Benchmark Papers in Electrical Engineering and Computer Science

Series Editor: John B. Thomas
Princeton University

PUBLISHED VOLUMES

SYSTEM SENSITIVITY ANALYSIS
J. B. Cruz, Jr.
RANDOM PROCESSES, PART I: Multiplicity Theory and Canonical Decompositions
Anthony Ephremides and John B. Thomas
ALGEBRAIC CODING THEORY: History and Development
Ian F. Blake
PATTERN RECOGNITION: Introduction and Foundations
Jack Sklansky
COMPUTER-AIDED CIRCUIT DESIGN: Simulation and Optimization
S. W. Director
ENVIRONMENTAL MODELING: Analysis and Management
Douglas Daetz and Richard H. Pantell
AUTOMATIC CONTROL: Classical Linear Theory
George J. Thaler
CIRCUIT THEORY: Foundations and Classical Contributions
M. E. Van Valkenburg
DATA COMMUNICATION: Fundamentals of Baseband Transmission
L. E. Franks
NONLINEAR SYSTEMS: Processing of Random Signals—Classical Analysis
A. H. Haddad
RANDOM PROCESSES, PART II: Poisson and Jump-Point Processes
Anthony Ephremides
DIGITAL FILTERS AND THE FAST FOURIER TRANSFORM
Bede Liu
DETECTION AND ESTIMATION: Applications to Radar
Simon S. Haykin
NONLINEAR SYSTEMS: Stability Analysis
J. K. Aggarwal and M. Vidyasagar
LINEAR LEAST-SQUARES ESTIMATION
Thomas Kailath
OPTICAL COMMUNICATION THEORY
R. O. Harger

Benchmark Papers
in Electrical Engineering
and Computer Science / 17
A BENCHMARK® Books Series

LINEAR LEAST-SQUARES ESTIMATION

Edited by
THOMAS KAILATH
Stanford University

STROUDSBURG, PENNSYLVANIA

Benchmark Papers in Electrical Engineering and Computer Science, Volume 18
Library of Congress Catalog Card Number: 77–7465
ISBN: 0–87933–098–8

79 78 77 1 2 3 4 5
Manufactured in the United States of America.

LIBRARY OF CONGRESS CATALOGING IN PUBLICATION DATA

Main entry under title:
Linear least-squares estimation.
(Benchmark papers in electrical engineering and computer science; v. 17)
Includes indexes.
CONTENTS: A survey of the field: Kailath, T. A view of three decades of linear filtering theory.—Mathematical foundations of least-squares prediction theory: Kolmogorov, A. N. Interpolation and extrapolation of stationary random sequences. Kolmogorov, A. N. Stationary sequences in Hilbert space. Krein, M. G. On a problem of extrapolation of A. N. Kolmogoroff. Krein, M. G. On a fundamental approximation problem in the theory of extrapolation and filtration of stationary processes. Karhunen, K. On the structure of stationary random functions. [etc.]
1. Stochastic processes—Addresses, essays, lectures. 2. Estimation theory—Addresses, essays, lectures. 3. Least squares—Addresses, essays, lectures. I. Kailath, Thomas.
QA274.L55 519 77–7465
ISBN 0–87933–098–8

Exclusive Distributor: **Halsted Press**
A Division of John Wiley & Sons, Inc.
ISBN: 0-470-99238-7

SERIES EDITOR'S FOREWORD

This Benchmark Series in Electrical Engineering and Computer Science is aimed at sifting, organizing, and making readily accessible to the reader the vast literature that has accumulated. Although the series is not intended as a complete substitute for a study of this literature, it will serve at least three major critical purposes. In the first place, it provides a practical point of entry into a given area of research. Each volume offers an expert's selection of the critical papers on a given topic as well as his views on its structure, development, and present status. In the second place, the series provides a convenient and time-saving means for study in areas related to but not contiguous with one's principal interests. Last, but by no means least, the series allows the collection, in a particularly compact and convenient form, of the major works on which present research activities and interests are based.

Each volume in the series has been collected, organized, and edited by an authority in the area to which it pertains. In order to present a unified view of the area, the volume editor has prepared an introduction to the subject, has included his comments on each article, and has provided a subject index to facilitate access to the papers.

We believe that this series will provide a manageable working library of the most important technical articles in electrical engineering and computer science. We hope that it will be equally valuable to students, teachers, and researchers.

This volume, *Linear Least-Squares Estimation*, has been edited by Thomas Kailath of Stanford University. It contains twenty papers which provide the foundations for modern linear least-squares estimation of random processes. Included are a number of works that have remained relatively inaccessible and, consequently, should be of particular interest. Professor Kailath needs no introduction, having played a major role in the development of this field.

JOHN B. THOMAS

PREFACE

Least-squares estimation is one of those topics that has the happy characteristic of being an apparently inexhaustible source of new problems, new applications, and new results. The reason for this vitality is that despite the seeming narrowness of its title, the subject is intimately connected with the basic structural properties not only of random and deterministic signals, but also of linear and nonlinear systems. This has the dual consequence of frequently bringing new insights from other fields to bear on problems of least-squares estimation and of often showing the usefulness of least-squares ideas in new problems and new applications. We might mention Wiener-Hopf equations, operator factorization, signal detection, martingale theory, matrix algebra, integral equations, scattering theory, and generalized orthogonal polynomial recursions as only some examples of topics that have had both classical and recent mutually fruitful interactions with least-squares theory.

This depth makes several of the classical papers in this field valuable not only for their historical significance but also because they contain insights and ideas that are still capable of further exploration and development. This potential for additional investigation has been one of the criteria used in selecting the papers gathered here. Another criterion has been the relative inaccessibility and sometimes even obscurity of several of the papers, even though this has meant the omission of other works because of their greater availability (some in other volumes in this series). Therefore, I hope that new students as well as those quite learned in this field will find some pleasure and some profit in this collection.

As I have said on numerous occasions, I am grateful to many friends in various countries and in a diversity of fields for shared pleasures in discussing and exploring this rich subject. The broad support of the mathematics department of the Air Force Office of

Scientific Research has been an important ingredient in these efforts. However, for specific assistance with the selections and commentaries in this volume, I am especially indebted to Professor A. M. Yaglom of the Institute of Atmospheric Physics, USSR Academy of Sciences, Moscow.

THOMAS KAILATH

CONTENTS

PART III: WIENER-HOPF EQUATIONS AND OPTIMUM FILTERS

PART IV: STATE-SPACE MODELS AND RECURSIVE FILTERS

CONTENTS BY AUTHOR

LINEAR LEAST-SQUARES ESTIMATION

INTRODUCTION

The topic of linear least-squares estimation, apparently a very special one, has ramifications in a large number of problems, both specific and general, stochastic and deterministic. Among these, we might mention the detection of Gaussian signals in Gaussian noise, quadratic minimization problems in modern control theory, polynomial factorization problems in network theory, computation of spline functions, solution of Fredholm integral equations and two-point linear boundary-value problems, Padé approximation problems, system identification techniques, the theory of invariant subspaces and Hilbert-space operators, inverse Sturm-Liouville problems, and so on. Perhaps the main reason for this broad scope is the close relationship between linear least-squares estimation and the structure of second-order random processes. A major contribution of engineers to this general area has been, and continues to be, the demonstration and exploitation of the close interplay between structural properties and the structure of deterministic linear systems. Some early examples are Nyquist's theorem on thermal noise, and the analysis of shot noise (Campbell's theorem, filtered Poisson processes, etc). Textbook presentations of these topics can be found in Davenport and Root [1], Parzen [2], and Blanc-Lapierre and Fortet [3]. However, the real fruits of such an interplay have come via least-squares estimation theory, following its introduction into radar and communication systems analysis through the work and influence of Norbert Wiener [4].

Bode and Shannon [5] and Zadeh and Ragazzini [6] used linear-systems concepts to illuminate the least-squares formulas of Wiener for the hard-pressed engineers of the time. Very significantly, it was also realized later that this engineering approach was in many ways the deeper mathematical one, a fact discovered as engineers began to study the works of Kolmogorov [7, 8], two of which appear in translation as Papers 2 and 3, and Karhunen [9], also partly covered in Doob's famous textbook [10]. Later, the work of Laning and Battin [11] and especially of Kalman (see Paper 16 and reference [12]) brought to the fore the state-space descriptions of linear systems, thus enabling easier extensions to multivariable and time-variant systems. In this connection, one should mention two extensive and insightful, but comparatively overlooked, papers by Doob [13, 14] that somewhat foreshadowed such developments.

The introduction of state-space models provided a major impetus to the theory and especially to its application, because such models led naturally to recursive computer-based algorithms. For obvious reasons, the need for algorithms of this type has been felt in many fields, and variants of the state-space based estimation algorithms have appeared in control theory (see Athans [15]), system identification and time-series analysis (see Kailath [16]), network theory (see Anderson and Vongpanitlerd [17]), and detection and modulation theory (see Van Trees [18]). However, in many applications, state-space models are not readily at hand, and efforts must be made to obtain recursive solutions for other forms of system and process description. Therefore, there is again a renewed interest in input-output (or external) as opposed to state-space (or internal) descriptions, both in the deterministic (linear system) and stochastic (least-squares) theories.

This is too large a panorama to be satisfactorily encompassed in a single volume and we have therefore envisaged two volumes. This first one covers development from the early period to the flowering of the state-space theory. The second volume will note some nonestimation-theoretical applications, as well as developments in the currently active field of recursive techniques for general input-output models, into which framework state-space methods can be naturally embedded.

The papers in this volume are divided into four groups—early mathematical foundations of prediction theory (Part II), Wiener–Hopf equations and optimal filters (Part III), and the introduction of recursive algorithms (Part IV). Part I is comprised of a 1974 survey paper providing a fairly detailed analysis of the topics of this volume and also helping to provide some background and perspective for the topics of the next volume.

In addition to their significance, factors contributing to the choice of these papers have been their potential for further development and their relative inaccessibility. In particular, a natural overlap with some papers in the Benchmark volume *Random Processes: Multiplicity Theory and Canonical Decompositions* [19] has been avoided. Furthermore, no selections have been included from Norbert Wiener's work because of its ready accessibility—the book *Extrapolation, Interpolation and Smoothing of Stationary Time Series, with Engineering Applications* [4] should be in the library of everyone interested in this subject. (A four volume edition of Norbert Wiener's collected works, edited by P. Masani and with commentaries by noted authorities, is in preparation [MIT Press].) I should also note that the availability of the survey paper in Part I has obviated the need for more comprehensive introductory remarks on the later parts; on the other hand, opportunities have been taken in those remarks to update some of the references and perspectives collected in Part I.

REFERENCES

1. W. Davenport and W. Root, *Introduction to the Theory of Random Signals and Noise*, New York: McGraw-Hill, 1958
2. E. Parzen, *Stochastic Processes*, San Francisco: Holden-Day, 1962.
3. A. Blanc-Lapierre and R. Fortet, *Theory of Random Functions*, Trans. J. Gani, New York: Gordon and Breach Science Publishers, 1968. (French original, 1953).
4. N. Wiener, *Extrapolation, Interpolation and Smoothing of Stationary Time Series, with Engineering Applications*, New York: Technology Press and J. Wiley, 1949. (Originally issued in Feb. 1942 as a classified Nat. Defense Res. Council Rept.) Available in paperback form as *Time Series*, Cambridge, Mass.: MIT Press, 1964.
5. H. W. Bode and C. E. Shannon, "A Simplified Derivation of Linear Least Square Smoothing and Prediction Theory," *Proc. IRE* **38:** 417–425, Apr. 1950.
6. L. A. Zadeh and J. R. Ragazzini, "An Extension of Wiener's Theory of Prediction," *J. Appl. Phys.* **21:**645–655, July 1950.
7. A. N. Kolmogorov, "Stationary Sequences in Hilbert Space," (in Russian), *Bull. Math. Univ. Moscow* **2**(6):1–40, 1941.
8. A. N. Kolmogorov, "Interpolation and Extrapolation of Stationary Random Sequences," *Bull. Acad. Sci USSR*, Ser. Math. **5:** 3–14, 1941; W. Doyle and I. Selin, Trans. RAND Corporation, Santa Monica, Ca., Memo, RM-3090-PR, Apr. 1962, pp. 1–14.
9. K. Karhunen, "Uber die Struktur Stationäere Zufälliger Funktionen," *Ark. Mat.* **1:**141–160, 1950; I. Selin, trans. RAND Corporation, Santa Monica, Ca., Memo RM-3091-PR, Apr. 1962, pp. 1–29.
10. J. L. Doob, *Stochastic Processes*, New York: J. Wiley, 1953.
11. H. Laning and R. Battin, *Random Processes in Automatic Control*, New York: McGraw-Hill, 1958.

12. R. E. Kalman, "New Methods of Wiener Filtering Theory," *Proc. 1st Symp. Engineering Applications of Random Function Theory and Probability*, J. L. Bogdanoff and F. Kozin, eds., New York: J. Wiley, 1963, pp. 270–388; also, RIAS, Baltimore, Md., Tech Rept. 61-1, 1961.
13. J. L. Doob, "The Elementary Gaussian Processes," *Ann. Math. Statist.* **15**:229–282, 1944.
14. J. L. Doob, "Time Series and Harmonic Analysis," *Proc. Berkeley Symp. Mathematics, Statistics, and Probability*, Berkeley, Ca.: University of California Press, 1949, pp. 303–343.
15. M. Athans, ed., Special Issue on Linear-Quadratic Gaussian Problem, *IEEE Trans. Automat. Contr.* **AC–16**, Dec. 1971.
16. T. Kailath, ed., Special Issue on System Identification and Time-Series Analysis, *IEEE Trans. Automat. Contr.* **AC–19**(6), Dec. 1974.
17. B. D. O. Anderson and S. Vongpanitlerd, *Network Analysis and Synthesis—A Modern Systems Theory Approach*, Englewood Cliffs, N. J.: Prentice-Hall, 1973.
18. H. L. Van Trees, *Detection, Estimation, and Modulation Theory, Part I*, New York: J. Wiley, 1968.
19. A. Ephremides and J. B. Thomas, eds., *Random Processes: Multiplicity Theory and Canonical Decompositions*, Benchmark Papers in Electrical Engineering and Computer Science, Stroudsburg, Pa.: Dowden, Hutchinson & Ross, 1973.

Part I

A SURVEY OF THE FIELD

Editor's Comments on Paper 1

1 KAILATH
A View of Three Decades of Linear Filtering Theory

Paper 1 provides a review of the field from its effective beginnings in the mid-forties to about the mid-seventies. Writing now, two years after Paper 1 was submitted, the main theoretical developments have come roughly as predicted.

There has been progress in developing and applying the fast algorithms described in Section VI (see [1, 2, 3, and 4]). A deeper, theoretical understanding of these algorithms has been obtained. For example, the parameter, α, introduced via the highly state-space dependent relations (39) to (40) of Paper 1, can now be defined for arbitrary nonstationary processes [5, 6]. It is unity for any scalar stationary process. Otherwise, it provides a measure of "distance from stationarity" reflected in the increase of computational effort required over the stationary case. The significance of the constant-parameter state-space models is chiefly that the state dimension then provides a finite upper bound on α.

Another important development has been in the further recognition that recursive methods are not restricted to the availability of state-space models. The Levinson algorithm (see Eqs. (73) to (84)) provides recursive algorithms for calculating the impulse response of the whitening filter for arbitrary stationary processes. In recent work [7, 8, 9], this has been extended to nonstationary processes [7, 8] by using the concepts of Kailath, et al. [5] and Friedlander et al. [6]; it has also been shown in this work that if there is an underlying constant-parameter state-space model, then the Levinson-type equations reduce to the Chandrasekhar-type equations (see Eqs. (37) to (41)). Thus, state-space models have been embedded naturally into a more general input-output framework.

Also, one can pass naturally from state-space to transfer-function models (see the remarks at the end of Section IX and [1]).

Progress has also been made in studying the Riccati equation itself. It was discovered that one of the earliest occurrences of this equation was in transmission line and scattering theory. An algebra of linear two point boundary-value problems was developed by Redheffer [10]. These results have been adapted to least-squares estimation in Ljung et al. [11] and Friedlander et al. [12]. Among the benefits gained have been new and powerful results on the much-studied problem of the asymptotic behaviour of the Riccati equation (see the discussion on page **16** and References [126, 158, 167, 151, 188] of Paper 1), new doubling formulas, new insights into the continuous- and discrete-Chandrasekhar equations, new families of smoothing formulas, and so on. As noted above, these developments start with an algebra for two point boundary-value problems which were first introduced into estimation problems by control theorists (see the remarks and references in the second paragraph on page **16** of Paper 1). Another application in which control ideas (this time from the singular control problem) have led to new developments is a new algorithm for multivariate spectral factorization (see [14, 15]). This algorithm is also closely related to some (Brune) synthesis concepts in network theory [16, 17].

Another topic first developed in a control application is the problem of square-root techniques, briefly noted in the paragraph below Eq. (60) in Paper 1. Comprehensive analyses from different points of view can now be found in Morf and Kailath [18], Silverman [19], Lawson and Hanson [20], and Bierman [21].

Reference has already been made to generalizations of the Levinson algorithm. The connections with orthogonal polynomials indicated in the survey paper (see the end of Section VII) have been further pursued [22] and seem likely to lead to new connections and extensions of classical results. This is also true for the material in Section VIII on Hardy functions and canonical kernels that is explored further in system-theoretical language in Helton [23] and Baras and Dewilde [24]. This work again promises to lead to new connections between network theory methods of scattering function synthesis and spectral factorization. The system-theoretical formulation of stochastic realization problems briefly discussed at the end of Section X has now been further developed [25]. These papers are further evidence of the remarkable range and power of the fundamental ideas of linear least-squares estimation, and one can be confident that many more interesting and useful results remain to be discovered.

REFERENCES

1. B. Dickinson, T. Kailath, and M. Morf, "Canonical Matrix Fraction and State-Space Description for Deterministic and Stochastic Linear Systems," *IEEE Trans. On Automat. Control*, Special Issue on System Identification and Time-Series Analysis, **AC–19**(6):656–667, Dec. 1974.
2. A. A. Desalu, L. A. Gould, and F. C. Schweppe, "Dynamic Estimation of Air Pollution," *IEEE Trans. on Automat. Control*, Special Issue on System Identification and Time-Series Analysis, **AC–19**(6):904–910, Dec. 1974.
3. J. Casti and L. Ljung, "Some New Analytic and Computational Results for Operator Riccati Equations," *SIAM J. Contr.* **13**(4):817–826, July 1975.
4. M. Morf, B. Dickinson, and T. Kailath, "General Speech Models and Linear Estimation Theory," in D. R. Reddy, ed., *Speech Processing*, New York: Academic Press, 1975, pp. 157–182.
5. T. Kailath, L. Ljung, and M. Morf, "Generalized Krein-Levinson Equations for the Efficient Computation of Fredholm Resolvents of Nondisplacement Kernels," in *Advances in Mathematics*, 1978.
6. B. Friedlander, M. Morf, T. Kailath and L. Ljung, "New Inversion Formulas for Matrices Classified in Terms of Their Distance From Toeplitz Matrices," *SIAM J. App.. Math.*, forthcoming.
7. T. Kailath, L. Ljung, and M. Morf, "Recursive Input-Output and State-Space Solutions for Continuous-Time Linear Estimation Problems," in *Surveys in Mathematical Analysis: Essays Dedicated to M. G. Krein*, New York: Academic Press, forthcoming.
8. B. Friedlander, T. Kailath, L. Ljung and M. Morf, "Levinson- and Chandrasekhar-Type Equations for a General Discrete-Time Linear Estimation Problem," *IEEE Trans. on Automat. Control*, forthcoming.
9. T. Kailath, "Some New Results and Insights in Linear Least-Squares Estimation Theory," in *Proc. 1st Joint IEEE-USSR Workshop on Information Theory*, New York: IEEE, 1976, pp. 97–104.
10. R. M. Redheffer, "On the Relation of Transmission-Line Theory to Scattering and Transfer," *J. Math. Phys.* **41:**1–41, 1962.
11. L. Ljung, T. Kailath, and B. Friedlander, "Scattering Theory and Linear Least-Squares Estimation, Part I: Continuous-Time Problems," *Proc. IEEE* **64**(1):131–138, Jan. 1976.
12. B. Friedlander, T. Kailath, and L. Ljung, "Scattering Theory and Linear Least Squares Estimation, Part II: Discrete-Time Problems," *J. Franklin Inst.* **301:**71–82, Jan. 1976.
13. B. Friedlander and T. Kailath, "Scattering Theory and Linear Least Squares Estimation, Part III: The Estimates," *Proc. 1977 IEEE Decision and Control Conf.*, New Orleans, La., Dec. 1977.
14. B. D. Anderson and P. Moylan, "Spectral Factorization of a Finite-Dimensional Nonstationary Matrix Covariance," *IEEE Trans. on Automat. Control*, Special Issue on System Identification and Time-Series Analysis, **AC–19**(6):680–692, Dec. 1974.
15. D. J. Clements and B. D. Anderson, "Transformational Solution to the Linear–Quadratic Singular Control Problem," *IEEE Trans. on Automat. Control*, **AC–22**(1):57–60, Feb. 1977.

16. P. J. Moylan, "Preliminary Simplifications in State-Space Impedance Synthesis," *IEEE Trans. Circuits and Systems* **CAS-21**:203–206, 1974.
17. B. D. Anderson and P. J. Moylan, "Synthesis of Linear Time-Varying Passive Networks," *IEEE Trans. Circuits and Systems* **CAS-21**:678–687, 1974.
18. M. Morf and T. Kailath, "Square-Root Algorithms for Least-Squares Estimation," *IEEE Trans. on Automat. Contr.* **AC-20**(4):487–497, Aug. 1975.
19. L. Silverman, "Discrete Riccati Equations: Alternative Algorithms Asymptotic Properties and System Theory Interpretations," in C. T. Leondes, ed., *Advances in Control and Dynamic Systems: Theory and Applications*, vol. 12, New York: Academic Press, 1975, pp. 313–386.
20. C. Lawson and R. J. Hanson, *Solving Least-Squares Problems*, Englewood Cliffs, N. J.: Prentice-Hall, 1974.
21. G. Bierman, *Factorization Methods for Discrete-Time Estimation*, New York: Academic Press, 1976.
22. T. Kailath, A. Vieira, and M. Morf, "Inverses of Toeplitz Operators, Innovations, and Orthogonal Polynomials," *SIAM Review*, forthcoming.
23. J. W. Helton, "Systems with Infinite-Dimensional State Space: The Hilbert Space Approach," *Proc. IEEE* **64**(1):145–160, Jan. 1976.
24. J. S. Baras and P. Dewilde, "Invariant Subspace Methods in Linear Multivariable-Distributed Systems and Lumped-Distributed Networks Synthesis," *Proc. IEEE* **64**(1):160–178, Jan. 1976.
25. G. Picci, "Stochastic Realization of Gaussian Processes, *Proc. IEEE* **64** (1):112–122, Jan. 1976.

1

Reprinted from *IEEE Trans. Inf. Theory* **IT–20**(2):146–181 (1974)

A View of Three Decades of Linear Filtering Theory

THOMAS KAILATH, FELLOW, IEEE

***Abstract*—Developments in the theory of linear least-squares estimation in the last thirty years or so are outlined. Particular attention is paid to early mathematical work in the field and to more modern developments showing some of the many connections between least-squares filtering and other fields.**

I. Introduction and Outline

THE SERIES of survey papers of which this is a part was begun largely to commemorate the twenty-fifth anniversary of the publication of Shannon's classic paper on information theory. However, 1974 is also twenty-five years after the publication in the open literature of Wiener's famous monograph, "Extrapolation, Interpolation and Smoothing of Stationary Time Series, with Engineering Applications" [1], so that it is appropriate this year to commemorate this event as well. [As noted elsewhere in this issue, this month is also the tenth anniversary of Wiener's death (March 18, 1964).] Not only was this work the direct cause for the great activity of the last three decades in signal estimation, but it was perhaps the greatest factor in bringing the statistical point of view clearly into communication theory and also control theory. It may suffice to quote Shannon's own major acknowledgment: "Credit should also be given to Professor N. Wiener, whose elegant solution of the problems of filtering and prediction of stationary ensembles has considerably influenced the writer's thinking in this field."

The subject of estimation is a vast one, and most of our attention will be devoted to the particular problems of *linear least-squares estimation*, or *linear filtering* as it has generally come to be called in the engineering literature. Even though least-squares estimation is clearly only a small part of the possible forms of estimation theory, in the author's opinion it is perhaps the most interesting and most important part. Least-squares theory not only provides useful solutions to certain specific estimation problems, but it also has connections to and implications for a surprisingly large number of other problems, both statistical and deterministic. As some examples we mention signal detection [93], [291]; the calculation of mutual information in certain channels [303]; the solution of integral equations [288], [292], two-point boundary value problems in many fields; problems of scalar and matrix polynomial factorization with applications in network theory and stability theory [135], [177]; the solution of linear equations, especially as they arise in constructing state-variable realizations from impulse-response or transfer-function data, which in turn is related to the Berlekamp–Massey algorithm for decoding BCH codes [289], [191]; and the inversion of multivariable linear systems [357], [361], [362]. There are also more purely mathematical ramifications in Hilbert-space theory, operator theory, and more generally in functional analysis [245], [250], [261].

The section headings give a quick idea of the scope of the paper.

I. Introduction.
II. A Key Linear Estimation Problem.
III. Wiener Filters and Early Generalizations.
IV. Kalman Filters.
V. Recursive Wiener Filters.
VI. New Algorithms for Time-Invariant Systems.
VII. Some Early Mathematical Work.
VIII. Canonical Representations of Continuous-Time Processes.
IX. Recent Results on Innovations Processes and Some Applications.
X. Karhunen–Loève Expansions, Canonical Correlations and State Models.
XI. Concluding Remarks.
XII. Bibliography.

Needless to say, the choice of material and emphasis in this paper are mine; the field is a vast one and can be surveyed in various ways. My main aims are to provide some perspective on presently used methods, to bring out the significance and relevance of some relatively early, but often neglected, work in this field, and to illustrate some of the connections between least-squares theory and other fields.

In Section II we formulate the problem of determining the causal linear least-squares estimate of a signal process corrupted by additive white noise. Although this is only one of a large number of possible estimation problems, it is a key one in the sense that its solution underlies that of many others.

Sections III–VI describe some of the current approaches to solution of this key problem. Information theorists and communication engineers have been more familiar with problems in which covariance information is given about

Manuscript received June 15, 1973; revised October 3, 1973. This work was supported in part by the Air Force Office of Scientific Research, Air Force Systems Command, under Contract AF 44-620-69-C-0101, and in part by the Joint Services Electronics Program under Contract N-00014-67-A-0112-0044.

The author is with Stanford University, Stanford, Calif. 94305.

signal and noise, which are usually called Wiener filtering problems. Control engineers deal more often with problems where the signal and noise are described by state-space models, usually called Kalman filtering problems. Because of differing backgrounds, mutual knowledge of these two general approaches is often small, and one of our aims is to bring out the close and useful relation that must exist between these two approaches. Of course no proofs can be given here, but the main results are stated and their significance and role explained. Appropriate references are given for the proofs.

The discussion in Sections III–VI is fairly self-contained, but at various points allusions are made to earlier results in the mathematical literature, especially of the forties. This work is explored in Sections VII and VIII, partly for the record but really because it contains ideas that in my opinion have still not yet been adequately appreciated and exploited. For example, the work of Krein (1944) and of Levinson (1947) is only now beginning to be rediscovered and extended. Limitations of space again prevent any detailed discussion, but I have tried to provide some guidance for a reader interested in further study. Moreover, even a casual reader might find some fascinating nuggets exposed here, although I should stress that the lode is really much richer.

In Section IX, I have described in somewhat more detail one vein that I have personally found to be very illuminating and powerful: the role of canonical or innovations representations of random processes. Again I have given only references to many results and applications, but I could not resist being a little more specific about one aspect; namely, the connections to spectral factorization and to the so-called positive real functions of network and complex variable theory. The aim is to show at least one explicit connection between stochastic and deterministic problems. A fact I attempt to stress in Sections VIII and IX is the importance of deterministic system structure in the theory of random processes. This is at the moment an active field of research, but one that is being largely carried out in control theory. There is scope for many more communication theorists to enter this field. Conversely, in the last section I have briefly referred to some recent results in information theory that have useful implications for estimation. The time seems to be ripe for a fruitful symbiosis.

Section X is a brief look at the role and significance of series expansions. These should be relatively more familiar to readers of this journal, and I have therefore concentrated on some special, but often overlooked, aspects of such expansions. The section concludes in fact with an indication of how certain random process ideas can illuminate the currently active system-theoretic problem of abstract state-space determination. Again there is scope for fruitful interaction with control and system theorists. The paper closes with the thought that it is the range and scope of such possibilities that has kept estimation theory vital and active, without, it seems to me, the above-average doubts and misgivings that the more self-contained field of strict-sense Shannon theory has been exposed to recently.

The bibliography is of necessity rather vast, although it could easily have been even larger. On several occasions, the availability of convenient bibliographies has led me to omit various references. This is undoubtedly an injustice to the authors of many fine papers, but it seems to be unavoidable. I have attempted to do more justice to papers published in this journal, and in fact all such linear filtering papers in the period 1968–1972 have been included in the bibliography, even though no explicit reference may have been made to them in the text. This has also been done for certain other papers appearing in other journals that I feel contain some ideas or approaches that may appeal to our readers. The choice is of necessity rather subjective and any omissions should be regarded as a measure of my ignorance rather than a conscious slight.

The bibliography is organized under five subheadings, though the division of papers between the five categories is on occasion somewhat arbitrary, partly because the fields are of course not completely exclusive. In retrospect, some reassignments would really have been desirable, but I have not had the courage or the time to attempt them.

I must repeat that the inevitable limitations of time, space, and personal knowledge are undoubtedly reflected in this survey. The only palliative I can offer is that perusal of the various references will enable the reader to learn many additional facts and results that could not be covered in the paper and to make his own judgment of any controversial matters.

II. A Key Linear Estimation Problem

Some Early History

From the earliest times, people have been concerned with interpreting observations and making estimates and predictions. Neugebauer [370] has noted that the Babylonians used a rudimentary form of Fourier series for such purposes. As with so much else, the beginnings of a "theory" of estimation in which attempts are made to minimize various functions of the errors can apparently be attributed to Galileo Galilei in 1632 [174]. Then came a whole series of illustrious investigators, including the young Roger Cotes (of whom Newton said "had he lived, we might have known something"), Euler, Lagrange, Laplace, Bernoulli, and others.

As is well known, the method of least squares was apparently first used by Gauss in 1795 [197], though it was first published by Legendre in 1805 [198]. (It is less well known that Adrain in America, unaware of these developments, independently developed the method in 1808 [196]). Since then, there has been a vast literature on various aspects of the least-squares method. A comprehensive annotated bibliography of least-squares estimation for random variables has been given in a report by Harter [375]. (See also a brief survey by Sorenson [376].) Therefore we shall not go into this here, but shall proceed to least-squares estimation in stochastic processes, the first studies of which were made by Kolmogorov [207], [209], Krein [214], [215], and Wiener [1].

The works of Kolmogorov and Krein were independent of Wiener's, and while there was some overlap in the results, their aims were rather different. Kolmogorov, inspired by some work of Wold [209], gave a comprehensive treatment of the prediction problem for discrete-time stationary processes. Krein noted the relationship of this work to some early work of Szegö [200], [201], on orthogonal polynomials, and extended the results to continuous time by clever use of a bilinear transformation. (We shall describe these results in more detail later (Section VII).)

However, no special attention was paid to explicit formulas for the optimum predictor itself. Such formulas are obviously necessary for applications, and in fact, certain anti-aircraft fire-control problems led Wiener to formulate independently the continuous-time linear prediction problem and derive an explicit formula for the optimum predictor. He also considered the "filtering" problem of estimating a process corrupted by a "noise" process. An interesting nontechnical account of this work and its background and development was given by Wiener in his autobiography [371, pp. 240–262].

Wiener constantly attempted to examine and stress the engineering significance of his ideas and results and his book [1] contains several explicit examples, which are still generally the only ones to be found in many textbooks on the subject. Wiener was also conscious of the problems of actually building circuits to implement the theoretical solutions. For example, he notes that "The detailed design of a filter involves certain choices of constants which must be justified economically. In general, it does not pay to eliminate a small error from a quantity when there is a large *irremovable error* in it." Another paragraph of his book is entitled "The Determination of Lag and Number of Meshes in a Filter." Partly because of such concerns, much of Wiener's work, despite its hard mathematics, has had a wide influence in engineering circles.

A Key Problem

As we begin to be more specific, we are immediately confronted with the fact that there is a large variety of estimation problems, even just linear ones. For example, as the reader knows, we can have prediction or filtering or smoothing problems, in state-variable form or transfer-function form, with additive white noise or colored noise, etc. However, in my opinion there is a *key* linear estimation problem in the sense that its solution can be shown to underlie that of many other problems (see [298], [181], for some examples): we have observations $y(\cdot)$ of a signal process $z(\cdot)$ in additive white noise $v(\cdot)$

$$y(s) = z(s) + v(s), \qquad t_0 \le s \le t_f \tag{1}$$

where

$$Ev(t)v'(s) = I_p\delta(t - s). \tag{2}$$ [1]

The usual assumption on $z(\cdot)$ is that it is uncorrelated with $v(\cdot)$

$$Ez(t)v'(s) \equiv 0. \tag{3}$$

[1] Some notational conventions: all random variables will be assumed to have zero mean. No special notation will be used to distinguish scalars and matrices; primes denote transposes; I_p is a $p \times p$ identity matrix. We could have assumed any strictly positive-definite matrix instead of I_p, but by a normalization we can return to the case described by (2).

However, this breaks down in feedback communication and feedback control problems where the signal $z(\cdot)$ may be influenced by past signal and noise. Therefore, a more general assumption is that

$$Ez(t)v'(s) = \begin{cases} \text{arbitrary}, & t \ge s \\ 0, & t < s. \end{cases} \tag{4}$$

It should be stressed that (4) can be introduced only with difficulty in many of the analyses found in the literature; for example, it cannot be directly handled by methods that rely on representing the signal and noise processes $z(\cdot)$ and $v(\cdot)$ by Karhunen–Loève expansions. (See Section X for further discussion of this point.)

To proceed with our description, let us define

$$K(t,s) = E[z(t)z'(s) + z(t)v'(s) + v(t)z'(s)]. \tag{5}$$

Note that $K(\cdot,\cdot)$ is generally not a covariance, unless $z(\cdot)$ and $v(\cdot)$ are uncorrelated. However $K(\cdot,\cdot)$ does determine the covariance function of the process $y(\cdot)$ as

$$R_y(t,s) = Ey(t)y'(s) = I_p\delta(t - s) + K(t,s). \tag{6}$$

We shall require that $R_y(t,s)$ be strictly positive definite on the square $[t_0,t_f] \times [t_0,t_f]$. Other assumptions will be that the signal process has finite expected energy and that $K(t,s)$ is continuous, though both of these assumptions can in fact be relaxed; the essential thing is really that $K(t,s)$ be "smoother" than $I_p\delta(t - s)$. (See also [327c].)

The problem is to determine a random variable $\hat{z}(t \mid t_f)$ of the form

$$\hat{z}(t \mid t_f) = \int_{t_0}^{t_f} H(t,\tau)y(\tau)\, d\tau, \qquad t_0 \le t \le t_f \tag{7}$$

so that

$$\operatorname{tr} E[z(t) - \hat{z}(t)][z(t) - \hat{z}(t)]' = \text{minimum}.$$

It is by now well known (see, e.g., [10]) that such an optimum linear least-squares estimate is characterized by the "orthogonality" property

$$E[z(t) - \hat{z}(t \mid t_f)]y'(s) = 0, \qquad t_0 \le s \le t_f \tag{8}$$

so that a simple calculation shows that the optimum filter $H(\cdot,\cdot)$ is determined by the solution of the integral equation

$$H(t,s) + \int_{t_0}^{t_f} H(t,\tau)K(\tau,s)\, d\tau = Ez(t)z'(s) + Ez(t)v'(s), \qquad t_0 \le t,\, s \le t_f. \tag{9}$$

Depending on whether $t_f < t$, $t_f = t$, or $t_f > t$, we have what are called *predicted*, *filtered*, or *smoothed* estimates, respectively. For convenience, we shall write $\hat{z}(t \mid t)$ as

$$\hat{z}(t) = \int_{t_0}^{t} h(t,s)y(s)\, ds.$$

Using (8) and (4), the relevant integral equation for *filtering* can be found as

$$h(t,s) + \int_{t_0}^{t} h(t,\tau)K(\tau,s)\, d\tau = K(t,s), \qquad t_0 \le s \le t \le t_f. \tag{10}$$

This is a key equation in linear system theory.

It is important to note that (10) is substantially more difficult to solve than (9). Equation (9) is a Fredholm integral equation of the second kind and a lot is known about its properties. Several general solution methods are available, including reduction in different ways to a set of linear algebraic equations and the use of various gradient methods, see, e.g., [51], [37], [18], [27]. On the other hand, (10) is a much more difficult equation because of the constraint that $s \leq t$. This is seen most dramatically by taking

$$p = 1, \quad t_0 = -\infty, \quad t_f = +\infty, \quad K(t,s) = K(|t - s|)$$

in which case it is easy to see that in both (9) and (10) the solution $h(t,s)$ depends only on $|t - s|$. Then with an obvious change of variables (9) can be reduced to

$$H(t) + \int_{-\infty}^{\infty} H(\tau)K(t - \tau)\, d\tau = \tilde{K}(t), \qquad -\infty < t < \infty$$

which can be readily solved by Fourier transformation. However, the corresponding form of (10) is

$$h(t) + \int_{0}^{\infty} h(\tau)K(t - \tau)\, d\tau = K(t), \qquad 0 < t < \infty \quad (11)$$

which is quite a different thing. Simple Fourier transformation does not work, and we must in general use the celebrated Wiener–Hopf spectral factorization technique [2]–[4], which in fact has given (11) its name.

The Wiener–Hopf equation (11) first arose in astrophysics in 1894 and has been widely studied; see especially two comprehensive papers [234], [235], which deserve to be more widely read by engineers. On the other hand, much less is known about the nonstationary version (10), which we shall describe as being of Wiener–Hopf type.[2] Several references are collected in [257], [51], [298], and some of these will be brought up as we proceed with this survey. However, it is appropriate to begin with (11), which was where Wiener started.

III. Wiener Filters and Early Generalizations

The first explicit solutions for least-squares estimates of stochastic processes were given by Wiener in 1942 [1] under the assumptions of a scalar observation process ($p = 1$), a semi-infinite observation interval ($t_0 = -\infty$), and jointly stationary signal and noise processes. Wiener used a variational argument to determine the optimum estimate and was delighted to find that what was required was the solution of the Wiener–Hopf equation (11), a problem to which Hopf and he had ten years earlier contributed an elegant solution [2]. The method applies to quite general kernels, but, as Wiener himself noted, it took its simplest form for processes with rational spectral densities, for which $K(t,s)$ had the form

$$K(t,s) = K(|t - s|) = \sum_1^n \alpha_i \exp(-\beta_i|t - s|)$$

$$= \begin{cases} \sum_1^n \alpha_i \exp(-\beta_i t)\exp(\beta_i s), & t \geq s \quad (12a)^3 \\ \sum_1^n \alpha_i \exp(-\beta_i s)\exp(\beta_i t), & t \leq s \quad (12b)^3 \end{cases}$$

where the $\{\beta_i\}$ are constants, possibly complex. For such kernels, the bilateral Laplace transform

$$S_y(s) = \int_{-\infty}^{\infty} [\delta(t) + K(t)] \exp(-st)\, dt \qquad (13)$$

is a ratio of polynomials in s^2, whose zeros display a quadrantal symmetry in the complex s-plane: every root of the form $\sigma + j\omega$ is accompanied by roots of the form $\sigma - j\omega$, $-\sigma \pm j\omega$. By virtue of this root distribution, we can make a *unique* factorization of $S_y(s)$ as

$$S_y(s) = S_y^+(s)S_y^+(-s) \qquad (14)$$

where the so-called *canonical factor* is

$$S_y^+(s) = \prod_1^n (s - z_i) \Big/ \prod_1^n (s - p_i) \qquad (15)$$

and $\{z_i\}$ and $\{p_i\}$ are the *left-half-plane* zeros and poles of $S_y(s)$, $\operatorname{Re} z_i < 0$, $\operatorname{Re} p_i < 0$. The Wiener–Hopf technique (see, e.g., [4], [20], [51]) shows that this factorization completely determines the Laplace transform of the optimum filter as

$$H_{\text{opt}}(s) = 1 - [S_y^+(s)]^{-1}. \qquad (16)$$

This expression, though implicit in Wiener's own examples, was first explicitly given by Yovits and Jackson [11] and by Krein [235].

Yovits and Jackson also gave a closed-form expression for the mean-square error in the special case of uncorrelated signal and noise

$$E[z(t) - \hat{z}(t)]^2 = \int_{-\infty}^{\infty} \ln[1 + S_z(i\omega)]\, d\omega. \qquad (17)$$

Since (17) does not require explicit knowledge of the optimum filter, it can be used to help decide if an optimum filter is worthwhile; several such applications in modulation theory have been discussed by Van Trees [40], [47], Stiffler [46], Lindsey [52], and the references therein. Recently Yao [48], [49], and Snyders [54], [54a], have extended this formula to cover certain problems with nonwhite noise, and signals or noise with nonrational spectra (see also Prouza [53]).

Wiener's theory has mainly been applied to the optimum choice of various components in modulation systems, as we have noted previously. We should explicitly mention

[2] One might argue that (10) is just a family of Fredholm equations, indexed by t. This is true, but the major problem lies in showing that the solutions $\{h(t,\cdot)\}$ can be satisfactorily fitted together, for example to make $h(\cdot,\cdot)$ square-integrable in both variables.

[3] This is not the most general form except when the poles $\{p_i\}$ in (15) are distinct; the generalization is not difficult but is notationally cumbersome and so has been avoided.

a notable early application to the design of loop filters in phase-locked loops [9]. Related applications were also made in optimal control (see [13], [24], and the references therein).

Some Generalizations

The Wiener–Hopf equation was soon extended to cover estimation of stationary processes given only over a finite observation interval ($t_0 > -\infty$), and more generally to cover the estimation of nonstationary processes. However, while it was not hard to discover that the general equation was of the form (10), there was no general method for solving such equations, and therefore a host of special results and techniques were developed. We especially mention an early paper by Zadeh and Ragazzini [6], which we shall encounter again in Section VIII. For processes with rational power-spectral densities, fairly explicit results were obtained by Yaglom [10], Hajek [243], Rozanov and Pisarenko [31], Whittle [32], Helstrom [33], and Slepian and Kadota [43].

Some solutions were also obtained for nonstationary processes [6], [7], [17], [21]. The most useful of these were by Shinbrot [14], [25], who however found it necessary to restrict attention to $K(\cdot,\cdot)$ of the form

$$K(t,s) = \begin{cases} \sum_1^n a_i(t)b_i(s), & t \geq s \quad (18a) \\ \sum_1^n a_i(s)b_i(t), & t \leq s. \quad (18b) \end{cases}$$

This is clearly a generalization of (12), but its true significance was not appreciated until later (see Section V). Despite such important contributions, however, too large a part of the literature dealt with minor variations and special cases, so much so that Elias felt compelled to editorialize in the IEEE TRANSACTIONS ON INFORMATION THEORY in 1958 that it was time to stop writing "two famous papers." One was "The Optimum Linear Mean-Square Filter for Separating Sinusoidally-Modulated Triangular Signals from Randomly-Sampled Stationary Gaussian Noise, with Applications to a Problem in Radar." (The other was "Information Theory, Photosynthesis, and Religion," a title suggested by D. Huffman.)

Furthermore, there were other reasons for being dissatisfied even with the most significant of the results of this period.

i) They were rather complicated, often requiring the solution of auxiliary differential and algebraic equations and the calculation of roots of polynomials.

ii) They were not easily updated with increases in the observation interval.

iii) They could not be conveniently adapted to the vector case ($p > 1$).

These last two difficulties came immediately to the fore in the late fifties in the problem of determining satellite orbits. Here there were generally vector observations of some combinations of position and velocity, and also there were large amounts of data sequentially accumulated with each pass over a tracking station. Swerling was one of the first to tackle this problem, and he presented some useful recursive algorithms [61], [80], [155], that were soon recognized and applied, especially by a group at the Bell Laboratories, who added various contributions of their own [84]. For different reasons, growing out of his successful application of state-space ideas in deterministic problems, Kalman developed [64], [68], [69], a somewhat more restricted algorithm than Swerling's, but it was one that seemed particularly matched to the dynamical state-estimation problems that were brought forward by the advent of the space age. Groups at the NASA Ames Laboratory [71], [72], and at the M.I.T. Draper Laboratories [70], [83], took up Kalman's ideas and developed them into programs that were successfully used in many space applications [139], [145], [138].

We shall examine the Kalman filter in some detail in Section IV. The reasons for our attention go beyond the specific algorithm and are more broadly connected with the importance of dynamical structure in data-processing algorithms. Unfortunately communication engineers and information theorists have lagged behind control engineers in appreciating this fact, though, as pointed out in Wong's recent survey [327], the gap is closing.

IV. KALMAN FILTERS

Kalman [64], [68], [69], changed the conventional formulation of the problem by giving, not the covariance of the signal process, but a "model" for it as the output of a dynamical linear system driven by white noise. Specifically, he assumed that the signal process $z(\cdot)$ could be described by

$$z(t) = H(t)x(t), \qquad t \geq t_0 \quad (19a)$$

$$\dot{x}(t) = F(t)x(t) + G(t)u(t), \qquad x(t_0) = x_0 \quad (19b)$$

where $x(\cdot)$ is an $n \times 1$ "state" vector and $u(\cdot)$ is an $m \times 1$ random input such that

$$Eu(t)u'(s) = Q(t)\delta(t-s) \quad (19c)$$

$$Ex_0x_0' = \Pi_0, \qquad Eu(t)x_0' \equiv 0, \qquad t \geq t_0. \quad (19d)$$

Also the matrices $F(\cdot)$, $G(\cdot)$, $H(\cdot)$, $Q(\cdot)$, and Π_0 are assumed known and continuous. In trajectory estimation (19b) could be the "linearized" equations of motion describing the evolution of the position and velocity vector $x(\cdot)$ subject to the wideband perturbations $u(\cdot)$ caused by random drag, gravitational uncertainties, etc., and the initial uncertainties x_0.

The assumption that x_0 and $u(\cdot)$ are uncorrelated not only is physically reasonable but also has the important consequence that the process $x(\cdot)$ is now a wide-sense Markov process [225]. Kalman also assumes that the "plant" noise $u(\cdot)$ and the "observation" noise $v(\cdot)$ in the observed process

$$y(t) = z(t) + v(t) = H(t)x(t) + v(t) \quad (20)$$

can be correlated, but he restricts the dependence to being of the form

$$Eu(t)v'(s) = C(t)\delta(t-s) \quad (21)$$

which is consistent with our more general earlier assumption (4) on the one-sided dependence between $z(\cdot)$ and $v(\cdot)$. The equations (19)–(21) describe the Kalman model for the estimation problem. The Kalman filter is not given by an explicit formula for the impulse response of the optimal filter, but as an algorithm suitable for direct evaluation by analog or digital computers

$$\hat{z}(t) = H(t)\hat{x}(t) \tag{22}$$

where

$$\dot{\hat{x}}(t) = F(t)\hat{x}(t) + K(t)\varepsilon(t), \qquad \hat{x}(t_0) = 0 \tag{23}$$

$$\varepsilon(t) = y(t) - \hat{z}(t) = y(t) - H(t)\hat{x}(t) \tag{24}$$

$$K(t) = P(t)H'(t) + G(t)C(t) \tag{25}$$

and the $n \times n$ matrix $P(\cdot)$ is the covariance matrix of the errors in the state estimates

$$P(t) = E\tilde{x}(t)\tilde{x}'(t), \qquad \tilde{x}(t) = x(t) - \hat{x}(t). \tag{26}$$

$P(\cdot)$ can be computed as the unique solution of the nonlinear differential equation

$$\dot{P}(t) = F(t)P(t) + P(t)F'(t) - K(t)K'(t) + G(t)Q(t)G'(t), \qquad P(t_0) = \Pi_0. \tag{27}$$

This equation is a matrix version of the familiar Riccati equation, first introduced by Francesco, Count Riccati, in 1724 [195] and since then often encountered in the calculus of variations. It seems surprising that a nonlinear equation should arise in a linear problem and be regarded as advantageous. However, the point is that it is a *differential equation* with known *initial* conditions, and such equations are comparatively easy to solve on a digital or analog computer because they involve only the iteration of relatively simple updating operations. This circumstance is indeed a very happy one, because Riccati equations can be introduced to solve general linear two-point boundary value problems, which arise often in various fields, see, e.g., [146], [247].

Discrete-Time Results

The discrete-time Kalman filter results were actually the first to be obtained, partly because the major system-theory activity in the mid-fifties was in the field of sampled-data systems, which arose when modern digital computers were put into control and communication links. Sampled-data filters for least-squares estimation were given by Franklin [8], Friedland [60], and others. While Friedland used infinite triangular matrices, Blum [59], [67], studied recursive filters for limiting the storage requirements of such algorithms. We have already noted the work of Swerling [61]. Kalman's contribution was to introduce state-space models. He assumed that

$$y_i = z_i + v_i, \qquad z_i = H_i x_i, \qquad i \geq 0 \tag{28a}$$

$$x_{i+1} = \Phi_i x_i + \Gamma_i u_i \tag{28b}$$

$$Eu_i x_0' \equiv 0 \equiv Ev_i x_0', \qquad Ex_0 x_0' = \Pi_0 \tag{28c}$$

$$Eu_i u_j' = Q_i \delta_{ij}, \qquad Ev_i v_j' = R_i \delta_{ij}, \qquad Eu_i v_j' = C_i \delta_{ij} \tag{28d}$$

where $\{\Phi,\Gamma,H,Q,R,C,\Pi_0\}$ are known matrices. The Kalman filter solution is

$$\hat{x}_{i+1|i} = \Phi_i \hat{x}_{i|i-1} + K_i (R_i^\varepsilon)^{-1} \varepsilon_i, \qquad \hat{x}_{0|-1} = 0 \tag{29a}$$

$$\varepsilon_i = y_i - \hat{z}_{i|i-1}, \qquad \hat{z}_{i|i-1} = H_i \hat{x}_{i|i-1} \tag{29b}$$

$$R_i^\varepsilon = E\varepsilon_i\varepsilon_i' = H_i P_{i|i-1} H_i' + R_i, \qquad P_{i|i-1} = E\tilde{x}_{i|i-1}\tilde{x}'_{i|i-1} \tag{29c}$$

$$K_i = \Phi P_{i|i-1} H_i' + \Gamma_i C_i \tag{29d}$$

where the $n \times n$ matrix

$$P_{i|i-1} \triangleq E[x_i - \hat{x}_{i|i-1}][x_i - \hat{x}_{i|i-1}]'$$

can be computed via the so-called Riccati difference equation

$$P_{i+1|i} = \Phi_i P_{i|i-1} \Phi_i' - K_i (R_i^\varepsilon)^{-1} K_i' + \Gamma_i Q_i \Gamma_i', \qquad P_{0|-1} = \Pi_0. \tag{30}$$

The similarity of this set of equations to (22)–(27) is clear; in fact, the latter can be obtained from (28), (29), by a limiting procedure [68]. Note that because of the presence of the $P_{i|i-1}$-dependent term R_i^ε, the discrete-time formulas are somewhat more complicated than the continuous-time ones, or even than discretized versions of the continuous-time formulas. Moreover, in discrete time, there is no particular need to assume that the covariance of the additive-noise is nonsingular, and we have therefore written it as R_i rather than as I. Note that we could even take R_i to be zero without affecting the formulas (29), (30). This is not possible in continuous time, where problems with no nonsingular white noise component require more care (cf. [90], [122], [181] and the references therein). The study of state-estimation problems where there is no additive noise has recently uncovered some interesting differences between discrete- and continuous-time estimation and control problems (cf. [170], [179]).

By now the Kalman filter is widely known and widely used, notably in aerospace engineering; see, for example, the papers and references in the survey volumes [139] and [145]. Furthermore, (19)–(30) have turned out to have a fundamental role in understanding the structure and properties of dynamical systems, in many stochastic and deterministic problems. We may refer to work in quadratic optimization, stability theory, network theory, covariance and spectral factorization, stochastic control, sensitivity analyses, signal detection, etc. (The factorization problems will be briefly discussed in Section IX.) We forebear from giving specific references, but shall merely note some recent books in which such topics are covered [106], [128], [129], [132], [135], [143], [153], [157], [177]; it should be noted that Kalman himself launched the study of several of these questions.

However, despite all the good that has come out of this theory, there have been many excesses and oversights in its pursuit, partly reflected in an incredible volume of papers. Although some of this literature was necessary and worthwhile, a good fraction of it must be attributed to the general expansion of technological and especially space activity that Sputnik and the Apollo project brought to America scene, in terms both of research and development contracts for industry and of the rapid expansion of graduate education in universities. Another factor was that this period coincided with the emergence of what is now called modern control theory, which was being built upon the rediscovery of the importance, emphasized in the mid-fifties by Bellman, of the notion of "state." The fact that the Kalman filter dealt largely with state-estimation made it comparatively easy to include it in books and courses in state-variable theory, without having to go very deeply into estimation theory or even into Wiener filtering.

Although several excellent examples of the clever and successful application of control theory ideas to estimation problems can be found, e.g., [74], [111], [128], [133], [148], [84a], [193a], the majority of contributions have suffered from having too narrow a base. I feel it is unfortunate that a whole generation of control engineers has grown up whose only knowledge of estimation theory is through Kalman filtering. The work of Kolmogorov, Krein, Wiener, Karhunen, Levinson, Lévy, Hida, and others (see Section VII) on many still important aspects of estimation has generally been neglected, not without loss. On the other hand, I should also state that in my opinion, the potential of the results and insights in the just-cited control-theoretic ideas has also not yet been fully exploited.

It may be useful to reinforce the previous comments by giving an illustration from control theory itself. As Kalman has often stressed [68] the major contribution of his work is not perhaps the actual filter algorithm, elegant and useful as it no doubt is, but the proof that under certain technical conditions called "controllability" and "observability," the optimum filter is "stable" or "robust" in the sense that the effects of initial errors and round-off and other computational errors will die out asymptotically. However, the known proofs of this result are somewhat difficult, and it is significant that only a small fraction of the vast literature on the Kalman filter deals with this problem. (Significant recent contributions have been made in [126], [158], [167], [151], [188].) The concepts of controllability and observability actually first arose as technical conditions in certain optimal control problems [119]. They also enter in a fundamental way [76], [119], in characterizing irreducible transfer functions and minimal state-space realizations of linear systems. Kalman isolated these notions and, for conceptual and other reasons, also defined them in terms of certain idealized but simple control problems; e.g., he observed that controllability is equivalent to being able to take an arbitrary initial state to the origin [62]. However, such definitions are only somewhat incidental, and their main justification lies in the theorems that can be proved with them. Nevertheless, many textbooks deal largely with examinations and elaborations of the definitions of controllability and observability, with hardly a mention of the associated theorems as being the reasons for this great attention. Information theorists may recognize similarities to the fate of the words information and entropy!

In my opinion, it was the peculiar atmosphere of the sixties, with its catchwords of "building research competence," "training more scientists," etc., that supported the uncritical growth of a literature in which quantity and formal novelty were often prized over significance and attention to scholarship. There was little concern for fitting new results into the body of old ones; it was important to have "new" results! Wiener had some interesting comments on the scene as early as 1956 [371, p. 271].

Despite this unfortunate historical context, one should not underestimate the significance of the Kalman filter, which, to repeat, is more than just a solution to a specific estimation problem. As a tribute to this work, I now attempt to add my view concerning the slight controversy that exists as to its origin.

Historical Notes on the Kalman Filter

Recursive solutions to least-squares problems are not of recent origin. Gauss was forced to invent them to handle the vast calculations he undertook in order to help astronomers locate the asteroid Ceres. His work dealt with the discrete-time model (28), where, however, the state x_i was constant (i.e., Φ and Γ were zero). Given hindsight one can generalize this work to handle dynamics and, for example, Rosenbrock has done this in his interesting note [92], (see also a note by Genin in [139]). Incidentally, Whittle in 1963 [32, p. 35] pointed out that the classical Wiener filter could be rewritten in a recursive form as a differential equation, and he also studied some nonstationary extensions [95].[4]

However, the general case was first studied by Kalman [64], who combined state-space descriptions and the notion of discrete-time innovations, as described for example in Doob [225, especially sects. XII.1 and XII.3], to give a complete and elegant solution. Kalman's solution also introduced a nonlinear recurrence equation (30) which was the discrete-time counterpart of a Riccati differential equation he had already encountered in studies on quadratic minimization problems in optimal control [63]. From this it was an easy step to obtain the continuous analog of the discrete-time equation for the least-squares estimate, especially since Kalman also recognized a "duality" between the filtering and control problems. An immediate bonus of his analysis of the steady-state behavior of the Riccati equation in optimal control was the important result that, under the previously mentioned technical conditions of "observability" and "controllability," the finite-time solution converges to a unique steady-state solution, independent of the initial condition and of errors introduced during the computation. [This stability question did not arise in the classical Wiener problem, which roughly speaking, corresponds to a Kalman filter problem with the $F(\cdot)$ matrix in the state-space signal model (19) constant and stable (i.e., having eigenvalues with negative real parts).

[4] Whittle [32], [95], used difference equation (autoregressive-moving average) models, and interestingly enough it is only recently that several advantages of such models have been fully appreciated (see the discussion at the end of Section IX).

Therefore, the signal variance goes to a steady-state value and it can be shown that so does the (always smaller) variance of the error. It is a striking fact, at least without further thought, that the error variance goes to a finite steady-state value under the structural conditions of controllability and observability even if F is unstable, so that the signal variance becomes unbounded.]

In view of these facts it seems fair to use the name Kalman filter for the continuous-time algorithm as well as for its discrete-time analog. The continuous-time filter is often also called the Kalman–Bucy filter, or sometimes the Bucy–Kalman filter (see Part I of [113] and also [136]). Bucy's coauthorship in Kalman and Bucy [69] grew out of some early work by Carlton and Follin [56] and Hanson [57], at the Applied Physics Laboratory of Johns Hopkins University, in which algorithms of the Kalman type were obtained for some special cases. Kalman's discovery of the general continuous-time formulas was apparently independent of this, being based as we have noted on analogies with optimal control [63]. Later Kalman obtained a direct derivation by applying a limiting argument to the discrete-time formulas [68]. Bucy's important contribution to the joint paper [69] was a derivation using the finite-time Wiener–Hopf equation (10). It should also be noted that Siegert in 1953–1955 [58] had already shown in a different context that finite-time Wiener–Hopf equations could be solved by reduction to a Riccati differential equation.

It is also not so widely known that, independently of all this, Stratonovich in the USSR had begun to study recursive solutions for nonlinear least-squares estimates of the states of a nonlinear dynamical system driven by white noise. In this connection, it is natural to consider the linearized problem and its solution, and in so doing Stratonovich in 1960 also obtained the Kalman filter equations ([65], [141, p. 675]). However, no stability analysis was undertaken.

We should also mention that with hindsight one can specialize certain recursive formulas obtained in 1958 by Swerling [61] for nondynamical systems to again obtain the Kalman filter. Swerling did not actually explicitly consider this special case, nor did he anywhere have a Riccati equation. However, as noted earlier, Swerling's papers [61], [155], contain several useful and interesting ideas, for linear and nonlinear filtering, many of which have been widely overlooked. [One such idea will be encountered in Section X.]

As a final comment on this topic, we may note that in a little known 1944 paper [213] (unfortunately not cited in his 1953 book, but added by Yaglom to the 1956 Russian translation) Doob made explicit and effective use of linear state-variable models to study processes with rational spectral density. This paper (see also [219]) contains several formulas and results that were rediscovered much later in the state-space literature.

V. Recursive Wiener Filters

Kalman replaced the conventional specification of the filtering problem in terms of signal and noise covariance functions by one in which state-space models were specified for the signal and noise, and it seemed to many that this difference in specification was the chief reason for the success of the Kalman filter. Therefore, it was thought that to obtain similar computationally efficient recursive solutions for problems with covariance information one should first deduce state-space models consistent with the given covariance specifications, to which the Kalman solution could then be applied. Unfortunately, the problem of determining such state-space models for nonstationary processes, which include stationary processes observed over finite time intervals, is quite difficult. Most known solutions require an amount of work roughly equivalent to that involved in solving the Riccati equation for a process with an already known signal model. Thus it appears in effect that the price paid for starting with covariance information rather than model information is essentially a doubling of work.

However, this is not true. With a proper formulation, the same amount of computation suffices to solve either problem. More specifically, suppose that we return to Shinbrot's covariance specification (18), which we shall rewrite more compactly in matrix notation as

$$K(t,s) = A(t)B(s)1(t-s) + B'(t)A'(s)1(s-t) \quad (31)$$

where $A(\cdot)$ and $B'(\cdot)$ are $p \times n$ matrices and $1(\cdot)$ is the Heaviside unit step function. The meaning of this assumption, which as stated earlier Shinbrot was forced to make for purely mathematical reasons, is that this is the form that $K(t,s) = E[z(t)z'(s) + z(t)v'(s) + v(t)z'(s)]$ must take for the processes $z(\cdot)$ and $v(\cdot)$ in a state-space model (19), (20).

Before proceeding, it will be convenient to rewrite $K(t,s)$ in the form

$$\begin{aligned} K(t,s) &= M(t)\Phi(t,s)N(s)1(t-s) \\ &\quad + N'(t)\Phi'(s,t)M'(s)1(s-t) \end{aligned} \quad (32)$$

where $\Phi(\cdot,\cdot)$ is a so-called state transition matrix defined [106] as the unique solution of the linear differential equation

$$\frac{d\Phi(t,s)}{dt} = F(t)\Phi(t,s), \qquad \Phi(s,s) = I \quad (33)$$

and $F(\cdot)$ is an arbitrary matrix that can be chosen conveniently for the problem at hand. There is no loss of generality in doing this because $\Phi(\cdot,\cdot)$ is nonsingular and obeys $\Phi(t,s) = \Phi(t,t_0)\Phi(t_0,s)$ for an arbitrary t_0. When $F(\cdot)$ is constant

$$\begin{aligned} \Phi(t,s) &= \exp F(t-s) \\ &\triangleq I + F(t-s) + F^2\frac{(t-s)^2}{2!} + \cdots . \end{aligned} \quad (34)$$

For a *given* $F(\cdot)$, the correspondence between (29) and (30) is established by the relations (with t_0 arbitrary)

$$A(t) = M(t)\Phi(t,t_0), \qquad B(t) = \Phi(t_0,t)N(t)$$

$$M(t) = A(t)\Phi(t_0,t), \qquad N(t) = \Phi(t,t_0)B(t).$$

Now it is shown in [149] (see [292], [105], [144], for earlier efforts) that the estimate $\hat{z}(\cdot)$ can be calculated by the following recursive algorithm:

$$\hat{z}(t) = M(t)\phi(t) \tag{35a}$$

where

$$\dot{\phi}(t) = F(t)\phi(t) + K(t)[y(t) - M(t)\phi(t)], \qquad \phi(t_0) = 0 \tag{35b}$$

$$K(t) = N(t) - \Sigma(t)M'(t) \tag{36a}$$

$$\dot{\Sigma}(t) = F(t)\Sigma(t) + \Sigma(t)F'(t) + K(t)K'(t), \qquad \Sigma(t_0) = 0. \tag{36b}$$

The equation for $\Sigma(\cdot)$ is again an $n \times n$ nonlinear matrix differential equation of Riccati type. It is different from the Riccati equation (26) of the Kalman filter, though it is closely related [149]. As expected, (32)–(35) reduce to the Wiener filter formulas (11)–(16) in the special case $p = 1$, $t_0 = -\infty$, and $K(t,s)$ a function only of $|t - s|$. Thus we have now found the recursive generalization of the Wiener filter. To put it another way, we can now see how to make Shinbrot's integral equation solution recursive. A proof of these results is outlined in (128)–(133) of Section IX. (The discrete-time version of these equations can be found in [180].)

The important point is that the equations for $\Sigma(\cdot)$ and $\hat{z}(\cdot)$ can be directly written down from the covariance specifications without first having to determine a state-space model. Thus both specifications, in terms of covariances or in terms of state-space models, are seen to be equivalent, not just in that they give the same final answer (because, of course, they must), but in that their solutions involve comparable amounts of work. The choice between them lies purely in whether state-space models or covariance specifications are more readily at hand. This fact is still not widely appreciated and the literature contains many discussions of attempts to "identify" state-space models from covariance data so as to be able to use a Kalman filter.

Nevertheless, "modeling" is, as in all subjects, a thorny problem and we should say a few more words about it here. State-space models are often at hand in aerospace problems, where we may have enough information to write down the equations of motion, whether they be time invariant or time variant. However, the choice of the proper number of states to model a given problem adequately is not always an easy one. In many problems of industrial process control and communications, it is generally impossible to write down state equations (as is clear if we try to do so for a large power grid, or chemical plant or a telephone-line channel) and recourse has to be had to terminal measurements, for example of the covariance function or power spectrum of the channel output. Now covariance estimation is itself a vast subject, but even if we assume that good estimates are available, $K(t,s)$ will be available only as a numerical function of t and s and not in the factored form (31) or (32); getting the functions $\{A(\cdot),B(\cdot)\}$ or $\{M(\cdot),\Phi(\cdot,\cdot),N(\cdot)\}$ involves a further step of approximation. How can this be done? In the stationary case, where $A(\cdot)$ and $B(\cdot)$ are exponential functions, we have basically the classical network-theory problem of approximating a time function by a sum of exponential functions (or, in frequency-domain terms, approximating a function by a ratio of polynomials). A modern version of this problem is that of obtaining minimal state-space realizations from measurements of transfer functions. There are now several methods available for doing this, and in fact research into more efficient methods is still going on (see, e.g., [129], [360], and the references therein). However, this is only for the stationary case. Although certain analogous procedures can be devised for the nonstationary case [112], in general it is a difficult matter even in network theory to obtain time-variant system realizations.

For this reason, and also because attention is shifting away from aerospace problems, there is renewed interest in time-invariant models. Such models have generally been the only ones studied by statisticians and communication engineers, and recent research in control and system theory has taken a big swing in this direction.

VI. New Algorithms for Time-Invariant Systems

When the parameters in the Kalman state-space model are constant (time invariant), it has recently been discovered that one can obtain recursive solutions without going through a Riccati equation, and in several problems it is possible to obtain significant computational advantages thereby (cf. [166], [183], [184]). One reason for presenting this result is that a special case of it is closely related to algorithms invented in 1943–1947 by the astrophysicists Ambartsumian (USSR) [211] and Chandrasekhar (USA) [217], [220], to solve a class of Wiener–Hopf equations by reduction to nonlinear differential equations. This reduction was actually sought to obtain better numerical procedures, the same fundamental motivation underlying the work a decade later of Swerling, Stratonovich, and Kalman.

We shall start with the general state-space model (19), (20), where now F,G,H,Q,C are assumed to be time invariant. Then it has been shown [183] that the linear least-squares estimate of $z(\cdot)$ can be computed via the equations

$$\hat{z}(t) = H\hat{x}(t) \tag{37a}$$

$$\dot{\hat{x}}(t) = F\hat{x}(t) + K(t)[y(t) - H\hat{x}(t)], \qquad \hat{x}(t_0) = 0 \tag{37b}$$

which are (cf. (22), (23)) as in the Kalman filter, except that now $K(\cdot)$ need not be computed via a Riccati-type equation (27), but through the equations

$$\dot{K}(t) = L(t)SL'(t)H', \qquad K(t_0) = \Pi_0 H' + GC \tag{38a}$$

$$\dot{L}(t) = [F - K(t)H]L(t), \qquad L(t_0) = L_0 \tag{38b}$$

where S and the initial condition matrix L_0 are found as follows. Let

$$D \triangleq F\Pi_0 + \Pi_0 F' + GQG' - (\Pi_0 H' + GC)(\Pi_0 H' + GC)' \tag{39}$$

and suppose that

$$\text{rank of } D = \alpha, \qquad \alpha \le n. \tag{40}$$

Since D is symmetric, we can write it via a standard numerical procedure (called the LDU decomposition, see e.g., [387]) as

$$D = L_0 S L_0' \tag{41}$$

where L_0 is an $n \times \alpha$ matrix and S is the $\alpha \times \alpha$ "signature" matrix of D

$$S = \text{diag}\{1,1,\cdots,1,-1,-1,\cdots,-1\}$$

with as many ones as D has positive eigenvalues. For reasons that will be given later, we shall say that the nonlinear differential equations (38) are of Chandrasekhar-type. This new algorithm determines $K(\cdot)$ directly via the solution of $n(p+\alpha)$ nonlinear differential equations for the $n \times p$ matrix $K(\cdot)$ and the $n \times \alpha$ matrix $L(\cdot)$. In the Kalman filter solution we have $n(n+1)/2$ equations for the components of the matrix $P(\cdot)$, from which $K(\cdot)$ must then be found as $P(\cdot)H' + GC$.

In this generality, there may not be much to gain by choosing one algorithm or the other. However, there are important special cases where there is a decided advantage. For example, if we are very certain about the value of the initial state $x(t_0)$, we might assume that $\Pi_0 = 0$, which can be seen to lead to certain simplifications in D, in particular that $S = I$ and $\alpha \le \min(n,m)$.

Another special case, of particular interest in communications and in system identification, is obtained by assuming that the state and signal processes $x(\cdot)$ and $z(\cdot)$ are in the statistical steady state, i.e., that they are stationary processes. To achieve this requires that we assume F to have eigenvalues with negative real parts and the initial state variance to be

$$\Pi_0 = \bar{\Pi}, \qquad 0 = F\bar{\Pi} + \bar{\Pi}F' + GQG'. \tag{42}$$

Now the matrix D reduces to

$$D = -(\bar{\Pi}H' + GC)(\bar{\Pi}H' + GC)' \text{ and rank } D \le p. \tag{43}$$

Assuming for definiteness that the rank is p, the number of outputs we can take $S = -I$ and $L_0 = \bar{\Pi}H' + GC$, so that no special factorization is needed to specify the equations, which now comprise $2np$ nonlinear equations compared to $n(n+1)/2$ for the Kalman filter. There can be a considerable computational saving when $n \gg p$.

If we assume not a state-space model for the *stationary* signal process $z(\cdot)$, but only knowledge of the covariance function of $z(\cdot)$, we shall be closer to the classical assumptions of the pre-Kalman theory, and will be able to bring out some interesting connections. Thus if (cf. (32))

$$R_y(t,s) = I_p\delta(t-s) + Me^{F(t-s)}N'1(t-s) + N'e^{F'(s-t)}M'1(s-t) \tag{44}$$

then it can be shown [183] that the algorithm is just (37) with (instead of (38))

$$\dot{K}(t) = -L(t)L'(t)H', \qquad K(t_0) = N \tag{45a}$$

$$\dot{L}(t) = [F - K(t)H]Y(t), \qquad L(t_0) = N \tag{45b}$$

which should be compared with the Riccati-type algorithm (35), (36) of the previous section.

We can now point out a close relationship to some famous equations obtained in astrophysics in connection with a Wiener–Hopf equation (10) with $K(t,s)$ of the form

$$K(t,s) = K(|t-s|) = \int_0^1 \exp(-|t-s|\alpha)w(\alpha)\,d\alpha \tag{46}$$

for a certain weighting function $w(\alpha)$. In 1947, Chandrasekhar [217] showed that the solution could be obtained in terms of two functions, now generally known as Chandrasekhar's X and Y functions, obeying the simultaneous nonlinear differential equations

$$\frac{\partial X(t,\alpha)}{\partial t} = -Y(t,\alpha)\int_0^1 Y(t,\beta)w(\beta)\,d\beta \tag{47}$$

$$\frac{\partial Y(t,\alpha)}{\partial t} = -\alpha Y(t,\alpha) - X(t,\alpha)\int_0^1 Y(t,\beta)w(\beta)\,d\beta \tag{48}$$

$$X(0,\alpha) = 1 = Y(0,\alpha), \qquad 0 \le \alpha \le 1. \tag{49}$$

These equations attracted considerable attention and have been studied, especially in transport theory and related fields, by Case [232], Noble [252], Wing [256], and Kalaba, Kagiwada, and Bellman (see [254] and the many references in the survey volume [146]). In 1972, Casti, Kalaba, and Murthy [161] used these results to show that the least-squares estimate $\hat{z}(\cdot)$ could be written

$$\hat{z}(t) = \int_0^1 L(t,\alpha)y(\alpha)\,d\alpha \tag{50}$$

where

$$\frac{\partial}{\partial t}L(t,\alpha) = -\alpha L(t,\alpha) + X(t,\alpha)[y(t) - \hat{z}(t)],$$
$$L(0,\alpha) = 0, \qquad 0 \le \alpha \le 1. \tag{51}$$

With some small effort, the reader should be able to see that these equations are essentially the same as (37), (45a), (45b) if we make the assumptions

$$w(\alpha) = \sum_1^n \alpha_i\,\delta(\alpha - \alpha_i), \quad \alpha_i \ge 0, \quad -F = \text{diag}\{\alpha_1,\cdots,\alpha_n\}. \tag{52}$$

This is why the nonlinear differential equations (38) are said to be of Chandrasekhar type. We should note that the X and Y functions were already introduced by Ambartzumian in [211]. Chandrasekhar [217] first gave the differential forms, which Bellman, Kalaba, and their colleagues began to numerically exploit in the early sixties.

Discrete-Time Models

Analogous results can be obtained for discrete-time problems but the formulas are somewhat more complicated. This happened also with the Kalman formulas but the difference is even more pronounced here. Once again, we use the same estimator equation as in the Kalman filter

$$\hat{x}_{i+1|i} = \Phi\hat{x}_{i|i-1} + K_i[R_i^\varepsilon]^{-1}\varepsilon_i, \qquad \hat{x}_{0|-1} = 0 \tag{53a}$$

$$\varepsilon_i = y_i - H\hat{x}_{i|i-1} \tag{53b}$$

but K_i is found not via the Riccati difference equation (30), but via the equations [184]

$$K_{i+1} = K_i + \Phi L_i[R_i^r]^{-1}L_i'H' \quad (54a)$$

$$L_{i+1} = [\Phi - K_{i+1}[R_{i+1}^e]^{-1}H]L_i \quad (54b)$$

$$R_{i+1}^e = R_i^e + HL_i[R_i^r]^{-1}L_i'H' \quad (55a)$$

$$R_{i+1}^r = R_i^r - L_i'H'[R_i^e]^{-1}HL_i \quad (55b)$$

where

$$K_0 = \Phi_0 H' + \Gamma C, \qquad R_0^e = R + H\Pi_0 H' \quad (56)$$

and L_0, R_0^r are found by factoring the matrix

$$D \triangleq \Phi\Pi_0\Phi' + \Gamma Q\Gamma' - \Pi_0 - K_0[R_0]^{-1}K_0' \quad (57)$$

as

$$D = L_0 \begin{bmatrix} M_+ & 0 \\ 0 & M_- \end{bmatrix} L_0', \qquad M_+ > 0, \qquad M_- < 0. \quad (58)$$

The matrix L_0 has dimension $n \times \alpha$, where α = rank of D. Then L_0 is the initial value for (54b), while

$$R_0^r = \begin{bmatrix} M_+^{-1} & 0 \\ 0 & M_-^{-1} \end{bmatrix}. \quad (59)$$

The form of (59) suggests that we define

$$M_i = [R_i^r]^{-1}$$

in which case (55b) gives the equation

$$M_{i+1} = M_i + M_i L_i'H'[R_i^e]^{-1}HL_iM_i. \quad (60)$$

We could obtain other variations by updating $[R_i^e]^{-1}$ directly. The best choices seem to be those we have given; a count of the number of operations, which is more significant in discrete time than the number of equations, shows that all forms of the new algorithm involve less computation than the Kalman filter. Recently, other variants have been found in which specific equations as in (54) or (60) are replaced by the specification of successive orthogonalizing transformations (e.g., of the Householder type) to be applied to certain data arrays [193c], [193d]. These forms are intimately related to square-root estimation algorithms (see [98a], [128a], [148], and the references therein), and to the ideas of canonical spectral factorization (shades of Wiener, again). This last topic is further discussed at the end of Section IX.

As in continuous time, in various special cases the algorithm can be simplified further, e.g., when $\Pi_0 = 0$ or when $\Pi_0 = \bar{\Pi} = \Phi\bar{\Pi}\Phi' + \Gamma Q\Gamma'$. In the latter case, the processes z and y are stationary, and it can be shown that the relevant equations are closely related to the work of Levinson in 1947 [218]. Curiously, the algorithm for the continuous-time stationary process case is also related to work done in 1947, namely that of Chandrasekhar [217]. While Chandrasekhar's paper was in astrophysics and therefore somewhat inaccessible, it is unfortunate that Levinson's paper, which was reprinted as an appendix in Wiener's monograph [1], has been somewhat overlooked in the literature of communication and control theory, though it has been widely used by geophysicists [79], [96], [110], and very recently in speech analysis [164], [173]. Since they solve similar problems, there must clearly be a close connection between Chandrasekhar's and Levinson's results. Actually, both can be derived by using certain "invariance" principles, which consist basically of "invariantly imbedding" the given problem in a family of similar problems. This will be explained in some detail in Section VII.

We should also note that the techniques used to obtain (38)–(41) and (54)–(58) can also be applied [183], [184], to other problems where Riccati equations arise, even to general two-point boundary value problems whose solution is well known to be obtainable via a nonsymmetric Riccati equation. Furthermore, by using the ideas discussed at the end of Section IX, the results can also be extended to certain classes of time-variant and even nonlinear models. Applications to infinite-dimensional (distributed parameter) problems seem to hold special computational promise since operator Riccati equations on Hilbert × Hilbert spaces are replaced by equations on Hilbert × R^n spaces.

As stressed in [184], [193d], a significant aspect of all the results of this section is that a reevaluation is timely of the almost total concentration on the Riccati equation in the sixties. The rest of this paper hopes to describe some of the concepts that will underlie such a reexamination. These concepts have actually been available for quite a while, but, as stated before, they seem to have been generally neglected, perhaps as historical curiosities. However, many of the results of Sections V and VI would probably not have been developed without awareness of the important role of the Wiener–Hopf equation and of spectral factorization in this field. Our survey of these various ideas can only be partial, but it is deliberately also somewhat tutorial so as to aid the interested reader in making a closer study of the many references that will be noted later.

VII. Some Early Mathematical Work on Linear Least-Squares Estimation

The adjectives in the title might seem strange to someone who has gone through the previous sections; it might also seem a bit presumptuous considering how often the work of Wiener has been mentioned so far. Nevertheless, as stated by Masani [255] in the special Wiener issue of the Bulletin of the American Mathematical Society, the portion of Wiener's work [1] that we have described does not have "the theoretical strength and completeness of that of Kolmogorov." Wiener became aware of this himself when he tackled the multivariate and nonlinear least-squares problems. Masani writes that "Wiener adopted the [Kolmogorov] Hilbertian approach in his later papers under the stimulus of his younger collaborators."

Before embarking upon our examination of the more mathematical work of Kolmogorov and his successors, we should perhaps reassure the reader that the mathematical level of our presentation is not going to take a big jump. "Deeper" mathematics, or even "more abstract" mathematics, does not necessarily entail more formidable mathematical "language." It is quite possible to present the basic ideas of the deeper mathematics in a physical way, and in fact it is no longer a novelty that often the deeper mathematics is closer to physical constructs. The

Schwartz theory of generalized functions is a good example, where the abstract topological notions of generalized equality and convergence are more closely tied to the mechanisms of physical measurements than are the classical pointwise or L^p definitions.

The Work of Wold (1938)

In 1938, just a few years after Kolmogorov had put the theory of probability and stochastic processes on a sound general footing [203], Wold presented a Ph.D. dissertation on discrete-time stationary processes. This dissertation, now available in book form [206], contains several interesting results, of which we mention only a very few, related to our theme. For example, Wold already used the idea proposed by Frechet in 1937 [205] of regarding random variables as elements of a metric space with the distance between two elements being the variance of their difference. This geometric interpretation made it natural to interpret least-squares estimation as projection onto a subspace. It took many years for this natural idea to penetrate the engineering literature, where even in the sixties, strenuous efforts were made in many papers to avoid using the so-called "orthogonality" condition for least-squares estimates. We may mention in passing that Wold was influenced by the work of Frisch [202], where as Wold states, "matrix calculus was for the first time systematically employed in statistics." In 1969 Frisch received the first Nobel Prize in Economics.

One of Wold's major observations was that it simplified calculations to replace a sequence of correlated random variables by an "equivalent" sequence of uncorrelated variables. He also noted that certain processes could be "singular" in that their future values could be predicted exactly from knowledge of their past values. Such processes are nowadays, following Doob [213], called *deterministic* processes. These various ideas were combined into the following fundamental result [206, p. 89].

> Let $y(t)$ be a finite-variance stationary discrete-time process. Then there exist three *jointly stationary* processes $\{\chi(t), \varepsilon(t), \psi(t)\}$
>
> i) $y(t) = \chi(\cdot) + \psi(\cdot)$;
> ii) $\psi(\cdot)$ and $\chi(\cdot)$ are uncorrelated;
> iii) $\psi(\cdot)$ is deterministic and unique;
> iv) $\varepsilon(\cdot)$ has uncorrelated components, $E\varepsilon(i)\varepsilon(j) = \delta_{ij}$;
> v) $\chi(t) = b_0\varepsilon(t) + b_1\varepsilon(t-1) + b_2\varepsilon(t-2) + \cdots$, where $\sum b^2 < \infty$.

This decomposition is now called the *Wold decomposition* and it has been widely used and generalized, especially in functional analysis [245], [250], [261].

The Work of Kolmogorov (1939–1941)

While Wold went on in his thesis to apply his ideas to economic time series, it was left to Kolmogorov to pick up and complete Wold's results on prediction, which he did in a brilliant and comprehensive fashion in the papers [207], [209], [210]. Later Wiener made similar, but less complete efforts, and Masani [255] comments that "so thorough had been Kolmogorov's treatment of univariate prediction in the discrete case that there was little left to do."

Kolmogorov first noted that, though Wold's decomposition was stated as an existence theorem, it becomes a prediction formula as soon as the deterministic process $\{\psi(\cdot)\}$ and the coefficients $\{b_i\}$ of the moving-average process (so named by Kolmogorov $\{\varepsilon(\cdot)\}$) are fixed. Thus suppose we know that $\psi(\cdot)$ is identically zero. Wold was aware, but did not explicitly state in his theorem, that $\varepsilon(t)$ could be uniquely determined by $\{y(t), y(t-1), \cdots\}$. In fact, Wold essentially constructed the $\varepsilon(\cdot)$ sequence by successive Gram–Schmidt orthogonalization, for which this property is obvious. (We say "essentially" because Wold was dealing with an infinite sequence $\{y(t), y(t-1), \cdots\}$ and a possibly nonzero deterministic $\psi(\cdot)$, so that a careful double limiting procedure had to be used.) This property was explicitly introduced and exploited by Kolmogorov as follows. By Wold's theorem

$$y(t) = b_0\varepsilon(t) + b_1\varepsilon(t-1) + b_2\varepsilon(t-2) + \cdots \quad (60a)$$

where the $\{\varepsilon(t), \varepsilon(t-1), \cdots\}$ are uncorrelated random variables that can be computed from $\{y(t), y(t-1), \cdots\}$ by linear operations. Also

$$y(t+1) = b_0\varepsilon(t+1) + [b_1\varepsilon(t) + b_2\varepsilon(t-1) + \cdots].$$

However, as just noted, the terms in the square bracket are completely determined by knowledge of past $y(\cdot)$; moreover, these terms are not correlated with $\varepsilon(t+1)$. Therefore, we have

$$\hat{y}(t+1 \mid t) \triangleq \text{linear least-squares estimate of } y(t+1) \text{ given } \{y(t), y(t-1), \cdots\}$$
$$= b_1\varepsilon(t) + b_2\varepsilon(t-1) + \cdots.$$

This solves the prediction problem. Here

$$b_0\varepsilon(t+1) = y(t+1) - \hat{y}(t+1 \mid t)$$

so that one may call $\varepsilon(t+1)$ the "new information" or the "innovation" in the process $y(\cdot)$ at time $t+1$, and the process $\varepsilon(\cdot)$ may be called the *innovations process* of $y(\cdot)$, a name that was apparently first used by Wiener and Masani in the mid-fifties (personal communication in 1968 from P. Masani). (See also a 1960 paper by Cramer [268].) We shall see that such processes play a fundamental role in our understanding of the process $y(\cdot)$ in both discrete and continuous time (Sections VIII and IX).

So far we have chiefly an easy application of Wold's theorem. Kolmogorov went on to deepen Wold's theorem by relating it to properties of the so-called integrated power spectrum of the process $y(\cdot)$. With the covariance

$$R(i-j) \triangleq Ey(i)y(j)$$

Wold had shown that there exists a nondecreasing function $F(\lambda)$ called the *integrated power spectrum* of $y(\cdot)$ such that

$$R(k) = \int_{-1/2}^{1/2} \exp(i2\pi k\lambda)\, dF(\lambda).$$

In general $F(\cdot)$ will consist of an absolutely continuous part, a jump part, and a singular part (continuous and nondecreasing but with zero derivative almost everywhere). Kolmogorov showed that the deterministic part of a process $y(\cdot)$ is identically zero if and only if $F(\lambda)$ is absolutely continuous and its derivative $\dot{F}(\lambda)$ satisfies

$$\int_{-1/2}^{1/2} \ln \dot{F}(\lambda)\, d\lambda > -\infty. \tag{61}$$

He also gave explicit formulas for the coefficients $\{b_i\}$ in terms of the Fourier series coefficients of $\ln \dot{F}(\lambda)$[5] and finally showed that the one-step prediction error has a simple form

$$\lim_{t\to\infty} E[y(t) - \hat{y}(t \mid t-1)]^2 = \lim_{t\to\infty} E\varepsilon^2(t) = \exp \tfrac{1}{2} \int_{-1/2}^{1/2} \ln \dot{F}(\lambda)\, d\lambda. \tag{62}$$

This last formula, which is valid for all processes (even those with a nonzero deterministic part), is closely related to the Yovits–Jackson formula (17) mentioned in Section III (cf. [48], [54]).

These remarkable results were obtained by connecting the study of stationary random processes with that of certain deterministic functions in the frequency domain. Kolmogorov did this via the relationship

$$Ey(k)y(l) = R(k-l) = \int_{-1/2}^{1/2} \exp i2\pi\lambda(k-l)\, dF(\lambda). \tag{63}$$

It will be useful later to set

$$z = \exp i2\pi\lambda$$

so that (63) can also be written

$$Ey(k)y(l) = R(k-l) = \oint z^k z^{-l} \frac{dF(z)}{z} \tag{64}$$

where the special symbol denotes integration around the unit circle. The left side of (63) can be regarded as an inner product between the random variables $y(k)$ and $y(l)$ in a Hilbert space of random variables formed from finite linear combinations of the $\{y(k)\}$ and their limits in the covariance norm. Similarly, because $F(\lambda)$ is nondecreasing, the right side can be regarded as an inner product between $\exp i2\pi k\lambda$ and $\exp i2\pi l\lambda$ in the space $L_2(dF)$ of functions square-integrable with respect to the measure $dF(\lambda)$. Similarly the right side of (64) defines an inner product between the powers z^k in the space of polynomials on the unit circle. Therefore, we have an obvious norm-preserving mapping (or isometry) between these spaces in which

$$y(k) \leftrightarrow \exp i2\pi k\lambda \leftrightarrow z^k \tag{65}$$

$$\sum_1^n a_k y(k) \leftrightarrow \sum_1^n a_k \exp i2\pi k\lambda \leftrightarrow \sum_1^n a_k z^k. \tag{66}$$

[5] Incidentally, this defines the so-called "cepstrum" of $y(\cdot)$, which has had a vogue in signal processing recently [364]–[368].

These λ-functions can be extended from the interval $(-\frac{1}{2},\frac{1}{2})$, or equivalently from the unit circle, into the complex plane, an idea of Hardy's [199] that had been extensively studied by Paley and Wiener [204]. Nevertheless, it was Kolmogorov who exploited these ideas in general prediction.

A Formula of Szegö's (1915)

Kolmogorov's explicit formula (62) for the mean-square error had already been discovered in a different, but isomorphic problem by G. Szegö in 1915 [200], [201]. The isomorphism was precisely the one introduced by Kolmogorov, according to which the problem of choosing $\{a_j\}$ to minimize

$$\sigma_n^2 \triangleq E\left[y(n) - \sum_0^{n-1} a_j y(j)\right]^2 \tag{67}$$

is the same as that of choosing them to minimize

$$\sigma_n^2 = \int_{-1/2}^{1/2} \left|\exp(i2\pi\lambda n) - \sum_0^{n-1} a_j \exp(i2\pi\lambda j)\right|^2 dF(\lambda) \tag{68}$$

or

$$\sigma_n^2 = \oint \left|z^n - \sum_0^{n-1} a_j z^j\right|^2 \frac{dF(z)}{z}. \tag{69}$$

Thus the problem of minimizing σ_n^2 by suitable choice of the $\{a_i\}$ is just a problem of *polynomial approximation* on the unit circle. This problem was solved for absolutely continuous $F(z)$ by Szegö, and rederived by Kolmogorov for general $F(z)$. The connection to Szegö's work was noted by Krein [214] and later by Grenander [223].

The Work of Krein (1944–1945)

In response to questions raised by Kolmogorov, Krein in 1944–1945 [214], [215], showed how Kolmogorov's results could be extended to continuous time by use of a simple bilinear transformation. To each *discrete-time* stationary process with integrated spectrum $F(\lambda)$ we can associate a stationary *continuous-time* process with integrated spectrum $S(f)$, where

$$F(\lambda) = S(f), \qquad \lambda = \frac{1}{\pi}\tan^{-1} f \tag{70}$$

$$\exp i2\pi\lambda = \frac{1+if}{1-if}, \qquad \tan \pi\lambda = \frac{\exp(2\pi i\lambda) - 1}{\exp(2\pi i\lambda) + 1} = if. \tag{71}$$

This transformation has the useful property of preserving causality, and therefore it is often used in digital signal processing (see, e.g., [367]).

Use of the bilinear transformation shows easily [225, ch. XII] that the necessary and sufficient condition for no deterministic part is (compare (64))

$$\int_{-\infty}^{\infty} \frac{\ln \dot{F}(f)}{1+f^2}\, df > -\infty. \tag{72}$$

The bilinear transformation can also be extended to multivariable systems [233] and to systems in state-space form

[163],[6] where it has been exploited to give a new technique for solving the steady-state Riccati equation. The bilinear transformation was also used in some generality by Masani and Robertson [246], [35], [45], who paid particular attention to the question of how the discrete-time innovations process goes over to continuous time. We may note that the discrete-time one-step prediction error formula has no continuous-time analog, but there is a continuous-time version of the Wold decomposition and of the innovations process $\varepsilon(\cdot)$. These are important results, which we shall discuss further in the next section.

Krein has made several other important contributions to filtering theory. In 1954, he discovered that the spectral analysis of a weighted string, in which he had been interested since 1940, enabled him to obtain [227] some deep results on estimation given a finite data segment. Krein's analysis also led him to several other results on the solution of integral equations, the so-called "inverse-scattering" problem (see, e.g., [224], [226]). Recently, Dym and McKean [259], [260] have pursued Krein's ideas even further.

The Work of Levinson (1947)

In the USA, work on least-squares estimation proceeded along different lines. Wiener's basic ideas, rather than his ingenious solution of the problem, influenced work on anti-aircraft devices, where the need to find computationally simple solutions led to some alternative approaches. Thus Phillips [212] began with the assumption that the optimum filter has a rational transfer function and used the mean-square-error criterion to solve for the coefficients. Others, apparently including Blackman, Bode, and Shannon [3], tried to incorporate the dynamical constraints on the targets into the prediction schemes.

Levinson [218] formulated the problem in discrete time. In his words, "A few months after Wiener's work appeared, the author, in order to facilitate computation procedure, worked out an approximate, and one might say, mathematically trivial procedure." Levinson's deprecatory comments notwithstanding, this work has had an important impact on the field, both directly and indirectly. His equations were rediscovered in 1960 by Durbin [239] in a scheme for recursive fitting of autoregressive models to scalar time-series data. Whittle [81], [32] extended these recursions to multivariate time series, and his work has been widely used by statisticians. Levinson's work was directly used and extended to multivariate series by workers in geophysics, especially Robinson, with various contributions by groups in the Geology Department at M.I.T. and in the oil industry (cf. [110] and the references therein). These algorithms are now being used in speech analysis (see, e.g., [164], [173]) and in spectral estimation [107], [156], [182]. There are also close relations to the theory of orthogonal polynomials [208], [236], [241] and to the algorithms of Section VI. These and other connections will be briefly discussed after we present the basic problem and its solution.

[6] See also Popov [87], [87a], papers that contain several important ideas on spectral factorization and innovations representations (see especially [87, sects. 5, 7, and appendix E, F).

Given a segment of a stationary time-series $\{y(0), y(1),\cdots, y(N-1)\}$, where the $\{y(i)\}$ are p-vectors, we wish to find the optimum one-step prediction

$$\hat{y}_{N|N-1} \triangleq -\sum_{i=0}^{N-1} A_{N,N-i}y(i). \tag{73}$$

Let

$$R_{i-j} \triangleq Ey(i)y'(j)$$

then by using the orthogonality property of least-squares estimates, we will have the equations

$$R_{N-j} = -\sum_{i=0}^{N-1} A_{N,N-i}R_{i-j}, \qquad j = N-1,\cdots,0. \tag{74}$$

The mean-square error is given by

$$R_N^\varepsilon \triangleq = E[y_N - \hat{y}_{N|N-1}]y_N' = R_0 + \sum_0^{N-1} A_{N,N-i}R_{i-N}. \tag{75}$$

Since R_N^ε is a nonincreasing function of N, its value can be used to decide whether it is necessary to collect more data (i.e., increase N) in order to achieve a desired mean-square error. As stressed by Levinson, this makes it important to find a way of successively calculating R_N^ε, $N = 0,1,\cdots$. The first step is rearrange the filter equations (74) and the error equation (75) in a single block-Toeplitz matrix equation

$$[I,A_{N,1},\cdots,A_{N,N}]\underbrace{\begin{bmatrix} R_0 & R_1 & \cdots & R_N \\ R_{-1} & R_0 & \ddots & \\ \vdots & \ddots & \ddots & R_1 \\ R_{-N} & \cdots & R_{-1} & R_0 \end{bmatrix}}_{\mathscr{R}_N} = [R_N^\varepsilon,0,\cdots,0] \tag{76}$$

where $\mathscr{R}_N$ is a block-Toeplitz matrix. The unknowns are the $\{A_{N,i}\}$ and R_N^ε. The aim is to determine R_{N+1}^ε and $\{A_{N+1,i}\}$ in a way that takes maximum advantage of the previous computations made to find R_N^ε and $\{A_{N,i}\}$. It takes almost as long to describe the result as it does to give a derivation, following Robinson [110, ch. 6]. We shall therefore do so here, partly also with the hope that readers may recognize analogies with similar procedures in other problems.

The method is first to try an "obvious" solution, pushing our luck the most by assuming that adding a zero to the previous solution may work. It will if in the resulting equation

$$[I,A_{N,1},\cdots,A_{N,N},0]\mathscr{R}_{N+1} = [R_N^\varepsilon,0,\cdots,0,\alpha_N] \tag{77}$$

the term $\alpha_N = R_{N+1} + \sum_1^{N-1} A_{N,i}R_{N-i}$ is zero. If this does not happen, we have to find a (simple) way of forcing α_N to zero. For this we introduce the "auxiliary" (adjoint or reversed) equations

$$[0,B_{N,N},\cdots,B_{N,1},I]\mathscr{R}_{N+1} = [\beta_N,0,\cdots,0,R_N^r] \tag{78}$$

where at stage N we assume that we know $\{A_{N,i},R_N^\varepsilon,\alpha_N\}$ and $\{B_{N,i},R_N^r,\beta_N\}$. [For $N = 1$, we take $B_{1,1} = I$, $R_0^r = R_0 = R_0^\varepsilon$.] Next we form a weighted combination of (77),

(78)

$$[I, A_{N,1} + K_N{}^\alpha B_{N,N}, \cdots, K_N{}^\alpha]\mathscr{R}_{N+1} = [R_N{}^\varepsilon + K_N{}^\alpha \beta_N, 0, \cdots, 0, \alpha_N + K_N{}^\alpha R_N{}^r]$$

from which it is easy to see that choosing

$$K_N{}^\alpha = -\alpha_N [R_N{}^r]^{-1} \tag{79}$$

gives us a solution of the extended equation, i.e.,

$$[I, A_{N+1,1}, \cdots, A_{N+1,N+1}] = [I, A_{N,1}, \cdots, A_{N,N}, 0] + K_N{}^\alpha [0, B_{N,N}, \cdots, B_{N,1}, I]. \tag{80}$$

Similarly with the help of

$$K_N{}^\beta = -\beta_N [R_N{}^\varepsilon]^{-1}$$

we can update the $\{B_{N,i}\}$ as

$$[B_{N+1,N+1}, \cdots, B_{N+1,1}, I] = [0, B_{N,N}, \cdots, B_{N,1}, I] + K_N{}^\beta [I, A_{N,1}, \cdots, A_{N,N}, 0]. \tag{81}$$

These relations were independently discovered by Whittle [81] and Wiggins and Robinson [96]. However, the latter had the benefit of an observation of Burg that

$$\alpha_N = \beta_N{}'. \tag{82}$$

The recursions for $R_N{}^\varepsilon$ and $R_N{}^r$ can be seen to be

$$R_{N+1}^\varepsilon = R_N{}^\varepsilon - \alpha_N [R_N{}^r]^{-1} \beta_N \tag{83}$$

$$R_{N+1}^r = R_N{}^r - \beta_N [R_N{}^\varepsilon]^{-1} \alpha_N. \tag{84}$$

Some Connections and Extensions

We should point out that the method of using a partial solution and a subsequent correction using a "reversed-time" auxiliary solution is the same in spirit as the method of invariance introduced into astrophysics by Ambartsumian [211] and Chandrasekhar [217]. This is a powerful idea, whose main thrust can be recognized in many problems—e.g., in the method of adjoints (or influence functions) in optimization problems [128] or in the Berlekamp–Massey decoding algorithm for BCH codes [289], [354], to mention only a few problems that may be of interest to readers of this article (see also [191]).

Here we note the obvious similarity between (54), (55) and (83), (84). In [185] we have given a stochastic interpretation of the previous algorithm, in which the Burg identity (82) arises very naturally. These ideas are then extended to a general study of decompositions of a time-series (into *conditional innovations* and *residuals*) that yield a certain statistical shift-invariance property. This property, analogous in many ways to the natural shift-invariance of stationary processes, enables "imbedding" derivations of the discrete-time algorithms (53)–(60) of Section VI and of certain so-called "Fast-Cholesky" algorithms for triangularizing nonnegative-definite block-Toeplitz and certain related matrices [191], [193].

Orthogonal Polynomials

It is natural to describe the relations (80)–(83) in polynomial language. Let

$$A_N(z) = A_{N,N} + \cdots + A_{N,1} z^{N-1} + z^N \tag{85}$$

where z is an indeterminate and similarly define $B_N(z)$, $A_{N+1}(z)$, $B_{N+1}(z)$. Then the recursions (81)–(83) can be written compactly as

$$\begin{bmatrix} A_{N+1}(z) \\ B_{N+1}(z) \end{bmatrix} = \begin{bmatrix} I & zK_N{}^\alpha \\ zK_N{}^\beta & I \end{bmatrix} \begin{bmatrix} A_N(z) \\ B_N(z) \end{bmatrix} \tag{86}$$

which turn out to be exactly the recurrence formulas for orthogonal polynomials on the unit circle [236], [241]. These are polynomials $\{A_N(z), |z| = 1\}$ such that

$$\langle A_i(z), A_j(z) \rangle \triangleq \oint A_i(z) \frac{dF(z)}{z} A_j{}'(z) \tag{87}$$

$$= \delta_{ij} R_i{}^\varepsilon \tag{88}$$

where $F(\cdot)$ is the matrix integrated power spectrum (cf. (63)) defined by

$$\oint z^i z^{-j} \frac{dF(z)}{z} = \text{a covariance function, } R_{i-j}. \tag{89}$$

A striking thing about these polynomials is that, unlike the classical orthogonal polynomials on the line, to get a nice recursion one has to introduce "auxiliary" polynomials $\{B_i(z)\}$ and define a simultaneous recursion for the $\{A_i(z)\}$ and the $\{B_i(z)\}$.

This fact was first noticed by Szegö [200] who considered the scalar case in which the previous relations simplify because the Toeplitz matrix $\mathscr{R}_N$ will be symmetric rather than just block-symmetric. Now it can be seen that

$$B_N(z) = z^N A_N(z^{-1}) = z^N + A_{N,N} z^{N-1} + \cdots + A_{N,1}$$

$$= \text{the "reverse" of the polynomial } A_N(z). \tag{90}$$

In 1961, Baxter [240] extended Szegö's formulas to scalar nonsymmetric Toeplitz matrices, of which the symmetric block matrix form is a special case. In related work, Baxter [242], [247], Devinatz [249], Ibragimov [251], and others have studied the asymptotic behavior of the error in Wiener filtering for quite general classes of stationary processes. These general techniques can also be usefully adapted to Kalman filters.

To close this section, we shall point out a connection with innovations that can be useful in generalizing several of the previous results to certain classes of nonstationary processes. For this we recall the isometric mapping (64)–(66) between the space of random variables and the space of polynomials on the unit circle under which $y(k) \leftrightarrow z^k$. Now, given $\{y(0), y(1), \cdots\}$, it is natural, following Wold [206], to consider the orthogonalized innovations sequence, say,

$$\varepsilon(0) = y(0) = I \cdot y(0)$$

$$\varepsilon(1) = y(1) - \hat{y}(1 \mid 0) = y(0) + A_{1,1} y(0).$$

$$\vdots \qquad \vdots$$

By our mapping each of the innovations $\varepsilon(N)$ corresponds to a polynomial $A_N(z) = I + A_{N,N} z + \cdots + A_{N,1} z^N$, and these polynomials must be orthogonal. In other words, Szegö's orthogonal polynomials are the images of the innovations under the isometric mappings (65), (66). Many properties of orthogonal polynomials can be interpreted as properties of innovations and vice versa. For example,

Szegö's classic result that the polynomials $A_N(z)$ have all their roots inside the unit circle, so that $A_N^{-1}(z)$ can be regarded as the transfer function of a bounded filter, is the analog of the fact that the transformation between a discrete-time process $y(\cdot)$ and its innovations is boundedly reversible.

So far, we have only noted connections with known results. On the other hand, innovations can be defined for large classes of processes besides stationary discrete-time sequences. Therefore, we have the possibility of obtaining various generalizations of the previous results and in particular of the classical theory of orthogonal polynomials. (See also [248a].)

These possibilities provide additional motivation for going on to a deeper study of innovations and the Wold representation.

VIII. Canonical Representations of Continuous-Time Processes

The concept of innovations processes was introduced for stationary discrete-time processes via the Wold representation theorem. The continuous-time decomposition, which was obtained by different methods by Krein [215], Karhunen [222], and Hanner [221], is a natural generalization of the discrete-time representation. It contains a regular part plus a deterministic part, which will be absent if and only if condition (72) is met, as we shall henceforth assume for convenience of writing. Then the Wold representation is

$$y(t) = \int_{-\infty}^{t} g(t-u)\, dE(u) \tag{91}$$

where $E(\cdot)$ is a process with uncorrelated increments, and the integral is the so-called Wiener stochastic integral, cf. Doob [225, ch. IX]. There are many kernels $g(\cdot)$ that can be used in the previous representation, as we shall see later, but there is always a particular one $g_0(\cdot)$ such that $y(\cdot)$ and $E(\cdot)$ are *causally equivalent* in the sense that any finite-variance random variable linearly dependent on $\{E(u), u \leq t\}$ can also be calculated by linear operations on $\{y(s), s \leq t\}$ and vice versa. In this case, we shall say that the Wold representation is *canonical*, or that

$$y(t) = \int_{-\infty}^{t} g_0(t-u)\, dE(u) \tag{92}$$

is an *innovations representation* of $y(\cdot)$, and we shall call $E(\cdot)$ the *innovations process* of $y(\cdot)$. This can be rewritten in a form closer to Wold's (cf. [60a])

$$y(t) = \int_{-\infty}^{t} g_0(t-u)e(u)\, du, \qquad e(u) = dE(u)/du.$$

However, $e(\cdot)$ is now a continuous time "white noise" process, so that the differentiation of $E(\cdot)$ is not valid in a classical sense, but only in the sense of generalized functions. However, as long as one is at most concerned with a white noise process (and not any of its generalized derivatives), there is no need to introduce the machinery of generalized processes; it suffices to work with the integrated process of uncorrelated increments $E(\cdot)$ and the notion of Wiener stochastic integrals. [The situation is analogous to that in deterministic system theory, where we can use Heaviside functions and Stieltjes integrals to handle singularities that are at worst delta functions.] Even when the second approach is taken, the concept of white noise is still very convenient and useful, as engineers have long known. A thought-provoking example is provided by the development of general series expansions for random processes, which were known to engineers much before they were formally discovered by mathematicians (cf. Section X).

However, as the mathematical sophistication of the field increases, there is a tendency to uncritically reject the use of white noise. This I believe is a serious mistake, which delays for many potential users the appreciation of several general and useful results that may have been originally derived in a more abstract context. Moreover, at the very least it can be a powerful guide to our intuition and a hedge against many unfruitful investigations. It may be apt to note that, according to Doob [253a], it was Wiener who first "showed, and applied repeatedly, that the [process $E(\cdot)$] acts as though [its] derivative process exists, is stationary and has a constant spectral density." Doob's book [225, pp. 435–436, p. 533, pp. 546–547] has some nice examples of how white noise can be usefully employed even in a rigorous exposition of stochastic processes addressed to mathematicians. A recent book of Hida [295] stresses the importance of white noise and this has been further reinforced by recent work of Hida and of McKean [323]. Rao has developed perhaps the most general results to date [305], studying more white noise processes than those defined as the derivatives of processes with independent increments [225].

Properties of Canonical Kernels

Karhunen and others have studied the kernels $g(\cdot)$ and $g_0(\cdot)$ in some detail. Karhunen [222] showed that all $g(\cdot)$ that could serve to define the Wold representation (91) are of the form

$$g(t) = \int \exp(i\omega t)G(\omega)\frac{d\omega}{2\pi}$$

$$G(\omega) = \lim_{\sigma \to 0} G(s), \qquad s = \sigma + i\omega$$

$$G(s) = A(s)G_0(s) \tag{93}$$

where

$$G_0(s) = \exp\left[\frac{1}{\pi}\int_{-\infty}^{\infty} \frac{1 - i\lambda s}{s - i\lambda}\frac{\log \dot{F}(\lambda)}{1 + \lambda^2}\, d\lambda\right]$$

$$\dot{F}(\lambda) = \text{the power spectral density of the process } y(\cdot) \tag{94}$$

and

$$A(s) = \text{transfer function of an "all-pass" system.}$$

Any such all-pass transfer function can be further decomposed into

$$A(s) = A_0A_1(s)A_2(s)A_3(s) \tag{95}$$

where

A_0 = constant of magnitude 1 (called a trivial all-pass function)

$$A_1(s) = \prod_k \frac{s_k - s}{\bar{s}_k + s} \cdot \frac{|s_k - 1|}{s_k - 1} \cdot \frac{|s_k + 1|}{s_k + 1} \quad \text{(Blaschke product)}$$

$$\text{Re } s_k > 0, \qquad \Sigma \text{ Re } s_k/1 + |s_k|^2 < \infty \tag{96}$$

$$A_2(s) = e^{-\alpha s} \quad \text{(pure delay)} \tag{97}$$

$$A_3(s) = \exp\left[-\frac{1}{\pi}\int_{-\infty}^{\infty} \frac{1 - i\lambda s}{s - i\lambda} d\beta(\lambda)\right] \quad \text{(singular part)} \tag{98}$$

where $\beta(\cdot)$ is a nondecreasing function whose derivative vanishes almost everywhere, and $0 < \beta(\infty) - \beta(-\infty) < \infty$. The $\{A_i\}$ all have unit modulus along the $i\omega$ axis, so that

$$|G(i\omega)| = |G_0(i\omega)|. \tag{99}$$

Also, for all but A_0 we have

$$|A_i(s)| < 1, \qquad \text{Re } s > 0, \qquad i = 1,2,3. \tag{100}$$

If we restrict $G(s)$ to being rational, then $A_2(s)$ and $A_3(s)$ will not appear and the so-called Blaschke product in $A_1(s)$ will be finite ($k < \infty$). Note that all the poles of $A_1(s)$ must be in the left half-plane (corresponding to causality, since the region of definition of all our functions here includes the $i\omega$-axis), though there can be zeros in the right half-plane. These results were obtained by using classical results on bilateral Laplace transforms (cf. Paley and Wiener [204]) and the so-called Hardy functions (see Hoffman [244] and Duren [258] for recent accounts). It may be of interest that the decomposition for $G(s)$ was obtained before Karhunen and Krein by Krylov [262] in a purely mathematical study of the transforms of "one-sided" (causal) time functions. Such results have again become of interest in recent studies on infinite-dimensional realization theory [307].

Karhunen proved that

$$G(s) = A_0 G_0(s) \tag{101}$$

is a necessary and sufficient condition for

$$L_2(y; t) = L_2(E; t), \qquad -\infty \le t \le \infty \tag{102}$$

where

$L_2(y; t)$ = the linear space of all finite linear combinations of $\{y(s), s \le t\}$, and their limits in the covariance norm. (103)

For all other $G(s)$, we have

$$L_2(y; t) \subseteq L_2(E; t), \qquad -\infty \le t \le \infty.$$

This explains the names canonical representation and innovations representation for (92). We note that $G_0(s)$ and $A(s)$ are sometimes, following Beurling [264], [312], called "outer" and "inner" functions, respectively.

There are various other interesting properties of the canonical kernel $g_0(\cdot)$, some of which would seem to be useful in developing adaptive filters [38], [173], [55].

1) Let $f(\cdot)$ be a square-integrable function on $(-\infty,\infty)$. Then

$$\int_{-\infty}^{t} g_0(t - u) f(u)\, du = 0, \qquad t \le 0,$$

$$\text{if and only if } f(t) = 0, \qquad t \le 0. \tag{104}$$

This property was discovered by Karhunen [222].

2) The canonical kernel has maximum partial energy in the sense that

$$\int_0^t g_0^2(u)\, du \ge \int_0^t g^2(u)\, du, \qquad \text{all } t > 0 \tag{105}$$

and all causal or noncausal $g(\cdot)$ with $|G(i\omega)| = |G_0(i\omega)|$. This property was given in 1962 by Robinson [271], though closely related results were also noted by Lévy in the mid-fifties (see, e.g., [266, p. 140]). See also [317].

3) The set $\{g_0(t - r), r \ge 0\}$ spans $L_2(0,\infty)$. This last result is due to Beurling [264] and provides a version for L_2 of a famous theorem of Wiener's that the Banach space L_1 can be spanned by the translates of a function with positive Fourier transform.

4) Let $G(s)$ be rational and let

$$G(i\omega) = |G(i\omega)| \exp i\phi(i\omega). \tag{106}$$

Then $\phi(\cdot)$ is called the *phase lag* of the filter $G(\cdot)$ and $-d\phi(i\omega)/d\omega$ is called the *group delay*. Of all filters with the same "gain" $|G(i\omega)|$, $G_0(\cdot)$ has the smallest phase lag and group delay. This is often labeled the *minimum-phase property* of $g_0(\cdot)$ (see [263]).

Applications to Prediction

It should be clear (as noted by Krein and Karhunen) that the innovations representation (92) can be used to solve the prediction problem just as in the discrete-time case. In fact we have

$$\hat{x}(t + \alpha \mid t) = \int_{-\infty}^{t} g_0(t - u)\, dE(u), \qquad \alpha > 0. \tag{107}$$

Of course, the difficult thing is to explicitly calculate $\{E(u), u \le t\}$ from $\{x(u), u \le t\}$. For rational $G_0(\cdot)$ the solution is simple: just pass $x(\cdot)$ through the filter with transfer function $[G_0(s)]^{-1}$. (In principle, this is what we do in general, but care has to be taken in defining the inverse transfer function.) Thus we now have an alternative basically probabilistic approach to the continuous-time prediction problem, without bringing in any Wiener–Hopf equations.

These ideas were rediscovered by Bode and Shannon [5], and Zadeh and Ragazzini [6], though not in as much mathematical generality (only for rational spectra). These authors recognized very clearly that what was involved in the innovations approach was 1) replacing, causally and without loss of information, the given process by a "simpler" process, and 2) solving the estimation problem for the simpler process. So far we have used white noise as a simpler process, but it is not necessarily the only such process. For example, Zadeh and Ragazzini [6] show how we may use a process with spectral density equal to the numerator of the spectral density of the original process. This idea was rediscovered for discrete-time autoregressive-moving average processes by Rissanen and Barbosa [131] (see also Whittle [32]) and has been exploited in several recent papers [175], [192], [184]. In principle, there are many other possibilities as well, and studies of "generalized innovations processes" have been made by Kailath and Duttweiler [311], Kallianpur and Oodaira [321], and Rozanov [314], [325].

Nonstationary Processes

The cited work up to 1950 was all for stationary processes and used frequency-domain methods. In 1950 Hanner [221] gave a purely time-domain derivation of the continuous-time Wold decomposition, and this raised the possibility of extensions to nonstationary processes. The first results in this direction were perhaps those of Yaglom and Pinsker [265], who studied nonstationary processes that had stationary increments of order n. The simplest example is the Wiener process, which has stationary increments of first order, and can be represented as the integral of white noise. The general case was first studied by Lévy [260], [274], who sought representations of the form

$$y(t) = \int_0^t g(t,u)\, dE(u), \qquad 0 \le t \le T \le \infty. \tag{108}$$

When

$$L_2(y;\, t) = L_2(E;\, t) \tag{109}$$

he called the representations *proper canonical*, using the word *canonical* for the case $L_2(y;\, t) \subseteq L_2(E;\, t)$. This terminology has now been abandoned. The covariance of a process as in (108) is

$$R(t,s) = \int_0^{t\wedge s} g(t,u)g(s,u)\, du, \qquad t\wedge s = \min(t,s). \tag{110}$$

Now Lévy asked whether, given a covariance $R(t,s)$, one could find a suitable function $g(t,u)$. He did not obtain a general solution to this problem, but among other results, he did obtain the following useful one.

Let

$$R(t,s) = t\wedge s + \int_0^t \int_0^s K(u,v)\, du\, dv, \qquad 0 \le t,\, s \le T \tag{111a}$$

$$= \int_0^t \int_0^s [\delta(u - v) + K(u,v)]\, du\, dv \tag{111b}$$

where $K(\cdot,\cdot)$ is a continuous symmetric function of two variables. Assume that the eigenvalues of $K(\cdot,\cdot)$ on the square $[0,T] \times [0,T]$ are greater than -1, or equivalently that $R(\cdot,\cdot)$ is strictly positive definite on this square. Next determine a function $h(\cdot,\cdot)$ as the unique solution of the Wiener–Hopf type of equation

$$h(t,s) + \int_0^t h(t,u)K(u,s)\, du = K(t,s), \qquad 0 \le s \le t \le T. \tag{112}$$

Let $k(\cdot,\cdot)$ be the so-called resolvent function of $h(\cdot,\cdot)$, defined by the equation

$$h(t,s) - k(t,s) = \int_0^t k(t,u)h(u,s)\, du. \tag{113}$$

Then the innovations representation is

$$y(t) = \int_0^t \left[1 + \int_u^t k(\tau,u)\, d\tau\right] dE(u). \tag{114}$$

This is a useful result, though it is a bit difficult to see the rationale for the steps (see some following comments by Hida). We shall give an explanation in Section IX.

Although Lévy [266] noted that his method applied to certain more general kernels, e.g., those that were an α-fold double integral of $[\delta(t - s) + K(t,s)]$ (cf. (111)) with α not necessarily integral, he was unable to prove that (110) always has a solution, or, equivalently, that an innovations representation always exists for a (nondeterministic) process. The reason for this failure appeared only a few years later, when in 1960 Cramér [268] and Hida [269] independently discovered a new dimension to this problem, finding that a single kernel $g(\cdot,\cdot)$ does not suffice in general. The proper Wold decomposition for a finite-variance nonstationary process is

$$y(t) = \sum_1^N \int_0^t g_i(t,u)\, dE_i(u) + \psi(t) \tag{115}$$

where $\psi(\cdot)$ is a deterministic process, and the $\{E_i(\cdot)\}$ are orthogonal-increment processes, uncorrelated with each other. The number N is uniquely determined by the covariance of $y(\cdot)$, though the $\{g_i\}$ and $\{E_i(\cdot)\}$ are not. N is called the *multiplicity* of the process $y(\cdot)$, and if $N > 1$ the representation (115) is called a generalized canonical representation. The multiplicity N can be infinite [273], even though all presently known examples of processes with $N > 1$ have rather pathological kernel functions $\{g_i(t,u)\}$. Hitsuda has very recently discovered [320] that a process of the form

$$w_1(t) + f(t)w_2(t), \qquad t \ge 0 \tag{116}$$

where $w_1(\cdot)$ and $w_2(\cdot)$ are independent Wiener processes, will have multiplicity 1 if $f(\cdot)$ is absolutely continuous with a square-integrable derivative. However, if the derivative is not square-integrable on every open interval $(1,m) \subset [0,\infty)$, then the multiplicity is 2. The multiplicity is also 2 if $f(\cdot)$ has unbounded variation everywhere.

Hida (269] developed some of Lévy's ideas more clearly and obtained several new results. In his words "[Lévy's] pioneering works contain some points difficult for us to follow. The main aim of this paper is to establish his theory systematically and to prove some new facts." Among other things, Hida proved in great generality that if a process has a representation of the form (108), then it always has a canonical (innovations) representation obeying (109). He also gave the following extension of the criterion (104) of Karhunen: the representation (108) is an innovations representation if and only if

$$\int_0^t g(t,u)f(u)\,du = 0, \qquad \text{for every } t \le T \tag{117a}$$

implies

$$f(u) = 0, \qquad \text{a.e. on } [0,T]. \tag{117b}$$

These results are for processes of multiplicity one and are difficult to extend to the general case (115), because of the nonuniqueness of the $\{g_i\}$ and $\{E_i\}$.

Processes of Multiplicity One

It is useful to identify processes whose multiplicity is 1, since statistical applications will be easier in this case.

As noted earlier, the results of Krein, Karhunen, and Hanner show that nondeterministic stationary processes have multiplicity one. Cramér [268], [273], following the ideas of Hanner, obtained several interesting results. For example, he showed that nondeterministic discrete-time processes, stationary or not, always have multiplicity one, which is in sharp contrast to what we have just noted for continuous-time nonstationary processes. This result shows that some care must be exercised in studying continuous-time problems via discretization. Some studies of this question have recently been made [318]. Some other special examples of processes of multiplicity one (e.g., wide-sense Markov processes and certain harmonizable processes) have been given by Cramér [275], [300], who was apparently unaware of the class of processes identified by Lévy [cf. (111)].

In the engineering literature, the great appeal of the innovations approach as described by Bode and Shannon and by Zadeh and Ragazzini led to various attempts to discover innovations representations for nonstationary processes. A survey of the results achieved up to 1961 is given in a paper by Zadeh [270] and a further review was given by Stear [94] in 1965. However, the results achieved until then did not seem particularly exciting. In 1966, Anderson [98] extended some of Stear's work to show that a canonical representation for covariances of the separable form (27) could be obtained if a certain Riccati equation could be solved. This was nice because such equations were familiar from the Kalman filter. However, the Riccati equation developed by Anderson was of a different form from that used in the Kalman theory, and therefore the nontrivial question of existence of a solution to this nonlinear equation had to be settled differently [122], [127]. Related results were obtained by Brandenburg and Meadows [134] and Geesey [281].

However, the work in [98], [134], [122], [127], proceeded in apparent ignorance of the fundamental results of Hida and Cramér. It is perhaps noteworthy that even Lévy's result, though it was described in Zadeh's 1961 survey paper, escaped attention until it was rediscovered in a different way by the present author in 1968. This will be described in the next section, where we shall show that the Lévy result and the occurrence of the Riccati equation in [98], [134], can be easily explained by means of some estimation results.

IX. Recent Results on Innovations Processes and Some Applications

The innovations process was derived in discrete time by use of the Gram–Schmidt technique, according to which vectors are successively orthogonalized by projecting each new vector on the manifold spanned by the previous vectors. For a random sequence $\{(y_i)\}$ this means forming

$$\varepsilon_i = y_i - \hat{y}_{i|i-1}.$$

The analogous procedure for a continuous-time process is to form

$$\varepsilon(t) = y(t) - \hat{y}(t \mid t-)$$

but this will be identically zero for any process with continuous paths. However, if

$$y(t) = z(t) + v(t), \qquad 0 \le t \le T$$

and

$$E\int_0^T |z(t)|\,dt < \infty \tag{118}$$

then we will have a nontrivial innovations process

$$\varepsilon(t) = y(t) - \hat{y}(t \mid t-) = y(t) - \hat{z}(t). \tag{119}$$

Notice that (118), (119) is the time-domain analog of the Wiener filter formula (16). The process $\varepsilon(\cdot)$ has already been encountered in the Kalman filter (cf. (24)), and in that context it was recognized by a number of people ([109], [111], [114]) that $\varepsilon(\cdot)$ is a white noise process with

$$E\varepsilon(t)\varepsilon'(s) = Ev(t)v'(s) = I_p\delta(t-s). \tag{120}$$

However, the proofs of (120) in all these references were based on knowledge of the Kalman formulas for $\hat{x}(\cdot)$ and the mean-square error $P(\cdot)$. It was apparently not suspected that (120) holds quite generally; in fact my belief in this was the object of much skepticism at first. However, in April 1967, I obtained a proof initially via the Wiener–Hopf equation (10), and then by a simple use of the fundamental property that $z(t) - \hat{z}(t) \perp \{y(\tau), \tau < t\}$. This proof is described in [116, appendix I] (see also [298] and [291]). Thus we now had the opportunity to use the innovations method for *nonstationary* continuous-time processes.

During 1966–1967, the author, his students, Duncan and Frost, and Clark were studying nonlinear filtering. In that context it had become clear by a theorem of Lévy's (Doob [225, p. 384]) that if (120) could be established, one could obtain the striking fact that if $v(\cdot)$ were Gaussian (but not necessarily $z(\cdot)$), and if $\hat{z}(\cdot)$ was the least-squares estimate (not necessarily linear) then $\varepsilon(\cdot)$ would be not only white, but also Gaussian! This general result thus followed from our proof in April 1967; Frost simultaneously established it [279] in a different way using the Itô differential rule. Shiryaev later informed the author that he had found a similar result in 1966 [101, p. 22] in trying to make the evolution equations for nonlinear filtering mathematically meaningful. Namely, one had stochastic integrals with respect to the process $\varepsilon(\cdot)$ and these would not make sense unless $\varepsilon(\cdot)$ was a martingale process. The same reason led to an independent discovery by Kallianpur (personal communication, 1969). By now, several proofs of (119), (120) have been given under increasingly general assumptions, with the most recent results being in Kailath [304] and Meyer [324]. Specific extensions to discontinuous processes with independent increments have been made by Frost [294], [302], Snyder [172], Bremaud [308], and in some generality by Segall [326]. By now, the role of martingale theory in this question has become much clearer, e.g., the innovations theorem is just the Meyer–Doob decomposition for semimartingales with respect to the sigma fields of the observation process [326]. However, this is not the place for an exposition of this topic.

The first application of the previous result was to deduce a general formula for the likelihood ratio in terms of the nonlinear estimate $\hat{z}(\cdot)$ (cf. [291], [297]). Such formulas had been obtained by Schweppe [93] for Gaussian $z(\cdot)$ and, independently and in more generality, namely, for Markov $z(\cdot)$ by Stratonovich and Sosulin in a series of papers beginning in 1964 ([89], [102], [141]). However, the arguments of these papers were rather cumbersome and, at the author's suggestion and with his help, Duncan succeeded in streamlining and clarifying Stratonovich's proof (cf. [108], [115]).

A popular method for attacking such detection problems (and related nonlinear estimation problems) is via the use of a generalized Bayes' rule; the likelihood ratio is first written for a fixed sample of the signal process, and then one averages over all possible signal paths. (The function-space version of this familiar technique was first introduced into the engineering literature by Bucy [91], and it is sometimes called Bucy's representation theorem. Bucy has given proofs under various conditions (cf. [131], [136]), apparently unaware that the most general form appeared in a 1968 paper by Kallianpur and Striebel [118].) This was essentially the method used by Stratonovich and Sosulin and then clarified and extended by Duncan [108], [115], [293]. However, it has a serious limitation. If the signal process can depend upon past observations of signal plus noise, as happens in feedback communication and control problems, then clearly the conditioning-and-averaging procedure of the generalized Bayes' rule cannot be followed, because fixing the signal may also fix the observation. It seems that the only way to overcome this difficulty is to use the innovations process. This is done for additive Gaussian noise in [296], [297], with a heavy use of martingale theory. It may be of interest that the general likelihood ratio formula was conjectured in May 1967, some time before much of the machinery for a rigorous proof was made available through a now famous paper of Kunita and Watanabe [278], published in late 1967, and a related December 1968 paper [282] of Hitsuda.

Applications of the innovations process to linear problems actually came after the application to nonlinear detection problems and were partly stimulated by the paper of Schweppe [93], just cited. Let

$$\hat{z}(t \mid T) = \int_0^T H(t,s)y(s)\, ds$$

and

$$\hat{z}(t) = \int_0^t h(t,s)y(s)\, ds \tag{121}$$

and assume that the signal $z(\cdot)$ and the noise $v(\cdot)$ are completely uncorrelated. Then from Schweppe's paper one can deduce that

$$H(t,s) = h(t,s) + h(s,t) - \int_0^T h(\tau,t)h(\tau,s)\, d\tau. \tag{122}$$

It turned out that this relation could be proved directly by using an integral equation identity of Siegert, Krein, and Bellman [58], [228], [231]. A proof via innovations was desired, but for this it was necessary to show that the process $\varepsilon(\cdot)$ spans the same space as the observation process $y(\cdot)$. The author succeeded in doing this in November 1967 by the following argument [116], [165].

We can write the relation

$$\varepsilon(t) = y(t) - \hat{z}(t) = y(t) - \int_0^t h(t,\tau)y(\tau)\, d\tau \tag{123}$$

symbolically as

$$\varepsilon = (I - h)y. \tag{124}$$

Now we can find y from ε if $(I - h)^{-1}$ exists, a sufficient condition for which is (Smithies [386]) that h be Volterra and square-integrable on $[0,T] \times [0,T]$. Causality is equivalent to the Volterra property for such functions (but not necessarily otherwise (cf. [311]), and it is easy to see that the square-integrability will follow from the assumption that $z(\cdot)$ has finite expected energy

$$E\int_0^T z^2(t)\, dt < \infty. \tag{125}$$

Therefore in the linear case, we know that under quite general circumstances $\varepsilon(\cdot)$ is a true innovations process: for every t, $\{(\varepsilon(\tau), \tau < t)\}$ spans the same linear space of random variables as $\{(y(\tau), \tau < t)\}$. With this fact, the Kalman filter can be readily derived [116], [165], and several related linear problems solved [117], [181]. Unfortunately, the analogous result for the nonlinear problem has still not been proved, except for uniformly bounded $z(\cdot)$ [283] (see also [301, appendix I]). However Fujisaki, *et al.* [309] have elegantly circumvented this lack, by proving that $\hat{z}(\cdot)$ can still be written as a stochastic integral with respect to the innovations process, after which the remaining calculations are straightforward. Nevertheless, a proof of the equivalence would still have various benefits, since it can also be regarded as a very general existence result for solutions of functional stochastic equations. Moreover, there is a feeling, not unshared by the author, that the term "innovations process" should only be used for $\varepsilon(\cdot)$ when it can be proved to contain the same information as the original observation process, i.e., whenever $\{y(s), s < t\}$ and $\{\varepsilon(s), s < t\}$ generate the same sigma fields. Nevertheless, many important results, e.g., those on signal detection [297] and on nonlinear filtering [309]

can be proved without using this property. Therefore, the process $\varepsilon(\cdot)$ certainly deserves a special name, and we use *innovations* in the hope that the equivalence question will be settled (sooner or later) in some generality for non-Gaussian $z(\cdot)$. The term "residual" might be used, but this term is already used in statistics to denote (in discrete-time) the process $r_i = y_i - H\hat{x}_{i|i}$, while the innovations process is $\varepsilon_i = y_i - H\hat{x}_{i|i-1}$.

The name innovations was apparently first used for the discrete-time sequence $\{y_i - \hat{y}_{i|i-1}\}$ by Wiener, Masani, and Kallianpur in their studies in the mid-fifties (personal communication from Masani). Some studies of its properties were made in the unpublished report [267], which is partially elaborated in Masani's survey [255, pp. 92–93]. Masani [255] and Loéve [248, p. 386] have pointed out that Lévy used the centered sequence $\{y_i - \hat{y}_{i|i-1}\}$ for various problems, beginning in the mid thirties. As noted earlier, Wold was the first to use it for linear estimation. Wiener and Kallianpur attempted to establish conditions for non-Gaussian $\{y_i\}$ under which the sigma field of the innovations would be causally equivalent to the observations. Masani [255] mentions this as one important problem "bequeathed to posterity" by Wiener. He continues "Also left to us is the extension of this theory to the continuous-parameter case. Here the absence of an atomic time-unit makes the problem of defining nonlinear innovations extremely hard; obviously all we may expect are virtual or differential innovations". As noted earlier [304], [324], [326], martingale theory shows how to do this for semi-martingales (processes of bounded variation plus a martingale, e.g., a Wiener or centered Poisson process).

The proof of equivalence in the linear case immediately gives a factorization result for the covariance of the process y. Let us define a Volterra function k via

$$(I - h)^{-1} = I + k \tag{126a}$$

or equivalently via the Volterra equation

$$h + hk = k. \tag{126b}$$

Then

$$\begin{aligned} I + K = Eyy' &= (I - h)^{-1}E\varepsilon\varepsilon'(I - h')^{-1} \\ &= (I - h)^{-1}(I - h')^{-1} \\ &= (I + k)(I + k'). \end{aligned} \tag{127}$$

Equivalently, we have a canonical representation for $y(\cdot)$, viz.,

$$y = (I + k)\varepsilon. \tag{128}$$

This factorization, so natural now, took some time to be recognized. Shepp had given a noncausal factorization, but could not solve the problem of causal factorization [276, pp. 332]. This is now accomplished by (127). This result having been obtained, it became clear that for continuous K we had just rediscovered the Lévy factorization (c.f. (111)–(114) of Section VIII, where y there is actually the integral of the previous y). However, our result was somewhat more general; assumption (125) does not require $K(\cdot,\cdot)$ to be continuous, only that $K(t,t)$ be integrable. Then in May 1968, we happened to discover a book by Gohberg and Krein [280] that showed that the square-integrability of $K(\cdot,\cdot)$ was sufficient. Later Hitsuda [282] gave a proof of this fact using martingale theory (cf. the discussion in [296]). Combining these results with a decomposition formula of Shepp's [276, theorem 7] for differential processes gives innovations representations for smooth processes as well [285], [181].

Besides these generalizations, however, a significant aspect of our approach was its provision of a neat stochastic interpretation for Lévy's calculations, namely that $h(t,s)$ is a least-squares filter, so that we can interpret Lévy's process $\varepsilon(t)$ as $y(t) - \hat{z}(t) = y(t) - \hat{y}(t \mid t-)$. Moreover, this interpretation suggests that when $z(\cdot)$ is generated by a lumped system, then a Riccati equation can be used to compute $\hat{z}(\cdot)$ and therefore also to solve the covariance factorization problem. This fact explains (see [149], [181]) the occurrence of the Riccati equation in the previously cited work of Anderson and Moore [98], [127], [140], Brandenburg [134], and Geesey [281, ch. II]. Moreover, it shows that other algorithms for computing estimates can also be used when appropriate, e.g., algorithms of Chandrasekhar type (Section VI).

Briefly, suppose that we have a process $y(\cdot)$ with covariance function as in (32)

$$\begin{aligned} &I\delta(t - s) + M(t)\Phi(t,s)N(s)1(t - s) \\ &\qquad + N'(t)\Phi'(s,t)M'(s)1(s - t). \end{aligned} \tag{128a}$$

Suppose also that the process $y(\cdot)$ arises from some state-space model of the form

$$\begin{aligned} \dot{x}(t) &= F(t)x(t) + G(t)u(t), \qquad x(0) = x_0 \\ \dot{y}(t) &= H(t)x(t) + v(t) \end{aligned} \tag{129}$$

with the usual assumptions (19)–(21) on $u(\cdot)$, $v(\cdot)$, x_0. The least-squares filter (22)–(24) for this state-space model can be rewritten as

$$\begin{aligned} \dot{\hat{x}}(t) &= F(t)\hat{x}(t) + K(t)\varepsilon(t), \qquad \hat{x}(0) = 0 \\ y(t) &= H(t)\hat{x}(t) + \varepsilon(t) \end{aligned} \tag{130}$$

where the gain function $K(\cdot)$, which is defined by

$$K(t) = P(t)H'(t) + G(t)C(t), \qquad P(t) = E\tilde{x}(t)\tilde{x}'(t) \tag{131}$$

can be calculated via the Riccati equation (27) for $P(\cdot)$, or directly via the Chandrasekhar-type equations of Section VI if F, G, H are constant.

Now, since $\varepsilon(\cdot)$ is known to be white, (130) can be regarded as another causal model for $y(\cdot)$ driven by a single white noise. Also, it is easy to calculate $\varepsilon(\cdot)$ from $y(\cdot)$: replace ε by $y - H\hat{x}$ in the differential equation, calculate $\hat{x}$, and then form ε as $y - H\hat{x}$. Therefore, for a process with a *known* state model (129), (130) defines the innovations representation (IR). (This simple fact, widely known by now, was to our knowledge first pointed out in [116, appendix D] and explicitly restated in [284].)

Now the general studies of Hida and Cramér on IRs (cf. Section VIII) have shown that the IR of a process $y(\cdot)$ can depend only upon the covariance of $y(\cdot)$. Therefore, F, K, H in (129) *must be determinable* (up to state-space transformations) from the covariance of $y(\cdot)$, no matter what state model we initially assumed. To do this we can calculate the covariance of $y(\cdot)$ as given by the particular model (129) and compare with the given expression (128) to make the nonunique identifications

$$M(t) = H(t), \qquad F(t) = \frac{d\Phi(t,s)}{dt}\Phi^{-1}(t,s)$$

$$N(t) = \Pi(t)H'(t) + G(t)C(t).$$

We now use these relations to express the parameters (F,H,K) of the IR (130) in terms of (M,Φ,N). Let

$$\Sigma(t) \triangleq E\hat{x}(t)\hat{x}'(t).$$

Then by the orthogonality of $\hat{x}$ and $\tilde{x}$ we have

$$Ex(t)x'(t) \triangleq \Pi(t) = \Sigma(t) + P(t)$$

so that we can write $K(\cdot)$, as given by (131), as

$$K(t) = \Pi(t)H'(t) + G(t)C(t) - \Sigma(t)H'(t)$$
$$= N(t) - \Sigma(t)M'(t). \qquad (132)$$

Moreover, using (130) and the fact that $\varepsilon(\cdot)$ is white readily yields

$$\dot{\Sigma}(t) = F(t)\Sigma(t) + \Sigma(t)F'(t) + [N(t) - \Sigma(t)M'(t)]$$
$$\cdot [N(t) - \Sigma(t)M'(t)]', \qquad \Sigma(0) = 0. \qquad (133)$$

The stochastic interpretation of $\Sigma(\cdot)$ guarantees its existence, and we see that (132) and (133) enable us to calculate $K(\cdot)$ from knowledge only of $M(\cdot)$, $N(\cdot)$, and $\Phi(\cdot,\cdot)$. Therefore, (130), (132), and (133) determine the canonical or innovations factorization of the covariance (128). The state model (129) was only used here to motivate the development of the IR, as defined by (130), (132), (133). The result can be deduced just from the assumption that the covariance in (128) is strictly positive definite (cf. [285], [181, appendix III]). However, the present derivation does have the advantage that it clearly displays the intimate relationship between the filtering problem and the factorization problem. Thus we see that the state vector of the IR is $\hat{x}(\cdot)$, so that $H(t)\hat{x}(t) = M(t)\hat{x}(t)$ can also be computed directly from knowledge of the covariance function; this gives us a proof of the result presented in Section V. Other applications are described in [181], [285], [292], [310].

As one example, we shall show how to obtain some matrix spectral factorization algorithms.

Multivariate Spectral Factorizations

Consider a covariance function $R_y(\cdot)$ of the form (128a) where M and N are constant $p \times n$ and $n \times p$ matrices, respectively, and

$$\Phi(t,s) = \exp F(t-s), \qquad F = \text{a stability matrix}. \qquad (134)$$

The power spectral density matrix is defined by the values for $s = i\omega$ of the function

$$S_y(s) = \int_{-\infty}^{\infty} R(t)e^{-st}\,dt$$
$$= I + M(sI - F)^{-1}N + N'(-sI - F')^{-1}M'. \qquad (135)$$

This is not generally the form in which the power spectral density will be given, but let us postpone this aspect for a while. The problem is to factor $S_y(s)$ as [cf. (14)]

$$S_y(s) = S_y^+(s)S_y^+(-s)$$

where $S_y^+(s)$ is the transfer-function matrix of a causal and causally invertible system with p inputs and p outputs.

Now, for finite time, we have such a system in the IR (130), (132), (133). However, when F is stable, it is not hard to prove (see, e.g., [126], [135]) that as $t \to \infty$, $\Sigma(t)$ in (133) will tend to a constant matrix Σ and hence that $K(t)$ will tend to a constant matrix $\bar{K}$

$$\bar{K} = N - \bar{\Sigma}M' \qquad (136)$$

where $\bar{\Sigma}$ is the unique nonnegative-definite solution of the algebraic Riccati equation

$$0 = F\bar{\Sigma} + \bar{\Sigma}F' + [N - \bar{\Sigma}M'][N - \bar{\Sigma}M']'. \qquad (137)$$

Then the innovations representation (130) has the transfer function

$$S_y^+(s) = I + M(sI - F)^{-1}\bar{K} \qquad (138)$$

which we have designated $S_y^+(s)$ because in the limit it continues to provide the canonical (causal and causally convertible) factor of the power spectral density. From (135) and (138), we obtain the useful spectral factorization formula

$$I + M(sI - F)^{-1}N + N'(-sI - F')^{-1}M'$$
$$= [I + M(sI - F)^{-1}\bar{K}][I + \bar{K}'(-sI - F')^{-1}M']. \qquad (139)$$

We clearly have several procedures for computing $\bar{K}$ and thereby factoring $S_y(s)$. We can find the unique nonnegative-definite solution $\bar{\Sigma}$ of the nonlinear algebraic equation (137), or we can find $\bar{\Sigma}$ as the limiting solution of the Riccati differential equation (133); another more direct method is to find $\bar{K}$ as the limiting solution of the Chandrasekhar-type equations of Section VI. The differential equation procedures would be preferred because of their simplicity and automatic production of the right $\bar{\Sigma}$ or $\bar{K}$, but it is sometimes difficult to control the accumulation of computational errors until steady state is reached. However, the Chandrasekhar-type algorithms seem to behave quite well in this regard. The solution of the quadratic algebraic equation (137) involves choosing among the several possible solutions for $\bar{\Sigma}$ and is generally more laborious, though efficient eigenvalue–eigenvector methods have recently been proposed [78], [100], [160].

We now examine the problem of how to handle $S_y(s)$ that are not given in the form (135). The properties of power spectra show that we can easily write $S_y(s)$ in the form

$$S_y(s) = \frac{\text{a polynomial matrix}}{\Psi(s)\Psi(-s)} \tag{140}$$

where $\psi(s)\psi(-s)$ is a scalar polynomial that is the greatest common divisor of the denominators of all the entries in $S_y(s)$.

Now by a partial fraction expansion (or other means) we can decompose $S_y(s)$ as

$$S_y(s) = Z(s) + Z'(-s) \tag{141}$$

where $Z(s)$ contains all the terms corresponding to the left half-plane zeros of $\psi(s)\psi(-s)$. We can identify $Z(s)$ as the Laplace transform of the positive-time part of $R(t)$ [cf. (134), (135)]

$$Z(s) = \int_0^\infty [\tfrac{1}{2}I\delta(t) + Me^{Ft}N]e^{-st}\,dt \tag{142}$$

$$= \tfrac{1}{2}I + M(sI - F)^{-1}N. \tag{143}$$

Now $Z(s)$ can be regarded as the transfer function of a system with impulse response function $[\frac{1}{2}I + Me^{Ft}N]$, and therefore given $Z(s)$ we can find M, F, N by using one of several algorithms [129], [306], [360], going from rational transfer functions to state-variable realizations.

Thus we have some new methods for the classical multivariate spectral factorization problem. This problem has a long history and several different algorithms have been proposed by Wiener and Masani [12], Youla [26], Davis [28], Yaglom [23], Rozanov [22], Masani [35], Csaki and Fischer [36], Tuel [125], Strintzis [55], and others. The method based on solution of the nonlinear algebraic equation (137) was first derived in a different way by Anderson [104], who used certain connections, initially noted by Youla [26] and Kalman [75], [77], between spectral factorization and certain functions long familiar in network theory.

The point is that the $Z(s)$ in (142), (143) is not an arbitrary transfer function, but a "driving-point" impedance function, a property that Brune showed, in a 1931 dissertation that essentially founded network theory, is equivalent to $Z(s)$ being *positive real*, viz., that it obeys the conditions 1) all elements of $Z(s)$ are analytic in Re $s > 0$; 2) $Z(s)$ is real when s is real and positive; 3) $Z(s) + Z'(-s) \geq 0$, if Re $s > 0$. It can be shown (see, e.g. [177]) that equivalent conditions are that $Z(s) + Z'(-s)$ is a power spectral density matrix or that

$$Z(s) = \int_0^\infty R(t)e^{-st}\,dt$$

where $R(\cdot)$ is a nonnegative-definite function (i.e., a covariance function). These equivalences enable a considerable interplay between results in network theory, stability theory, control theory, and estimation (see, e.g., [77], [87], [158], [177]).

When $Z(s)$ has rational elements, an important tool in such studies has been the so-called *positive-real lemma*, first given by Yakubovic [73] and Kalman [75] for scalar $S_y(s)$, and extended to the matrix case by Popov [87] and Anderson (see references in [177]).

The positive-real lemma starts with a (nonunique) *minimal realization* of $Z(s)$ in the form

$$Z(s) = J + M(sI - F)^{-1}N \tag{144}$$

where minimality means that the square matrix F has lowest dimension among all possible F that could be used. Then it states that $Z(s)$ will be positive real if and only if there exists a real symmetric matrix $\Pi \geq 0$ such that

$$\mathcal{M} = \begin{bmatrix} F\Pi + \Pi F' & N - \Pi M' \\ (N - \Pi M')' & J + J' \end{bmatrix} \geq 0. \tag{145}$$

Since the $\mathcal{M}$ matrix is nonnegative definite, it can be factored as

$$\mathcal{M} = \begin{bmatrix} L \\ W \end{bmatrix} [L' \quad W'] \tag{146}$$ [7]

where the column size of L and W is arbitrary. Then an alternative statement is clearly that $Z(s)$ will be positive real if and only if there exist matrices L and W such that

$$F\Pi + \Pi F' = LL' \tag{147a}$$

$$N - \Pi M' = LW \tag{147b}$$

$$J + J' = WW'. \tag{147c}$$

The significance of L and W is that they immediately give a factorization of $S_y(s)$. In fact, we can check that

$$\begin{aligned} S_y(s) &= Z(s) + Z'(-s) \\ &= [W + M(sI - F)^{-1}L][W' + L'(-sI - F')^{-1}M']. \end{aligned} \tag{148}$$

There are many matrices Π that will satisfy (145), and consequently there will be many factorizations. The family of all such solutions has been studied by Anderson [177], Willems [158], Kucera [167], and Canabal [171]. The maximum and minimum Π, when they exist, play a significant role in the analysis, with the minimum being the one that gives the innovations factorization; the maximum relates similarly to a certain dual system.

We cannot pursue such discussions any further here, but it may be interesting to note that related and in fact somewhat more general minimality properties were discovered by Krein in 1945 (cited in [22]) and by Masani [237]. The multivariate estimation problem has many fascinating aspects that are not generally known in the engineering literature, but we must content ourselves here to calling attention to the book [30] and two fine surveys by Masani [35], [45].

We conclude this section in a more engineering vein.

[7] Such factorizations of the discrete-time time-invariant generalization of the matrix in (145) are at the heart of the square-root algorithms mentioned briefly at the end of Section VI.

Transfer-Function Models Versus State Models

The many successes of the state model in recent years have led to an unwise neglect of more traditional methods and problems. Some examples were discussed in Sections VI and VII. Another example, briefly noted in Section IX, is the use of multivariate transfer-function models and frequency-domain analysis. Such models are essentially the only ones used in the statistical literature, where they are known as ARMA (autoregressive-moving average models). There are several interesting features associated with algorithms that work directly with such models, including the possibilities of fewer computations, easier proofs of stability, and more insight into certain structural aspects.

Briefly, suppose we have a process $y(\cdot)$ such that

$$y(n) + A_1 y(n-1) + \cdots + A_n y(0) = w(n)$$

where

$$w(n) = B_0 u(n) + B_1 u(n-1) + \cdots + B_m u(n-m)$$

and $u(\cdot)$ is a white-noise sequence. Then $w(\cdot)$ is a *moving-average* process. If $B_i = 0$, $i \geq 1$, then $y(\cdot)$ is an *autoregressive* process; otherwise we have a *mixed* or *autoregressive-moving average* process. The interesting point is that the innovations for $y(\cdot)$ are just the innovations for $w(\cdot)$

$$\hat{y}(n \mid n-1) = -A_1 y(n-1) - \cdots - A_n y(0) - \hat{w}(n \mid n-1)$$

so that

$$\varepsilon(n) = y(n) - \hat{y}(n \mid n-1) = w(n) - \hat{w}(n \mid n-1).$$

The process $w(\cdot)$ is often much simpler than $y(\cdot)$, e.g., m may be much less than n, or the $\{B_i\}$ may be constant while the $\{A_i\}$ are time variant or even nonlinear. These possibilities have already been exploited in [131], [175], [191], but more can be done. A useful stimulus to such researches is also provided by the relatively recent work of Popov [355], [359], Rosenbrock [356], Wang [358], Morf [191], Forney [361], Wolovich [363], and others, which has uncovered the close relationships between linear ARMA models and state-space models, thus enabling a fruitful combination of time- and frequency-domain methods.

X. Karhunen–Loève Expansions: Canonical Correlations and State Models

Those familiar with the textbooks on statistical communication theory in the last decade or so may be surprised that we have not referred to series expansions of random processes, and more particularly to the Karhunen–Loève (K–L) expansions (see, e.g., [341], [342], [40]). Consider a scalar Gaussian process $z(\cdot)$ with a continuous covariance $R_z(t,s)$ such that

$$\int_0^T \int_0^T R_z^2(t,s)\, dt\, ds < \infty.$$

The assumption of Gaussianness is made for terminological convenience; all statements can be translated in a standard way to apply to just "second-order" processes.

The K–L expansion of $z(\cdot)$ is

$$z(t) = \sum_1^\infty z_i \Psi_i(t) \tag{149}$$

where the $\{\Psi_i(\cdot)\}$ are eigenfunctions of $R_z(\cdot,\cdot)$, viz.,

$$\int_0^T R_z(t,s)\Psi_i(s)\, ds = \lambda_i \Psi_i(t), \qquad 0 \leq t \leq T$$

and

$$z_i = \int z(t)\Psi_i(t)\, dt, \qquad i = 1,2,\cdots.$$

It is known from the theory of integral equations that the $\{\Psi_i(\cdot)\}$ are orthonormal

$$\int_0^T \Psi_i(t)\Psi_j(t)\, dt = \delta_{ij}$$

and that (Mercer's formula)

$$R_z(t,s) = \sum_1^\infty \lambda_i \Psi_i(t)\Psi_i(s). \tag{150}$$

A simple calculation shows that the coefficients $\{z_i\}$ are uncorrelated

$$Ez_i z_j = \lambda_i \delta_{ij}.$$

If we temporarily write the random process as $z(t,\omega)$, ω being the probability-space variable, then the K–L expansion is a decomposition of a function of two variables into a sum of products of functions of one variable

$$z(t,\omega) = \sum_1^\infty z_i(\omega)\Psi_i(t).$$

Such decompositions are familiar from the partial differential equations of physics and their significance in random-process theory is basically the same. Since the $\{\Psi_i(\cdot)\}$ are deterministic, we can replace study of the uncountable family of random variables $z(t,\omega)$ by that of the countable family $\{z_i(\omega)\}$. The K–L expansions have the further useful property that the $\{z_i(\omega)\}$ are independent because $z(\cdot)$ is Gaussian, which simplifies many probabilistic calculations, e.g., determination of moments and convergence of sums. The K–L expansion (Karhunen [331], Loève [329]) was independently introduced by Kac and Siegert [330], [331], to simplify the calculation of the distribution of the output power from a nonlinear circuit (limiter-squarer-filter) driven by noise. On more abstract grounds, it had already been introduced in 1943 by Kosambi [328], an Indian statistician and Marxist philosopher, and also by Obukhov (in a 1946 dissertation, cited in [346]), Pugachev [339], and perhaps many others. The popularity of the K–L expansion (this terminology is now well entrenched) grew from its use in the 1950 Ph.D. dissertation [333] of Grenander to extend to stochastic processes the classical theories of statistical estimation and hypothesis testing, which had been developed for finite families of random variables.

The K–L expansion was soon used in estimation problems by Davis [335], Slepian [336], and Youla [337]. Its presentation in 1958 in the pioneering textbook of Davenport and Root [341] and the many applications in the widely used 1960 textbook of Helstrom [342] gave the K–L expansion a major place in the literature of the sixties.

Despite these many successes, however, the use of such expansions is diminishing for several reasons. One is of course that K–L expansions really apply only to Gaussian (or second-order) processes, while recently martingale theory has enabled significant headway to be made with non-Gaussian processes (see, e.g., the survey [327] by Wong). However, this deficiency does not apply to linear filtering, the main concern of our survey. Here a common complaint is that the K–L expansion does not lend itself to recursive calculation because the $\{z_i\}$ and $\{\Psi_i(\cdot)\}$ are not easily updated as T increases. Also, since the $\{z_i\}$ depend upon the values $z(t)$, for all $t \in [0,T]$, K–L expansions would seem to be more appropriate for smoothing problems (data over $[0,T]$) rather than causal filtering problems.

For example, consider the smoothing integral equation (9) for *uncorrelated* signal $z(\cdot)$ and noise $v(\cdot)$, say (in an obvious notation)

$$H(t,s) + \int_0^T H(t,\tau)R_z(\tau,s)\,d\tau = R_z(t,s), \qquad 0 \le t,s \le T. \tag{151}$$

The Mercer formula (150) for $R_z(t,s)$ now shows readily that the solution can be written

$$H(t,s) = \sum_1^\infty \frac{\lambda_i}{1+\lambda_i}\Psi_i(t)\Psi_i(s), \qquad 0 \le t,s \le T. \tag{152}$$

The computational value of such a solution is debatable because the $\{\lambda_i\}$ and $\{\Psi_i(\cdot)\}$ are difficult to compute, but at least its explicitness is often convenient. No similar solution appears possible for the filtering (or Wiener–Hopf) equation (10) because of the causality constraint $0 \le s < t \le T$. Nevertheless, recently a number of authors [352], [353], have shown that, with proper interpretations, series-expansion techniques can also be exploited in causal filtering problems. The key to these results can be found in an old device of Swerling's (cited in [155]). Swerling calculates the filtered estimate $\hat{z}(t \mid t)$ as

$$\hat{z}(t \mid t) = \sum_1^\infty \hat{z}_{i|t}\Psi_i(t)$$

where the $\{\hat{z}_{i|t}\}$ are *smoothed* estimates of the coefficients $\{z_i\}$ given data on $[0,t]$, and therefore can be determined by solving the smoothing equation(7). The filtered estimate can then be put together as an infinite combination of smoothed estimates. (This reflects in a different way our comment in Section II that (10) is a family (but not the obvious one) of equations of the form (9).) The recursive Kalman filter can be derived along these lines.

While one common criticism of the K–L approach can thus be partly met, there is another more serious difficulty. The ease of solving the smoothing equation (151) is heavily dependent upon the assumption that $z(\cdot)$ and $v(\cdot)$ are uncorrelated, which makes the right side equal to $R_z(t,s)$. Otherwise we would also have [cf. (9)] the term $Ez(t)v(s)$, and now there is no obvious solution. Some reflection will show that what really yielded the solution (152) of (151) was the *simultaneous expansions*

$$Ez(t)z(s) = R_z(t,s) = \sum_1^\infty \lambda_i\Psi_i(t)\Psi_i(s)$$

$$Ev(t)v(s) = \delta(t-s) = \sum_1^\infty 1\cdot\Psi_i(t)\Psi_i(s)$$

$$z(t) = \sum z_i\Psi_i(t)$$

$$Ez_iz_j = \lambda_i\delta_{ij}$$

$$Ev_iv_j = \delta_{ij}$$

and the fact that

$$Ez(t)v(s) \equiv 0 \Rightarrow Ez_iv_j = 0, \qquad \text{for all } i,j.$$

But how can dependence between $z(\cdot)$ and $v(\cdot)$ be reflected into the $\{z_i\}$ and the $\{v_i\}$, especially a one-sided dependence as in (4)? This question does not seem to have been raised in the engineering literature, rigorous or formal, even though dependence is needed to model many problems, e.g., when feedback is present.

The answer is that one should not treat $z(\cdot)$ and $v(\cdot)$ separately but should work with the observed process $y(\cdot) = z(\cdot) + v(\cdot)$, which has covariance

$$Ey(t)y(s) = \delta(t-s) + K(t,s).$$

$K(t,s)$ will not generally be a covariance when $z(\cdot)$ and $v(\cdot)$ are correlated, but since $Ey(t)y(s)$ is a covariance one can show that $K(t,s)$ (assumed to be continuous in t and s) has only a finite number of negative eigenvalues, and the Mercer formula extends to such functions as well (Riesz–Nagy [385, p. 242]). The emphasis on the observed process $y(\cdot)$, as against the signal and noise processes separately, is also the key to the development of the recursive Wiener filters; cf. the discussion of (128)–(133) in Section IX.

We have passed quickly over the above point that the white noise $v(\cdot)$ or its covariance $\delta(t)$ do not meet the conditions for the validity of the K–L expansion (149) or the Mercer formula (150). However, the validity of the expansion

$$\delta(t-s) = \sum_1^\infty \phi_i(t)\phi_i(s)$$

$$\{\phi_i(\cdot)\} = \text{any complete orthonormal family on } L_2[0,T]$$

is usually argued on the grounds that any L_2-function $f(\cdot)$ can be correctly calculated as

$$f(t) = \int_0^T f(s)\delta(t-s)\,ds = \sum_1^\infty f_i\phi_i(t)$$

$$f_i = \int_0^T f(t)\phi_i(t)\,dt, \qquad i = 1,2,\cdots.$$

Therefore the $\{\phi_i(\cdot)\}$ can be chosen as, say, the eigenfunctions $\{\phi_i(\cdot)\}$ of some other covariance, say $R_z(\cdot,\cdot)$, and it is usually argued that we can write

$$v(t) = \sum_1^\infty v_i\Psi_i(t) + v_{rem}(t)$$

where the quantity $v_{rem}(\cdot)$ needed to make both sides "equal" is orthogonal to the signal process $z(\cdot)$ and can therefore be ignored. We do not wish to push this argument too far, though it has been quite successfully used in the literature. However, lest one get too scornful of what some people call "engineering nonsense," we shall show how it can be used to obtain (or let us say conjecture) a result that appeared only much later in the mathematical literature (not that this is a new phenomenon).

Expansions of a Wiener Process

White Gaussian noise can be thought of as the formal derivative of a Wiener process $w(\cdot)$ with continuous covariance function

$$Ew(t)w(s) = \min(t,s).$$

Now the eigenfunctions of this covariance are readily calculated

$$\Psi_n(t) = (2/T)^{1/2} \sin[(2n-1)\pi t/2T], \qquad n = 1,\cdots$$

so that the K–L expansion is

$$w(t) = \sum_1^\infty w_n\Psi_n(t)$$

where the $\{w_n\}$ are uncorrelated random variables with variances $\{4T^2/(2n-1)^2\pi^2\}$. However, since formally

$$w(t) = \int_0^t v(\tau)\,d\tau, \qquad v(\cdot) = \text{white noise}$$

we can also write

$$w(t) = \sum_1^\infty v_i\Phi_i(t)$$

where the $\{v_i\}$ are uncorrelated unit-variance random variables

$$\Phi_i(t) = \int_0^t \varphi_i(\tau)\,d\tau$$

and the $\{\varphi_i(\cdot)\}$ are *any* complete orthonormal set in $L_2[0,T]$.

Thus we can get many expansions for $w(\cdot)$, each with uncorrelated random variables $\{w_i\}$, but with *different* families of deterministic functions $\Phi_i(\cdot)$. The $\{w_i\}$ can be calculated as

$$w_i = \int_0^T \varphi_i(t)v(t)\,dt = \int_0^T \varphi_i(t)\,dw(t).$$

For most purposes these expansions are almost as useful as the K–L expansion (149). The previous result was first obtained by Shepp [276] and has since been extended to fairly general (non-Wiener) processes [350]. These general results can also be heuristically explained by using white noise, but we shall give a slightly different explanation.

Reproducing Kernel Hilbert Spaces

The time functions $\{\Psi_i(\cdot)\}$ in the K–L expansion are orthogonal over $[0,T]$ in the sense that

$$\langle\Psi_i,\Psi_j\rangle = \int_0^T \Psi_i(t)\Psi_j(t)\,dt = \delta_{ij}.$$

This property does not hold for the $\{\Phi_i\}$

$$\langle\Phi_i,\Phi_j\rangle \neq \delta_{ij}$$

but if we define a new inner product as

$$\langle a(\cdot),b(\cdot)\rangle_{H(w)} = \int_0^T \dot{a}(t)\dot{b}(t)\,dt + a(0)b(0)$$

then it is easy to see that

$$\langle\Phi_i,\Phi_j\rangle_{H(w)} = \delta_{ij}$$

for all choices of the $\{\Phi_i(\cdot)\}$. This inner product is appropriate for the Wiener process $w(\cdot)$. For other processes, we can determine suitable inner products that will make the corresponding expansions have simultaneously orthogonal random variables and time functions. The only special feature of the K–L expansion is that the time functions are orthogonal with respect to the $L_2(dt)$ inner product. However, there is nothing sacred about L_2 or about Lebesgue measure dt; we could, for example, use a measure $p(t)dt$, where $p(\cdot)$ is a weight function such that

$$\int_I\int_I R^2(t,s)p(t)p(s)\,dt\,ds < \infty.$$

This leads to expansions that are orthogonal in $L_2(p(t)dt)$ norm. In fact, to avoid dependence on such arbitrary weight functions $p(\cdot)$ or on the arbitrary family $\{\varphi_i(\cdot)\}$, it is desirable to try to seek an "intrinsic" norm associated just with the covariance of the process $w(\cdot)$. The norm $\langle\ ,\ \rangle_{H(w)}$ is just such a norm, and it is called a *reproducing kernel norm* because of the property

$$\langle R_w(t,s),m(s)\rangle_{H(w)} = m(t)$$

for all $m(\cdot)$, such that

$$\|m\|_{H(w)}^2 = \langle m(\cdot),m(\cdot)\rangle_{H(w)} < \infty.$$

Such reproducing kernel Hilbert spaces (RKHS) were introduced into stochastic process theory by Loève in 1948 (cf. [274, appendix I]), and their usefulness in statistical applications has been made clear, notably by Parzen [347]. Many other references as well as some tutorial explanations and applications are discussed in [347], [349], [311]. However, it should be noted here that engineers have generally tended to think only of the space L_2 whenever Hilbert space is mentioned, perhaps on the grounds that L_2 is isomorphic to any Hilbert space. But the norm is the most important feature of a Hilbert space, and *isometries* (norm-preserving isomorphisms) are more significant than additive isomorphisms. The theory of RKHS shows that there are many other quite different and quite useful Hilbert spaces besides L_2.

There are many other aspects of series expansions that could be discussed (some are noted in Wong's survey [327]), but we shall conclude with a discussion of a somewhat different type of process representation.

Canonical Correlations and State Models

Since series expansions generally contain an infinite number of terms, they cannot be used directly without truncation to a finite number, and this brings up the question of which terms to keep. It is usually suggested that we keep the coefficients that have maximum variance. However, calculations of least-squares estimates on this basis have not been very satisfactory, especially for processes with rational spectral density. The reason is *roughly* that the whole data interval enters equally into the determination of all the series coefficients, whereas, for example in prediction, the most recent observations should make a larger contribution. The state-space description of random processes reflects this circumstance better, but there is another quite general statistical technique that is roughly equivalent. This is the so-called theory of *canonical correlations*, which was independently developed in the mid-thirties by Hotelling and Obukhov (cf. [346]). [Incidentally, Hotelling was apparently the first to use eigenvalue-eigenvector decompositions (the finite analogs of the K-L expansion) in statistics (in a 1933 paper in the *Journal of Educational Psychology*).]

These authors proposed the following method for studying the interrelations between two (full-rank) families of random variables $\{X_1,X_2,\cdots,X_n\}$ and $\{Y_1,Y_2,\cdots Y_m\}$. First find the linear combinations

$$U_1 = \sum_1^n \alpha_{1i}X_i, \qquad V_1 = \sum_1^m \beta_{1i}Y_i$$

that have the largest cross-correlation coefficient

$$\rho_1 = EU_1V_1/\sqrt{EU_1^2EV_1^2}.$$

Next find linear combinations U_2 and V_2 that are uncorrelated with (U_1,V_1) and have maximum correlation coefficient, and so on until a new set of random variables $(U_1,U_2,\cdots,U_n, V_1,\cdots,V_m)$ has been found that span the same space as the $\{X_1,X_2,\cdots,X_n, Y_1,\cdots,Y_m\}$ and are pairwise uncorrelated, except for the pairs (U_i,V_i), $i = 1,\cdots$, min (n,m). The actual calculation can be shown to be equivalent to the solution of a certain eigenvalue problem (see, e.g., [340]).

It is reasonable that in making inferences about the $\{Y_i\}$ from the $\{X_i\}$, the first few canonical variables $\{U_i\}$ should give us more information than the first few coefficients of the discrete-time analog of the K-L expansion. This observation was made by Yaglom [346], who with Gelfand generalized the theory to the case of continuous-time random processes $\{x(s), s \in S\}$ and $\{y(t), t \in T\}$, and used it for the calculation of mutual information [338]. Some especially interesting results are obtained when the two families of random variables are the past and future of the same random process. Yaglom has shown that the past and future (and in fact any two disjoint segments) of a nondeterministic continuous-time process have a finite number of canonical variables if and only if the process has a rational spectral density.

Recent studies in automata theory and algebraic system theory [119], [129], have shown that the analysis of a system in terms of past and future leads naturally to a "state-space" description. Related ideas can be recognized in an interesting but somewhat obscure paper by Levinson and McKean [345]. For a stationary process $y(\cdot)$, they introduce, among others, the subspaces

$B_{(a,b)}$ = the linear (Hilbert) space spanned by the random variables $\{y(\tau), a \leq \tau \leq b\}$

$B_+ = B_{(0,\infty)}$ = the future

$B_- = B_{(-\infty,0)}$ = the past

$B_{+|-}$ = the projection of B_+ on B_-

$$B_{0+} = \bigcap_{\delta>0} B_{(0,\delta)}.$$

B_{0+} is called the *germ field* [343] and $B_{+|-}$ is the *minimal splitting field* of past and future, viz., it is the smallest field such that, given $B_{+|-}$, B_+ is independent of B_-. [There are clearly many splitting fields, e.g., B_-; the proof that $B_{+|-}$ is the minimal field was given by McKean in a fascinating paper [344] on multidimensional Brownian motion.] These concepts are clearly related to the notion of state, and with this in mind we should not be surprised that $B_{+|-}$ is finite-dimensional if and only if $y(\cdot)$ has a rational spectral density or that $B_{+|-} = B_{0+}$ if and only if the spectral density has no zeros [345], [261a]. Recall that the state of a system can be determined from the output and its derivatives, without knowledge of the input, if and only if the transfer function has no zeros.

It can be proved that the canonical variables for the sets $\{y(\tau), \tau \geq 0\}$ and $\{y(\tau), \tau \leq 0\}$ are a useful basis for the state-space $B_{+|-}$, and this fact has recently been cleverly exploited by Akaike [316] to study state-space modeling. In particular he obtains a stochastic interpretation of the Ho-Youla-Silverman and other algorithms for determining a minimal state-space realization of an impulse response. We cannot pursue these matters any further here, though as a final comment we may note that the canonical correlations would seem to be useful in more problems than those to which they have been explicitly applied so far.

XI. Concluding Remarks

In this survey, we have described several well established and widely used results and on occasion we have also indicated some areas for further work and some new directions.[8]

One important trend is the growing interplay between linear system theory and linear filtering theory. This is reinforced by the growing realization in both fields of the importance of understanding the underlying structural features and invariants of dynamical linear systems and of the stochastic processes that can be generated from them. This structural knowledge is bound to be useful in all applications where linear systems and stochastic processes arise, ranging from long-distance communication problems to the analysis of single circuits.

Another obvious direction is into nonlinear filtering. Actually the last decade has seen a considerable effort to extend the Kalman filter to nonlinear problems. This vast area will have to be surveyed separately, but some important aspects should be mentioned here.

Recursive formulas for updating the least-squares estimate (the conditional mean) were first obtained by Stratonovich [65], [66], and Kushner [86]. However, it was found that in general the formulas involve all the conditional moments, so that an infinite set of simultaneous equations (or equivalently, a *partial* differential equation for the conditional probability density or the conditional characteristic functional) is necessary. There is still no consensus as to a satisfactory way of "truncating" this set of equations, which incidentally is also encountered in other fields such as fluid mechanics and quantum mechanics [388]-[390]. Furthermore, no other computationally satisfactory approaches to solving directly, even approximately, the partial differential equations seem to be at hand, though spline function approaches do hold out some promise.

At the moment, one of the chief benefits of having attacked the nonlinear problem was that it brought to the fore certain difficulties associated with the proper definition of the differential and integral equations used to describe nonlinear operations on white noise. For nearly the first time, as far as engineers were concerned, the use of different definitions of integrals made a difference to the answer, rather than just to the "rigor" of the proofs. It also affected the proper modeling of physical nonlinear problems. Wong and Zakai [277], McShane [277a], [277b], and others have made important contributions to this subject, but there are still many unresolved questions. Nevertheless, this work pointed the way to the introduction of martingale theory into communication and control problems, a fact whose significance will probably far outreach any specific nonlinear filtering problems. This point has been discussed at greater length in Wong's survey [327]; see also [298].

It should be noted that the previously mentioned unexpected difficulties with models and integrals lent the nonlinear recursions more than academic interest. Wonham gave the first rigorous proof, but only for signals taking finitely many values (e.g., finite Markov chains) [97]. Other proofs have since been constructed for more general signals, but under rather complicated and physically obscure conditions. In particular, dependence of the signal on feedback observations was excluded. As noted in Section IX, Fujisaki *et al.* have recently used the innovations approach to overcome most of these difficulties.

However, although such rigorous proofs are of interest, their main contribution is to highlight the fundamental difficulties in the way of practical optimum filtering. Despite major efforts (see, e.g., the proceedings of many recent symposia) the field is in some disarray. A good account of some of the more successful efforts is given by Jazwinski [138]. I believe that the situation is somewhat analogous to that of linear filtering in the mid-fifties, when the field was rapidly grinding to a halt amid a welter of numerous attempts at direct extensions of the Wiener filter. The Kalman filter provided a new impulse that moved things out of the doldrums into a new and fruitful direction. Similarly in nonlinear filtering it may be that attempts to solve the nonlinear filtering problem along the lines of the successful Kalman linear filter are misdirected. Some new approach needs to be uncovered.

Perhaps the way to begin is by lowering our sights by restricting ourselves to parameter estimation rather than to estimation of rather general stochastic processes. Even this is a difficult subject, which will not even be outlined here. However, I bring it up because recently information-theoretic[9] ideas have been found to be useful in getting error bounds for such finite-parameter problems (see, e.g., [377]-[381], [384]), and to a small extent for certain infinite-parameter (or stochastic-process) problems as well, see [383]. Furthermore, recently Blahut [382] and others have begun to show how the basic results of information theory can also be illuminated by the use of some simple hypothesis-testing and parameter-estimation problems. This interchange will be valuable and is in fact somewhat overdue, which brings me to my final point.

At many times in the twenty-five years of information theory there has been a not inconsiderable dissatisfaction with the scope, development, and application of the theory. This may or may not (probably not) have been justified, but it is interesting, at least in my opinion, that the fields of signal detection, estimation, and stochastic processes have not experienced such traumas. It seems to me that the reason lies in the actively pursued connections between these subjects and many other topics. "Strict-sense" information theory, although a beautiful and important subject, has suffered by its partially deliberate insularity and isolation. The numerous interconnections of statistical

[8] It perhaps goes without saying that we have confined ourselves for many reasons to the *theory* of linear filtering and in particular to the *probabilistic* theory, where knowledge of all statistical parameters (e.g., means and covariances) is assumed. The development of a *statistical* theory of filtering will introduce several new dimensions, although it might be noted that the lack of a complete statistical theory does not seem to have significantly limited the successful use of the ideas of the probabilistic theory.

[9] In the Shannon sense.

signal processing with other fields, as richly displayed even in the small subdomain of linear filtering that we have surveyed, is perhaps the surest guarantee of its continued vitality. It has never been other than a pleasure for me to have worked in such a field.

ACKNOWLEDGMENT

The last sentence really says it all: I am grateful to all who have contributed to this field. Several friends have generously provided comments on drafts of this paper.

BIBLIOGRAPHY

A. Wiener Filtering and Related Topics

[1] N. Wiener, *Extrapolation, Interpolation and Smoothing of Stationary Time Series, with Engineering Applications.* New York: Technology Press and Wiley, 1949. (Originally issued in February 1942, as a classified Nat. Defense Res. Council Rep.)
[2] N. Wiener and E. Hopf, "On a class of singular integral equations," *Proc. Prussian Acad., Math.-Phys. Ser.*, p. 696, 1931.
[3] R. B. Blackman, H. W. Bode, and C. E. Shannon, "Data smoothing and prediction in fire-control systems," Research & Development Board, Washington, D.C., Aug. 1944.
[4] N. Levinson, "A heuristic exposition of Wiener's mathematical theory of prediction and filtering," *J. Math. Phys.*, vol. 25, pp. 110–119, July 1947; reprinted as an Appendix in [1].
[5] H. W. Bode and C. E. Shannon, "A simplified derivation of linear least square smoothing and prediction theory," *Proc. IRE*, vol. 38, pp. 417–425, Apr. 1950.
[6] L. A. Zadeh and J. R. Ragazzini, "An extension of Wiener's theory of prediction," *J. Appl. Phys.*, vol. 21, pp. 645–655, July 1950.
[7] C. L. Dolph and H. A. Woodbury, "On the relation between Green's functions and covariances of certain stochastic processes and its applications to unbiased linear prediction," *Trans. Amer. Math. Soc.*, vol. 72, pp. 519–550, 1952.
[8] G. F. Franklin, "The optimum synthesis of sampled-data systems," Electron. Res. Lab., Columbia Univ., New York, Tech. Rep. T-6/B, May 1955.
[9] R. Jaffe and E. Rechtin, "Design and performance of phase-lock circuits capable of near-optimum performance over a wide range of input signal and noise levels," *IRE Trans. Inform. Theory*, vol. IT-1, pp. 66–76, Mar. 1955.
[9a] P. Elias, "Predictive coding," *IRE Trans. Inform. Theory*, vol. IT-1, pp. 16–33, Mar. 1955.
[9b] R. Price, "On entropy equivalence in the time- and frequency-domains," *Proc. IRE*, vol. 43, p. 484, Apr. 1955.
[10] A. M. Yaglom, *Theory of Stationary Random Functions*, translated from the Russian by R. A. Silverman. Englewood Cliffs, N.J.: Prentice-Hall, 1962. (Originally published as a survey paper in 1955.)
[11] M. C. Yovits and J. L. Jackson, "Linear filter optimization with game theory considerations," in *IRE Nat. Conv. Rec.*, pt. 4, pp. 193–199, 1955.
[12] N. Wiener and P. Masani, "The prediction theory of multivariate stochastic processes, Pt. I," *Acta Math.*, vol. 98, pp. 111–150, 1957; Pt. II, *ibid.*, vol. 99, pp. 93–137, 1958.
[13] G. C. Newton, L. A. Gould, and J. F. Kaiser, *Analytical Design of Linear Feedback Controls.* New York: Wiley, 1957.
[14] M. Shinbrot, "A generalization of a method for the solution of the integral equation arising in optimization of time-varying linear systems with nonstationary inputs," *IRE Trans. Inform. Theory*, vol. IT-3, pp. 220–225, Dec. 1957.
[15] F. J. Beutler, "Prediction and filtering for random parameter systems," *IRE Trans. Inform. Theory*, vol. IT-4, pp. 166–171, Dec. 1958.
[16] H. Laning and R. Battin, *Random Processes in Automatic Control.* New York: McGraw-Hill, 1958.
[17] S. Darlington, "Nonstationary smoothing and prediction using network theory concepts," *IRE Trans. Inform. Theory* (Special Suppl.), vol. IT-5, pp. 1–14, May 1959.
[18] P. Leonov, "On an approximate method for synthesizing optimal linear systems for separating signals from noise," *Automat. Remote Contr.*, vol. 20, pp. 1039–1048, 1959.
[19] A. V. Balakrishnan, "On a characterization of processes for which the optimal mean-square systems are of specified form," *IRE Trans. Inform. Theory*, vol. IT-6, pp. 490–500, Sept. 1960.
[20] Y. W. Lee, *Statistical Theory of Communication.* New York: Wiley, 1960.
[21] V. S. Pugachev, *Theory of Random Functions and Its Applications in Automatic Control.* Moscow: Goztekhizdat, 1960; 2nd ed., 1964 (Transl.: Reading, Mass.: Addison-Wesley, 1966).
[22] Yu. A. Rozanov, "Spectral properties of multivariate stationary processes and boundary properties of analytic matrices," *Theory Prob. Appl.* (USSR), vol. 5, pp. 362–376, 1960.
[23] A. M. Yaglom, "Effective solutions of linear approximation problems for multivariate stationary processes with a rational spectrum," *Theory Prob. Appl.* (USSR), vol. 5, pp. 239–264, 1960.
[24] S. S. L. Chang, *Synthesis of Optimum Control Systems.* New York: McGraw-Hill, 1961.
[25] E. L. Peterson, *Statistical Analysis and Optimization of Systems.* New York: Wiley, 1961.
[26] D. C. Youla, "On the factorization of rational matrices," *IRE Trans. Inform. Theory*, vol. 7, pp. 172–189, July 1961.
[27] A. V. Balakrishnan, "An operator-theoretic formulation of a class of control problems and a steepest descent method of solution," *SIAM J. Contr.*, vol. 1, pp. 109–127, 1963.
[27a] H. C. Hsieh and R. A. Nesbit, "Functional analysis and its applications to mean-square estimation problems," in *Modern Control Systems Theory*, C. Leondes, Ed. New York: McGraw-Hill, 1965, pp. 97–120.
[28] M. C. Davis, "Factoring the spectral matrix," *IEEE Trans. Automat. Contr.*, vol. AC-8, pp. 296–305, Oct. 1963.
[29] E. Parzen, "A new approach to the synthesis of optimal smoothing and prediction systems," in *Mathematical Optimization Techniques*, R. Bellman, Ed. Berkeley, Calif.: Univ. California Press, 1963, pp. 75–108.
[30] Yu. A. Rozanov, *Stationary Random Processes.* Moscow: Fizmatgiz, 1963 (Transl.: A. Feinstein, San Francisco: Holden-Day, 1967).
[31] V. F. Pisarenko and Yu. A. Rozanov, "On some problems for stationary processes reducing to equations of the Wiener–Hopf type," *Probl. Pered. Inform.* (in Russian), vol. 14, pp. 113–135, 1963.
[32] P. Whittle, *Prediction and Regulation.* New York: Van Nostrand Reinhold, 1963.
[33] C. W. Helstrom, "Solution of the detection integral equation for stationary filtered white noise," *IEEE Trans. Inform. Theory*, vol. IT-11, pp. 335–339, July 1965.
[34] K. Steiglitz, "The equivalence of digital and analog signal processing," *Inform. Contr.*, vol. 8, pp. 455–467, 1965.
[35] P. Masani, "Recent trends in multivariate prediction theory," in *Multivariate Analysis*, P. R. Krishnaiah, Ed. New York: Academic Press, 1966.
[36] F. Csaki and P. Fischer, "On the spectrum factorization," *Acta Tech. Acad. Sci. Hung.*, vol. 58, pp. 145–168, 1967.
[37] S. G. Mikhlin and K. L. Smolitsky, *Approximate Methods for Solution of Differential and Integral Equations.* New York: American Elsevier, 1967.
[38] B. Widrow, P. E. Mantey, L. J. Griffiths, and B. Goode, "Adaptive antenna systems," *Proc. IEEE*, vol. 55, pp. 2143–2159, Dec. 1967.
[39] J. F. Claerbout, "A summary, by illustrations, of least squares filters with constraints," *IEEE Trans. Inform. Theory*, vol. IT-14, pp. 269–272, Mar. 1968.
[40] H. L. Van Trees, *Detection, Estimation, and Modulation Theory, Part I.* New York: Wiley, 1968. Reviewed by R. A. Scholtz, *IEEE Trans. Inform. Theory* (Book Rev.), vol. IT-14, pp. 612–613, July 1968.
[41] I. F. Blake, "Linear filtering and piecewise linear correlation functions," *IEEE Trans. Inform. Theory*, vol. IT-15, pp. 345–349, May 1969.
[42] W. M. Brown and R. B. Crane, "Conjugate linear filtering," *IEEE Trans. Inform. Theory*, vol. IT-15, pp. 462–465, July 1969.
[43] D. Slepian and T. T. Kadota, "Four integral equations of detection theory," *SIAM J. Appl. Math.*, vol. 17, pp. 1102–1117, 1969.
[44] D. L. Snyder, *The State-Variable Approach to Continuous Estimation, with Applications to Analog Communication Theory.* Cambridge, Mass.: M.I.T. Press, 1969. Reviewed by E. C. Posner, *IEEE Trans. Inform. Theory* (Book Rev.), vol. IT-18, p. 314, Mar. 1972.
[45] P. Masani, "Review of *Stationary Random Processes*, by Yu. A. Rozanov," *Ann. Math. Statist.*, vol. 42, pp. 1463–1467, 1971.
[46] J. J. Stiffler, *Theory of Synchronous Communications.* Englewood Cliffs, N.J.: Prentice-Hall, 1971. Reviewed by R. A. Scholtz, *IEEE Trans. Inform. Theory* (Book Rev.), vol. IT-18, pp. 218–219, Jan. 1972.
[47] H. L. Van Trees, *Detection, Estimation, and Modulation Theory, Part II—Nonlinear Modulation Theory.* New York: Wiley, 1971. Reviewed by J. B. Thomas, *IEEE Trans. Inform. Theory* (Book Rev.), vol. IT-18, pp. 450–451, May 1972.
[48] K. Yao, "On the direct calculations of MMSE of linear realizable estimator by Toeplitz form method," *IEEE Trans. Inform. Theory* (Corresp.), vol. IT-17, pp. 95–97, Jan. 1971.
[49] ——, "An alternative approach to the linear causal least-square filtering theory," *IEEE Trans. Inform. Theory*, vol. IT-17, pp.

232–240, May 1971.
[50] Yu. A. Rozanov, "Some approximation problems in the theory of stationary processes," *J. Multivariable Anal.*, vol. 2, pp. 135–144, June 1972.
[51] J. A. Cochran, *Analysis of Linear Integral Equations*. New York: McGraw-Hill, 1972.
[52] W. C. Lindsey, *Synchronization Systems in Communication and Control*. Englewood Cliffs, N.J.: Prentice-Hall, 1972.
[53] L. Prouza, "On generalized linear discrete inversion filters," *Kybernetika*, vol. 8, pp. 264–267, 1972; also, *ibid.*, vol. 6, pp. 225–240, 1970.
[54] J. Snyders, "Error expressions for optimal linear filtering of stationary processes," *IEEE Trans. Inform. Theory*, vol. IT-18, pp. 574–582, Sept. 1972.
[54a] J. Snyders, "Error formulae for optimal linear filtering, prediction and interpolation of stationary time series," *Ann. Math. Statist.*, vol. 43, pp. 1935–1943, 1972.
[54b] M. G. Strintzis, "A solution to the matrix factorization problem," *IEEE Trans. Inform. Theory*, vol. IT-18, pp. 225–232, Mar. 1972.
[55a] D. G. Messerschmitt, "A geometric theory of intersymbol interference, Part I: Zero-forcing and decision-feedback equalization," *Bell Syst. Tech. J.*, vol. 52, pp. 1483–1519, 1973.
[55b] R. W. Lucky, "A survey of the communication theory literature: 1968–1973," *IEEE Trans. Inform. Theory*, vol. IT-19, pp. 725–739, Nov. 1973.
[55c] J. Salz, "Optimum mean-square decision feedback equalization," *Bell Syst. Tech. J.*, 1974.

B. Recursive Wiener and Kalman Filtering

[56] A. G. Carlton and J. W. Follin, Jr., "Recent developments in fixed and adaptive filtering," NATO Advanced Group for Aerospace R&D, AGARDograph 21, 1956.
[57] J. E. Hanson, "Some notes on the application of the calculus of variations to smoothing for finite time, etc.," Appl. Phys. Lab., Johns Hopkins Univ., Baltimore, Md., Internal Memo BBD-346, 1957.
[58] A. J. F. Siegert, "A systematic approach to a class of problems in the theory of noise and other random phenomena, Pt. II," *IRE Trans. Inform. Theory*, vol. IT-13, pp. 38–43, Mar. 1957; Pt. III, *ibid.*, vol. IT-4, pp. 4–14, Mar. 1958.
[59] M. Blum, "Recursion formulas for growing memory digital filters," *IRE Trans. Inform. Theory*, vol. IT-4, pp. 24–30, Mar. 1958.
[60] B. Friedland, "Least-squares filtering and prediction of nonstationary sampled data," *Inform. Contr.*, vol. 1, pp. 297–313, 1958.
[61] P. Swerling, "First-order error propagation in a stagewise smoothing procedure for satellite observations," *J. Astronaut. Sci.*, vol. 6, pp. 46–52, Autumn 1959; see also "A proposed stagewise differential correction procedure for satellite tracking and prediction," RAND Corp. Rep. P-1292, Jan. 1958.
[62] R. E. Kalman, "On the general theory of control," in *Proc. 1st IFAC Cong.* London: Butterworth, 1960.
[63] ——, "Contributions to the theory of optimal control," *Bol. Soc. Mat. Mex.*, vol. 5, pp. 102–119, 1960.
[64] ——, "A new approach to linear filtering and prediction problems," *J. Basic Eng.*, vol. 82, pp. 34–45, Mar. 1960.
[65] R. L. Stratonovich, "Application of the theory of Markov processes for optimum filtration of signals," *Radio Eng. Electron. Phys.* (USSR), vol. 1, pp. 1–19, Nov. 1960.
[66] ——, "Conditional Markov process theory," *Theory Prob. Appl.* (USSR), vol. 5, pp. 156–178, 1960.
[67] M. Blum, "A stagewise parameter estimation procedure for correlated data," *Numer. Math.*, vol. 3, pp. 202–208, 1961.
[68] R. E. Kalman, "New methods of Wiener filtering theory," in *Proc. 1st Symp. Engineering Applications of Random Function Theory and Probability*, J. L. Bogdanoff and F. Kozin, Eds. New York: Wiley, 1963, pp. 270–388; also, RIAS, Baltimore, Md., Tech. Rep. 61-1, 1961.
[69] R. E. Kalman and R. S. Bucy, "New results in linear filtering and prediction theory," *Trans. ASME, Ser. D, J. Basic Eng.*, vol. 83, pp. 95–107, Dec. 1961.
[70] R. H. Battin, "A statistical optimizing navigation procedure for space flight," *J. Amer. Rocket Soc.*, vol. 32, pp. 1681–1692, 1962.
[71] J. D. McLean, S. F. Schmidt, and L. A. McGee, "Optimal filtering and linear prediction applied to a midcourse navigation system for the circumlunar mission," NASA Rep. TND-1208, 1962.
[72] S. F. Schmidt, "State-space techniques applied to the design of a space navigation system," in *Proc. 1962 Joint Automatic Control Conf.*, Paper 11-3.
[73] V. A. Yakubovic, "The solution of certain matrix inequalities in automatic control theory," *Dokl. Akad. Nauk SSSR*, vol. 143, pp. 1304–1307, 1962.
[74] A. E. Bryson and M. Frazier, "Smoothing for linear and nonlinear dynamic systems," Aeronaut. Syst. Div., Wright-Patterson AFB, Ohio, Tech. Rep. ASD-TDR-63-119, Feb. 1963.
[75] R. E. Kalman, "Lyapunov functions for the problem of Lur'e in automatic control," *Proc. Nat. Acad. Sci.*, vol. 49, pp. 201–205, Feb. 1963.
[76] ——, "Mathematical description of linear dynamical systems," *SIAM J. Contr.*, vol. 1, pp. 152–192, 1963.
[77] ——, "On a new characterization of linear passive systems," in *Proc. 1st Annu. Allerton Conf. Circuit and System Theory*, Nov. 1963, pp. 456–470.
[78] A. G. J. MacFarlane, "An eigenvector solution of the optimal linear regulator," *J. Electron. Contr.*, vol. 14, pp. 643–654, June 1963.
[79] E. A. Robinson, "Mathematical development of discrete filters for detection of nuclear explosions," *J. Geophys. Res.*, vol. 68, pp. 5559–5567, 1963.
[80] P. Swerling, "Comment on 'A statistical optimizing navigation procedure for space flight'," *AIAA J.* vol. 1, p. 1968, Aug. 1963.
[81] P. Whittle, "On the fitting of multivariate autoregressions and the approximate canonical factorization of a spectral density matrix," *Biometrika*, vol. 50, pp.129–134, 1963.
[82] K. Astrom and S. Wensmark, "Numerical identification of stationary time-series," in *Proc. 6th Int. Instrumentation and Measurements Congr.*, Sept. 1964.
[83] R. H. Battin, *Astronautical Guidance*. New York: McGraw-Hill, 1964.
[84] R. B. Blackman, "Methods of orbit refinement," *Bell Syst. Tech. J.*, vol. 43, pp. 885–909, May 1964.
[84a] H. Cox, "On the estimation of state variables and parameters for noisy dynamic systems," *IEEE Trans. Automat. Contr.*, vol. AC-9, pp. 5–12, Jan. 1964.
[85] R. E. Kalman, "When is a linear control system optimal?," *Trans. ASME, Ser. D, J. Basic Eng.*, vol. 86, pp. 51–60, June 1964.
[86] H. J. Kushner, "On differential equations satisfied by conditional probability densities of Markov processes," *SIAM J. Contr.*, vol. 2, pp. 106–119, 1964.
[87] V. M. Popov, "Hyperstability and optimality of automatic systems with several control functions," *Rev. Roum. Sci. Tech.*, vol. 9, pp. 629–690, 1964.
[87a] ——, "Incompletely controllable positive systems and applications to optimization and stability of automatic control systems," *Rev. Roum. Sci. Tech., Electrotech. Energ.*, vol. 12, pp. 337–357, 1967.
[88] G. L. Smith, "Multivariable linear filter theory applied to space vehicle guidance," *SIAM J. Contr.*, vol. 2, pp. 19–32, 1964.
[89] R. L. Stratonovich and Yu. G. Sosulin, "Optimal detection of a Markov process in noise," *Eng. Cybern.*, vol. 6, pp. 7–19, Oct. 1964.
[90] A. E. Bryson and D. E. Johansen, "Linear filtering for time-varying systems using measurements containing colored noise," *IEEE Trans. Automat. Contr.*, vol. AC-10, pp. 4–10, Jan. 1965.
[91] R. S. Bucy, "Nonlinear filtering theory," *IEEE Trans. Automat. Contr* (Corresp.), vol. AC–10, p. 198, Apr. 1965.
[92] H. H. Rosenbrock, "On the connection between discrete linear filters and some formulae of Gauss," in *Act. Congr. Automatique Théorique*. Paris: Dunod, 1965.
[93] F. C. Schweppe, "Evaluation of likelihood functions for Gaussian signals," *IEEE Trans. Inform. Theory*, vol. IT-11, pp. 61–70, 1965.
[94] E. B. Stear, "Shaping filters for stochastic processes," in *Modern Control Systems Theory*, C. T. Leondes, Ed. New York: McGraw-Hill, 1965, pp. 121–155.
[95] P. Whittle, "Recursive relations for predictors of non-stationary processes," *J. Roy. Statist. Soc., Ser. B*, vol. 27, pp. 525–532, 1965.
[96] R. A. Wiggins and E. A. Robinson, "Recursive solution to the multichannel filtering problem," *J. Geophys. Res.*, vol. 70, pp. 1885–1891, Apr. 1965.
[97] W. M. Wonham, "Some applications of stochastic differential equations to optimal nonlinear filtering," *SIAM J. Contr.*, vol. 2, pp. 347–369, 1965.
[98] B. D. O. Anderson, "Time-varying spectral factorization," Stanford Electron. Lab., Stanford, Calif., Tech. Rep. SEL-66-107, Oct. 1966.
[98a] P. Businger and G. H. Golub, "Linear least-squares solutions by Householder transformation," *Math. Comput.*, vol. 20, pp. 325–328, 1966.
[99] R. E. Mortensen, "Optimal control of continuous-time stochastic systems," Ph.D. dissertation, Univ. California, Berkeley, 1966.
[100] J. E. Potter, "Matrix quadratic solutions," *SIAM J. Appl. Math.*, vol. 14, pp. 496–501, May 1966.
[101] A. N. Shiryaev, "Stochastic equations of nonlinear filtration for purely discontinuous Markov processes," *Probl. Peredach. Inform.*, vol. 2, pp. 3–22, 1966.

[102] R. L. Stratonovich and Yu. G. Sosulin, "Optimum reception of signals in nonGaussian noise," *Radio Eng. Electron.*, vol. 11, pp. 497–507, Apr. 1966.

[103] D. C. Youla, "The synthesis of linear dynamical systems from prescribed weighting patterns," *SIAM J. Appl. Math.*, vol. 14, pp. 527–549, May 1966.

[104] B. D. O. Anderson, "An algebraic solution to the spectral factorization problem," *IEEE Trans. Automat. Contr.*, vol. AC-12, pp. 410–414, Aug. 1967.

[105] B. D. O. Anderson and J. B. Moore, "Solution of a time-varying Wiener filtering problem," *Electron. Lett.*, vol. 3, pp. 562–563, Dec. 1967.

[106] K. J. Astrom, *Introduction to Stochastic Control Theory.* New York: Academic Press, 1967.

[106a] A. Bjorck and G. H. Golub, "Iterative refinement of linear least-squares solutions by Householder transformation," *BIT*, vol. 7, pp. 322–337, 1967.

[107] J. Burg, "Maximum entropy spectral analysis," in *Proc. 37th Annu. Meet. Soc. Explor. Geophys.*, 1967.

[108] T. E. Duncan, "Probability densities for diffusion processes with applications to nonlinear filtering theory and detection theory," Ph.D. dissertation, Dep. Elec. Eng., Stanford Univ., Stanford, Calif., June 1967.

[109] H. J. Kushner, *Stochastic Stability and Control.* New York: Academic, 1967.

[110] E. A. Robinson, *Multichannel Time-Series Analysis with Digital Computer Programs.* San Francisco, Calif., Holden-Day, 1967.

[111] W. M. Wonham, "Lecture notes on stochastic optimal control," Div. Appl. Math., Brown Univ., Providence, R.I., Rep. 67-1, 1967.

[112] C. Bruni, A. Isidori, and A. Ruberti, "A method of factorization of the impulse-response matrix," *IEEE Trans. Automat. Contr.* (Corresp.), vol. AC-13, pp. 739–741, Dec. 1968.

[113] R. S. Bucy and P. D. Joseph, *Filtering for Stochastic Processes with Applications to Guidance.* New York: Wiley, 1968.

[114] L. D. Collins, "Realizable whitening filters and state-variable realizations," *Proc. IEEE* (Lett.), vol. 56, 100–101, Jan. 1968.

[115] T. E. Duncan, "Evaluation of likelihood functions," *Inform. Contr.*, vol. 13, pp. 62–74, July 1968.

[116] T. Kailath, "An innovations approach to least-squares estimation—Part I: Linear filtering in additive white noise," *IEEE Trans. Automat. Contr.*, vol. AC-13, pp. 646–655, Dec. 1968.

[117] T. Kailath and P. Frost, "An innovations approach to least-squares estimation, Part II: Linear smoothing in additive white noise," *IEEE Trans. Automat. Cont.*, vol. AC-13, pp. 655–660, Dec. 1968.

[118] G. Kallianpur and C. Striebel, "Estimation of stochastic systems: arbitrary system process with additive white observation errors," *Ann. Math. Statist.*, vol. 39, pp. 785–801, 1969.

[119] R. E. Kalman, "Lectures on controllability and observability," Lecture Notes, CIME, Bologna, 1968.

[120] R. S. Lipt'ser and A. N. Shiryaev, "Nonlinear filtering of Markov diffusion processes," *Proc. Steklov Inst.* Math. (English Transl.), vol. 104, pp. 163–218, 1968.

[121] ——, "Nonlinear interpolation of Markov diffusion processes," *Theory Prob. Appl.* (USSR). vol. 13, pp. 564–583, 1968.

[122] J. B. Moore and B. D. O. Anderson, "Extensions of quadratic minimization theory, I," *Int. J. Contr.*, vol. 7, pp. 465–472, 1968.

[123] F. C. Schweppe and H. K. Knudsen, "The theory of amorphous cloud trajectory prediction," *IEEE Trans. Inform Theory*, vol. IT-14, pp. 415–427, May 1968.

[124] E. Stear and A. Stubberud, "Optimal filtering for Gauss-Markov noise," *Int. J. Contr.*, vol. 8, pp. 123–130, 1968.

[125] W. G. Tuel, "Computer algorithm for spectral factorization of rational matrices," *IBM J. Res. Develop.*, vol. 12, pp. 163–170, Mar. 1968.

[126] W. M. Wonham, "On a matrix Riccati equation of stochastic control," *SIAM J. Contr.*, vol. 6, pp. 681–697, Nov. 1968.

[127] B. D. O. Anderson, J. B. Moore, and S. G. Loo, "Spectral factorization of time-varying covariance functions," *IEEE Trans. Inform. Theory*, vol. IT-15, pp. 550–557, Sept. 1969.

[128] A. E. Bryson and Y. C. Ho, *Applied Optimal Control.* Waltham, Mass.; Blaisdell, 1969.

[128a] P. Dyer and S. McReynolds, "Extension of square-root filtering to include process noise," *J. Optimiz. Theory Appl.*, vol. 3, pp. 444–459, 1969.

[129] R. E. Kalman, P. Falb, and M. A. Arbib, *Topics in Mathematical System Theory.* New York: McGraw-Hill, 1969.

[129a] N. Morrison, *Introduction to Sequential Smoothing and Prediction.* New York: McGraw-Hill, 1969.

[130] N. E. Nahi, "Optimum recursive estimation with uncertain observation," *IEEE Trans. Inform. Theory*, vol. IT-15, pp. 457–462, July 1969.

[131] J. Rissanen and L. Barbosa, "Properties of infinite covariance matrices and stability of optimum predictors," *Inform. Sci.*, vol. 1, pp. 221–236, 1969.

[132] W. M. Wonham, "Random differential equations in control theory," in *Probabilistic Methods in Applied Mathematics*, A. T. Bharucha-Reid, Ed. New York: Academic Press, 1969, ch. 2.

[133] L. E. Zachrisson, "On optimal smoothing of continuous-time Kalman processes," *Inform. Sci.*, vol. 1, pp. 143–172, 1969.

[134] L. H. Brandenburg and M. E. Meadows, "Shaping filter representation of nonstationary colored noise," *IEEE Trans. Inform. Theory*, vol. IT-17, pp. 26–31, Jan. 1971; see also L. Brandenburg, Ph.D. dissertation, Columbia Univ., New York, June 1970.

[135] R. W. Brockett, *Finite-Dimensional Linear Systems.* New York: Wiley, 1970.

[136] R. S. Bucy, "Linear and nonlinear filtering," *Proc. IEEE*, vol. 58, pp. 854–864, June 1970.

[137] A. Gersho and D. J. Goodman, "Projecting filters for recursive prediction of discrete-time processes," *Bell Syst. Tech. J.*, vol. 49, pp. 2377–2403, Nov. 1970.

[138] A. H. Jazwinski, *Stochastic Processes and Filtering Theory.* New York: Academic Press, 1970.

[139] C. T. Leondes, Ed., "Theory and applications of Kalman filtering," NATO Advanced Group for Aerospace R&D, AGARDOgraph 139, Feb. 1970.

[140] J. B. Moore and B. D. O. Anderson, "Spectral factorization of Time-varying covariance functions," *Math. Syst. Theory*, vol. 4, pp. 10–23, 1970.

[141] R. L. Stratonovich, "Detection and estimation of signals in noise when one or both are non-Gaussian," *Proc. IEEE*, vol. 58, pp. 670–679, May 1970.

[142] H. Akaike, "Autoregressive model fitting for control," *Ann. Inst. Statist. Math.*, vol. 23, pp. 163–180, 1971.

[143] B. D. O. Anderson and J. B. Moore, *Linear Optimal Control.* Englewood Cliffs, N.J., Prentice-Hall, 1971.

[144] ——, "The Kalman-Bucy filter as a true time-varying Wiener filter," *IEEE Trans. Syst., Man, Cybern.*, vol. SMC-1, pp. 119–128, Apr. 1971.

[145] M. Athans, Ed., *Special Issue on Linear-Quadratic-Gaussian Problem, IEEE Trans. Automat. Contr.*, vol. AC-16, Dec. 1971.

[146] R. Bellman and E. D. Denman, Eds., "Invariant imbedding," in *Lecture Notes in Operations Research*, vol. 52. New York: Springer, 1971.

[147] G. Epstein, "On finite-memory, recursive filters," *IEEE Trans. Inform. Theory* (Corresp.), vol. IT-17, pp. 486–487, July 1970; see also *ibid.*, p. 614, Sept. 1971, p. 753, Nov. 1971.

[148] P. G. Kaminski, "Square-root filtering and smoothing for discrete processes," Ph.D. dissertation, Stanford Univ., Stanford, Calif., Sept. 1971.

[148a] P. G. Kaminski, A. E. Bryson, Jr., and S. F. Schmidt, "Discrete square root filtering: A survey of current techniques," *IEEE Trans. Automat. Contr.*, vol. AC-16, pp. 727–735, Dec. 1971.

[149] T. Kailath and R. A. Geesey, "An innovations approach to least-squares estimation—Part IV: Recursive estimation given lumped covariance functions," *IEEE Trans. Automat. Contr.*, vol. AC-16, pp. 720–727, Dec. 1971.

[150] D. G. Lainiotis, "Optimal nonlinear estimation," *Int. J. Contr.*, vol. 14, pp. 1137–1148, 1971.

[150a] ——, "Optimal linear smoothing: Continuous-data case," *Int. J. Contr.*, vol. 17, pp. 921–930, May 1973.

[151] K. Martensson, "On the matrix Riccati equation," *Inform. Sci.*, vol. 3, pp. 17–49, 1971.

[151a] J. K. Omura, "Optimal receiver design for convolutional codes and channels with memory via control theoretical concepts," *Inform. Sci.*, vol. 3, pp. 243–266, 1971.

[152] I. B. Rhodes, "A tutorial introduction to estimation and filtering," *IEEE Trans. Automat. Contr.*, vol. AC-16, pp. 688–707, Dec. 1971.

[153] A. P. Sage and J. L. Melsa, *Estimation Theory with Applications to Communication and Control.* New York: McGraw-Hill, 1971. Reviewed by K. Yao, *IEEE Trans. Inform. Theory* (Book Rev.), vol. IT-19, pp. 374–376, May 1973.

[154] M. D. Srinath and P. K. Rajasekaran, "Estimation of randomly occurring stochastic signals in Gaussian noise," *IEEE Trans. Inform. Theory*, vol. IT-17, p. 206, Mar. 1971.

[155] P. Swerling, "Modern state estimation methods from the viewpoint of the method of least squares," *IEEE Trans. Automat. Contr.* vol. AC-16, pp. 707–720, Dec. 1971.

[156] A. van den Bos, "Alternative interpretation of maximum entropy spectral analysis," *IEEE Trans. Inform. Theory* (Corresp.), vol. IT-17, pp. 493–494, July 1971.

[157] H. L. Van Trees, *Detection, Estimation, Modulation Theory, Pt. III—Radar-Sonar Signal Processing and Gaussian Signals in Noise.* New York: Wiley, 1971.

[158] J. C. Willems, "Least squares stationary optimal control and the algebraic Riccati equation," *IEEE Trans. Automat. Contr.*, vol. AC-16, pp. 621–634, Dec. 1971.

[158a] ——, "Dissipative dynamical systems, Part II: Linear systems with quadratic supply rates," *Arch. Ration. Mech. Anal.*, vol. 45, pp. 352–393, 1972.

[159] M. G. Wood, J. B. Moore, and B. D. O. Anderson, "Study of an

integral equation arising in detection theory," *IEEE Trans. Inform. Theory*, vol. IT-17, pp. 677–686, Nov. 1971.
[160] A. V. Balakrishnan, "System theory and stochastic optimization," in *Proc. NATO 1972 Advanced Study Institute on Network and Signal Theory*, Bournemouth, England, 1972.
[160a] A. F. Fath, "Computational aspects of the linear optimal regulator problem," *IEEE Trans. Automat. Contr.*, vol. AC-14, pp. 547–550, Oct. 1969; also, A. E. Bryson and W. E. Hall, "Optimal control and filter synthesis by Eigenvector decomposition," Dep. Aeron. and Astron., Stanford Univ., Stanford, Calif., Rep. 436, Dec. 1971.
[161] J. L. Casti, R. E. Kalaba, and V. K. Murthy, "A new initial-value method for on-line filtering and estimation," *IEEE Trans. Inform. Theory*, vol. IT-18, pp. 515–518, July 1972.
[162] P. E. Gill, G. H. Golub, W. Murray, and M. A. Saunders, "Methods for modifying matrix factorizations," Comput. Sci. Dep., Stanford Univ., Stanford, Calif., Rep. CS-72-322, Nov. 1972.
[163] K. L. Hitz and B. D. O. Anderson, "Iterative method of computing the limiting solution of the matrix Riccati differential equation," *Proc. Inst. Elec. Eng.* vol. 119, pp. 1402–1406, Sept. 1972.
[164] F. Itakura, "Extraction of feature parameters of speech by statistical methods," in *Proc. 8th Symp. Speech Information Processing*, Feb. 1972.
[165] T. Kailath, "A note on least-squares estimation by the innovations method," *J. SIAM Contr.*, vol. 10, pp. 477–486, Aug. 1972.
[166] ——, "Some Chandrasekhar-type algorithms for quadratic regulator problems," in *Proc. IEEE Conf. Decision and Control and 11th Symp. Adaptive Processes*, Dec. 1972, pp. 219–223.
[167] V. Kučera, "A contribution to matrix quadratic equations," *IEEE Trans. Automat. Contr.*, vol. AC-17, pp. 344–346, June 1972.
[167a] ——, "A review of the matrix Riccati equation," *Kybernetika*, vol. 9, pp. 42–61, 1973.
[168] H. Kwakernaak and R. Sivan, *Linear Optimal Control Systems*. New York: Wiley, 1972.
[169] J. T. H. Lo, "Finite-dimensional sensor orbits and optimal nonlinear filtering," *IEEE Trans. Inform. Theory*, vol. IT-18, pp. 583–588, Sept. 1972.
[170] D. Rappaport, "Constant directions of the Riccati equation," *Automatica*, vol. 8, pp. 175–186, Mar. 1972. See also *IEEE Trans. Automat. Contr.*, vol. AC-15, pp. 535–540, Oct. 1970.
[171] J. M. Rodriguez-Canabal, "The geometry of the Riccati equation," Ph.D. dissertation, Univ. Southern California, Los Angeles, June 1972.
[172] D. L. Snyder, "Filtering and detection for doubly stochastic Poisson processes," *IEEE Trans. Inform. Theory*, vol. IT-18, pp. 91–101, Jan. 1972.
[173] H. Wakita, "Estimation of the vocal tract shape by optimal inverse filtering," Speech Commun. Res. Lab., Inc., Santa Barbara, Calif., Mono. 9, July 1972. See also *IEEE Trans. Audio Electroacoust.*, vol. AU-21, pp. 417–427, Oct. 1973.
[174] G. T. Wilson, "The factorization of matrical spectral densities," *SIAM J. Appl. Math.*, vol. 23, pp. 420–426, Dec. 1972.
[175] H. Aasnaes and T. Kailath, "An innovations approach to least-squares estimation—Pt VII: Some applications of vector autoregressive-moving average models," *IEEE Trans. Automat. Contr.*, vol. AC-18, pp. 601–607, Dec. 1973.
[176] H. Aasnaes and T. Kailath, "Robustness of linear-least squares filtering algorithms," in *Proc. 1973 Joint Automatic Control Conf.* See also *IEEE Trans. Automat. Contr.*, vol. AC-19, June 1974.
[177] B. D. O. Anderson and S. Vongpanitlerd, *Network Analysis and Synthesis—A Modern Systems Theory Approach*. Englewood Cliffs, N.J.: Prentice-Hall, 1973.
[178] J. A. Edward and M. M. Fitelson, "Notes on maximum-entropy processing, *IEEE Trans. Inform. Theory*. (Corresp.), vol. IT-19, pp. 232–234, Mar. 1973.
[179] M. Gevers and T. Kailath, "Constant, predictable and degenerate directions of the discrete-time Riccati equation," *Automatica*, vol. 9, pp. 699–712, Nov. 1973.
[180] ——, "An innovations approach to least-squares estimation—Part VI: Discrete-time innovations representations and recursive estimation," *IEEE Trans. Automat. Contr.*, vol. AC-18, pp. 588–600, Dec. 1973.
[181] T. Kailath and R. Geesey, "An innovations approach to least-squares estimation—Part V: Innovations representations and recursive estimation in colored noise," *IEEE Trans. Automat. Contr.*, vol. AC-18, pp. 435–453, Oct. 1973.
[182] R. H. Jones, "Autoregressive spectrum estimation," in *Proc. Amer. Meteorol. Soc. 3rd Conf. Probability and Statistics in Atmospheric Science*, 1973.
[183] T. Kailath, "Some new algorithms for recursive estimation in constant linear systems," *IEEE Trans. Inform. Theory*, vol. IT-19, pp. 750–760, Nov. 1973.
[184] T. Kailath, M. Morf, and G. S. Sidhu, "Some new algorithms for recursive estimation in constant discrete-time linear systems," in *Proc. 7th Princeton Symp. Information and System Science*, 1973. See also *IEEE Trans. Automat. Contr.*, vol. AC-19, Aug. 1974.
[185] G. S. Sidhu, T. Kailath, and M. Morf, "Development of fast algorithms via innovations decompositions," in *Proc. 7th Hawaii Int. Conf. Syst. Sci.*, Honolulu, Hawaii, Jan. 1974.
[186] R. Sh. Lipt'ser and A. N. Shiryaev, "Statistics of conditionally Gaussian random sequences," in *Proc. 6th Berkeley Symp. Mathematics, Statistics, and Probability*, vol. III. Berkeley, Calif.: Univ. California Press, 1973, pp. 389–422.
[187] B. P. Molinari, "The stabilizing solution of the algebraic Riccati equations," *SIAM J. Contr.*, vol. 11, pp. 262–272, May 1973.
[188] ——, "Equivalence relations for the algebraic Riccati equation," *SIAM J. Contr.*, vol. 11, pp. 272–286, May 1973.
[189] J. B. Moore and P. Hetrakul, "Optimal demodulation of PAM signals," *IEEE Trans. Inform. Theory*, vol. IT-19, pp. 188–197, Mar. 1973.
[190] H. J. Payne and L. M. Silverman, "On the discrete time algebraic Riccati equation," *IEEE Trans. Automat. Contr.*, vol. AC-18, pp. 226–234, June 1973.
[190a] M. Aoki, "On the subspaces associated with partial reconstruction of state vectors, the structure algorithm and the predictable directions of Riccati equations," *IEEE Trans. Automat. Contr.*, vol. AC-18, pp. 399–400, Aug. 1973.
[191] M. Morf, "Discrete-time multivariable systems," Ph.D. dissertation, Stanford Univ., Stanford, Calif., 1974.
[192] J. Rissanen, "A fast algorithm for optimum predictors," *IEEE Trans. Automat. Contr.*, vol. AC-18, p. 555, Oct. 1973.
[193] ——, "Algorithms for triangular decomposition of block Hankel and Toeplitz matrices with application to factorizing positive matrix polynomials," *Math. Comput.*, vol. 27, pp. 147–154, 1973.
[193a] F. C. Schweppe, *Uncertain Dynamic Systems*. Englewood Cliffs, N.J.: Prentice-Hall, 1973.
[193b] A. V. Balakrishnan, "Stochastic differential systems, Pt. I," in *Lecture Notes in Economics and Mathematical Systems*, vol. 84. New York: Springer, 1973.
[193c] H. J. Payne and L. Silverman, "Matrix Riccati equations and system structure," in *Proc. IEEE Conf. on Decision and Control*, pp. 558–563, Dec. 1973.
[193d] M. Morf and T. Kailath, "Square-root algorithms for linear least-squares estimation and control," in *Proc. 8th Princeton Symp. Information and System Science*, 1974.

C. Some Early Mathematical Work

[194] G. Galileo, "Dialogo Sopra i due Massimi Sestemi del Mondo: Tolemaico e Copernicano. Florence: Landini, 1632 (Transl.: Berkeley, Calif.: Univ California Press, 1953.)
[195] Jacōpo Francesco, Count Riccati, "Animadversationes in aequationes differentiales secundi gradus," in *Actorum Eruditorum quae Lipsiae publicantur*, Suppl. 8, pp. 66–73, 1724.
[196] R. Adrain, "Research concerning the probabilities of the errors which happen in making observations," *Analyst*, vol. 1, pp. 193–209, 1808.
[197] C. F. Gauss, *Theoria Motus Corporum Coelestium in Sectionibus Conicis Solem Ambientum*, Hamburg, 1809 (Transl.: Dover, New York: 1963).
[198] A. M. Legendre, "Methode des moindres quarrés, pour trouver le milieu le plus probable entre les résultats de differentes observations," *Mem. Inst. France*, pp. 149–154, 1810.
[199] G. H. Hardy, "The mean value of the modulus of an analytic function," *Proc. London Math. Soc.*, vol. 14, pp. 269–277, 1915.
[200] G. Szegö, "Ein Grenszwertsatz uber die Toeplitzschen Determinanten einer reellen positiven Funktion," *Math. Ann.*, vol. 76, pp. 490–503, 1915.
[201] ——, "Beitrage zur Theorie der Toeplitzschen Formen," *Math. Z.*, vol. 6, pp. 167–202, 1920.
[202] R. Frisch, "Correlation and scatter in statistical variables," *Nord. Stat. Tidsk.*, vol. 8, pp. 36–102, 1928.
[203] A. N. Kolmogorov, *Foundations of the Theory of Probability*, New York: Springer, 1933 (Transl.: New York: Chelsea, 1950).
[204] R. E. A. C. Paley and N. Wiener, "Fourier transforms in the complex domain," *Amer. Math. Soc. Colloq. Publ.*, vol. 19, 1934.
[205] M. Frechet, *Recherches théoriques modernes sur la théorie des Probabilities*, vol. 1. Paris: Gauthier-Villars, 1937; 2nd ed., 1950.
[206] H. Wold, *A Study in the Analysis of Stationary Time Series*. Uppsala, Sweden: Almqvist and Wiksell, 1938; 2nd. ed., 1954.
[207] A. N. Kolmogorov, "Sur l'interpolation et extrapolation des suites stationnaires," *C. R. Acad. Sci.*, vol. 208, p. 2043, 1939.
[208] G. Szegö, "Orthogonal polynomials," *Amer. Math. Soc. Colloq. Publ.*, vol. 23, 1939; 2nd ed., 1958; 3rd ed. 1967.
[209] A. N. Kolmogorov, "Stationary sequences in Hilbert space" (in Russian), *Bull. Math. Univ. Moscow*, vol. 2, no. 6, 1941 (A transl. by N. Artin is available in many libraries).
[210] ——, "Interpolation and extrapolation of stationary random sequences," *Bull. Acad. Sci. USSR, Ser Math.*, vol. 5, 1941; Transl.: RAND Corp., Santa Monica, Calif., Memo. RM-3090-PR, Apr. 1962).

[211] V. A. Ambartsumian, "Diffuse reflection of light by a foggy medium," *Dokl. Akad. Sci. SSSR*, vol. 38, pp. 229–322, 1943.
[212] R. S. Phillips, "Servomechanisms," M.I.T. Radiation Lab., Cambridge, Mass.: Rep. 372, May 1943; also in *Theory of Servomechanisms*, H. M. James, N. B. Nichols, and R. S. Phillips, Eds. New York: McGraw-Hill, 1957, ch. 7.
[213] J. L. Doob, "The elementary Gaussian processes," *Ann. Math. Statist.*, vol. 15, pp. 229–282, 1944.
[214] M. G. Krein, "On a generalization of some investigations of G. Szegö, W. M. Smirnov, and A. N. Kolmogorov," *Dokl. Akad. Nauk SSSR*, vol. 46, pp. 91–94, 1945.
[215] ——, "On a problem of extrapolation of A. N. Kolmogorov," *Dokl. Akad. Nauk SSSR*, vol. 46, pp. 306–309, 1945.
[216] W. T. Reid, "A matrix differential equation of the Riccati type," *Amer. J. Math.*, vol. 68, pp. 237–246, 1946.
[216a] ——, *Riccati Differential Equations*. New York: Academic Press, 1972.
[217] S. Chandrasekhar, "On the radiative equilibrium of a stellar atmosphere, Pt XXI," *Astrophys. J.*, vol. 106, pp. 152–216, 1947; Pt XXII, *ibid*, vol. 107, pp. 48–72, 1948.
[218] N. Levinson, "The Wiener rms (root-mean-square) error criterion in filter design and prediction," *J. Math. Phys.*, vol. 25, pp. 261–278, Jan. 1947; reprinted as appendix in [1].
[219] J. L. Doob, "Time series and harmonic analysis," in *Proc. Berkeley Symp. Mathematics, Statistics, and Probability*. Berkeley, Calif.: Univ. California Press, 1949, pp. 303–343.
[220] S. Chandrasekhar, *Radiative Transfer*. Oxford, England: Oxford Univ. Press, 1950; also New York: Dover, 1960.
[221] O. Hanner, "Deterministic and nondeterministic stationary random processes," *Ark. Mat.*, vol. 1, pp. 161–177, 1950.
[222] K. Karhunen, "Über die Struktur Stationärer Zufälliger Funktionen," *Ark. Mat.*, vol. 1, pp. 141–160, 1950 (Transl.: RAND Corp., Memo. RM-3091-PR, Apr. 1962).
[223] U. Grenander, "On Toeplitz forms and stationary processes," *Ark. Mat.*, vol. 1, pp. 555–571, 1952.
[224] I. M. Gel'fand and B. H. Levitan, "On the determination of a differential equation from its spectral function," *Izv. Akad. Nauk SSSR*, vol. 15, pp. 309–360, 1951 (Transl.: *Amer. Math. Soc. Transl.*, Ser. 2, vol. 1, pp. 253–304, 1955).
[225] J. L. Doob, *Stochastic Processes*. New York: Wiley, 1953.
[226] M. G. Krein, "Some problems of the effective determination of a nonhomogeneous string by means of its spectral function," *Dokl. Akad. Nauk SSSR*, vol. 93, pp. 617–620, 1953.
[227] ——, "On a fundamental approximation problem in the theory of extrapolation and filtration of stationary processes," *Dokl. Akad. Nauk SSSR*, vol. 94, pp. 13–16, 1954 (Transl.: *Select. Transl. Probl. Math. Statist.*, vol. 4, pp. 127–131, 1964).
[228] ——, "On integral equations governing differential equations of second order," *Dokl. Akad. Nauk SSSR*, vol. 97, pp. 21–24, 1954.
[229] ——, "On a new method of solving linear integral equations of the first and second kinds," *Dokl. Akad. Nauk SSSR*, vol. 100, pp. 413–416, 1955.
[230] ——, "The continuous analogues of theorems on polynomials orthogonal on the unit circle," *Dokl. Akad. Nauk SSSR*, vol. 104, pp. 637–640, 1955.
[231] R. E. Bellman, "Functional equations in the theory of dynamic programming, VII: A partial differential equation for the Fredholm resolvent," *Proc. Amer. Math. Soc.*, vol. 8, pp 435–440, 1957.
[232] K. M. Case, "On Wiener-Hopf equations," *Ann. Phys.*, vol. 2, pp. 384–405, 1957.
[233] E. G. Gladyshev, "On multidimensional stationary random processes," *Theory Prob. Appl.* (in Russian), vol. 3, pp. 425–428, 1958.
[234] I. C. Gohberg and M. G. Krein, "Systems of integral equations on a half-axis with kernels depending on the difference of the arguments," *Usp. Mat. Nauk*, vol. 13, pp. 3–72, 1958.
[235] M. G. Krein, "Integral equations on a half-axis with kernel depending on the difference of the arguments," *Usp. Mat. Nauk*, vol. 13, pp. 3–120, 1958.
[236] U. Grenander and G. Szegö, *Toeplitz Forms and Their Applications*. Berkeley, Calif.: Univ. California Press, 1958.
[237] P. Masani, "Cramér's theorem on monotone matrix-valued functions and the Wold decomposition," in *Probability and Statistics*, U. Grenander, Ed. New York: Wiley, 1959, pp. 175–189.
[238] S. Sandor, "Sur l'equation differentielle matricielle de type Riccati," *Bull. Math. Soc. Sci. Math. Phys.*, Revue Physique Roumaine (New Series), vol. 3, pp. 229–249, 1959.
[239] J. Durbin, "The fitting of time-series models," *Rev. Intern. Statist. Inst.*, vol. 28, pp. 233–244, 1960.
[240] G. Baxter, "Polynomials defined by a difference system," *J. Math. Anal. Appl.*, vol. 2, pp. 223–263, 1961.
[241] L. Ya. Geronimus, *Orthogonal Polynomials* (Transl. from the Russian). New York: Consultant's Bureau, 1961.
[242] G. Baxter, "An asymptotic result for the finite predictor," *Math. Scand.*, vol. 10, pp. 137–144, 1962.
[243] J. Hajek, "On linear statistical problems in stochastic processes," *Czech. Math. J.*, vol. 12, pp. 404–444, 1962.
[244] K. Hoffman, *Banach Spaces of Analytic Functions*. Englewood Cliffs, N.J.: Prentice-Hall, 1962.
[245] P. Masani, "Shift-invariant spaces and prediction theory," *Acta Math.*, vol. 107, pp. 275–290, 1962.
[246] P. Masani and J. B. Robertson, "The time-domain analysis of continuous-parameter, weakly stationary stochastic processes," *Pac. J. Math.*, vol. 12, pp. 1361–1378, 1962.
[247] R. M. Redheffer, "On the relation of transmission-line theory to scattering and transfer," *J. Math. Phys.*, vol. 41, pp. 1–41, 1962.
[247a] G. Baxter, "A norm inequality for a 'finite-section' Wiener-Hopf equation," *Ill. J. Math.*, vol. 7, pp. 97–103, 1963.
[248] M. Loève, Probability Theory, 3rd ed. New York: Van Nostrand Reinhold, 1963.
[248a] F. B. Atkinson, *Discrete and Continuous Boundary Problems*. New York: Academic Press, 1964.
[249] A. Devinatz, "Asymptotic estimates for the finite predictor," *Math. Scand.*, vol. 15, pp. 111–120, 1964.
[250] H. Helson, *Lectures on Invariant Subspaces*. New York: Academic Press, 1964.
[251] I. A. Ibragimov, "On the asymptotic behavior of the prediction error," *Theory Prob. Appl. USSR*, vol. 9, pp. 627–633, 1964.
[252] B. Noble, "The numerical solution of nonlinear integral equations and related topics," in *Nonlinear Integral Equations*, P. M. Anselone, Ed. Madison, Wis.: Wisconsin Univ. Press, 1964.
[253] C. M. Deo, "Prediction theory of nonstationary random processes," *Sankhya*, Ser. A, vol. 27, pp. 113–132, 1965.
[253a] J. L. Doob, "Wiener's work in probability theory," *Bull. Amer. Math. Soc.*, vol. 72, no. 1, pt. II, pp. 69–72, Jan. 1966.
[254] H. Kagiwada and R. E. Kalaba, "An initial-value method for Fredholm integral equations of convolution type," RAND Corp. Memo RM-5186-PR, 1966.
[255] P. Masani, "Wiener's contribution to generalized harmonic analysis, prediction theory and filter theory," *Bull. Amer. Math. Soc.*, vol. 72, no. 1, pt. II, pp. 73–125, Jan. 1966.
[256] G. M. Wing, "On certain integral equations reducible to initial value problems," *SIAM Rev.*, vol. 9, pp. 655–670, 1967.
[257] A. Devinatz and M. Shinbrot, "General Wiener-Hopf operators," *Trans. Amer. Math. Soc.*, vol. 145, pp. 467–494, Nov. 1967.
[258] P. L. Duren, *Theory of* H^p *Spaces*. New York: Academic Press, 1970.
[259] H. Dym and H. P. McKean, "Application of de Branges spaces of integral functions to the prediction of stationary Gaussian processes," *Ill. J. Math.*, vol. 14, pp. 299–343, 1970.
[260] H. Dym and H. P. McKean, "Extrapolation and interpolation of stationary Gaussian processes," *Ann. Math. Statist.*, vol. 41, pp. 1817–1844, Dec. 1970.
[261] B. Sz. Nagy and C. Foias, *Harmonic Analysis of Operators on Hilbert Space*. New York: Academic Press, 1970.
[261a] L. D. Pitt, "On problems of trigonometrical approximation from the theory of stationary Gaussian processes," *J. Multivariable Anal.*, vol. 2, pp. 145–161, June 1972.

D. Canonical Representations, Innovations, Martingales, and All That

[262] V. I. Krylov, "On functions regular in a half-plane," *Mat. Sb.*, vol. 6, pp. 95–138, 1939 (Transl.: *Amer. Math. Soc. Transl.* (2), vol. 32, pp. 37–81, 1963.
[263] H. W. Bode, *Network Analysis and Feedback Amplifier Design*, Princeton, N.J.: Van Nostrand Reinhold, 1945.
[264] A. Beurling, "On two problems concerning linear transformations in Hilbert space," *Acta Math.*, vol. 81, pp. 239–255, 1949.
[265] A. M. Yaglom and M. S. Pinsker, "Random processes with stationary increments of order *n*," *Dokl. Akad. Nauk SSSR*, vol. 90, pp. 731–733, 1953.
[266] P. Lévy, "Sur une classe de courbes de l'espace de Hilbert et sur une équation intégrale non linéaire," *Ann. Sci. Ec. Norm. Super.*, vol. 73, pp. 121–156, 1956.
[267] N. Wiener and G. Kallianpur, "Nonlinear prediction," Office of Naval Research, Tech. Rep. 1, Cu-2-56-NONR-266, (39)-CIRMIP, Project NR-047-015, 1956.
[268] H. Cramér, "On some classes of nonstationary stochastic processes," in *Proc. 4th Berkeley Symp. Mathematics, Statistics, and Probability*. Berkeley, Calif.: Univ. California Press, 1960, pp. 57–78.
[269] T. Hida, "Canonical representations of Gaussian processes and their applications," *Mem. College Sci., Univ. Kyoto, Ser. A*, vol. 33, pp. 109–155, 1960.
[270] L. A. Zadeh, "Time-varying networks, I," *Proc. IRE*, vol. 49, pp. 1488–1502, Oct. 1961.
[271] E. A. Robinson, "Extremal representation of stationary stochastic processes," *Ark. Mat.*, vol. 4, pp. 379–384, 1962.
[272] ——, *Random Wavelets and Cybernetic Wavelets*. New York: Hafner, 1962.
[273] H. Cramér, "Stochastic processes as curves in Hilbert space," *Teor. Veroyat. Primen.*, vol. 9, pp. 169–179, 1964.
[274] P. Lévy, *Processus Stochastiques et Mouvement Brownien*, 2nd ed. Paris: Gauthier-Villars, 1964.

[275] H. Cramér, "A contribution to the multiplicity theory of stochastic processes," in *Proc. 5th Berkeley Symp. Mathematics, Statistics, and Probability*, vol. 2. Berkeley, Calif.: Univ. California Press, 1965, pp. 215–224.

[276] L. A. Shepp, "Radon-Nikodym derivatives of Gaussian measures," *Ann. Math. Statist.*, vol. 37, pp. 321–354, Apr. 1966.

[277] E. Wong and M. Zakai, "On the relation between ordinary and stochastic differential equations and applications to stochastic problems in control theory," in *Proc. 3rd IFAC Congr.* London: Butterworth, 1966.

[277a] E. J. McShane, "Stochastic functional equations: Continuity properties and relation to ordinary equations," in *Control Theory and the Calculus of Variations*, A. V. Balakrishnan, Ed. New York: Academic Press, 1969.

[277b] ——, "Stochastic differential equations and models of random processes," in *Proc. 6th Berkeley Symp. Mathematics, Statistics, and Probability*. Berkeley: Univ. California Press, 1973.

[278] H. Kunita and S. Watanabe, "On square-integrable martingales," *Nagoya Math. J.*, vol. 30, pp. 209–245, Aug. 1967.

[279] P. A. Frost, "Nonlinear estimation in continuous-time systems," Ph.D. dissertation, Dep. Elec. Eng., Stanford Univ., Stanford, Calif., June 1968.

[280] I. C. Gohberg and M. G. Krein, "Theory and Applications of Volterra Operators in Hilbert Space," (in Russian). Moscow: Nauka, 1967. (Transl.: *Amer. Math. Soc.*, Providence, R.I., 1970.)

[281] R. Geesey, "Canonical representations of second-order processes with applications," Ph.D. dissertation, Dep. Elec. Eng., Stanford Univ., Stanford, Calif., 1968.

[282] M. Hitsuda, "Representation of Gaussian processes equivalent to Wiener processes," *Osaka J. Math.*, vol. 5, pp. 299–312, 1968.

[283] J. M. C. Clark, "Conditions for one-to-one correspondence between an observation process and its innovations," Cent. Comput. Automat., Imperial College, London, Tech. Rep. 1, 1969.

[284] R. Geesey and T. Kailath, "Comments on 'The relationship of alternate state-space representations in linear problems'," *IEEE Trans. Automat. Contr.* (Corresp.), vol. AC-14, pp. 113–114, Feb. 1969.

[285] R. Geesey and T. Kailath, "Applications of the canonical representation to estimation and detection of colored noise," in *Proc. Symp. Computer Processing in Communications*. Brooklyn, N.Y.: Polytechnic Inst. Brooklyn Press, Apr. 1969.

[286] I. M. Gel'fand and N. Ya. Vilenkin, *Generalized Functions*. New York: Academic Press, 1968.

[287] T. Kailath and R. Geesey, "Covariance factorization—an explication via examples," in *Proc. 2nd Asilomar Conf. Systems Science*, 1968.

[288] A. Schumitzky, "On the equivalence between matrix Riccati equations and Fredholm resolvents," *J. Comp. Syst. Sci.*, vol. 2, pp. 76–87, June 1968.

[288a] A. McNabb and A. Schumitzky, "Factorization of operators, Part III: Initial value methods for linear two-point boundary value problems," *J. Math. Anal. Appl.*, vol. 31, pp. 391–405, Aug. 1970.

[289] E. R. Berlekamp, *Algebraic Coding Theory*. New York: McGraw-Hill, 1969.

[290] T. Kailath, "Application of a resolvent identity to a linear smoothing problem," *SIAM J. Contr.*, vol. 7, pp. 68–74, Feb. 1969.

[291] ——, "A general likelihood-ratio formula for random signals in Gaussian noise," *IEEE Trans. Inform. Theory*, vol. IT-15, pp. 350–361, May 1969.

[292] ——, "Fredholm resolvents, Wiener-Hopf equations, and Riccati differential equations," *IEEE Trans. Inform. Theory*, vol. IT-15, pp. 665–672, Nov. 1969.

[293] T. E. Duncan, "On the absolute continuity of measures," *Ann. Math. Statist.*, vol. 41, pp. 30–38, 1970.

[294] P. A. Frost, "The innovations process and its application to nonlinear estimation and detection of signals in additive white noise," in *Proc. Univ. Missouri, Rolla–M. J. Kelly Communications Conf.*, Oct. 1970, pp. 7.3.1–7.3.6. See also *Proc. 4th Princeton Symp. on Information and System Science*, 1970.

[295] T. Hida, *Stationary Stochastic Processes*. Princeton, N.J.: Princeton Univ. Press, 1970.

[296] T. Kailath, "Likelihood ratios for Gaussian processes," *IEEE Trans. Inform. Theory*, vol. IT-16, pp. 276–288, May 1970.

[297] ——, "A further note on a general likelihood formula for random signals in Gaussian noise," *IEEE Trans. on Inform. Theory*, vol. IT-16, pp. 393–396, July 1970.

[298] ——, "The innovations approach to detection and estimation theory," in *Proc. IEEE*, vol. 58, pp. 680–695, May 1970.

[299] B. D. O. Anderson and T. Kailath, "The choice of signal process models in Kalman-Bucy filtering," *J. Math. Anal. Appl.*, vol. 35, pp. 659–668, Sept. 1971.

[300] H. Cramér, *Structural and Statistical Problems for a Class of Stochastic Processes*. Princeton, N.J.: Princeton Univ. Press, 1971.

[301] P. A. Frost and T. Kailath, "An innovations approach to least-squares estimation—Part III: Nonlinear estimation in white Gaussian noise," *IEEE Trans. Automat. Contr.*, vol. AC-16, pp. 217–226, June 1971.

[302] P. A. Frost, "Estimation and detection for a simple class of conditionally independent-increment processes," in *Proc. IEEE Decision and Control Conf.*, 1971. Also published as lecture notes for Washington University summer course on current trends in automatic control, St. Louis, Mo., 1970.

[303] T. T. Kadota, M. Zakai, and J. Ziv, "Mutual information of the white Gaussian channel with and without feedback," *IEEE Trans. Inform. Theory*, vol. IT-17, pp. 368–371, July 1971.

[304] T. Kailath, "Some extensions of the innovations theorem," *Bell Syst. Tech. J.*, vol. 50, pp. 1487–1494, Apr. 1971.

[305] M. M. Rao, "Local functionals and generalized random fields with independent values," *Theory Prob. Appl.* (in Russian), vol. 16, pp. 457–473, 1971.

[306] J. Rissanen, "Recursive identification of linear systems," *SIAM J. Contr.*, vol. 9, pp. 420–430, Aug. 1971.

[307] J. Baras and R. Brockett, "H^2-functions and infinite-dimensional realization theory," in *Proc. IEEE Decision and Control Conf.*, 1972, pp. 355–360; also, Ph.D. dissertation, Harvard Univ., Cambridge, Mass., Sept. 1973.

[308] P. Bremaud, "A Martingale approach to point processes," Ph.D. dissertation, Univ. Calif., Berkeley, Aug. 1972; also Electron. Res. Lab., Rep. M345.

[309] M. Fujisaki, G. Kallianpur, and H. Kunita, "Stochastic differential equations for the nonlinear filtering problem," *Osaka J. Math.*, vol. 9, pp. 19–40, 1972.

[310] T. Kailath, R. Geesey, and H. Weinert, "Some relations between RKHS norms, and Fredholm equations innovation representations," *IEEE Trans. Inform. Theory*, vol. IT-18, pp. 341–348, May 1972.

[311] T. Kailath and D. Duttweiler, "An RKHS approach to detection and estimation problems—Part III: Generalized innovations representations and a likelihood-ratio formula," *IEEE Trans. Inform. Theory*, vol. IT-18, pp. 730–745, Nov. 1972.

[312] H. Dym and H. McKean, *Fourier Series and Integrals*. New York: Academic Press, 1972.

[313] J. Rissanen and T. Kailath, "Partial realization of stochastic processes," *Automatica*, vol. 8, pp. 380–386, July 1972.

[314] Yu. A. Rozanov, "On nonanticipative linear transformations of Gaussian processes with equivalent distributions," *Nagoya Math. J.*, vol. 47, pp. 227–235, 1972.

[315] E. Wong, *Stochastic Processes in Information and Communication Systems*. New York: McGraw-Hill, 1971. Reviewed by F. J. Beutler, *IEEE Trans. Inform. Theory* (Book Rev.), vol. IT-18, pp. 827–828, Nov. 1972.

[316] H. Akaike, "Markovian representation of stochastic processes by canonical variables," *SIAM J. Contr.*, to be published, 1973.

[316a] ——, "Stochastic theory of minimal realizations," *IEEE Trans. Automat. Contr.*, vol. AC-19, to be published.

[317] B. D. O. Anderson, "Algebraic properties of minimal degree spectral factors," *Automatica*, vol. 9, pp. 491–500, 1973.

[318] A. Ephremides and L. Brandenburg, "On the reconstruction error of sampled data estimates," *IEEE Trans. Inform. Theory* (Corresp.), vol. IT-19, pp. 365–367, May 1973.

[319] A. Ephremides and J. B. Thomas, *Random Processes—Multiplicity Theory and Canonical Decompositions*. Stroudsburg, Pa.: Dowden, Hutchinson and Ross, 1973.

[320] M. Hitsuda, "Multiplicity of some classes of Gaussian processes," *Nagoya J. Math.*, to be published, 1974.

[321] G. Kallianpur and H. Oodaira, "Nonanticipative representations of equivalent Gaussian processes," *Ann. Prob.*, vol. 1, pp. 104–122, 1973.

[322] T. Kailath and A. Segall, "A further note on innovations, martingales and nonlinear estimation," in *Proc. IEEE Decision and Control Conf.*, 1973.

[323] H. P. McKean, "Geometry of differential space," *Ann. Prob.*, vol. 1, pp. 197–206, Apr. 1973.

[324] P. A. Meyer, "Sur un problème de filtration," Séminaire de probabilitiés, Pt VII, *Lecture Notes in Mathematics*, vol. 321. New York: Springer, 1973, pp. 223–247.

[325] Yu. A. Rozanov, "Innovations and nonanticipative processes," in *Multivariate Analysis*, vol. 3, P. R. Krishnaiah, Ed. New York: Academic Press, 1973.

[326] A. Segall, "A Martingale approach to modeling, estimation and detection of jump processes," Ph.D. dissertation, Dep. Elec. Eng., Stanford Univ., Stanford, Calif., Aug. 1973.

[327] E. Wong, "Recent progress in stochastic processes—a survey," *IEEE Trans. Inform. Theory*, vol. IT-19, pp. 262–275, May 1973.

[327a] J. H. Van Schuppen, "Estimation theory for continuous-time processes, a martingale approach," Ph.D. dissertation, Dep. Elec. Eng., Univ. California, Berkeley, Sept. 1973.

[327b] R. Boel, P. Varaiya, and E. Wong, "Martingales on jump processes, I: Representation results, II: Applications," Univ. of California, Berkeley, ERL Memos 407 and 409, Oct. 1973; also, submitted to *SIAM J. Contr.*

[327c] L. H. Brandenburg, "Covariance factorization: Some unified results encompassing both stationary and nonstationary processes," to appear in *IEEE Trans. Inform. Theory.*

E. Miscellaneous

1) *Series Expansion (further references can be found in [327])*

[328] D. D. Kosambi, "Statistics in function space," *J. Indian Math. Soc.* vol. 7, pp. 76–88, 1943.
[329] M. Loève, "Sur les functions aléatories stationnaires de second order," *Rev. Sci.*, vol. 83, pp. 297–310, 1945; see also *Comptes Rénd.*, vol. 220, p. 380, 1945, and vol. 222, p. 489, 1946.
[330] M. Kac and A. J. F. Siegert, "On the theory of noise in radio receivers with square-law detectors," *J. App. Phys.*, vol. 18, pp. 383–397, Apr. 1947; see also, *Ann. Math. Statist.*, vol. 18, pp. 438–442, 1947.
[331] K. Karhunen, "Über Lineare Methoden in der Wahrscheinlichkeitsrechnung," *Amer. Acad. Sci.*, Fennicade, Ser. A, I, vol. 37, pp. 3-79, 1947; (Transl.: RAND Corp., Santa Monica, Calif., Rep. T-131, Aug. 1960).
[332] N. Aronszajn, "Theory of reproducing kernels," *Trans. Amer. Math. Soc.*, vol. 63, pp. 337–404, May 1950.
[333] U. Grenander, "Stochastic processes and statistical inference," *Ark. Mat.*, vol. 1, pp. 195–277, Oct. 1950.
[334] A. Ya. Povzner, "A class of Hilbert function spaces," *Dokl. Akad. Nauk SSSR*, vol. 68, pp. 817–820, 1949; see also, *ibid.*, vol. 74, pp. 13–17, 1950.
[335] R. C. Davis, "On the theory of prediction of nonstationary stochastic processes," *J. Appl. Phys.*, vol. 23, pp. 1047–1053, Sept. 1952.
[336] D. Slepian, "Estimation of signal parameters in the presence of noise," *IRE Trans. Inform. Theory*, vol. P6IT-3, pp. 68–89, Mar. 1954.
[337] D. C. Youla, "The use of the method of maximum likelihood in estimating continuous-modulated intelligence which has been corrupted by noise," *IRE Trans. Inform. Theory*, vol. P6IT-3, pp. 90–105, Mar. 1954.
[338] I. M. Gel'fand and A. M. Yaglom, "Calculation of the amount of information about a random function contained in another such function," *Usp. Mat. Nauk*, vol. 12, pp. 3–52, 1956.
[339] V. S. Pugachev, "Application of canonic expansions of random functions in determining an optimum linear system," *Automat. Remote Contr.*, vol. 17, pp. 489–499, 1956.
[340] T. W. Anderson, *An Introduction to Multivariate Statistical Analysis.* New York: Wiley, 1958.
[341] W. Davenport and W. L. Root, *An Introduction to the Theory of Random Signals and Noise.* New York: McGraw-Hill, 1958.
[342] C. W. Helstrom, *Statistical Theory of Signal Detection.* London: Pergamon Press, 1960; 2nd ed., 1968.
[343] V. N. Tutubalin and M. I. Freidlin, "On the structure of the infinitesimal σ-algebra of a Gaussian process," *Teor. Veroyat. Primen.*, vol. 7, pp. 196–199, 1962.
[344] H. P. McKean, "Brownian motion with a several dimensional time," *Teor. Veroyat. Primen.*, vol. 8, pp. 357–378, 1963.
[345] N. Levinson and H. P. McKean, "Weighted trigonometrical approximation on R^1 with application to the germ field of a stationary Gaussian noise," *Acta Math.*, vol. 112, pp. 99–143, 1964.
[346] A. M. Yaglom, "Outline of some topics in linear extrapolation of stationary random processes," in *Proc. 5th Berkeley Symp. Mathematics, Statistics, and Probability*, vol. II. Berkeley, Calif.: Univ. California Press, 1970, pp. 259–278.
[346a] ——, "Strong limit theorems for stochastic processes and orthogonality conditions for probability," in *Proc. Bernoulli, Bayes, Laplace Symp.* Berlin: Springer, 1965, pp. 253–262.
[347] E. Parzen, *Time Series Analysis Papers.* San Francisco: Holden-Day, 1967.
[348] T. T. Kadota, "Optimum estimation of nonstationary Gaussian signals in noise," *IEEE Trans. Inform. Theory*, vol. IT-15, pp. 253–257, Mar. 1969.
[349] T. Kailath, "RKHS approach to detection and estimation problems—Part I: Deterministic signals in Gaussian noise," *IEEE Trans. Inform. Theory*, vol. IT-17, pp. 530–549, Sept. 1971.
[350] R. D. LePage, "Note relating Bochner integrals and reproducing kernels to series expansions on a Gaussian Banach space," *Proc. Amer. Math. Soc.*, vol. 32, pp. 285–289, Mar. 1972. N. Jain and G. Kallianpur, *Proc. Amer. Math. Soc.*, vol. 25, pp. 890–895, 1970.
[350a] E. Lyttkens, "Regression aspects of canonical correlation," *J. Multivariate Anal.*, vol. 2, pp. 418–439, 1972.
[351] S. Cambanis, "A general approach to linear mean-square estimation problems," *IEEE Trans. Inform. Theory*, (Corresp.), vol. IT-19, pp. 110–114, Jan. 1973.
[352] W. A. Gardner, on "A general approach to linear mean-square estimation problems," *IEEE Trans. Inform. Theory*, (Corresp.), vol. IT-19, pp. 114–115, Jan. 1973. See also this issue, pp. 271–274.
[353] T. E. Fortmann and B. D. O. Anderson, "On the approximation of optimal realizable linear filters using a Karhunen–Loeve expansion," *IEEE Trans. Inform. Theory*, (Corresp.), vol. IT-19, pp. 561–564, July 1973.
[353a] V. Belevitch, "On network analysis by polynomial matrices," in *Recent Developments in Network Theory*, S. R. Deards, Ed. Oxford: Pergamon, 1963, pp. 19–30.
[353b] ——, *Classical Network Theory.* San Francisco: Holden-Day, 1968.
[353c] C. Gueguen, A. Fossard, and M. Gauvrit, "Une représentation intermédiaire des systèmes multi-dimensionnels," in *Proc. 1st IFAC Symp. on Multivariable Control*, Dusseldorf, Germany, Oct. 1968.

2) *Linear System Structure*

[354] J. L. Massey, "Shift-register synthesis and BCH decoding," *IEEE Trans. Inform. Theory*, vol. IT-15, pp. 122–127, Jan. 1969.
[355] V. M. Popov, "Some properties of control systems with matrix transfer functions," in *Lecture Notes in Mathematics*, vol. 144. Berlin: Springer 1970, pp. 250–261.
[356] H. H. Rosenbrock, *State Space and Multivariable Theory.* New York: Wiley, 1970.
[357] L. Silverman, "Inversion of multivariable linear systems," *IEEE Trans. Automat. Contr.*, vol. AC-14, pp. 270–276, June 1969.
[357a] L. Silverman and H. J. Payne, "Input–output structure of linear systems," *SIAM J. Contr.*, vol. 9, pp. 199–233, May 1971.
[358] S. H. Wang, "Design of linear multivariable systems," Ph.D. dissertation, Univ. Calif., Berkeley, Dec. 1971; also Electron. Res. Lab. Memo. ERL-M309, Oct. 1971.
[358a] A. S. Morse and M. A. Wonham, "Status of noninteracting control," *IEEE Trans. Automat. Contr.*, vol. AC-16, pp. 568–581, Dec. 1971.
[359] V. M. Popov, "Invariant description of linear, time-invariant controllable system," *SIAM J. Contr.*, vol. 10, pp. 252–264, 1972.
[360] B. Dickinson, M. Morf, and T. Kailath, "A minimal realization algorithm for matrix sequences," *IEEE Trans. Automat. Contr.*, vol. AC-19, pp. 31–38, Feb. 1974.
[361] G. D. Forney, Jr., "Convolutional codes I: Algebraic structure," *IEEE Trans. Inform. Theory*, vol. IT-16, pp. 720–738, Nov. 1970.
[361a] ——, "Structural analysis of convolutional codes via dual codes," *IEEE Trans. Inform. Theory*, vol. IT-19, pp. 512–518, July 1973.
[361b] ——, "Minimal bases of rational vector spaces with applications to multivariable linear systems," *SIAM J. Contr.*, 1974.
[362] S. H. Wang and E. J. Davison, "A minimization algorithm for the design of linear multivariable systems," *IEEE Trans. Automat. Contr.*, vol. AC-18, pp. 220–223, June 1973.
[363] W. A. Wolovich, "The determination of state-space representations for linear multivariable systems," *Automatica*, vol. 9, pp. 97–106, 1973.

3) *Cepstral Analysis*

[364] B. Bogert, M. Healy, and J. Tukey, "The quefrency alanysis of time series for echoes," in *Proc. Symp. Time Series Analysis*, M. Rosenblatt, Ed. New York: Wiley, 1963, ch. 15, pp. 209–243.
[365] B. P. Bogert and J. F. Ossanna, "The heuristics of cepstrum analysis of a stationary complex echoed Gaussian signal in stationary Gaussian noise," *IEEE Trans. Inform. Theory*, vol. IT-2, pp. 373–380, July 1966.
[366] R. C. Kemerait and D. G. Childers, "Signal detection and extraction by cepstrum techniques," *IEEE Trans. Inform. Theory*, vol. IT-18, pp. 745–759, Nov. 1972.
[367] L. R. Rabiner and C. M. Rader, Eds., *Digital Signal Processing.* New York: IEEE Press, 1972.
[368] T. J. Cohen, "Source-depth determinations using spectral, pseudoautocorrelation and cepstral analysis," *Geophys. J. Roy. Astron. Soc.*, vol. 20, pp. 223–231, 1970.

4) *Historical Surveys*

[369] R. L. Plackett, "A historical note on the method of least squares," *Biometrika*, vol. 36, pp. 458–460, 1949.
[370] O. Neugebauer, *The Exact Sciences in Antiquity.* Princeton, N.J.: Princeton Univ. Press, 1952.
[371] N. Wiener, *I Am a Mathematician.* Cambridge, Mass.: M.I.T. Press, 1956.
[372] C. Eisenhart, "The meaning of 'least' in least squares," *J. Wash. Acad. Sci.*, vol. 54, pp. 24–33, 1964.
[373] H. L. Seal, "The historical development of the Gauss linear model," *Biometrika*, vol. 54, pp. 1–23, 1967.
[374] H. W. Sorenson, "Least-squares estimation: from Gauss to Kalman," *IEEE Spectrum*, vol. 7, pp. 63–68, July 1970.
[375] H. L. Harter, "The method of least squares and some alternatives," Aerospace Res. Lab., Air Force Systems Command, Wright-Patterson AFB, Ohio, Rep. ARL 72-0129, Sept. 1972.
[376] H. W. Sorenson, "Estimation theory: a historical perspective," in *Proc. SW IEEE Conf.*, 1972.

5) *Information-Theoretic Analyses*

[377] I. Vajda, "A contribution to the informational analysis of patterns," in *Methodologies of Pattern Recognition*, M. S. Watanabe, Ed. New York: Academic Press, 1969, pp. 509–519.
[378] S. Arimoto, "Information-theoretical considerations on estimation problems," *Inform. Contr.*, vol. 19, pp. 181–194, 1971.
[379] J. Ziv and M. Zakai, "Some lower bounds on signal parameter estimation," *IEEE Trans. Inform. Theory*, vol. IT-15, pp. 386–391, May 1969.
[380] L. P. Seidman, "Performance limitations and error calculations for parameter estimation," *Proc. IEEE*, vol. 58, pp. 644–652, May 1970.
[381] J. Seidler, "Bounds on the mean-square error and the quality of domain decisions based on mutual information," *IEEE Trans. Inform. Theory*, vol. IT-17, pp. 655–665, Nov. 1971.
[382] R. E. Blahut, "An hypothesis-testing approach to information theory," Ph.D. dissertation, Cornell Univ., Ithacar, N.Y., Aug. 1972; Abstract in *IEEE Trans. Inform. Theory* (Dissertation Abstr.), vol. IT-19, p. 253, Mar. 1973.
[383] M. Zakai and J. Ziv, "Lower and upper bounds on the optimal filtering of certain diffusion processes," *IEEE Trans. Inform. Theory*, vol. IT-18, pp. 325–331, May 1972.
[384] J. Ziv and M. Zakai, "On functionals satisfying a data-processing theorem," *IEEE Trans. Inform. Theory*, vol. IT-19, pp. 275–283, May 1973.

6) *Others*

[385] F. Riesz and B. Sz.-Nagy, *Functional Analysis*. New York: Ungar, 1955.
[386] F. Smithies, *Integral Equations*. New York: Cambridge Univ. Press, 1962.
[387] J. H. Wilkinson and C. Reinsch, *Linear Algebra, Handbook of Automatic Computation*, vol. 2. Berlin: Springer, 1971.
[388] J. R. Klauder and E. G. G. Sudarshan, *Quantum Optics*. New York: Benjamin, 1969.
[389] E. Hopf, "Statistical hydro-dynamics and functional calculus," *J. Ration. Mech. Anal.*, vol. 2, pp. 587–591, 1953.
[390] R. H. Kraichnan, "The closure problem of turbulence theory," in *Proc. Symp. Applied Mathematics*, vol. 13, 1962, pp. 199–225.

Part II

MATHEMATICAL FOUNDATIONS OF LEAST-SQUARES PREDICTION THEORY

Editor's Comments on Papers 2 Through 6

2 **KOLMOGOROV**
Interpolation and Extrapolation of Stationary Random Sequences

3 **KOLMOGOROV**
Stationary Sequences in Hilbert Space

4 **KREIN**
On a Problem of Extrapolation of A. N. Kolmogoroff

5 **KREIN**
On a Fundamental Approximation Problem in the Theory of Extrapolation and Filtration of Stationary Processes

6 **KARHUNEN**
On the Structure of Stationary Random Functions

Kolmogorov was the first to consider least-squares approximation problems for stochastic processes, though he was partly inspired by some special cases studied by H. Wold in 1938 [1]. Paper 2 is an elaboration of a very brief note in the *Comptes Rendus*, Paris, 1939 [2]. The main results are the formulation of the problems of least-squares extrapolation (or prediction) and interpolation of discrete-time processes, and the development of elegant expressions for the asymptotic error in the two cases. These results are simple applications of the studies in Paper 3 that present a general theory of second-order processes regarded as a collection of vectors in a Hilbert space. The geometric viewpoint is fully exploited, as are the powerful Hilbert-space techniques of M. Stone, especially the spectral decomposition theorem. The classic book by Doob describes some of these results [3, Chapters 9–12] but, in fact, not all of them have been fully exposed in the literature. This is perhaps even more true of the results of M. G. Krein, who, in a different context, found some results [4] closely related to Kolmogorov's. At the latter's suggestion, Krein pursued these problems further. Thus in Paper 4, Krein shows how to use a bilinear transformation to transfer Kolmogorov's analysis to continuous-time. In a related note [5], he showed the connection of these problems to the early work (ca. 1920) of Szegö and Smirnov on the theory of orthogonal polynomials on a circle (see Section VII of the survey Paper 1). More than two decades later, these ideas and techniques (which are also partly exposed in the book by Akhiezer [6, App. B]) began to reappear in digital filtering [7], geophysical data processing [8, 9], and speech analysis [10, 11] (see also the discussion of the work of Krein and Levinson in Paper 1, Section VII). Moreover, these ideas seem to have even greater potential and are rewarding objects for further study even in the context of nonstationary processes (see Dickinson et al. [12], Kailath et al. [13], and Morf et al. [14]). Another important paper is a short note of 1954, in which Krein relates his extensive work on inverse Sturm-Liouville problems to finite-interval prediction. Much remains to be done to fully understand and exploit the ideas in this work, reprinted here as Paper 5. We might remark that a recent 352 page book by Dym and McKean [15] is devoted to some mathematical investigations essentially growing out of this four page Doklady note of Krein's.

Returning to the work of Kolmogorov, we include as the final paper in this section, a work by Karhunen (Paper 6) exploring further the structure of stationary processes and making heavy use of the results of Paley and Wiener and others on Hardy functions, bilateral Laplace transforms, and so on. A modern exposition of these topics and some of Karhunen's results appears in the book by Hoffman [16, Chapter 4].

The basic ideas of Wold and Kolmogorov have also had many purely mathematical ramifications, especially in operator theory—see, for example, the books of Helson [17], Nagy and Foias [18], and the paper of Halmos [19], and Masani [20, 21]. Moreover, to reemphasize the striking fact that it is hard to clearly demarcate the realm of pure mathematics, we should note that these ideas have been relevant in system-theoretical analyses of (infinite-dimensional) distributed-parameter systems (see Baras and Dewilde [22] and Helton [23]).

REFERENCES

1. H. Wold, *A Study in the Analysis of Stationary Time Series*, Uppsala, Sweden: Almqvist and Wiksell, 1938; 2nd ed., 1954.
2. A. N. Kolmogorov, "Sur l'Interpolation et Extrapolation des Suites Stationnaires," *C. R. Acad. Sci.* **208:**2043–2045, 1939.
3. J. L. Doob, *Stochastic Processes*, New York: J. Wiley, 1953.

4. M. G. Krein, "On the Problem of Continuation of Helical Arcs in Hilbert Space," *Dokl. Akad. Nauk SSSR* **45:**139–142, 1944.
5. M. G. Krein, "On a Generalization of Some Investigations of G. Szegö, V. I. Smirnov, and A. N. Kolmogorov," *Dokl. Akad. Nauk SSSR* **46:**306–309, 1945.
6. N. I. Akhiezer, *Theory of Approximation*, Trans. C. J. Hyman, New York: F. Ungar, 1956.
7. K. Steiglitz, "The Equivalence of Digital and Analog Signal Processing," *Inf. Control,* **8:** 455–467, 1965; also reprinted in L. R. Rabiner and C. M. Rader, eds., *Digital Signal Processing*, New York: IEEE Press, 1972, pp. 7–19.
8. E. A. Robinsin, "Mathematical Development of Discrete Filters for Detection of Nuclear Explosions," *J. Geophys. Res.* **68:**5559–5567, 1963.
9. L. C. Wood and S. Treitel, "Seismic Signal Processing," *Proc. IEEE* **63** (4):649–661, Apr. 1975.
10. B. S. Atal and S. L. Hanauer, "Speech Analysis and Synthesis by Linear Prediction of the Speech Wave," *J. Acoust. Soc. Amer.* **50:**637–655, 1971; also reprinted in J. L. Flanagan and L. R. Rabiner, eds., *Spe ech Synthesis,* Benchmark Papers in Acoustics, Stroudsburg, Pa.: Dowden, Hutchinson & Ross, 1973, pp. 270–288.
11. F. Itakura and S. Saito, "Digital Filtering Techniques for Speech Analysis and Synthesis," *Proc. 7th International Congress on Acoustics*, Budapest, Hungary, Paper 25Cl, pp. 261–264, 1971.
12. B. W. Dickinson, T. Kailath, and M. Morf, "Canonical Matrix Fraction and State-Space Descriptions for Deterministic and Stochastic Linear Systems," *IEEE Trans. Automat. Control, Special Issue on System Identification and Time-Series Analysis,* **AC–19**(6):656–667, Dec. 1974.
13. T. Kailath, A. Vieira, and M. Morf, "Inverses of Toeplitz Operators, Innovations, and Orthogonal Polynomials," *SIAM Review,* forthcoming.
14. M. Morf, B. Dickinson, and T. Kailath, "General Speech Models and Linear Estimation Theory," in D. R. Reddy, ed., *Speech Processing*, New York: Academic Press, 1975, pp. 157–182.
15. H. Dym and H. McKean, *Gaussian Processes*, New York: Academic Press, 1975.
16. K. Hoffman, *Banach Spaces of Analytic Functions*, Englewood Cliffs, N. J.: Prentice-Hall, 1962.
17. H. Helson, *Lectures on Invariant Subspaces*, New York: Academic Press, 1964.
18. B. Sz. Nagy and C. Foias, *Harmonic Analysis of Operators on Hilbert Space*, New York: Academic Press, 1970.
19. P. R. Halmos, "Shifts on Hilbert Spaces," *Crelle's Journal* **208:** 102–112, 1961.
20. P. Masani, "Shift-Invariant Spaces and Prediction Theory," *Acta Math.,* **107:**275–290, 1962.
21. P. Masani and J. B. Robertson, "The Time-Domain Analysis of Continuous-Parameter, Weakly Stationary Stochastic Processes," *Pac. J. Math.* **12:**1361–1378, 1962.

22. J. S. Baras and P. Dewilde, "Invariant-Subspace Methods in Linear Multivariable-Distributed Systems and Lumped-Distributed Network Synthesis," *Proc. IEEE* **64**(1):160–178, Jan. 1976.
23. J. W. Helton, "Systems with Infinite-Dimensional State Space: The Hilbert Space Approach," *Proc. IEEE* **64**(1):145–160, Jan. 1976.

2

Reprinted from RM-3090-PR, W. Doyle and I. Selin, trans., RAND Corporation, 1962, pp. 1–14

INTERPOLATION AND EXTRAPOLATION OF STATIONARY RANDOM SEQUENCES

A. N. Kolmogorov

Spectral conditions are established for the possibility of extrapolating and interpolating stationary random sequences by a sufficiently large number of terms with arbitrary given accuracy.

INTRODUCTION

Let there correspond to each integral $t(-\infty < t < +\infty)$ the real random variable $x(t)$ with finite mean square. We shall call the sequence $\{x(t)\}$ of random variables $x(t)$ stationary if the mathematical expectations*

$$m = Mx(t)$$

and

$$B(k) = M[(x(t + k) - m)\ (x(t) - m)]$$

do not depend on t. With no restriction on generality we may set

$$m = Mx(t) = 0. \tag{1}$$

Then

$$B(k) = M\,[x(t + k)x(t)]. \tag{2}$$

Since

$$B(-k) = B(k), \tag{3}$$

it suffices to consider the second moments $B(k)$ only for $k \geq 0$.

The problem of linear extrapolation of a stationary sequence satisfying condition (1) includes, in particular, the determination for given $n > 0$

---

*The mathematical expectation of the random variable y is denoted henceforth by M_y.

and $m \geq 0$ of real coefficients a_s, such that the linear combination

$$L = a_1 x(t-1) + a_2 x(t-2) + \dots + a_n x(t-n)$$

of random variables

$$x(t-1),\ x(t-2),\ \dots,\ x(t-n)$$

achieves the most accurate possible approximation to the random variables $x(t+m)$. As a measure of the accuracy of such an approximation, it is natural to take the mathematical expectation

$$\sigma^2 = M(x(t+m) - L)^2 =$$

$$= B(0) - 2 \sum_{s=1}^{n} B(m+s)a_s + \sum_{p=1}^{n} \sum_{q=1}^{n} B(p-q)a_p a_q .$$

If the second moments $B(k)$ are known, then it is easy to solve the problem of finding values of the coefficients a_s for which σ^2 achieves a minimum value. This minimum value σ^2 we shall denote by $\sigma_E^2(n,m)$.

Obviously, when n increases the magnitude of $\sigma_E^2(n,m)$ cannot increase. Hence, the limit

$$\lim_{n\to\infty} \sigma_E^2\,(n,m) = \sigma_E^2(m) \tag{4}$$

exists. The determination of this limit is the first problem solved in the present work.

As far as the problems of interpolation are concerned, we shall consider only the case of estimation of $x(t)$ from the quantities

$$x(t+1),\ x(t+2),\ \dots,\ x(t+n),$$
$$x(t-1),\ x(t-2),\ \dots,\ x(t-n).$$

For this case, we shall denote by $\sigma_I^2(n)$ the minimum value of the mathematical expectation

$$\sigma^2 = M(x(t) - Q)^2,$$

where Q is the linear form

$$Q = a_1 x\,(t+1) + a_2 x\,(t+2) + \ldots + a_n x\,(t+n) +$$

$$+ a_{-1}x(t-1) + a_{-2}x(t-2) + \ldots + a_{-n}x(t-n)$$

with constant real coefficients a_s.

For an increase of n the magnitude $\sigma_I^2(n)$ does not increase. Hence, the limit

$$\lim_{n \to \infty} \sigma_I^2(n) = \sigma_I^2 \tag{5}$$

exists.

Our second problem is the determination of σ_I^2. The solution put forth below of both the above formulated problems was reported without proof in my note.[(1)*] It is based on concepts from the spectral theory of stationary random processes.

The spectral theory of stationary random processes was developed by A. Ya. Khinchin for the case of continuously varying time parameter, t.[(2)]

---

*In Eq. (1) of this note, a misprint occurred. The correct form of Eq. (1) is this:

$$\lim_{n \to \infty} \sigma_n^2(I) = \sigma^2(I) = \frac{\pi}{\displaystyle\int_0^{\pi} \frac{d\lambda}{s(\lambda)}}$$

For a discrete, stationary sequence, the case of interest to us here, a complete exposition of the theory is given in the book by H. Wold.[(3)] Of basic significance here is the following theorem.*

THEOREM 1. For an arbitrary stationary sequence $\{x(t)\}$, the second moments can be represented in the form

$$B(k) = \frac{1}{\pi}\int_0^{\pi} \cos k\lambda \, dW(\lambda), \tag{6}$$

where $W(\lambda)$ is a non-decreasing real function determined by the formula

$$W(\lambda) = B(0)\lambda + 2\sum_{k=1}^{\infty} \frac{B(k)}{k} \sin k\lambda. \tag{7}$$

The derivative

$$w(\lambda) = \frac{dW(\lambda)}{d\lambda}$$

of the non-decreasing function $W(\lambda)$ exists almost everywhere and is non-negative and summable. Since

$$\log w(\lambda) \leq w(\lambda),$$

it follows from the summability of $w(\lambda)$ that the integral

$$P = \frac{1}{\pi}\int_0^{\pi} \log w(\lambda) d\lambda \tag{8}$$

is either finite or equal to $-\infty$.** Later, we prove the following theorem:

THEOREM 2. If $P = -\infty$, then $\sigma_E^2(m) = 0$ for all $m \geq 0$. But if the

---

*Cf. Ref. 3, §17.

**If $w(\lambda) = 0$ on a set of positive measure, we put $P = -\infty$.

integral P is finite, then

$$\sigma_E^2(m) = e^P\,(1+r_1^2+r_2^2+\dots+r_m^2), \tag{9}$$

where the r_s are determined from the relations

$$e^{a_1\zeta + a_2\zeta^2+\dots} = 1 + r_1\zeta + r_2\zeta^2 + \dots, \tag{10}$$

$$a_k = \frac{1}{\pi}\int_0^\pi \cos k\lambda \log w(\lambda)\, d\lambda. \tag{11}$$

Since $w(\lambda) \geq 0$, the integral

$$R = \frac{1}{\pi}\int_0^\pi \frac{d\lambda}{w(\lambda)} \tag{12}$$

is either finite or equals $+\infty$.* We shall later prove the following theorem:

<u>THEOREM 3.</u> If $R = +\infty$, then $\sigma_I^2 = 0$. But if the integral R is finite, then

$$\sigma_I^2 = \frac{1}{R}. \tag{13}$$

In my work[4] the theory of stationary sequences of elements of a complex Hilbert space is developed. In §1 of the present paper I show that stationary random sequences, in the sense defined above, may be considered a special case of the stationary sequences considered in Ref. 4. This permits the above formulated theorems 1, 2, and 3 to be obtained as simple consequences of the results of the work in Ref. 4.

In what follows, equations with numbers divided into two parts, separated by a point [for example (8.44)], refer to formulas from Ref. 4.

---

*If $w(\lambda) = 0$ on a set of positive measure, we put $R = +\infty$.

1. STATIONARY RANDOM SEQUENCES AND THE GEOMETRY OF HILBERT SPACE

We shall start from axiomatics and the constructions of the basic concepts of probability theory, presented in my book,[(5)] with the variation that the random quantities considered below may take on not only real but also complex values.*

Let us consider the set H of all random variables x, of an arbitrary Borel probability field (F,P), with finite mean square magnitude. We take equivalent random variables (i.e., random variables differing from one another only with probability zero) to be identical.

We introduce into H the scalar product

$$(x,y) = M(x\bar{y}). \tag{14}$$

The norm in H is determined by the formula

$$\|x\|^2 = (x,x) = M|x|^2. \tag{15}$$

We shall understand the addition of elements of H and their multiplication by complex numbers in the usual sense.

It is easy to verify that the set H with the above-defined operations satisfies postulates A, B, and E of Stone's book,[(6)] i.e., all the postulates of an abstract unitary space.

Now let $\{x(t)\}$ be a stationary sequence of real random variables x(t) in the probability field (F,P) in the sense given in the introduction, satisfying the additional condition (1). Then, by virtue of (2) and the realness

---

*A complex function $x(\zeta)$, determined on a set E of elementary events ζ, is called a random variable if for any choice of real numbers a and b the set of all ζ such that the real and imaginary parts of x(t) satisfy the respective inequalities $Rx(\zeta) < a$, $Ix(\zeta) < b$, belongs to the system F.

of x(t)

$$B(k) = M\,[x(t+k)x(t)] = (x(t+k),x(t)).$$

Since, by definition, B(k) is independent of t, $\{x(t)\}$ is a stationary sequence of elements of the space H in the sense of Ref. 4.

In Ref. 4, I consider stationary sequences lying in a Hilbert space, i.e., in a space satisfying, besides postulates A, B, and E, also postulates C and D of Stone's book.[6] This restriction, however, is unnecessary. In fact, let us denote by H_x the smallest closed linear subspace of the space H containing all elements of the sequence $\{x(t)\}$. It is easy to show that H_x is separable, i.e., satisfies postulate D. A separable unitary space is either itself a Hilbert space (i.e., satisfies besides A, B, D, and E, also postulate C) or is finite dimensional and, in the latter case, may be extended to some Hilbert space H.

Thus, to the sequence $\{x(t)\}$, it is possible to apply, letting

$$B(k) = B_{xx}(k) = (x(t+k),x(t)), \tag{16}$$

all the results obtained in Ref. 4.

2. PROOF OF THEOREM 1

By virtue of Eqs. (3) and (16), in the case of real random variables x(t)

$$B_{xx}(-k) = B_{xx}(k). \tag{17}$$

Therefore, from Eq. (3.10) we get

$$W_{xx}(\lambda) = B_{xx}(0)\lambda - \sum_{k\neq 0} \frac{B_{xx}(k)}{ik} e^{-ik\lambda} =$$

$$= B_{xx}(0)\lambda + 2\sum_{k=1}^{\infty} \frac{B_{xx}(k)}{k} \sin k\lambda. \tag{18}$$

From (18) it follows that

$$W_{xx}(-\lambda) = - W_{xx}(\lambda). \tag{19}$$

Finally, from (3.1), (3.9), and (19) we get

$$B(k) = B_{xx}(k) = \int_{-\pi}^{+\pi} e^{ik\lambda} dF_{xx}(\lambda) =$$

$$= \frac{1}{2\pi}\int_{-\pi}^{+\pi} e^{ik\lambda} dW_{xx}(\lambda) = \frac{1}{\pi}\int_{0}^{\pi} \cos k\lambda \, dW_{xx}(\lambda). \tag{20}$$

From Eqs. (18) and (3.9) and theorem 2 of Ref. 4, it is clear that $W_{xx}(\lambda)$ is a real, non-decreasing function. Together with Eqs. (18) and (20) this shows that the function

$$W(\lambda) = W_{xx}(\lambda)$$

satisfies the conditions of theorem 1.

3. $\sigma_E^2(m)$ IN THE GENERAL CASE

We shall denote, following Ref. 4, by $H_x(t-1)$ the smallest closed linear subspace of the space H_x containing the elements

$$x(t-1),\ x(t-2),\ \ldots,\ x(t-n),\ \ldots.$$

The element $x(t+m)$ for arbitrary $m \geq 0$ is uniquely representable in the form

$$x(t+m) = \xi(t-1,m) + \Delta(t-1,m), \tag{21}$$

where $\xi(t-1,m)$ lies in $H_x(t-1)$, while $\Delta(t-1,m)$ is orthogonal to $H_x(t-1)$.

It is easy to show that in the case of a stationary sequence of real random variables*

$$\sigma_E^2(m) = \|\Delta(t-1,m)\|^2. \tag{22}$$

In the general case of stationary sequences in the sense of Ref. 4, we shall consider (22) as the definition of the quantity $\sigma_E^2(m)$.

If the sequence $\{x(t)\}$ is singular, then $H_x(t-1) = H_x$, and consequently,

$$\sigma_E^2(m) = 0. \tag{23}$$

If the sequence $\{x(t)\}$ is not singular, then by formula (7.8)

$$x(t+m) = s_x(t+m) + \sum_{n=0}^{\infty} c_n^{(x)} u_x(t+m-n). \tag{24}$$

---

*As is known, $\|\Delta(t-1,m\|$ is the "distance" of the point $x(t+m)$ from the space $H_x(t-1)$, i.e., the lower bound of the distances $\|x(t+m)-y\|$ for all y in $H_x(t-1)$. Since elements of the form $L = a_1x(t-1) + a_2x(t-2) + \ldots + a_nx(t-n)$ are everywhere dense in $H_x(t-1)$, $\|\Delta(t-1,m\|$ also equals the lower bound of the distances

$$\|x(t+m) - L\| = \sqrt{M|x(t+m) - L|^2}. \tag{*}$$

If all $x(s)$ are real random variables, then the lower bound of expression (*) is not changed by restricting to real coefficients a_k, in which case it obviously coincides with $\sigma_E(m)$.

Since $s_x(t+m)$ and $u_x(t+m-n)$ for $n > m$ lie in $H_x(t-1)$, while the $u_x(t+m-n)$ for $n \leq m$ are orthogonal to $H_x(t-1)$, comparison of (21) with (24) gives

$$\Delta(t-1,m) = c_0^{(x)}u_x(t+m) + c_1^{(x)}u_x(t+m-1) + \dots + c_m^{(x)}u_x(t). \qquad (25)$$

Since the elements $u_x(t+i)$ are pairwise orthogonal and normed, it follows from (25) that

$$\sigma_E^2(m) = \|\Delta(t-1,m)\|^2 = (c_0^{(x)})^2 + (c_1^{(x)})^2 + \dots + (c_m^{(x)}). \qquad (26)$$

4. $\sigma_E^2(m)$ IN THE REAL CASE

For stationary sequences of real random variables it is easy to deduce theorem 2 from formula (26). The present section is devoted to this derivation.

By virtue of formula (3.9) we have

$$f_{xx}(\lambda) = \frac{dF_{xx}(\lambda)}{d\lambda} = \frac{1}{2\pi}\frac{dW_{xx}(\lambda)}{d\lambda} = \frac{1}{2\pi} w(\lambda). \qquad (27)$$

From (19) it follows that

$$w(-\lambda) = w(\lambda). \qquad (28)$$

Using (27) and (28) we get

$$\int_{-\pi}^{+\pi} \log f_{xx}(\lambda)d\lambda = 2\int_0^{\pi} \log w(\lambda)d\lambda - 2\pi \log 2\pi. \qquad (29)$$

Formula (29) together with theorem 23 of Ref. 4 shows that the equality

$$P = \frac{1}{\pi}\int_0^{\pi} \log w(\lambda)d\lambda = -\infty$$

is necessary and sufficient for the singularity of the sequence $\{x(t)\}$. We have already seen in §3 that in this and only this case

$$\sigma_E^2(m) = 0. \qquad (30)$$

If the sequence $\{x(t)\}$ is not singular then from (8.44) and (29) follows

$$(c_0^{(x)})^2 = 2\pi \exp\left(\frac{1}{2\pi}\int_{-\pi}^{+\pi} \log f_{xx}(\lambda)d\lambda\right) =$$

$$= \exp\left(\frac{1}{\pi}\int_0^{\pi} \log w\lambda \, d\lambda\right) = e^P. \qquad (31)$$

Under this same assumption, that the sequence $\{x(t)\}$ is not singular,* from (8.31), (27), and (28) we conclude that for $k \geq 0$

$$\left.\begin{aligned} a_k^{(x)} &= \frac{1}{2\pi}\int_{-\pi}^{+\pi} \cos k\lambda \log f_{xx}(\lambda)d\lambda = \frac{1}{\pi}\int_0^{\pi} \cos k\lambda \log w(\lambda)d\lambda, \\ b_k^{(x)} &= \frac{1}{2\pi}\int_{-\pi}^{+\pi} \sin k\lambda \log f_{xx}(\lambda)d\lambda = \frac{1}{2\pi}\int_{-\pi}^{+\pi} \sin k\lambda \log w(\lambda)d\lambda = 0. \end{aligned}\right\} \qquad (32)$$

*As demonstrated in theorem 23 of Ref. 4, formulas (8.16), (8.29), (8.30), and (8.31) are applicable to an arbitrary non-singular sequence.

From (8.16), (8.29), (8.30), and (32) it follows that

$$\sum_{n=0}^{\infty} c_n^{(x)} \zeta^n = \Gamma_x(\zeta) = \Gamma_x(0) \frac{\Gamma_x(\zeta)}{\Gamma_x(0)} = \Gamma_x(0) e^{Q_x(\zeta)-Q_x(0)} =$$

$$= c_0^{(x)} \exp\left(\sum_{k=1}^{\infty} a_k^{(x)} \zeta^k\right). \tag{33}$$

Putting

$$\exp\left(\sum_{k=1}^{\infty} a_k^{(x)} \zeta^k\right) = 1 + r_1\zeta + r_2\zeta^2 + \dots, \tag{34}$$

we have from a comparison of (33) with (34)

$$\frac{c_n^{(x)}}{c_0^{(x)}} = r_n. \tag{35}$$

From (35) and (26) follows

$$\sigma_E^2 (m) = (c_0^{(x)})^2 (1+r_1^2+r_2^2+ \dots + r_m^2). \tag{36}$$

Equations (30), (31), and (36) completely prove theorem 2.

5. DETERMINATION OF σ_I^2

Let us denote, following Ref. 4, by $\hat{H}_x(t)$ the smallest linear subspace of the space H_x containing the elements

$$x(t+1),\ x(t+2),\ \dots,\ x(t+n),\ \dots,$$

$$x(t-1),\ x(t-2),\ \dots,\ x(t-n),\ \dots.$$

The element x(t) is uniquely representable in the form (10.3):

$$x(t) = \upsilon(t) + \delta(t),$$

where $\upsilon(t)$ lies in $\hat{H}_x(t)$, while $\delta(t)$ is orthogonal to $\hat{H}_x(t)$. It is easy to show that for the case of a stationary sequence of real random variables

$$\sigma_I^2 = \|\delta(t)\|^2 = d_x^2. \tag{37}$$

By theorem 24 of Ref. 4,

$$d_x^2 = \frac{(2\pi)^2}{\int_{-\pi}^{+\pi} \frac{d\lambda}{f_{xx}(\lambda)}}. \tag{38}$$

Moreover, if the integral in the denominator of the right-hand side is infinite,

$$d_x^2 = 0.$$

From (37), (38), (27), and (28) we conclude that

$$\sigma_x^2 = d_x^2 = \frac{\pi}{\int_0^{\pi} \frac{d\lambda}{w(\lambda)}} = \frac{1}{R}$$

which immediately proves theorem 3.

REFERENCES

1. Kolmogorov, A. N., "Sur l'Interpolation et Extrapolation des Suites Stationnaires," Comptes Rendus de l'Acad. Sci., Paris, 208 (1939), pp. 2043-2045.

2. Khinchin, A., "Korrelationstheorie der Stationären Stochastischen Prozesse," Math. Annalen, 109 (1934), pp. 604-615.

3. Wold, H., A Study in the Analysis of Stationary Time Series, Uppsala, 1938.

4. Kolmogorov, A. N., "Stationary Sequences in a Hilbert Space," (Russian), Bulletin M.G.U., 2 (1941), No. 6.

5. Kolmogorov, A. N., Foundations of the Theory of Probability, English Translation, Chelsea, New York, 1950.

6. Stone, M., "Linear Transformations in Hilbert Space," American Mathem. Soc. Colloquium Publications, 15 (1932).

3

Reprinted from CN/74/2, J. F. Barrett, trans., Department of Engineering, Cambridge University, 1974, pp. 1–24

STATIONARY SEQUENCES IN HILBERT SPACE

A.N. Kolmogorov

Introduction

DEFINITION 1 The sequence $\{x(t)\}$ of elements of a complex Hilbert space* H, where t runs over all whole values from $-\infty$ to $+\infty$, is called stationary if the scalar product

$$B_{xx}(k) = (x(t+k), x(t)) \tag{0.1}$$

does not depend on t.

DEFINITION 2 Two stationary sequences $\{x(t)\}$ and $\{y(t)\}$ are called stationarily related if the scalar product

$$B_{yx}(k) = (y(t+k), x(t)) \tag{0.2}$$

does not depend on t.

From the definition of the quantities $B_{xy}(k)$ and $B_{xx}(k)$ it immediately follows that

$$B_{xy}(k) = \bar{B}_{yx}(-k) \tag{0.3}$$

$$B_{xx}(k) = \bar{B}_{xx}(-k) \tag{0.4}$$

Such types of stationary and stationarily related sequences have a great significance in the theory of probability and mathematical statistics. In terms of the theory of probability they are described in detail in the book of H. Wold and in further articles of H. Cramér.

We set out the fundamental results of H. Wold and H. Cramér in terms of the geometry of Hilbert space in †3 and in †7.

All new problems which are investigated and solved in the present work, originally sprung out of the theory of probability and mathematical statistics. The application of the results obtained by us to the problem of extrapolation and interpolation of stationary sequences of quantities is set out in my article(7).

The following theorem shows the very simple connexion which exists between stationary and stationarily connected sequences and unitary operators.

THEOREM 1 Suppose that the sequences

$$\{x_1(t)\}, \{x_2(t)\}, \ldots, \{x_n(t)\}$$

are stationary and stationarily connected and let $H_{x_1x_2\ldots x_n}$ be the minimal linear closed subspace of the space H containing all elements of this sequence.

Then the equation

$$Ux_\mu(t) = x_\mu(t+1), \quad \mu=1,2,\ldots,n \qquad -\infty < t < \infty$$

uniquely defines a unitary operator

$$U = U_{x_1x_2\ldots x_n}$$

transforming the space $H_{x_1x_2\ldots x_n}$ on to itself.

This theorem leads to the deduction of a series of basic properties of stationary and stationarily connected sequences as immediate consequences of known results of the spectral theory of unitary operators. This is the character of all theorems of †3-6.

The more original and the essentially new results presented by us are in ††8-10. ††1-2 have an auxiliary character.

Theorem 1 immediately follows from definitions 1 and 2 and lemma 1 proved in †1.

†1 Two Lemmas

Suppose there is defined on a sub-

* For the definition of Hilbert space see the book of M. Stone (3).

space M of a space H an operator

$$Tx = y$$

with values in H. The operator T is called isometric if for arbitrary x and y from M

$$(x,y) = (Tx,Ty) \tag{1.1}$$

Let H_M be the minimal closed linear subspace of the space H containing M. Then the following lemma holds

LEMMA 1 The operator T may be extended with preservation of isometry on the whole space H_M. Such an extension is unique.

For the proof of the lemma we select from M an everywhere dense sequence of elements

$$z_1, z_2, \ldots, z_n, \ldots \tag{1.2}$$

From this sequence we eliminate all elements z_n, which are linearly dependent on previous z_k. There is obtained a sequence

$$z_{n_1}, z_{n_2}, \ldots, z_{n_k}, \ldots \tag{1.3}$$

for which the set R of all elements of the form

$$z = a_1 z_{n_1} + a_2 z_{n_2} + \ldots + a_p z_{n_p} \tag{1.4}$$

contains all elements of the sequence (1.2) and is everywhere dense on H_M. Consequently for the elements of the set R put in the form (1.4), in a unique way, (because of the independence of the elements of the sequence (1.3)), the formula

$$T^*z = a_1 T z_{n_1} + a_2 T z_{n_2} + \ldots + a_p T z_{n_p} \tag{1.5}$$

defines in a single valued way some operator T* on R. In this, for arbitrary z' and z" from R

$$(z', z'') = (T^*z', T^*z'')$$

i.e. the operator T is isometric on R. By continuity it may be extended with preservation of isometry on the whole space H_M.

As all elements of the sequence (1.2) belong to R, it follows from formula (1.5) that for all elements (1.2)

$$T^*z_n = Tz_n$$

by continuity this equality is kept also on the whole set M.

The existence of the required extension has been proved.

An arbitrary other isometric extension of the operator T in view of the formula (1.5) agrees with T on the set R. By continuity such agreement is conserved also on the whole of H_M. Consequently, the isometric extension T on the whole space H_M is unique. The lemma is proved.

LEMMA 2 In order that in the space H there should exist a sequence of elements

$$u_1, u_2, \ldots, u_n, \ldots$$

satisfying the conditions

$$(u_m, u_n) = c_{mn};\ m,n = 1,2,\ldots \tag{1.6}$$

it is necessary and sufficient that for arbitrary $k, m_1, m_2, \ldots, m_k$ the matrix $\|c_{m_i m_j}\|$ should be Hermitian nonnegative, i.e. that

$$S = \sum_{i,j=1}^{k} c_{m_i m_j} \xi_i \bar{\xi}_j \geq 0 \text{ [1]} \tag{1.7}$$

for arbitrary complex $\xi_1, \xi_2, \ldots, \xi_k$.

In order to prove the necessity of condition (1.7), we consider the elements of the space H

$$u = \xi_1 u_{m_1} + \xi_2 u_{m_2} + \ldots + \xi_k u_{m_k}$$

It is easy to calculate that

$$\|u\|^2 = (u,u) = S \tag{1.8}$$

Since always $\|u\|^2 \geq 0$ it follows that (1.8) implies (1.7). Let us go to the proof of the sufficiency of condition (1.7). We remark in connexion with these conditions that for each n,

$$S_n = \sum_{i,j=1}^{n} c_{ij} \xi_i \bar{\xi}_j \geq 0$$

We introduce the transformation

$$\eta_i = \sum_{j=1}^{n} a_{ij}^{(n)} \xi_j$$

which puts the form S_n as

1) Here as in all that follows, a ≥ 0 means that a is real and nonnegative.

$$S_n = \sum_{i=1}^{m} |\eta_i|^2$$

We select in H an orthogonal normed system of vectors

$$z_1^{(n)}, z_2^{(n)}, \ldots, z_m^{(n)}$$

and we put

$$u_i^{(n)} = \sum_{j=1}^{m} a_{ji}^{(n)} z_j \qquad i = 1,2,\ldots,n$$

It is easy to calculate that then

$$(u_i^{(n)}, u_j^{(n)}) = c_{ij}$$

Carrying out this construction for every positive whole n, we construct a sequence

$$u_1, u_2, \ldots, u_n, \ldots$$

by the following inductive process:

1) we put
$$u_1 = u_1^{(1)}$$

2) assuming that $u_1, u_2, \ldots u_N$ are already found and satisfy condition (1.6) for all $m,n \leq N$ we put

$$T_N u_n^{(N+1)} = u_n \qquad n = 1,2,\ldots,N$$

On the set $M_N = \{u_1^{(N+1)}, \ldots, u_N^{(N+1)}\}$, the operator T_N is isometric. According to lemma 1 it is possible to extend with retention of isometry on the space H_M. Since the spaces H_{M_N} and $T_N H_{M_N}$ are finite dimensional, the operator T_N may be extended with retention of isometry on the whole space H (see for example theorem 2.49 of the book of M. Stone[3]). Choosing such an extension, we put

$$u_{N+1} = T_N u_{N+1}^{(N+1)}$$

It is clear that the elements $u_1, u_2 \ldots, u_{N+1}$ satisfy condition (1.6) for all $m,n \leq N + 1$.

2. Operational Calculations in the Space H_x

In connexion with theorem 1 we denote by H_x the minimal closed linear subspace of the space H containing all elements of the stationary sequence $\{x(t)\}$ and we consider the unitary operator U_x defined by the equality

$$U_x x(t) = x(t+1) \tag{2.1}$$

Let

$$U_x = \int_{-\pi}^{\pi} e^{i\lambda} dE_x(\lambda) \tag{2.2}$$

be the spectral representation[1)] of the operator U_x. We put for $-\pi \leq \lambda \leq \pi$

$$F_{xx}(\lambda) = |E_x(\lambda)x(0)|^2 \tag{2.3}$$

Following the exposition of M. Stone (3) (†1, chapter 6), we denote by L_x^2 the class of all complex functions measurable with respect to F_{xx} defined on the segment $-\pi \leq \lambda \leq \pi$ for which the integral

$$\int_{-\pi}^{\pi} |\varphi(\lambda)|^2 dF_{xx}(\lambda) \tag{2.4}$$

is finite. Two functions φ_1 and φ_2 will be considered equivalent if

$$\varphi_1(\lambda) = \varphi_2(\lambda)$$

almost everywhere with respect to F_{xx}.

To every function φ of class L_x^2 corresponds the operator

$$T(\varphi) = \int_{-\pi}^{\pi} \varphi(\lambda) dE_x(\lambda) \tag{2.5}$$

the region of definition $\partial(\varphi)$ which consists of all elements z of the space H_x, for which the integral

$$\int_{-\pi}^{\pi} |\varphi(\lambda)|^2 d|E(\lambda)z|^2 \tag{2.6}$$

is finite. From the finiteness of the integral (2.4) follows that x(0) belongs to

---

1) Cf. in connexion with the spectral representation of linear operators the book of M. Stone (3). We remark here that everything that follows demands only the resolution of the identity $E(\lambda)$ corresponding to a unitary operator for which $E(-\pi) = 0$ and $E(+\pi) = J$. To apply to it all the theorems of M. Stone, it is demanded further, which is formulated in this book for the more general case of the resolution of unity, the corresponding spectral representation of the self-adjoint operators in the form

$$A = \int_{-\infty}^{\infty} \lambda dE(\lambda)$$

$\partial(\varphi)$ for an arbitrary function φ from the class L_x^2. Because to every function φ of the class L_x^2 corresponds a definite element,

$$z\varphi = T(\varphi)x(0) = \int_{-\pi}^{\pi} \varphi(\lambda)dE_{xx}(\lambda)x(0) \qquad (2.7)$$

of the space H_x.

The introduction of the notation allows the formulation of the following two lemmas which are almost immediate deductions of theorem 6.1 and 6.2 of M. Stone (3).

LEMMA 3 The transformation

$$\varphi \to T(\varphi)$$

transforms the class L_x^2 in a one-one way on the class $\mathfrak{T}_x$ of all operators T, representable in the form (2.5) where φ belongs to L_x^2.

In this

$$a\varphi \to aT(\varphi) \qquad (3a)$$

$$\varphi_1 + \varphi_2 \to T(\varphi_1) + T(\varphi_2) \qquad (3b)$$

if $\varphi_1\varphi_2$ belongs to L_x^2 then

$$\varphi_1\varphi_2 \to T(\varphi_1)T(\varphi_2) = T(\varphi_2)T(\varphi_1) \qquad (3c)$$

$$e^{it\lambda} \to U_x^t \qquad (3d)$$

if $\chi_\lambda(\mu)=1$ for $\mu\le\lambda$ and $\chi_\lambda(\mu)=0$ for $\mu>\lambda$ then

$$\chi_\lambda \to E_x(\lambda) \qquad (3e)$$

LEMMA 4 The transformation

$$\varphi \to z\varphi$$

transforms the class L_x^2 in a one-one way on the whole space H_x.

In this

$$a\varphi \to az\varphi \qquad (4a)$$

$$\varphi_1 + \varphi_2 \to z\varphi_1 + z\varphi_2 \qquad (4b)$$

if $\varphi_1\varphi_2$ belongs to L_x^2 then

$$\varphi_1\varphi_2 \to T(\varphi_1)z\varphi_2 + T(\varphi_2)z\varphi_1 \qquad (4c)$$

$$e^{it\lambda}\varphi \to U_x^t z\varphi \qquad (4d)$$

$$\chi_\lambda(\varphi) \to E_x(\lambda)z\varphi \qquad (4e)$$

$$(z\varphi_1, z\varphi_2) = \int_{-\pi}^{\pi} \varphi_1(\lambda)\overline{\varphi}_2(\lambda)dF_{xx}(\lambda) \qquad (4f)$$

$$(E_x(\lambda)z\varphi_1, z\varphi_2) = \int_{-\pi}^{\lambda} \varphi_1(\lambda)\overline{\varphi}_2(\lambda)dF_{xx}(\lambda) \qquad (4g)$$

The properties (3a), (3b) and (3c) of the transformation $\varphi \to T\varphi$ follow from theorem 6.1 of M. Stone (3). The properties (3d) and (3e) may be concluded from the comparison of formula (2.5) with the formulae

$$U_x^t = \int_{-\pi}^{\pi} e^{it\lambda}dE_x(\lambda) \qquad (2.8)$$

$$E_x(\lambda) = \int_{-\pi}^{\lambda} dE_x(\lambda) = \int_{-\pi}^{\pi} \chi_\lambda(\mu)dE_x(\mu) \qquad (2.9)$$

The one-one nature of the transformation $\varphi \to T(\varphi)$ follows from the fact that in the case of $\varphi_1 \ne \varphi_2$ (in the sense of the definition of equivalence relative to L_x^2), the difference

$$T(\varphi_1)x(0) - T(\varphi_2)x(0) = \int_{-\pi}^{\pi} [\varphi_1(\lambda) - \varphi_2(\lambda)]dE_x(\lambda)x(0)$$

has a norm

$$\int_{-\pi}^{\pi} |\varphi_1(\lambda) - \varphi_2(\lambda)|^2 d|E_x(\lambda)x(0)|^2 = \int_{-\pi}^{\pi} |\varphi_1(\lambda) - \varphi_2(\lambda)|^2 dF_{xx}(\lambda) \ne 0$$

and consequently

$$T(\varphi_1)x(0) \ne T(\varphi_2)x(0)$$

For the proof of lemma 4 we remark the following: the set $\mathfrak{M}$ of all elements of H_x, represented in the form (2.7) where φ belongs to L_x^2 is according to theorem 6.2 of M.Stone[x] (3) a closed linear subspace of the space H_x; since

$$x(t) = U_x^t x(0) = \int_{-\pi}^{\pi} e^{it\lambda}dE_x(\lambda)x(0) \qquad (2.10)$$

will be real and nondecreasing (see M. Stone (3), page 189). By this the corresponding assertion is justified in the formulation of theorem 2.

From the elementary properties of the Fourier series for functions with bounded variation (see for example A. Zygmund (4), page 13) it follows that

$$F_{xy}(\lambda) = \frac{1}{2\pi} W_{xy}(\lambda+0) + C \qquad (3.9)$$

where

$$W_{xy}(\lambda) = B_{xy}(0)\lambda - \sum_{k\neq 0} \frac{B_{xy}(k)\, e^{-ik\lambda}}{ik} \qquad (3.10)$$

and the constant C is chosen from the condition

$$F_{xy}(-\pi) = 0 \qquad (3.11)$$

Formulae (3.9) and (3.10) show that the function $F_{xy}(\ )$ is uniquely defined by the quantities B_{xy}. Thus theorems 2 and 3 are completely proven.

We remark again that from (3.8) immediately follows

$$F_{xy}(\lambda) = \overline{F_{yx}(\lambda)} \qquad (3.12)$$

THEOREM 3 (H. Cramér) If the sequences

$$\{x_1(t)\},\ \{x_2(t)\},\ldots,\{x_n(t)\}$$

are stationary and are stationarily connected, then for arbitrary $-\pi\leq\alpha\leq\beta\leq\pi$ the increments

$$\Delta F_{x_\mu x_\nu} = F_{x_\mu x_\nu}(\beta) - F_{x_\mu x_\nu}(\alpha) \qquad (3.13)$$

form a positive definite matrix, i.e.

$$\sum_{\mu,\nu=1}^{n} \Delta F_{x_\mu x_\nu}\, \xi_\mu \overline{\xi}_\nu \geq 0 \qquad (3.14)$$

whatever are the complex $\xi_1,\xi_2,\ldots,\xi_n$.

Conversely if the functions, continuous on the right

$$F_{\mu\nu}(\lambda) \qquad \mu,\nu = 1,2,\ldots,n,\ldots$$

defined on the segment $-\pi\leq\lambda\leq\pi$ satisfy the conditions

$$\sum_{\mu,\nu=1}^{n} \Delta F_{\mu\nu}\, \xi_\mu\overline{\xi}_\nu \geq 0 \qquad F_{\mu\nu}(-\pi) = 0 \qquad (3.15)$$

then in H exist a system of stationary and stationarily connected sequences

$$\{x_1(t)\},\{x_2(t)\},\ldots,\{x_n(t)\}$$

for which

$$F_{x_\mu x_\nu}(\lambda) = F_{\mu\nu}(\lambda) \qquad (3.17)$$

For the proof of the first part of the theorem we consider the spectral representation

$$U = \int_{-\pi}^{\pi} e^{i\lambda} dE(\lambda)$$

of the operator

$$U = U_{x_1 x_2 \ldots x_n}$$

of theorem 1 and, putting

$$z = \sum_{\mu=1}^{n} \xi_\mu x_\mu(0)$$

we consider on the segment $-\pi\leq\lambda\leq\pi$ the expression

$$\varphi(\lambda) = |E(\lambda)z|^2 = (E(\lambda)z,z)$$

$$= (E(\lambda) \sum_{\mu=1}^{n} \xi_\mu x_\mu(0),\ \sum_{\nu=1}^{n} \xi_\nu x_\nu(0))$$

$$= \sum_{\mu,\nu=1}^{n} (\xi_\mu E(\lambda) x_\mu(0),\ \xi_\nu x_\nu(0))$$

$$= \sum_{\mu,\nu=1}^{n} \xi_\mu \overline{\xi}_\nu (E(\lambda)x_\mu(0), x_\nu(0))$$

$$= \sum_{\mu,\nu=1}^{n} F_{x_\mu x_\nu}(\lambda)\xi_\mu\overline{\xi}_\nu$$

Since the function $\varphi(\lambda)$ is real and monotonic increasing (see M. Stone (3) page 189), then for $\alpha < \beta$

$$\varphi(\beta) - \varphi(\alpha) = \Delta\varphi = \sum_{\mu,\nu=1}^{n} \Delta F_{x_\mu x_\nu} \xi_\mu \overline{\xi}_\nu \geq 0$$

which it was required to prove.

Going to the proof of the second part of the theorem, we put

$$B_{\mu\nu}(k) = \int_{-\pi}^{\pi} e^{ik\lambda} dF_{\mu\nu}(\lambda)$$

and the function $e^{it\lambda}$ belongs to L_x^2 then $\mathfrak{M}$ contains all x(t) from which it follows that

$$\mathfrak{M} = H_x$$

After this remark, lemma 4 as regards the single-valuedness of the correspondence $\varphi \to z\varphi$ and its properties (4a), (4b) and (4f) become immediate deductions of the theorem 6.1 of Stone (3).

The properties (4c), (4d) and (4e) of the correspondence $z \to z\varphi$ follow from the corresponding properties (3c), (3d) and (3e) of the transformation $\varphi \to T(\varphi)$. Finally from (4e) and (4f) follows (4g):

$$(E_x(\lambda)z\varphi_1, z\varphi_2) = \int_{-\pi}^{\pi} [\chi_\lambda(\mu)\varphi_1(\mu)]\overline{\varphi}_2(\mu)dF_{xx}(\mu)$$

$$= \int_{-\pi}^{\lambda} \varphi_1(\lambda)\overline{\varphi}_2(\lambda)dF_{xx}(\lambda)$$

Thus lemmas 3 and 4 are proved.

† 3. Basic Spectral Properties

THEOREM 2 For an arbitrary stationary sequence $\{x(t)\}$, the quantities $B_{xx}(k)$ may be put in the form

$$B_{xx}(k) = \int_{-\pi}^{\pi} e^{ik\lambda}dF_{xx}(\lambda) \tag{3.1}$$

where F_{xx} is a real continuous on the right non-decreasing function defined on the interval $-\pi\leq\lambda\leq\pi$ for which

$$F_{xx}(-\pi) = 0 \tag{3.2}$$

The function $F_{xx}(\lambda)$ satisfying the enumerated conditions is defined in one valued relation with the quantities B_{xx}.

We remark here that the function $F_{xx}(\lambda)$, which we shall call in the sequel the spectral function of the sequence $\{x(t)\}$, as will appear from the following, coincides with the function defined in †2, formula (2.3).

THEOREM 3 If two stationay sequences $\{x(t)\}$ and $\{y(t)\}$ are stationarily related to each other, then

$$B_{xy}(k) = \int_{-\pi}^{\pi} e^{ik\lambda}dF_{xy}(\lambda) \tag{3.3}$$

where $F_{xy}(\lambda)$ is a (generally speaking, complex) continuous to the right function of bounded variation, defined on the interval $-\pi\leq\lambda\leq\pi$ for which

$$F_{xy}(-\pi) = 0 \tag{3.4}$$

The function F( ) satisfying the enumerated conditions, and being defined uniquely by the quantities B_{xy}.

Since every stationary sequence is stationarily related to itself, theorem 2 with confirmation that the function $F_{xx}(\lambda)$ is real and nondecreasing, may be considered as a particular case of theorem 3.

For the proof of theorem 3 we consider the spectral representation

$$U_{xy} = \int_{-\pi}^{\pi} e^{i\lambda}dE_{xy}(\lambda) \tag{3.5}$$

of the operator U_{xy}, the existence of which follows from theorem 1 for

$$n = 2,\ x_1 = x, x_2 = y$$

We put for $-\pi\leq\lambda\leq\pi$

$$(E_{xy}(\lambda)x(0), y(0)) = F_{xy}(\lambda) \tag{3.6}$$

As is known (see M. Stone (3), page 177), $F_{xy}(\lambda)$ is a function continuous to the right, of bounded variation. From

$$E_{xy}(-\pi) = 0 \tag{3.7}$$

follows that

$$F_{xy}(-\pi) = 0$$

Finally

$$B_{xy}(k) = (x(k), y(0)) = (U^k x(0), y(0))$$

$$= (\int_{-\pi}^{\pi} e^{ik\lambda}dE(\lambda)x(0), y(0))$$

$$= \int_{-\pi}^{\pi} e^{ik\lambda}d(E(\lambda)x(0), y(0))$$

$$= \int_{-\pi}^{\pi} e^{ik\lambda}dF_{xy}(\lambda)$$

Thus the function $F_{xy}(\lambda)$, defined by formula (3.6) satisfied all demands of the theorem.

In the special case of a single sequence $\{x(t)\} = \{y(t)\}$

$$F_{xx}(\lambda) = (E_x(\lambda)x(0), x(0)) = |E_x(\lambda)x(0)|^2 \tag{3.8}$$

According to lemma 2 in order that there exist in the space H elements x(t) corresponding to an arbitrary whole t and arbitrary $\mu = 1,2,\dots n$ for which

$$(x_\mu(t'),x_\nu(t'')) = B_{\mu\nu}(t' - t'')$$

it is sufficient that

$$B = \sum_{p,q=1}^{n} B_{\mu_p\mu_q}(t_p - t_q)\,\xi_p\bar{\xi}_q \geq 0$$

whatever are n,μ_p,t_p,ξ_p. This requirement is always satisfied since putting

$$\xi_p(\lambda) = e^{it_p\lambda}\xi_p,\quad \zeta_\mu(\lambda) = \Sigma^{(\mu)}\xi_p(\lambda)$$

where $\Sigma^{(\mu)}$ extends over all p for which $\mu_p=\mu$ we have owing to inequality (3.15):

$$B = \sum_{p,q=1}^{n}\int_{-\pi}^{\pi} e^{i(t_p-t_q)}dF_{\mu_p\mu_q}(\lambda)\xi_p\bar{\xi}_q$$

$$= \sum_{p,q=1}^{n}\int_{-\pi}^{\pi}\xi_p(\lambda)\bar{\xi}_q(\lambda)dF_{\mu_p\mu_q}(\lambda)$$

$$= \int_{-\pi}^{\pi}\sum_{\mu,\nu=1}^{n}\zeta_\mu(\lambda)\bar{\zeta}_\nu(\lambda)dF_{\mu\nu}(\lambda) \geq 0$$

For the elements $x_\mu(t)$ of the space H, the existence of which is thus justified

$$B_{x_\mu x_\nu}(k) = (x_\mu(t+k),\ x_\nu(t)) = B_{\mu\nu}(k)$$

Consequently the sequences

$$\{x_1(t)\},\ \{x_2(t)\},\dots,\{x_n(t)\}$$

are stationary and stationarily connected. It is easy to see that in this

$$F_{x_\mu x_\nu}(\lambda) = F_{\mu\nu}(\lambda)$$

Theorem 4 is completely proved.

In the case of n = 1 from the second part of theorem 4 follows:

THEOREM 5 For an arbitrary real continuous to the right, nondecreasing function $F(\lambda)$ defined on the segment $-\pi\leq\lambda\leq\pi$ and satisfying the condition $F(-\pi) = 0$ there exists a stationary sequence $\{x(t)\}$ with

$$F_{xx}(\lambda) = F(\lambda)$$

In the case n = 2 the condition (3.15) is equivalent to the collection of conditions

$$\Delta F_{x_1x_1}\geq 0,\Delta F_{x_2x_2}\geq 0,|\Delta F_{x_1x_2}|^2\leq\Delta F_{x_1x_1}\Delta F_{x_2x_2}$$

(see H. Cramér (2), page 227). In this case, theorem 4 gives:

THEOREM 6 If the sequences $\{x(t)\}$ and $\{y(t)\}$ are stationary and stationarily related, then for arbitrary $-\pi\leq\alpha<\beta\leq\pi$,

$$|\Delta F_{xy}|^2 \leq \Delta F_{xx}\Delta F_{yy} \tag{3.18}$$

Conversely, if the functions $F_{11}(\lambda),F_{22}(\lambda)$, $F_{12}(\lambda)$ defined on the interval $-\pi\leq\lambda\leq\pi$ are continuous to the right and satisfy the conditions

$$\Delta F_{11} \geq 0,\ \Delta F_{22} \geq 0,\ |\Delta F_{12}|^2 \leq \Delta F_{11}\Delta F_{22}$$

$$F_{11}(-\pi) = F_{22}(-\pi) = F_{12}(-\pi) = 0 \tag{3.19}$$

then there exist stationary and stationarily connected sequences $\{x(t)\}$ and $\{y(t)\}$ for which

$$F_{xx}(\lambda)=F_{11}(\lambda),F_{yy}(\lambda)=F_{22}(\lambda),F_{xy}(\lambda)=F_{12}(\lambda)$$

In the sequel we shall need the following strengthening of theorem 6:

THEOREM 7 If the functions $F_{11}(\lambda)$, $F_{22}(\lambda)$ and $F_{12}(\lambda)$ satisfy all conditions of the second part of theorem 6, the stationary sequence $\{x(t)\}$ is such that $F_{xx}(\lambda) = F_{11}(\lambda)$ and the orthogonal complement $H\ominus H_x$ of the space H_x to the space H is infinite dimensional then in H there exists a stationary and stationarily connected with $\{x(t)\}$ sequence $\{y(t)\}$ for which

$$F_{yy}(\lambda) = F_{22}(\lambda),\ F_{xy}(\lambda) = F_{12}(\lambda)$$

In order to prove theorem 7, we remark that according to theorem 6 with suitable conditions on the functions $F_{11}(\lambda),F_{22}(\lambda)$ and $F_{12}(\lambda)$ follows the existence of stationary and stationarily connected sequences $\{x^*(t)\}$ and $\{y^*(t)\}$ for which

$$F_{x^*x^*}(\lambda) = F_{11}(\lambda),\ F_{y^*y^*}(\lambda) = F_{22}(\lambda)$$

$$F_{x^*y^*}(\lambda) = F_{12}(\lambda)$$

We put for all whole t:

$$Tx^*(t) = x(t)$$

It is easy to verify that the operator T is isometric on the set $\{x^*(t)\}$. According to lemma 1 the operator is extendable with preservation of isometry on the whole space H_{x^*}. Evidently in this case

$$TH_{x^*} = H_x$$

As the orthogonal complement $H \ominus H_x$ is infinite dimensional, the operator T may be extended with preservation of isometry with H_x on all the space H_{xy} (see M. Stone (3), theorem 2.49). Choosing such an extension, we put

$$Ty^*(t) = y(t)$$

The sequence $\{y(t)\}$ possesses, as is easy to see, all properties demanded in theorem 7.

† 4. Subordinate Sequences

DEFINITION 4 A stationary sequence $\{y(t)\}$ is called a subordinate stationary sequence to $\{x(t)\}$ if sequences $\{x(t)\}$ and $\{y(t)\}$ are stationarily connected and all elements of the sequence $\{y(t)\}$ lie in the space H_x.

Evidently, if the sequence $\{y(t)\}$ is a subordinate sequence to $\{x(t)\}$ then

$$H_{xy} = H_x \qquad (4.1)$$

$$U_{xy} = U_x \qquad (4.2)$$

Using the idea introduced in †2 we prove the following theorem:

THEOREM 8 Every sequence $\{y(t)\}$ subordinate to $\{x(t)\}$ corresponds to a unique function $\varphi_y^{(x)}(\lambda)$ of class L_x^2 for which

$$F_{yy}(\lambda) = \int_{-\pi}^{\lambda} |\varphi_y^{(x)}(\lambda)|^2 dF_{xx}(\lambda) \qquad (4.3)$$

$$F_{yx}(\lambda) = \int_{-\pi}^{\lambda} \varphi_y^{(x)}(\lambda) dF_{xx}(\lambda) \qquad (4.4)$$

The transformation

$$\{y(t)\} \to \varphi_y^{(x)}(\lambda)$$

transforms one-one the class Y_x of all sequences $\{y(t)\}$ subordinate to $\{x(t)\}$ to the class L_x^2 of functions $\varphi(\lambda)$.

In this, for two functions $\{y_1(t)\}$ and $\{y_2(t)\}$ from Y_x

$$F_{y_1y_2}(\lambda) = \int_{-\pi}^{\lambda} \varphi_{y_1}^{(x)}(\lambda)\overline{\varphi}_{y_2}^{(x)}(\lambda) dF_{xx}(\lambda) \qquad (4.5)$$

The assertion about the uniqueness of the functions $\varphi_y^{(x)}(\lambda)$ is understood in theorem 8 in the sense of the correspondence given in 2 defining equality for functions $\varphi(\lambda)$ of class L_x^2. The uniqueness of the definition of $\varphi_y^{(x)}(\lambda)$ by $\{y(t)\}$ in this sense immediately follows from formula (4.4) which gives

$$\varphi_y^{(x)}(\lambda) = dF_{yx}(\lambda)/dF_{xx}(\lambda) \qquad (4.6)$$

The existence of the function $\varphi_y^{(x)}(\lambda)$ for an arbitrary sequence $\{y(t)\}$ of class Y_x we prove on the basis of lemma 4 from †2. If $\{y(t)\}$ comes from Y_x, then $y(0)$ belongs to H_x and according to lemma 4 the existence of a function $\varphi_y^{(x)}(\ )$ of class L_x^2 for which

$$y(0) = z\varphi_y^{(x)} \qquad (4.7)$$

According to the formula (3.6) and the assertion (4g) of lemma 4 we have, whatever are the sequences $\{y_1(t)\}$ and $\{y_2(t)\}$ from Y_x:

$$F_{y_1y_2}(\lambda) = (E_{y_1y_2}(\lambda)y_1(0), y_2(0))$$

$$= (E_x(\lambda)y_1(0), y_2(0))$$

$$= \int_{-\pi}^{\lambda} \varphi_{y_1}^{(x)}\overline{\varphi}_{y_2}^{(x)} dF_{xx}(\lambda)$$

which proves formula (4.5). In the special case $y_1 = y, y_2 = x$ we obtain (4.4) and for $y_1 = y_2 = x$ formula (4.3).

The two way singlevalueness of the correspondence between Y_x and L_x^2 also follows from lemma 4.

An essential supplement to Theorem 8 is:

THEOREM 9 If the sequences $\{x(t)\}$ and $\{y(t)\}$ are stationary and stationarily connected, then in order that the sequence $\{y(t)\}$ should be a subordinate sequence to $\{x(t)\}$ it is necessary and sufficient for the existence of such functions $\varphi(\lambda)$ from

L_x^2 for which

$$F_{yy}(\lambda) = \int_{-\pi}^{\lambda} |\varphi(\lambda)|^2 dF_{xx}(\lambda) \qquad (4.8)$$

$$F_{yx}(\lambda) = \int_{-\pi}^{\lambda} \varphi(\lambda) dF_{xx}(\lambda) \qquad (4.9)$$

The necessity of the condition follows from theorem 8. We prove its sufficiency. If the function $\varphi(\lambda)$ from L_x^2 satisfies the equalities (4.8) and (4.9), then according to theorem 8 there exists a sequence $\{x(t)\}$ subordinate to $\{y^*(t)\}$ for which

$$\varphi_{y^*}^{(x)}(\lambda) = \varphi(\lambda)$$

and consequently

$$F_{y^*y^*}(\lambda) = F_{yy}(\lambda), \quad F_{y^*x}(\lambda) = F_{yx}(\lambda)$$

According to Lemma 1 on H_{xy^*} may be defined an isometric operator T for which for arbitrary whole t,

$$Tx(t) = x(t), \quad Ty^*(t) = y(t)$$

It is clear that

$$T(H_{xy^*}) = H_{xy}, \quad T(H_x) = H_x \qquad (4.10)$$

Since $\{y^*(t)\}$ is subordinate to $\{x(t)\}$, then

$$H_{xy^*} = H_x \qquad (4.11)$$

From (4.10) and (4.11) follows

$$H_{xy} = H_x$$

which also implies that $\{y(t)\}$ is subordinate to $\{x(t)\}$.

DEFINITION 4 Two stationary sequences $\{x(t)\}$ and $\{y(t)\}$ are called equivalent if each of them is subordinate to the other, i.e. if they are stationarily connected and

$$H_x = H_y$$

THEOREM 10 In order that the sequence $\{y(t)\}$ subordinate to $\{x(t)\}$ should be equivalent to $\{x(t)\}$ it is necessary and sufficient that the function $\varphi_y^{(x)}(\lambda)$ should be different from zero almost everywhere relative to $F_{xx}(\lambda)$. If this condition is satisfied then

$$\varphi_x^{(y)}(\lambda) = \frac{1}{\varphi_y^{(x)}(\lambda)} \qquad (4.12)$$

almost everywhere relative to $F_{xx}(\lambda)$ and relative to $F_{yy}(\lambda)$.

1. We prove the necessity of the condition. We suppose that $\{y(t)\}$ is subordinate to $\{x(t)\}$ and that $\{x(t)\}$ is subordinate to $\{y(t)\}$. Then from formula (4.3) and from the formula obtained by permutation of indices,

$$F_{xx}(\lambda) = \int_{-\pi}^{\lambda} |\varphi_x^{(y)}(\lambda)|^2 dF_{yy}(\lambda) \qquad (4.13)$$

it follows that the functions $F_{xx}(\lambda)$ and $F_{yy}(\lambda)$ are absolutely continuous one with the other. Consequently the notions "almost everywhere relative to F_{xx}" and "almost everywhere relative to F_{yy}" coincide. By (4.3) and (4.13)

$$F_{xx}(\lambda) = \int_{-\pi}^{\lambda} |\varphi_x^{(y)}(\lambda)|^2 dF_{yy}(\lambda)$$

$$= \int_{-\pi}^{\lambda} |\varphi_x^{(y)}(\lambda)|^2 |\varphi_y^{(x)}(\lambda)|^2 dF_{xx}(\lambda)$$

and consequently, almost everywhere relative to F_{xx}

$$|\varphi_x^{(y)}(\lambda)|^2 |\varphi_y^{(x)}(\lambda)|^2 = 1 \qquad (4.14)$$

Formula (4.14) shows that $\varphi_y^{(x)}(\lambda) \neq 0$ almost everywhere relative to F_{xx} which it was required to prove.

2. We prove the sufficiency of the condition. We suppose that $\{y(t)\}$ is subordinate to $\{x(t)\}$ and that $\varphi_y^{(x)}(\lambda) \neq 0$ almost everywhere relative to F_{xx}.

Since by formula (4.3) the function $F_{yy}(\lambda)$ is absolutely continuous relative to $F_{xx}(\lambda)$ then $\varphi_y^{(x)}(\lambda) \neq 0$ also almost everywhere and relative to $F_{yy}(\lambda)$. Consequently it is permissible to write

$$\int_{-\pi}^{\lambda} \frac{dF_{yy}(\lambda)}{|\varphi_y^{(x)}(\lambda)|^2} = \int_{-\pi}^{\lambda} \frac{|\varphi_y^{(x)}(\lambda)|^2 dF_{xx}(\lambda)}{|\varphi_y^{(x)}(\lambda)|^2} \qquad (4.15)$$

$$= \int_{-\pi}^{\lambda} dF_{xx}(\lambda) = F_{xx}(\lambda)$$

$$\int_{-\pi}^{\lambda} \frac{dF_{xx}(\lambda)}{\varphi_y^{(x)}(\lambda)} = \int_{-\pi}^{\lambda} \frac{|\varphi_y^{(x)}(\lambda)|^2 dF_{xx}(\lambda)}{\varphi_y^{(x)}(\lambda)}$$

$$= \int_{-\pi}^{\lambda} \overline{\varphi_y^{(x)}(\lambda)} dF_{xx}(\lambda)$$

$$= \overline{\int_{-\pi}^{\lambda} \varphi_y^{(x)}(\lambda)dF_{xx}(\lambda)}$$

$$= \overline{F_{yx}(\lambda)} = F_{xy}(\lambda) \qquad (4.16)$$

Formulae (4.15) and (4.16) show that the function

$$\varphi(\lambda) = \frac{1}{\varphi_y^{(x)}(\lambda)} \qquad (4.17)$$

satisfies the demands

$$F_{xx}(\lambda) = \int_{-\pi}^{\lambda} |\varphi(\lambda)|^2 dF_{yy}(\lambda) \qquad (4.18)$$

$$F_{xy}(\lambda) = \int_{-\pi}^{\lambda} \varphi(\lambda)dF_{yy}(\lambda) \qquad (4.19)$$

According to theorem 9 it hence follows that the sequence $\{x(t)\}$ is subordinate to $\{y(t)\}$ which was also required to be proved.

3. According to theorem 8, from formulae (4.15) and (4.16) it follows that

$$\varphi_x^{(y)}(\lambda) = \varphi(\lambda) = \frac{1}{\varphi_y^{(x)}(\lambda)}$$

† 5. Decomposition of a Stationary Process in Orthogonal Components

DEFINITION 5 Two sequences $\{y_1(t)\}$ and $\{y_2(t)\}$ are called orthogonal one with the other if

$$(y_1(t+k), y_2(t)) = 0 \qquad (5.1)$$

for all whole k and t.

It is clear that in order that two stationary and stationarily connected sequences $\{y_1(t)\}$ and $\{y_2(t)\}$ should be orthogonal one with another, it is necessary and sufficient the condition

$$B_{y_1y_2}(k) \equiv 0 \qquad (5.2)$$

or the equivalent condition

$$F_{y_1y_2}(\lambda) \equiv 0 \qquad (5.3)$$

should hold.

THEOREM 11 If two sequences $\{y_1(t)\}$ and $\{y_2(t)\}$ are stationary and orthogonal one with the other, then:

a) the sequence

$$\{x(t)\} = \{y_1(t)\} + \{y_2(t)\}$$

is stationary

b) the sequences $\{y_1(t)\}$ and $\{y_2(t)\}$ are stationarily connected with each other and with the sequence $\{x(t)\}$:

$c_1)$ $B_{xx}(k) = B_{y_1y_1}(k) + B_{y_2y_2}(k)$

$c_2)$ $B_{y_1x}(k) = B_{y_1y_1}(k)$

$c_3)$ $B_{y_2x}(k) = B_{y_2y_2}(k)$

$d_1)$ $F_{xx}(\lambda) = F_{y_1y_1}(\lambda) + F_{y_2y_2}(\lambda)$

$d_2)$ $F_{y_1x}(\lambda) = F_{y_1y_1}(\lambda)$

$d_3)$ $F_{y_2x}(\lambda) = F_{y_2y_2}(\lambda)$

The proof of theorem 11 depends on the help of simple calculations on the basis of formulae (0.2), (3.9), (3.10), (3.11).

In the conditions of theorem 10 in view of formula $d_1)$ and the nonnegativity of the increments $\Delta F_{xx}, \Delta F_{y_1y_1}$ and $\Delta F_{y_2y_2}$ (for nonnegative $\Delta\lambda$)

$$\left|\frac{\Delta F_{y_1y_1}}{\Delta F_{xx}}\right| \leq 1, \quad \left|\frac{\Delta F_{y_2y_2}}{\Delta F_{xx}}\right| \leq 1. \qquad (5.4)$$

Consequently the functions

$$\psi_{y_1}^{(x)}(\lambda) = \frac{dF_{y_1y_1}(\lambda)}{dF_{xx}(\lambda)}, \quad \psi_{y_2}^{(x)}(\lambda) = \frac{dF_{y_2y_2}(\lambda)}{dF_{xx}(\lambda)} \qquad (5.5)$$

are bounded and uniquely defined (in the sense appropriate to L_x^2) by the equations

$$F_{y_1y_1}(\lambda) = \int_{-\pi}^{\lambda} \psi_{y_1}^{(x)}(\lambda)dF_{xx}(\lambda)$$

$$F_{y_2y_2}(\lambda) = \int_{-\pi}^{\lambda} \psi_{y_2}^{(x)}(\lambda)dF_{xx}(\lambda) \qquad (5.6)$$

THEOREM 12 In the conditions of theorem 10 either both sequences $\{y_1(t)\}$ and $\{y_2(t)\}$ are subordinate to $\{x(t)\}$; or neither is subordinate to $\{x(t)\}$. In order that the first of these two cases should hold it is necessary and sufficient that the condition should be observed

$$\psi_{y_1}^{(x)}(\lambda)\psi_{y_2}^{(x)}(\lambda) = 0 \qquad (5.7)$$

almost everywhere relative to F_{xx}.

Proof. 1. If $\{y_1(t)\}$ is subordinate to $\{x(t)\}$, then in view of the relation

$$y_2(t) = x(t) - y_1(t)$$

$\{y_2(t)\}$ is also subordinate to $\{x(t)\}$. In exactly the same way, if $\{y_2(t)\}$ is subordinate to $\{x(t)\}$ then also $\{y_1(t)\}$ is subordinate to $\{x(t)\}$.

2. If both sequences $\{y_1(t)\}$ and $\{y_2(t)\}$ are subordinate to $\{x(t)\}$ then from the comparison of (5.6) with (4.3) it follows that almost everywhere relative to F_{xx}

$$\psi_{y_1}^{(x)} = |\varphi_{y_1}^{(x)}|^2 \ , \ \psi_{y_2}^{(x)} = |\varphi_{y_2}^{(x)}|^2 \qquad (5.8)$$

if there is taken into consideration equations (d_2) and (d_3) then from the comparison of (5.6) with (4.4) it follows that almost everywhere relative to F_{xx},

$$\psi_{y_1}^{(x)} = \varphi_{y_1}^{(x)} \ , \ \psi_{y_2}^{(x)} = \varphi_{y_2}^{(x)} \qquad (5.9)$$

The comparison of equations (5.8) and (5.9) leads to the deduction that almost everywhere relative to F_{xx} the functions $\varphi_{y_1}^{(x)}$ and $\varphi_{y_2}^{(x)}$ are equal to zero or unity. Since moreover, almost everywhere relative to F_{xx}

$$\psi_{y_1}^{(x)} + \psi_{y_2}^{(x)} = \frac{dF_{y_1y_1}}{dF_{xx}} + \frac{dF_{y_2y_2}}{dF_{xx}} = \frac{dF_{xx}}{dF_{xx}} = 1 \qquad (5.10)$$

there follows also the justification of the condition (5.7).

3. Suppose the condition (5.7) fulfilled, then by (5.10) the functions $\psi_{y_1}^{(x)}$ and $\psi_{y_2}^{(x)}$ almost everywhere relative to F_{xx} are equal to zero or unity.

Consequently, almost everywhere relative to F_{xx} are justified the equalities

$$\psi_{y_1}^{(x)} = |\psi_{y_1}^{(x)}|^2 \ , \ \psi_{y_2}^{(x)} = |\psi_{y_2}^{(x)}|^2 \qquad (5.11)$$

From (5.11), (5.6), (d_2) and (d_3) follows that

$$F_{y_1y_1}(\lambda) = \int_{-\pi}^{\lambda} |\psi_{y_1}^{(x)}(\lambda)|^2 \, dF_{xx}(\lambda)$$

$$F_{y_1x}(\lambda) = \int_{-\pi}^{\lambda} \psi_{y_1}^{(x)}(\lambda) dF_{xx}(\lambda) \qquad (5.12)$$

$$F_{y_2y_2}(\lambda) = \int_{-\pi}^{\lambda} |\psi_{y_2}^{(x)}(\lambda)|^2 dF_{xx}(\lambda)$$

$$F_{y_2x}(\lambda) = \int_{-\pi}^{\lambda} \psi_{y_2}^{(x)}(\lambda) dF_{xx}(\lambda)$$

In view of theorem 9 on the existence of functions $\psi_{y_1}^{(x)}$ and $\psi_{y_2}^{(x)}$ satisfying the equalities (5.12), it follows that the sequences $\{y_1(t)\}$ and $\{y_2(t)\}$ are subordinate to $\{x(t)\}$.

THEOREM 13 Suppose that for the stationary sequence $\{x(t)\}$ the orthogonal complement $H \ominus H_x$ of the space H_x to the space H is infinite dimensional. Then to an arbitrary representation of $F_{xx}(\lambda)$ in the form of a sum

$$F_{xx}(\lambda) = F_1(\lambda) + F_2(\lambda) \qquad (5.13)$$

of two real nondecreasing continuous to the right functions $F_1(\lambda)$ and $F_2(\lambda)$ for which $F_1(-\pi) = F_2(-\pi) = 0$ there corresponds at least one representation of the sequence $\{x(t)\}$ in the form

$$x(t) = y_1(t) + y_2(t) \qquad (5.14)$$

where the sequences $\{y_1(t)\}$ and $\{y_2(t)\}$ are orthogonal one to another and possess spectral functions

$$F_{y_1y_1}(\lambda) = F_1(\lambda) \ , \ F_{y_2y_2}(\lambda) = F_2(\lambda) \qquad (5.15)$$

For the proof we put

$$F_{11}(\lambda)=F_{xx}(\lambda), F_{22}(\lambda)=F_1(\lambda), F_{12}(\lambda)=F_2(\lambda)$$

The functions $F_{11}(\lambda)$, $F_{22}(\lambda)$ and $F_{12}(\lambda)$ satisfy the conditions of the second part of theorem 6. Consequently, according to theorem 7 there exists in the space H a stationary and stationarily connected with $\{x(t)\}$ sequence $\{y_1(t)\}$ for which

$$F_{y_1y_1}(\lambda)=F_{22}(\lambda)=F_1(\lambda),\ F_{xy_1}(\lambda)=F_{12}(\lambda)=F_2(\lambda)$$

Putting

$$y_2(t) = x(t) - y_1(t)$$

it is possible by simple calculations to convince oneself that the sequences $\{y_1(t)\}$ and $\{y_2(t)\}$ are orthogonal one with another and satisfy the equalities (5.15).

In theorem 12 it is already established what sort of conditions must be put on the functions $F_1(\lambda)$ and $F_2(\lambda)$ in the representation (5.13) in order that the sequences $\{y_1(t)\}$ and $\{y_2(t)\}$ should be subordinate to $\{x(t)\}$ i.e. lie in the space H_x. In the case of these supplementary conditions the demand for the infinite dimensionality of the complement $H \ominus H_x$ becomes superfluous,[1)] and the conclusion of theorem 13 can be strengthened in the respect that the representation (5.14) defines the representation (5.13) uniquely[2)] and we thus obtain the theorem:

THEOREM 14 Whatever is the stationary sequence $\{x(t)\}$ every representation of the spectral function $F_{xx}(\lambda)$ in the form

$$F_{xx}(\lambda) = F_1(\lambda) + F_2(\lambda)$$

where $F_1(\lambda)$ and $F_2(\lambda)$ are real nondecreasing continuous to the right functions of λ for which $F_1(-\pi) = F_2(-\pi) = 0$ and almost everywhere relative to F_{xx}

$$\frac{dF_1}{dF_{xx}} \cdot \frac{dF_2}{dF_{xx}} = 0 \qquad (5.16)$$

corresponds one unique representation of $\{x(t)\}$ in the form

$$x(t) = y_1(t) + y_2(t)$$

for which the sequences $\{y_1(t)\}$ and $\{y_2(t)\}$ are orthogonal one to another and

$$F_{y_1y_1}(\lambda) = F_1(\lambda)\ ,\ F_{y_2y_2}(\lambda) = F_2(\lambda)$$

In this the sequence $\{y_1(t)\}$ and $\{y_2(t)\}$ are subordinate to $x(t)$.

†6. Sliding Summation

DEFINITION 6 It will be said that the sequence $\{x(t)\}$ is obtained by sliding summation from the stationary sequence $\{u(t)\}$ if it is possible to pick out such coefficients a_n that

$$x(t) = \sum_{n=-\infty}^{\infty} a_n u(t-n) \qquad (6.1)$$

where the series on the right side converges in norm.

It is clear that a sequence $\{x(t)\}$ obtained by sliding summation from a stationary sequence $\{u(t)\}$ is always stationary and subordinate to $\{u(t)\}$. We analyze in this paragraph a special case in which the converse assertion is also true.

DEFINITION 7 The sequence $\{u(t)\}$ of elements of the space H is called fundamental if it is stationary and satisfied the conditions

$$B_{uu}(0) = 1\ ,\ B_{uu}(k) = 0 \text{ for } k \neq 0 \qquad (6.2)$$

By formulae (3.9) and (3.10) it is easy to calculate that for each fundamental sequence $\{u(t)\}$

$$F_{uu}(\lambda) = \frac{\lambda + \pi}{2\pi} \qquad (6.3)$$

Conversely, every stationary sequence $\{u(t)\}$ with a spectral function of the form (6.3) is fundamental.

The elements of the fundamental sequence $\{u(t)\}$ form a complete normed orthogonal system in the space H_u.

---

1) In fact we consider the enlargement H^* of the space H for which $H^* \ominus H$ is infinite dimensional. In H^* according to theorem 13 there are found sequences $\{y_1(t)\}$ and $\{y_2(t)\}$ of interest to us for which in the case of the additional conditions of theorem 12, the sequences $\{y_1(t)\}$, $\{y_2(t)\}$ lie in H_x.

2) This follows from the fact that according to theorem 8 the functions $\varphi^{(x)}_{y_1} = \frac{dF_1}{dF_{xx}}$ and $\varphi^{(x)}_{y_2} = \frac{dF_2}{dF_{xx}}$ uniquely define the sequences $\{y_1(t)\}$ and $\{y_2(t)\}$subordinate to $\{x(t)\}$.

Consequently every element z of the space H_u is uniquely representable in the form

$$z = \sum_{t=-\infty}^{\infty} c_t u(t) \qquad (6.4)$$

$$\sum_{t=-\infty}^{\infty} c_t^2 = \|z\|^2 < +\infty \qquad (6.5)$$

The coefficients c_t in the formula (6.4) are defined by the equations

$$c_t = (z, u(t)) \qquad (6.6)$$

THEOREM 15 In order that the stationary sequence $\{x(t)\}$ may be obtained by sliding summation from the fundamental sequence $\{u(t)\}$ it is necessary and sufficient that the sequence $\{x(t)\}$ should be subordinate to $\{u(t)\}$. If this condition is satisfied, then the unique representation of $\{x(t)\}$ in the form (6.1) is given by the formula

$$x(t) = \sum_{n=-\infty}^{\infty} B_{xu}(n) u(t-n) \qquad (6.7)$$

The necessity of the conditions of the theorem was already shown above. If this condition is fulfilled then each element $x(t)$ of the sequence $\{x(t)\}$ belongs to H_u and in view of formulae (6.4) and (6.6) we represent it in the form of putting $t-s = n$, in the form (6.7).

Taking into consideration the uniqueness of the representation of an arbitrary element in the form (6.4), we get convinced that every representation of the sequence $\{x(t)\}$ in the form (6.1) coincides with (6.7).

We remark further that from the comparison of (6.7) with (6.4) and (6.5) follows

$$\sum_{n=-\infty}^{\infty} |B_{xu}(n)|^2 = \|x(t)\|^2 = B_{xx}(0) \qquad (6.8)$$

$$F_{xx}(\lambda) = \frac{1}{2\pi} \int_{-\pi}^{\lambda} |\varphi_x^{(u)}(\lambda)|^2 d\lambda \qquad (6.9)$$

$$F_{xu}(\lambda) = \frac{1}{2\pi} \int_{-\pi}^{\lambda} \varphi_x^{(u)}(\lambda) d\lambda \qquad (6.10)$$

Formula (6.9) shows that $\varphi_x^{(u)}(\lambda)$ is a function with integrable square. From (3.3) and (6.10) there is obtained

$$B_{xu}(n) = \frac{1}{2\pi} \int_{-\pi}^{\pi} e^{in\lambda} \varphi_x^{(u)}(\lambda) d\lambda \qquad (6.11)$$

Thus the quantities $B_{xu}(n)$ are the Fourier coefficients of the function $\varphi_x^{(u)}(\lambda)$ i.e.

$$\varphi_x^{(u)}(\lambda) \sim \sum_{n=-\infty}^{\infty} B_{xu}(n) e^{-in\lambda} \qquad (6.12)$$

We remark in connection with the formulae just introduced, that in view of (6.3), "almost everywhere relative to F_{uu}" means simply "with the exception of a set of Lebesgue measure zero" and the class L_u^2 coincides with the class L^2 of the functions with integrable square in the usual Lebesgue sense of the words.

THEOREM 16 In order that the stationary sequence $\{x(t)\}$ can be obtained by sliding summation from some fundamental sequence it is necessary that the spectral function $F_{xx}(\lambda)$ should be absolutely continuous.

If the orthogonal complement $H \ominus H_x$ is infinite dimensional then the condition mentioned is also sufficient.

The necessity of the condition is clear from formula (6.9). In order to prove its sufficiency we consider an arbitrary function $\gamma(\lambda)$ from L^2 for which almost everywhere on $-\pi \le \lambda \le \pi$

$$|\gamma(\lambda)|^2 = 2\pi \frac{dF_{xx}}{d\lambda} \qquad (6.13)$$

and we define functions $F_{11}(\lambda), F_{22}(\lambda)$ and $F_{12}(\lambda)$ by the formulae

$$F_{11}(\lambda) = F_{xx}(\lambda), \quad F_{22}(\lambda) = \frac{\lambda + \pi}{2\pi},$$

$$F_{12}(\lambda) = \frac{1}{2\pi} \int_{-\pi}^{\lambda} \gamma(\lambda) d\lambda$$

It is easy to verify that these functions satisfy the conditions of the second part of theorem 6. According to theorem 7 under the condition of infinite dimensionality of $H \ominus H_x$ there follows the existence of a stationary and stationarily connected with $\{x(t)\}$ sequence $\{u(t)\}$ for which

$$F_{uu}(\lambda) = F_{22}(\lambda) = \frac{\lambda + \pi}{2\pi} \qquad (6.14)$$

$$F_{xu}(\lambda) = F_{12}(\lambda) = \frac{1}{2\pi}\int_{-\pi}^{\lambda} \gamma(\lambda)d\lambda \qquad (6.15)$$

Formula (6.15) shows that $\{u(t)\}$ is a fundamental sequence. From the absolute continuity of $F_{xx}(\lambda)$ and the formula (6.13) it follows that

$$F_{xx}(\lambda) = F_{11}(\lambda) = \frac{1}{2\pi}\int_{-\pi}^{\lambda} |\gamma(\lambda)|^2 d\lambda \quad (6.16)$$

From (6.15) and (6.16) by theorem 9 we conclude that the sequence $\{x(t)\}$ is subordinate to the sequence $\{u(t)\}$. According to theorem 15, it follows that $\{x(t)\}$ can be obtained by sliding summation from $\{u(t)\}$.

THEOREM 17 In order that the stationary sequence $\{x(t)\}$ may be obtained from sliding summation from some fundamental sequence, subordinate to $\{x(t)\}$ it is necessary and sufficient that the function $F_{xx}(\lambda)$ should be absolutely continuous and the function

$$f_{xx}(\lambda) = \frac{dF_{xx}(\lambda)}{d\lambda} \qquad (6.17)$$

almost everywhere on $-\pi \leq \lambda \leq \pi$ positive.

If both these conditions are fulfilled then:

A) every fundamental sequence from which $\{x(t)\}$ may be obtained by sliding summation is subordinate to $\{x(t)\}$.

B) the sequence $\{x(t)\}$ may be obtained by sliding summation from an arbitrary sequence subordinate to it.

C) in order that the sequence $\{u(t)\}$ subordinate to $\{x(t)\}$ should be fundamental, it is necessary and sufficient that almost everywhere on $-\pi \leq \lambda \leq \pi$ should be

$$|\varphi_u^{(x)}(\lambda)|^2 = \frac{1}{2\pi f_{xx}(\lambda)} \qquad (6.18)$$

In connextion with formula (6.18) we remark that in the case of absolute continuity of $F_{xx}(\lambda)$ and the positivity almost everywhere of $f_{xx}(\lambda)$, the class of sets of zero measure relative to F_{xx} coincides with the class of sets of zero Lebesgue measure. Consequently the functions of class L_x^2 in particular the functions $\varphi_u^{(x)}$ are defined to an accuracy up to a set of zero measure in the usual Lebesgue sense of the word.

Proof of Theorem 17:

1. The necessity of the condition of absolute continuity is already established in theorem 16. We prove the necessity of the positivity almost everywhere of the function f_{xx}. If the fundamental sequence $\{u(t)\}$ is subordinate to $\{x(t)\}$ then

$$\frac{\lambda + \pi}{2\pi} = F_{uu}(\lambda) = \int_{-\pi}^{\lambda} |\varphi_u^{(x)}(\lambda)|^2 dF_{xx}(\lambda) \qquad (6.19)$$

from which after differentiation with respect to λ is obtained that almost everywhere

$$\frac{1}{2\pi} = |\varphi_u^{(x)}(\lambda)|^2 f_{xx}(\lambda) \qquad (6.20)$$

From (6.20) it is clear that f_{xx} is positive almost everywhere.

2. We assume now that the function F_{xx} is absolutely continuous and the function f_{xx} almost everywhere positive. From the absolute continuity of F_{xx} follows according to theorem 16 that the sequence $\{x(t)\}$ may be obtained by sliding summation from some fundamental sequence $\{u(t)\}$, lying in some extension H' of the space H_x. According to formula (6.9) there is, almost everywhere,

$$|\varphi_x^{(u)}(\lambda)|^2 = 2\pi f_{xx}(\lambda) \neq 0$$

According to theorem 10 it follows that, that the sequence $\{u(t)\}$ is subordinate to $\{x(t)\}$ (i.e. lies in fact, in the space H_x itself). We thus prove the sufficiency of the condition of the theorem, and apart from that, the additional condition A.

The sequence $\{u(t)\}$ subordinate to $\{x(t)\}$ is fundamental if and only if

$$F_{uu}(\lambda) = \int_{-\pi}^{\lambda} |\varphi_u^{(x)}(\lambda)|^2 dF_{xx}(\lambda)$$

$$= \int_{-\pi}^{\lambda} |\varphi_u^{(x)}(\lambda)|^2 f_{xx}(\lambda)d\lambda = \frac{\lambda + \pi}{2\pi}$$

i.e. almost everywhere

$$|\varphi_u^{(x)}(\lambda)|^2 f_{xx}(\lambda) = \frac{1}{2\pi} \qquad (6.21)$$

Thus the assertion C is proved.

In view of (6.21) for each fundamental sequence $\{u(t)\}$ subordinate to $\{x(t)\}$, almost everywhere, $|\varphi_u^{(x)}(\lambda)|^2 \neq 0$. According to theorem 10 it follows from that that the sequence $\{x(t)\}$ is subordinate to $\{u(t)\}$ by sliding summation. By this is proved assertion B.

7. The Wold Decomposition

For an arbitrary stationary sequence $\{x(t)\}$ we shall denote by $H_x(t)$ the minimal linear closed subspace of the space H_x containing all $x(s)$ for which $s \leq t$ and by S_x the intersection of all $H_x(t)$. It is clear that

$$U_x^k H_x(t) = H_x(t+k) \tag{7.1}$$

$$U_x^k S_x = S_x \tag{7.2}$$

DEFINITION 8 The stationary sequence $\{x(t)\}$ is called singular if

$$S_x = H_x \tag{7.3}$$

It is clear that for a singular sequence

$$H_x(t) = H_x \tag{7.4}$$

whatever is t. Conversely, if equality (7.4) is true for a single t then the sequence $\{x(t)\}$ is singular. In fact from the justification of the equality (7.4) for some single t, it follows in view of (7.1) its justification also for all other t from which it follows that $S_x = H_x$.

We denote by $s_x(t)$ the projection of $x(t)$ in the space S_x. It is easy to see that $\{s_x(t)\}$ is a singular stationary sequence subordinate to $x(t)$ for which

$$S_{sx} = H_{sx} = S_x \tag{7.5}$$

The sequence $\{s_x(t)\}$ is called the singular component of the sequence $\{x(t)\}$.

We assume now that the sequence $\{x(t)\}$ is not singular. Then $\{x(t)\}$ is uniquely representable in the form

$$x(t) = \xi(t) + \Delta(t) \tag{7.6}$$

where $\xi(t)$ belongs to $H_x(t-1)$ and $\Delta(t) \neq 0$ is orthogonal to $H_x(t-1)$ we put

$$u_x(t) = \frac{\Delta(t)}{\|\Delta(t)\|} \tag{7.7}$$

It is easy to see that the elements

$$u_x(t), u_x(t-1), \ldots, u_x(t-n), \ldots$$

form a complete normed orthogonal system in the space

$$H_x(t) \ominus S_x$$

Since $x(t)$ belongs to $H_x(t)$ then $x(t)$ is uniquely representable in the form

$$x(t) = s_x(t) + \sum_{n=0}^{\infty} c_n^{(x)} u_x(t-n) \tag{7.8}$$

Since the sequence $\{u_x(t)\}$ is stationary and subordinate to $\{x(t)\}$ the coefficients

$$c_n^{(x)} = (x(t), u_x(t-n)) = B_{xu}(n) \tag{7.9}$$

in formula (7.8) do not depend on t. From (7.6) and (7.7) it follows that

$$c_0 = \|\Delta(t)\| > 0 \tag{7.10}$$

The representation of a non-singular sequence in the form (7.8) we shall call the Wold decomposition. From the preceding account it follows that

$$s(t) = s_x(t), \quad u(t) = u_x(t), \quad c_n = c_n^{(x)}$$

possesses the following properties:

W_1. The sequence $\{s(t)\}$ is singular and subordinate to $\{x(t)\}$.

W_2. $\{u(t)\}$ is a fundamental sequence subordinate to $\{x(t)\}$.

W_3. $u(t)$ lies in the space $H_x(t)$.

W_4. The sequences $\{s(t)\}$ and $\{u(t)\}$ are orthogonal one to the other.

W_5. $c_0 > 0$.

The following theorem shows that the decomposition (7.8) is uniquely defined by the five properties written down.

THEOREM 18 If the stationary sequence $\{x(t)\}$ is put in the form

$$x(t) = s(t) + \sum_{n=0}^{\infty} c_n u(t-n) \tag{7.11}$$

the sequences $\{x(t)\}$ and $\{u(t)\}$ and the coefficients c_n satisfying the conditions W_{1-5} then the sequence $\{x(t)\}$ is not singular and

$$s(t) = s_x(t), \quad u(t) = u_x(t), \quad c_n = c_n^{(x)}$$

Proof: Since $x(t')$ and all $u(t'-n)$ for $n = 0,1,2\ldots$ (according to condition W_3)

belong to $H_x(t')$ then

$$s(t') = x(t') - \sum_{n=0}^{\infty} c_n u(t'-n)$$

also belongs to $H_x(t')$ for $t' \leq t$ lies in $H_x(t)$ then all $s(t')$ for $t' \leq t$ lie in $H_x(t)$. Consequently $H_s(t)$ lies in $H_x(t)$. Since according to condition W_1, $S_s = H_s(t)$ then S_s belongs to all $H_x(t)$ and consequently to S_x. From that it follows that all $s(t')$ belongs to S_x and consequently to all $H_x(t)$.

Let us rewrite formula (7.11) in the form

$$x(t) = \zeta(t) + c_0 u(t)$$

where

$$\zeta(t) = s(t) + \sum_{n=1}^{\infty} c_n u(t-n) \tag{7.12}$$

Since $s(t)$ and all $u(t-n)$ for $n = 1,2,3,\ldots$ lie in $H_x(t-1)$, then $\zeta(t)$ lies in $H_x(t-1)$. Since all $x(t')$ for $t' < t$ is expressible linearly through the elements $s(t'')$ and $u(t'')$ with $t'' < t$ and $u(t)$ is orthogonal with all $s(t'')$ and with $u(t'')$ for $t'' < t$ then $u(t)$ is orthogonal to all those $x(t')$ for which $t' < t$ and consequently to space $H_x(t-1)$.

From the fact that $\zeta(t)$ lies in $H_x(t-1)$ and $u(t)$ is orthogonal to $H_x(t-1)$ we conclude comparing (7.12) with (7.6) that

$$\zeta(t) = \xi(t)$$
$$c_0 u(t) = \Delta(t) \tag{7.13}$$

From the comparison of (7.13) with (7.7) in view of condition W_5 it follows that

$$u(t) = u_x(t) \tag{7.14}$$

In view of W_2 and W_4 from (7.11) it follows:

$$c_n = [x(t), u(t)] \tag{7.15}$$

from (7.14), (7.15) and (7.9) we obtain

$$c_n = c_n^{(x)} \tag{7.16}$$

From the comparison of (7.8) with (7.11) taking into consideration (7.15) and (7.16) we obtain finally

$$s(t) = s_x(t)$$

†8. Regular Sequences

THEOREM 19 In order that the stationary sequence $\{x(t)\}$ may be represented in the form

$$x(t) = \sum_{n=0}^{\infty} c_n u(t-n) \tag{8.1}$$

where $\{u(t)\}$ is some fundamental sequence, it is necessary and sufficient for the condition

$$s_x(t) = 0 \tag{8.2}$$

If $x(t) = 0$, then the condition $s_x(t) = 0$ is fulfilled and the representation (8.1) is possible (with all coefficients c_n equal to zero). It remains to consider the case $x(t) \neq 0$. The sufficiency of the condition $s_x(t) = 0$ in this case is also clear: if $s_x(t) = 0$ and $x(t) \neq 0$ then the sequence $x(t)$ is not singular and according to formula (7.8) of the preceding paragraph is representable in the form

$$x(t) = \sum_{n=0}^{\infty} c_n^{(x)} u_x(t-n) \tag{8.3}$$

We prove the necessity of the condition (8.2). For this we remark that in the case of the existence of the representation (8.1), the space $H_x(t)$ is contained in the space $H_u(t)$ and the space S_x in the space S_u. Consequently from the equality

$$S_u = 0 \tag{8.4}$$

justified for each fundamental sequence, it follows that $S_x = 0$ and consequently that $s_x(t) = 0$.

DEFINITION 9 A stationary sequence $\{x(t)\}$ will be called regular if

$$x(t) \neq 0 \ , \ s_x(t) = 0$$

If $\{u(t)\}$ is a fundamental sequence, then (8.1) is a special case of the representation investigated in detail in †6. Consequently in connexion with formulae (6.7), (6.8), (6.12)

$$B_{xu}(n) = c_n \quad \text{for } n \geq 0$$
$$B_{xu}(n) = 0 \quad \text{for } n < 0 \tag{8.5}$$

$$\sum_{n=0}^{\infty} |c_n|^2 = |x(t)|^2 = B_{xx}(0) \qquad (8.6)$$

$$\varphi_x^{(u)}(\lambda) = \sum_{n=0}^{\infty} c_n e^{-in\lambda} \qquad (8.7)$$

In view of (8.6), the series

$$\Gamma_x^{(u)}(\zeta) = \sum_{n=0}^{\infty} c_n \zeta^n \qquad (8.8)$$

represents an analytic function in the circle $|\zeta| < 1$. Its boundary values for $|\zeta| = 1$ give in view of (8.7) the formula

$$\Gamma_x^{(u)}(e^{-i\lambda}) = \varphi_x^{(u)}(\lambda) \qquad (8.9)$$

From the assumed condition $x(t) \neq 0$ it follows that $\Gamma_x^{(u)}(\zeta)$ cannot be identically zero. Consequently[1] almost everywhere on $-\pi \leq \lambda \leq \pi$

$$\varphi_x^{(u)}(\lambda) \neq 0 \qquad (8.10)$$

Since according to formula (6.9) almost everywhere

$$\frac{dF_{xx}}{d\lambda} = f_{xx}(\lambda) = \frac{1}{2\pi} |\varphi_x^{(u)}(\lambda)|^2 \qquad (8.11)$$

then from (8.10) it follows that almost everywhere

$$f_{xx}(\lambda) > 0 \qquad (8.12)$$

According to theorem 17 we hence conclude that the sequence $\{u(t)\}$ is subordinate to $\{x(t)\}$.

THEOREM 20 If the regular sequence $\{x(t)\}$ is represented in the form (8.1) through a fundamental sequence $\{u(t)\}$ then the sequence $\{u(t)\}$ is subordinate to $\{x(t)\}$.

In order for the stationary sequence $\{x(t)\}$ subordinate to the fundamental sequence $\{u(t)\}$ to be represented through $\{u(t)\}$ in the form (8.1) it is necessary and sufficient that the function $\gamma_x^{(u)}(\zeta)$ defined for $|\zeta| = 1$ by the equality

$$\gamma_x^{(u)}(e^{-i\lambda}) = \varphi_x^{(u)}(\lambda) \qquad (8.13)$$

should agree almost everywhere on the circumference $|\zeta| = 1$ with boundary values

1) See for example the monograph of I.I. Privalov (5), p.39.

analytic functions $\Gamma_x^{(u)}(\zeta)$ defined for $|\zeta| < 1$ by the formula

$$\Gamma_x^{(u)}(\zeta) = \frac{1}{2\pi i} \int_{|\xi|=1} \frac{\gamma_x^{(u)}(\xi)d\xi}{\xi - \zeta} \qquad 14)$$

If this condition is fulfilled, then the coefficients c_n of formula (8.1) are defined from (8.8).

The first part of the theorem, the necessity of the condition of the second part of the theorem and the assertion on the form of the dependence between the coefficients c_n and the function $\Gamma_x^{(u)}(\zeta)$ is already proved before.

It remains to prove the sufficiency of the condition in the second part of the theorem.

If these conditions are satisfied, then for the boundary values of $\gamma_y^{(u)}(\zeta)$ we have the development in Fourier series:

$$\gamma_x^{(u)}(e^{-i\lambda}) = \varphi_x^{(u)}(\lambda) \sim \sum_{n=0}^{\infty} c_n e^{-in\lambda} \qquad (8.15)$$

From the comparison of (8.15) with (6.12) and (6.7) we conclude that $\{x(t)\}$ is represented in the form (8.1) with coefficients c_n defined from (8.15) or, which is the same thing, from (8.8). The proof of theorem 20 is finished.

THEOREM 21 For an arbitrary regular sequence $\{x(t)\}$ the function

$$\Gamma_x(\zeta) = \Gamma_x^{(u_x)}(\zeta) = \sum_{n=0}^{\infty} c_n^{(x)} \zeta^n \qquad (8.16)$$

does not have zeros outside the circle $|\zeta| < 1$.

We assume in spite of the assertion of the theorem, that $\Gamma_x(\zeta)$ has a zero ζ_0 lying in the circle $|\zeta| < 1$. We construct a sequence $\{u_x(t)\}$ subordinate to $\{u(t)\}$ for which

$$\varphi_u^{(u_x)}(\lambda) = \frac{e^{-i\lambda} - \zeta_0}{1 - \bar{\zeta}_0 e^{-i\lambda}} \qquad (8.17)$$

Such a sequence exists in view of theorem 8 since the function

$$\varphi(\lambda) = \frac{e^{-i\lambda} - \zeta_0}{1 - \bar{\zeta}_0 e^{-i\lambda}}$$

is bounded and consequently belongs

to class $L^2_{u_x} = L^2$.

It is easy to calculate that for arbitrary λ

$$|\varphi_u^{(u_x)}(\lambda)|^2 = 1 \tag{8.18}$$

Since

$$f_{u_x u_x}(\lambda) = \frac{d}{d\lambda} F_{u_x u_x}(\lambda) = \frac{d}{d\lambda} \frac{\lambda + \pi}{2\pi} = \frac{1}{2\pi}$$

then according to assertion C of theorem 17 from (8.18) it follows that $\{u(t)\}$ is a fundamental sequence.

According to formula (4.5),

$$F_{xx}(\lambda) = \int_{-\pi}^{\lambda} \varphi_x^{(u_x)}(\lambda)\bar{\varphi}_u^{(u_x)}(\lambda)dF_{u_x u_x}(\lambda)$$

$$= \frac{1}{2\pi} \int_{-\pi}^{\lambda} \varphi_x^{(u_x)}(\lambda)\bar{\varphi}_x^{(u_x)}(\lambda)d\lambda \tag{8.19}$$

From (4.6) and (8.19) it follows that almost everywhere

$$\varphi_x^{(u)}(\lambda) = \frac{dF_{xu}(\lambda)}{dF_{uu}(\lambda)} = 2\pi \frac{dF_{xu}(\lambda)}{d\lambda}$$

$$= \varphi_x^{(u_x)}(\lambda)\bar{\varphi}_u^{(u_x)}(\lambda) \tag{8.20}$$

In view of (8.18)

$$\bar{\varphi}_u^{(u_x)}(\lambda) = \frac{1}{\varphi_u^{(u_x)}(\lambda)}$$

Consequently formula (8.20) may be written in the form

$$\varphi_x^{(u)}(\lambda) = \frac{\varphi_x^{(u_x)}(\lambda)}{\varphi_u^{(u_x)}(\lambda)} \tag{8.21}$$

In view of (8.9) and (8.16) we have:

$$\varphi_x^{(u_x)}(\lambda) = \Gamma_x^{(u_x)}(e^{-i\lambda}) = \Gamma_x(e^{-i\lambda}) \tag{8.22}$$

From (8.21), (8.22) and (8.17) we conclude that

$$\varphi_x^{(u)}(\lambda) = \Gamma_x(e^{-i\lambda}) \frac{1 - \zeta_0 e^{-i\lambda}}{e^{-i\lambda} - \zeta_0} \tag{8.23}$$

It is easy to see that

$$\gamma_x^{(u)}(e^{-i\lambda}) = \varphi_x^{(u)}(\lambda)$$

coincides almost everywhere with boundary values for the function

$$\Gamma_x^{(u)}(\zeta) = \Gamma_x(\zeta) \frac{1 - \bar{\zeta}_0\zeta}{\zeta - \zeta_0} \tag{8.24}$$

Since ζ_0 is a zero of the function $\Gamma_x(\zeta)$, the function $\Gamma_x^{(u)}(\zeta)$ is analytic for $|\zeta| < 1$. According to theorem 20 it hence follows that[1)]

$$x(t) = \sum_{n=-\infty}^{\infty} c_n u(t-n) \tag{8.25}$$

where the coefficients c_n are defined from the formula (8.8). The function

$$\gamma_u^{(u_x)}(e^{-i\lambda}) = \varphi_u^{(u_x)}(\lambda) = \frac{e^{-i\lambda} - \zeta_0}{1 - \bar{\zeta}_0 e^{-i\lambda}}$$

coincides almost everywhere on the circumference with the boundary values analytic in the circle $|\zeta| < 1$ of the function

$$\Gamma_u^{(u_x)}(\zeta) = \frac{\zeta - \zeta_0}{1 - \bar{\zeta}_0\zeta} \tag{8.26}$$

Consequently according to theorem 20,

$$u(t) = \sum_{n=0}^{\infty} d_n u_x(t-n) \tag{8.27}$$

where d_n are the coefficients of the Taylor series of the function $\Gamma_x^{(u_x)}(\zeta)$. Formula (8.27) shows that $u(t)$ lies in $H_{ux}(t)$. Since according to property W_3 of the sequence $\{u_x(t)\}$ (see † 7), $H_{ux}(t)$ lies in $H_x(t)$ then $u(t)$ also lies in $H_x(t)$.

From only what has been said it fol-

1) In order to use theorem 20 it is essential that the function $\Gamma_x^{(u)}(\zeta)$ be represented in the form of a Cauchy integral through its boundary values $\gamma_x^{(u)}(e^{-i\lambda})$. That this is really so, follows, for example from remark C on page 94 of the work of F. Riesz (6) since clearly the function $\Gamma_x^{(u)}(\zeta)$ as well as the function $\Gamma_x(\zeta)$ belong to the class H_1 in the sense of this work of F. Riesz.

lows that all $u(t-n)$ for $n > 0$ lie in $H_x(t-1)$. Consequently, comparing formulae (8.25) and (7.6) we obtain:

$$c_0 u(t) = \Delta(t)$$

and in view of (7.7) and (7.10)

$$c_0 u(t) = c_0^{(x)} u_x(t)$$

It is easy to verify the sequence $\{u(t)\}$ defined by the formula

$$u(t) = \frac{c_0^{(x)}}{c_0} u_x(t)$$

must correspond to the function

$$\varphi_u^{(u_x)}(\lambda) = \frac{c_0^{(x)}}{c_0} \qquad (8.28)$$

The contradiction between (8.28) and (8.17) proves theorem 21.

THEOREM 22 In order that the stationary sequence $\{x(t)\}$ should be regular, it is necessary and sufficient that these set of conditions be fulfilled on $-\pi \le \lambda \le \pi$

1) the function $F_{xx}(\lambda)$ is absolutely continuous;

2) the function $\log f_{xx}(\lambda)$ is summable.

If these conditions are fulfilled then

$$\Gamma_x(\zeta) = \sqrt{2\pi}\, e^{Q_x(\zeta)} \qquad (8.29)$$

$$Q_x(\zeta) = \frac{a_0^{(x)}}{2} + \sum_{k=1}^{\infty} (a_k^{(x)} + ib_k^{(x)})\zeta^k \qquad (8.30)$$

where the coefficients a_k and b_k are defined from the development

$$\frac{1}{2}\log f_{xx}(\lambda) \sim \frac{a_0^{(x)}}{2} + \sum_{k=1}^{\infty} (a_k^{(x)}\cos k\lambda + b_k^{(x)}\sin k\lambda) \qquad (8.31)$$

Since every regular sequence $\{x(t)\}$ is obtained by sliding summation from a fundamental sequence $\{u_x(t)\}$ subordinate to it, the necessity of conditions 1 and 2 follow from condition 3. If the sequence $\{x(t)\}$ is regular then according to theorem 21 the function $\Gamma_x(\zeta)$ is analytic and different from zero in the circle $|\zeta| < 1$. We shall denote by $Q_x(\zeta)$ that value of $\log \frac{\Gamma_x(\zeta)}{\sqrt{2\pi}}$ which is obtained by continuity from the real value

$$\log \frac{\Gamma_x(0)}{\sqrt{2\pi}} = \log \frac{c_0^{(x)}}{\sqrt{2\pi}}$$

Thanks to the absence of zeros of the function $\Gamma_x(\zeta)$, the function $Q_x(\zeta)$ is uniquely defined at all points of the circle $|\zeta| < 1$. For the real part of the function $Q_x(\zeta)$ we have in view of (8.32):

$$\Re Q_x(\zeta) = \log|\Gamma_x(\zeta)| - \frac{1}{2}\log 2\pi \qquad (8.32)$$

We denote by $\Re^+ Q(\zeta)$ the function equal to $\Re Q_x(\zeta)$ for $\Re Q_x(\zeta) > 0$ and equal to zero for $\Re Q_x(\zeta) \le 0$. Since from (8.32) follows

$$\Re^+ Q_x(\zeta) < \log^+|\Gamma_x(\zeta)| \le |\Gamma_x(\zeta)|$$

then for arbitrary $\rho < 1$,

$$\int_{-\pi}^{\pi} \Re^+ Q_x(\rho e^{-i\lambda})d\lambda < \int_{-\pi}^{\pi} |\Gamma_x(\rho e^{-i\lambda})|d\lambda \le \int_{-\pi}^{\pi} |\varphi_x^{(u_x)}(\lambda)|d\lambda = K$$

Since

$$\int_{-\pi}^{\pi} \Re Q_x(\rho e^{-i\lambda})d\lambda = 2\pi \Re Q_x(0)$$

then

$$\int_{-\pi}^{\pi} |\Re Q_x(\rho e^{-i\lambda})|d\lambda = \int_{-\pi}^{\pi} [2\Re^+ Q_x(\rho e^{-i\lambda}) - \Re Q_x(\rho e^{-i\lambda})] \le 2K - 2\pi \Re Q_x(0)$$

From the preceeding inequalities it follows that the boundary value

$$\Re Q_x(e^{-i\lambda}) = \log\left|\frac{\Gamma_x(e^{-i\lambda})}{\sqrt{2\pi}}\right| = \log\left|\frac{\varphi_x^{(u_x)}(\lambda)}{\sqrt{2\pi}}\right| = \frac{1}{2}\log\left|\frac{\varphi_x^{(u_x)}(\lambda)}{\sqrt{2\pi}}\right| = \frac{1}{2}\log f_{xx}(\lambda) \qquad (8.33)$$

of the function $\Re Q_x(e^{-i\lambda})$ is summable with

respect to λ. Thus the necessity of the summability of $\log f_{xx}(\lambda)$ i.e. condition 3 of our theorem, is proved.

Before going to the proof of the sufficiency of the condition of the theorem, we prove that for an arbitrary regular sequence $\{x(t)\}$ the function $\Gamma_x(\zeta)$ is defined by formulae (8.29), (8.30) and (8.31). Since as was proved in the case of a regular sequence $\{x(t)\}$, the function $\frac{1}{2}\log f_{xx}(\lambda)$ is summable, then for this function there exists a Fourier expansion (8.31). From (8.33) and (8.31) it follows that

$$\Re Q_x(e^{i\lambda}) \sim \frac{a_0^{(x)}}{2} + \sum_{k=1}^{\infty} (a_k^{(x)}\cos k\lambda - b_k^{(x)}\sin k\lambda) \tag{8.34}$$

Since

$$Q_x(0) = \log\frac{\Gamma_x(0)}{\sqrt{2\pi}} = \log\frac{c_0^{(x)}}{\sqrt{2\pi}} \tag{8.35}$$

is real, the imaginary part $\Im Q_x(e^{i\lambda})$ is expressed by the formula

$$\Im Q_x(e^{i\lambda}) \sim \sum_{k=1}^{\infty} (a_k^{(x)}\sin k\lambda + b_k^{(x)}\cos k\lambda) \tag{8.36}$$

From (8.34) and (8.36) follows (8.30). Finally, formula (8.29) follows from the definition of $Q_x(\zeta)$.

We return now to the proof of the sufficiency of the condition of the theorem. For this we suppose that these conditions are fulfilled and we put

$$Q(\zeta) = \frac{a_0^{(x)}}{2} + \sum_{k=1}^{\infty} (a_k^{(x)} + ib_k^{(x)})\zeta^k \tag{8.37}$$

$$\Gamma(\zeta) = \sqrt{2\pi}\, e^{Q(\zeta)} \tag{8.38}$$

where the coefficients $a_k^{(x)}$ and $b_k^{(x)}$ are defined from the development (8.31). Then[1)]

$$\log\frac{|\Gamma(\rho e^{-i\lambda})|^2}{2\pi} = 2\Re Q(\rho e^{-i\lambda})$$

$$= a_0 + 2\sum_{k=1}^{\infty} \rho^k(a_k^{(x)}\cos k\lambda + b_k^{(x)}\sin k\lambda)$$

$$= \int_{-\pi}^{\pi} \log f_{xx}(\mu)P_\rho(\mu - \lambda)d\mu \tag{8.39}$$

where

$$P_\rho(\theta) = \frac{1}{2\pi}\;\frac{1-\rho^2}{1+\rho^2 - 2\rho\cos\theta} \tag{8.40}$$

From the inequality between the geometrical and arithmetical means[2)] it follows that

$$\frac{|\Gamma(\rho e^{-i\lambda})|^2}{2\pi} \le \int_{-\pi}^{\pi} f_{xx}(\mu)P_\rho(\mu-\lambda)d\lambda \tag{8.41}$$

The boundedness thus demonstrated of the integral in the left side of the equation (8.42) guarantees[3)] the representability of the function in the form of a Cauchy integral from its boundary values.

$$\Gamma(e^{-i\lambda}) = \varphi(\lambda)$$

From (8.38), (8.37) and (8.31) it follows that

$$|\varphi(\lambda)|^2 = |\Gamma(e^{-i\lambda})|^2 = 2\pi e^{2\Re Q(e^{-i\lambda})}$$

$$= 2\pi f_{xx}(\lambda) \tag{8.43}$$

Since

$$\int_{-\pi}^{\pi} \frac{1}{|\varphi(\lambda)|^2}\, dF_{xx}(\lambda) = 1$$

and consequently, the function $1/\varphi(\lambda)$ belongs to class L_x^2. We find a sequence $\{x(t)\}$ subordinate to $\{u(t)\}$ for which

$$\varphi_u^{(x)}(\lambda) = \frac{1}{\varphi(\lambda)}$$

From the equality (8.43) and theorem 17 C it follows, that $\{u(t)\}$ is a fundamental sequence equivalent to $\{x(t)\}$. By theorem 10

---

1) See (5), page 15.

2) This inequality we use in the following form: if

$$m = \int_a^b P(x)f(x)dx, \quad \log S = \int_a^b P(x)\log f(x)dx,$$

$$\int_a^b P(x)dx = 1\;, \quad P(x) \ge 0$$

then

$$S \le m$$

3) See (6) remark C on page 94.

$$\varphi_x^{(u)}(\lambda) = \frac{1}{\varphi_x^{(u)}(\lambda)} = \varphi(\lambda) = \Gamma(e^{-i\lambda})$$

Thus the sequence $\{u(t)\}$ satisfies all conditions of the second part of theorem 20. **Consequently** $\{x(t)\}$ is represented by $\{u(t)\}$ in the form (8.1) which also proves the sufficiency of the condition of theorem 22. It is easy to see that the sequence $\{u(t)\}$ constructed by us coincides in fact with $\{u_x(t)\}$.

We observe further, in conclusion, the formula following from (8.30) and (8.35).[1)]

$$c_0^{(x)} = \sqrt{2\pi}\, c^{Q_x(0)} = \sqrt{2\pi}\, e^{\frac{a_0^{(x)}}{2}}$$

$$= \sqrt{2\pi}\, \exp\{\frac{1}{4\pi} \int_{-\pi}^{\pi} \log f_{xx}(\lambda)d\lambda\} \qquad (8.44)$$

9. The Spectral Characteristic of a Singular Sequence and of the Singular Component

In view of the results of †7, every nonsingular stationary sequence $\{x(t)\}$ is representable in the form

$$x(t) = s_x(t) + r_x(t) \qquad (9.1)$$

where

$$r_x(t) = \sum_{n=0}^{\infty} c_n^{(x)} u_x(t-n) \qquad (9.2)$$

The sequence $\{r_x(t)\}$ we shall call the regular component[2)] of the sequence $\{x(t)\}$. It is clear that it is always subordinate to $\{x(t)\}$ and regular.

We put for an arbitrary stationary sequence $\{x(t)\}$:

$$A_{xx}(\lambda) = \int_{-\pi}^{\lambda} f_{xx}(\lambda)d\lambda \qquad (9.3)$$

$$D_{xx}(\lambda) = F_{xx}(\lambda) - A_{xx}(\lambda) \qquad (9.4)$$

We shall distinguish the following three cases:

1) $f_{xx}(\lambda) = 0$ on a set of positive **measure;**

2) $f_{xx}(\lambda) = 0$ only on a set of zero measure, but the function $\log f_{xx}(\lambda)$ is not summable.

3) $f_{xx}(\lambda) = 0$ only on a set of measure zero and the function $\log f_{xx}(\lambda)$ is summable.

THEOREM 23 In the cases 1 and 2 the sequence $\{x(t)\}$ is singular. In the case 3 the sequence $\{x(t)\}$ is not singular

$$F_{s_x s_x}(\lambda) = D_{xx}(\lambda) \quad F_{r_x r_x}(\lambda) = A_{xx}(\lambda) \qquad (9.5)$$

and the coefficients $c_0^{(x)}$ of the expansion (9.2) are defined from formula (8.16) where the function $\Gamma_x(\zeta)$ is defined by formulae (8.29), (8.30) and (8.31).

Proof: We assume that the sequence $\{x(t)\}$ is not singular. Then in view of the regularity of the sequence $\{r_x(t)\}$ according to theorem 22 the function $f_{r_x r_x}(\lambda)$ is almost everywhere positive and the function $\log f_{r_x r_x}(\lambda)$ is summable. Since the sequences $\{s_x(t)\}$ and $\{r_x(t)\}$ are orthogonal to each other (see †7), then according to theorem 11

$$F_{xx}(\lambda) = F_{s_x s_x}(\lambda) + F_{r_x r_x}(\lambda) \qquad (9.6)$$

and consequently, almost everywhere

$$f_{xx}(\lambda) = f_{s_x s_x}(\lambda) + f_{r_x r_x}(\lambda) \qquad (9.7)$$

From (9.7) it follows that almost everywhere

$$f_{xx}(\lambda) \geq f_{r_x r_x}(\lambda) > 0 \qquad (9.8)$$

since every set of positive Lebesgue measure has a positive measure also relative to $F_{xx}(\lambda)$.

Since $\{s_x(t)\}$ and $\{r_x(t)\}$ are subordinate to $\{x(t)\}$ then according to theorem 12 almost everywhere relative to $F_{xx}(\lambda)$ and consequently also in the usual Lebesgue sense we have:

$$\frac{dF_{s_x s_x}(\lambda)}{dF_{xx}(\lambda)} \cdot \frac{dF_{r_x r_x}(\lambda)}{dF_{xx}(\lambda)} = 0$$

---

1) This formula is true not only for regular but also for arbitrary nonsingular sequences as is clear from theorem 23 of the following paragraph.

2) Since a sequence identically equal to zero is singular but not regular, then in a regular sequence, the singular component is equal to zero and in a singular sequence there is no regular component.

Since in view of (9.8) almost everywhere

$$\frac{dF_{r_x r_x}(\lambda)}{dF_{xx}(\lambda)} = \frac{f_{r_x r_x}(\lambda)}{f_{xx}(\lambda)} > 0$$

then almost everywhere

$$\frac{dF_{s_x s_x}(\lambda)}{dF_{xx}(\lambda)} = \frac{f_{s_x s_x}(\lambda)}{f_{xx}(\lambda)} = 0$$

i.e.

$$f_{s_x s_x}(\lambda) = 0 \tag{9.9}$$

$$f_{xx}(\lambda) = f_{r_x r_x}(\lambda) \tag{9.10}$$

From formula (9.10) the positivity almost everywhere of $f_{r_x r_x}(\lambda)$ and the summability of $\log f_{r_x r_x}(\lambda)$ it follows that for nonsingular sequences almost everywhere case 3 holds.

We show that conversely in case 3 the sequence $\{x(t)\}$ is nonsingular. According to theorem 14 in the case 3 the sequence $\{x(t)\}$ uniquely decomposes into a sum

$$x(t) = s(t) + r(t)$$

of mutually orthogonal sequences subordinate to $\{x(t)\}$ for which

$$F_{ss}(\lambda) = D_{xx}(\lambda)\ ,\ F_{rr}(\lambda) = A_{xx}(\lambda)$$

According to theorem 22 the sequence $\{r(t)\}$ is regular.

Consequently

$$x(t) = s(t) + \sum_{n=0}^{\infty} c_n^{(r)} u_r(t-n) \tag{9.11}$$

Formula (9.11) shows that the space $H_x(t)$ lies in the space[1] $S_s \oplus H_r(t)$. Putting $x(t+1)$ in the form

$$x(t+1) = \alpha + \beta$$

where

$$\alpha = c_0^{(r)} u_r(t+1)\ ,\ \beta = s(t+1) + \sum_{n=1}^{\infty} c_n^{(r)} u_r(t-n+1)$$

1) $\mathcal{S}_1 \oplus \mathcal{S}_2$ denotes the minimal linear closed subspace of the space H containing $\mathcal{S}_1$ and $\mathcal{S}_2$.

it is easy to establish that β lies in the space $S_x \oplus H_{u_r}(t)$ but is orthogonal to this space. Consequently $x(t+1)$ does not belong in the space $S_x \oplus H_{u_r}(t)$ and what is more to the space $H_x(t)$. From this it follows that the sequence $\{x(t)\}$ is nonsingular.

For an arbitrary nonsingular sequence $\{x(t)\}$ we proved above formula (9.10). Consequently the function $\Gamma_x(\zeta)$ defined in connection with (8.29), (8.30) and (8.31) coincides with $\Gamma_{r_x}(\zeta)$. In order to prove that the coefficients $c_n^{(x)}$ are obtained from the expansion (8.16) of the function $\Gamma_x(\zeta)$ it remains only to establish that

$$c_n^{(x)} = c_n^{(r_x)} \tag{9.12}$$

For this it is sufficient to verify the expansion (9.2) satisfies the conditions W_{1-5} (with the singular component equal to zero). The fulfillment of the conditions W_1, W_2, W_4 and W_5 is clearly immediate. We prove the observance for the development (9.2) of the property W_3. For this we remark that in view of (9.8) and theorem 17A the sequence $\{u_x(t)\}$ is subordinate to $\{r_x(t)\}$. From (9.1) and (9.2) it is clear that $H_x(t)$ is contained in $S_x \oplus H_{r_x}(t)$. Since $u_x(t)$ belongs in $H_x(t)$ and is orthogonal to S_x (see †7) then $u_x(t)$ belongs to $H_{rx}(t)$ and this also means that the expansion (9.2) satisfies the condition W_3.

The formula (9.5) for a nonsingular sequence follows immediately from (9.1), (9.10) and the absolute continuity of $F_{r_x r_x}(\lambda)$ (the last is guaranteed by the regularity of the sequence $r_x(t)$ by theorem 16).

†10. Minimal Sequences

For an arbitrary stationary sequence $\{x(t)\}$ we shall denote by $\hat{H}_x(t)$ the minimal linear closed subspace of the space containing all $x(s)$ for which $s \neq t$.

Since

$$U_x^k \hat{H}_x(t) = \hat{H}_x(t+k) \tag{10.1}$$

then only one of two cases are possible: either the whole of $\hat{H}_x(t)$ coincides with H_x or the whole of $\hat{H}_x(t)$ is different from H_x.

DEFINITION 10 A stationary sequence is called minimal if

$$\hat{H}_x(t) \neq H_x \qquad (10.2)$$

Every element $x(t)$ of the sequence $\{x(t)\}$ is uniquely representable in the form of a sum:

$$x(t) = v(t) + \delta(t) \qquad (10.3)$$

where $v(t)$ lies in $\hat{H}_x(t)$ and $\delta(t)$ is orthogonal to $\hat{H}_x(t)$. We put

$$d_x = |\delta(t)| \qquad (10.4)$$

clearly the condition

$$d_x > 0 \qquad (10.5)$$

is necessary and sufficient for the minimality of the sequence $\{x(t)\}$.

THEOREM 24 The stationary sequence $\{x(t)\}$ is minimal in that and only that case when the function $f_{xx}(\lambda)$ is almost everywhere positive and the integral

$$\int_{-\pi}^{\pi} \frac{d\lambda}{f_{xx}(\lambda)} \qquad (10.6)$$

is finite.

If this condition is fulfilled then

$$d_x^2 = (2\pi)^2 / \int_{-\pi}^{\pi} \frac{d\lambda}{f_{xx}(\lambda)} \qquad (10.7)$$

We prove from the beginning the sufficiency of the condition of the theorem. Let

$$\varphi(\lambda) = \frac{1}{f_{xx}(\lambda)\left(\int_{-\pi}^{\pi} \frac{d\lambda}{f_{xx}(\lambda)}\right)^{1/2}}$$

if $f_{xx}(\lambda)$ is defined and finite and

$$\varphi(\lambda) = 0$$

if $f_{xx}(\lambda)$ is not defined or infinite. Then the integral

$$\int_{-\pi}^{\pi} |\varphi(\lambda)|^2 dF_{xx}(\lambda) = 1$$

is finite. Consequently $\varphi(\lambda)$ belongs to the class L_x^2 and the equality

$$\varphi_y^{(x)}(\lambda) = \varphi(\lambda)$$

defines some sequence $\{y(t)\}$ subordinate to $\{x(t)\}$ for which according to formula (4.4)

$$F_{yx}(\lambda) = \int_{-\pi}^{\lambda} \varphi_y^{(x)}(\lambda) dF_{xx}(\lambda)$$

$$= (\lambda+\pi) / \left(\int_{-\pi}^{\pi} \frac{d\lambda}{f_{xx}(\lambda)}\right)^{1/2}$$

According to (3.3) for $s \neq t$

$$(y(t), x(s)) = B_{yx}(t-s)$$

$$= \int_{-\pi}^{\pi} e^{i(t-s)\lambda} dF_{yx}(\lambda) = 0$$

Consequently the element $y(t)$ is orthogonal to the space.

Since

$$|y(t)|^2 = B_{yy}(0) = \int_{-\pi}^{\pi} dF_{yy}(\lambda)$$

$$= \int_{-\pi}^{\pi} |\varphi_y^{(x)}(\lambda)|^2 dF_{xx}(\lambda) \neq 0$$

then the space H_x containing $y(t)$ must not coincide with $H_x(t)$. The minimality of the sequence $\{x(t)\}$ is proved.

In order to prove the necessity of the condition of the theorem, we assume that the sequence $\{x(t)\}$ is minimal. It is easy to see that the sequence $\{\delta(t)\}$ is stationary and subordinate to $\{x(t)\}$. In this

$$B_{\delta_x}(k) = (\delta(t+k), x(t)) = 0 \text{ for } k \neq 0 \qquad (10.8)$$

$$B_{\delta_x}(0) = (\delta(t), x(t)) = |\delta(t)|^2 = d_x^2 \qquad (10.9)$$

Hence it follows by formulae (3.9), (3.10) and (3.11) that

$$F_{\delta_x}(\lambda) = d_x^2\left(\frac{\lambda+\pi}{2\pi}\right) \qquad (10.10)$$

At the same time according to formula (4.4)

$$F_{\delta_x}(\lambda) = \int_{-\pi}^{\pi} \varphi_\delta^{(x)}(\lambda) dF_{xx}(\lambda) \qquad (10.11)$$

From (10.10) and (10.11) it follows that

$$\lambda + \pi = \frac{2\pi}{d_x^2} \int_{-\pi}^{\lambda} \varphi_\delta^{(x)}(\lambda)dF_{xx}(\lambda) \qquad (10.12)$$

and consequently the derivative

$$\frac{d\lambda}{dF_{xx}(\lambda)} = \frac{2\pi}{d_x^2} \varphi_\delta^{(x)}(\lambda) \qquad (10.13)$$

is finite almost everywhere relative to $F_{xx}(\lambda)$. Since in view of (10.12) λ is absolutely continuous relative to $F_{xx}(\lambda)$, the derivative $d\lambda/dF_{xx}(\lambda)$ is finite almost everywhere relative to λ i.e. in the usual Lebesgue sense. Consequently the derivative

$$f_{xx}(\lambda) = \frac{dF_{xx}(\lambda)}{d\lambda}$$

is almost everywhere different from zero. After this remark there is obtained from (10.13) that almost everywhere

$$\varphi_\delta^{(x)}(\lambda) = \frac{d_x^2}{2\pi}\frac{d\lambda}{dF_{xx}(\lambda)} = \frac{d_x^2}{2\pi}\frac{dF_{xx}(\lambda)}{d\lambda}$$

$$= \frac{d_x^2}{2\pi f_{xx}(\lambda)}$$

According to theorem 8 the integral

$$\int_{-\pi}^{\pi} |\varphi_\delta^{(x)}(\lambda)|^2 dF_{xx}(\lambda) = \int_{-\pi}^{\pi} \left|\frac{d_x^2}{2\pi f_{xx}(\lambda)}\right|^2 dF_{xx}(\lambda)$$

$$= \frac{d_x^4}{(2\pi)^2} \int_{-\pi}^{\pi} \frac{d\lambda}{f_{xx}(\lambda)} \qquad (10.14)$$

is finite. The necessity of the condition of the theorem is proved.

Remarking that

$$d_x^2 = |\delta(t)|^2 = B_{\delta\delta}(0) = \int_{-\pi}^{\pi} dF_{\delta\delta}(\lambda)$$

$$= \int_{-\pi}^{\pi} |\varphi_\delta^{(x)}(\lambda)|^2 dF_{xx}(\lambda) \qquad (10.15)$$

from (10.14) and (10.15) we obtain formula (10.7).

Komarovka

7 September-21 October 1940

CITED LITERATURE

1. H. Wold. A study in the analysis of stationary time series, Uppsala, (1938).

2. H. Cramér. Annals of Mathematics, 41, (1940), 215-230.

3. H. Stone. Linear transformations in Hilbert space. American Mathematical Society, Colloquium Publications, 15, (1932).

4. A. Zygmund. Trigonometrical Series.

5. I.I. Privalov. Cauchy integral. Saratov, (1919).

6. F. Riesz. Mathematische Zeitschrift, 18, (1923), 87-95.

7. A.N. Kolmogorov. Izvestia A.N. SSSR (mathematical series) 5, (1941), 3-14.

4

Reprinted from *C. R. (Dokl.) Akad. Nauk SSSR* **46**(8):306–309 (1945)

ON A PROBLEM OF EXTRAPOLATION OF A. N. KOLMOGOROFF

By M. KREIN

(Communicated by A. N. Kolmogoroff, Member of the Academy, 18. VIII. 1944)

Let $\sigma(\lambda)=\sigma(\lambda-0)$ $(-\infty<\lambda<\infty,\ \sigma(-\infty)=0)$ be a certain bounded non-decreasing function, and L_σ, the Hilbert space of all σ-measurable and σ-integrable together with their squares functions $\varphi(\lambda)$ $(-\infty<\lambda<\infty)$ with the usual definition of the scalar product:

$$(\varphi,\psi)_\sigma=\int_{-\infty}^{\infty}\varphi(\lambda)\overline{\psi(\lambda)}\,d\sigma(\lambda)\qquad(\varphi,\psi\in L_\sigma)$$

By $L_\sigma(E)$, where E is a certain set of real numbers t, we denote the linear closed envelope in L_σ of all elements $\exp(it\lambda)\in L_\sigma$, where $t\in E$.

A. Kolmogoroff proposed * the following general problem from the theory of statinary random functions, which may be formulated in analytical form as follows:

(K) *Two sets E_1 and E_2 of real numbers are given. To find the conditions which must be satisfied by a non-decreasing bounded function $\sigma(\lambda)=\sigma(\lambda-0)$ $(-\infty<\lambda<\infty)$ in order that $L_\sigma(E_1)=L_\sigma(E_2)$.*

In the case when E_1 is the sequence of all integers (positive and negative) less than a given one, and E_2 is the sequence of all integers, as well as in the case when E_1 consists of a finite number of consecutive integers, and E_2 is the sequence of all other integers, the problem (K) was solved by A. Kolmogoroff in 1939 ([1]) (a detailed solution is to be found in ([2],[3]) **.

In the same year, solving the problem of prolongation of Hermite-positive functions, the author obtained accidentally (quite unconnected with the investigations of A. Kolmogoroff) the solution of the problem (K)in the case, where E_1 is a finite interval and E_2, the whole infinite real axis, but for want of space ([4]) omitted to formulate the result. At present the solution of the problem (K) in this case (and a somewhat more general problem) is stated by the author at the end of his note ([5]) on the problem of continuation of helics in the Hilbert space.

A. Kolmogoroff drew the attention of the author to the special interest, which the problem (K) represents in the case where E_1 is a half-axis, and E_2 $(\supset E_1)$, the whole axis.

In the present note we indicate the solution of the problem (K) in this case ***, which is obtained as a corollary of the solution of a more general question (cf. Theorem 1 and its corollary).

---

* In a letter to the author.

** In the second case A. Kolmogoroff published the solution under the condition that E_1 consists only of one integer ([2], [3]).

*** Without reducing the generality we may assume that E_1 is the positive half-axis $t\geqslant 0$.

1. Denote by H_ζ the class of functions $\varphi(\zeta)$ $(I\zeta > 0)$ holomorphic in the upper half-plane, such that $|\varphi(\zeta)|^2$ possesses in this half-plane a harmonical majorant. A holomorphic function $\varphi(\zeta)$ $(I\zeta > 0)$ belongs to the class H_ζ if and only if the function

$$f(z) = \varphi(\zeta), \quad \text{where} \quad \zeta = i\frac{1-z}{1+z}, \quad z = \frac{i-\zeta}{i+\zeta} (|z| < 1)$$

belongs to the class H_z, where by H_z is denoted the class of all holomorphic functions $f(z)$ $(|z| < 1)$ such that

$$f(z) = c_0 + c_1 z + c_2 z^2 + \ldots, \quad \text{where} \quad \sum_{k=0}^{\infty} |c_k|^2 < \infty$$

Another necessary and sufficient condition for the function $\varphi(\zeta)$ to belong to H_ζ consists in that $(\zeta + i)^{-1}\varphi(\zeta) \in \mathfrak{H}_\zeta$ *, where $\mathfrak{H}_\zeta$ denotes the class of all holomorphic functions $\psi(\zeta)$ $(I\zeta > 0)$ such that

$$\int_{-\infty}^{\infty} |\psi(\lambda + i\mu)|^2 \, d\lambda < M_\psi < \infty, \quad \mu > 0$$

Every function $\varphi(\zeta) \in \mathfrak{H}_\zeta$ possesses almost everywhere on the real axis limit values $\varphi(\lambda)$, to which it tends on all non-tangent paths. The same relates, of course, also to the functions $\psi(\zeta) \in \mathfrak{H}_\zeta$.

By H_λ (correspondingly, H_θ and $\mathfrak{H}_\lambda$) we denote the set of all boundary functions for the functions $\varphi(\zeta) \in H_\zeta$ (correspondingly, $f(z) \in H_z$ and $\psi(\zeta) \in \mathfrak{H}_\zeta$).

In the sequel L_λ denotes a certain linear set of functions from L_σ, possessing the following properties:

1) $1 \in L_\lambda$,

2) if $\varphi(\lambda) \in L_\lambda$, then $\exp(it\lambda)\varphi(\lambda) \in L_\lambda$ for $t \geqslant 0$.

Theorem 1. *The set $L_\lambda \subset L_\sigma$ is not dense in L_σ if and only if the following two conditions are satisfied*:

$$\text{a)} \qquad -\infty < \int_{-\infty}^{\infty} \frac{\log \sigma'(\lambda)}{1+\lambda^2} \, d\lambda \, (< \infty), \tag{1}$$

b) *there exists a function* $\Gamma(\lambda) \in H_\lambda$ *such that* $\int_{-\infty}^{\infty} \frac{|\Gamma(\lambda)|^2}{\sigma'(\lambda)} \frac{d\lambda}{1+\lambda^2} < \infty$ *and* $\Gamma(\lambda)\varphi(\lambda) \in H_\lambda$ *for any* $\varphi \in L_\lambda$.

Proof. Denote by $\overline{L}_\lambda$ the closure of L_λ in L_σ. In virtue of 2), if $\varphi \in \overline{L}$, then also $\exp(it\lambda)\varphi(\lambda) \in \overline{L}_\lambda$ for $t \geqslant 0$. And since every function (λ) $(-\infty < \lambda < \infty)$ of the form

$$p(\lambda) = c + \int_{-\infty}^{\infty} \exp(it\lambda) q(t) \, dt, \quad \text{where} \quad \int_{-\infty}^{\infty} |q(t)| \, dt < \infty \tag{2}$$

is the limit of a certain uniformly convergent in every finite interval and uniformly bounded sequence of trigonometrical polynomials $P_n(\lambda)$ $(n = 1, 2, \ldots)$ of the form

$$P_n(\lambda) = \sum a_k^{(n)} \exp(it_k^{(n)}\lambda) \qquad (t_k^{(n)} \geqslant 0),$$

we must have that together with $\varphi \in \overline{L}_\lambda$ also $p(\lambda)\varphi(\lambda) \in \overline{L}_\lambda$.

---

* This fact may be, for instance, deduced from Theorems VII, IX of the paper of Kryloff (9).

In particular, putting in (2) $c=-1$, $q(t)=\exp(-t)$ $(-\infty<t<\infty)$, we find $p(\lambda)=\frac{i-\lambda}{i+\lambda}$ and, consequently,

$$\frac{i-\lambda}{i+\lambda}\varphi(\lambda)\in L_\lambda, \quad \text{if} \quad \varphi(\lambda)\in \overline{L_\lambda} \tag{3}$$

Put

$$\tau(\theta)=\sigma\left(\operatorname{tg}\frac{\theta}{2}\right) \qquad (-\pi\leqslant\theta\leqslant\pi) \tag{4}$$

and consider the Hilbert space L_τ of all τ-measurable and τ-integrable together with their squares functions $g(\theta)$ $(-\pi\leqslant\theta\leqslant\pi)$. Evidently, the transformation

$$g(\theta)=\varphi(\lambda), \quad \lambda=\operatorname{tg}\frac{\theta}{2}=\frac{1}{i}\,\frac{e^{i\theta}-1}{e^{i\theta}+1}$$

establishes a linear isometrical correspondence T $(g=T\varphi)$ between L_σ and L_τ.

Let $L_\theta=TL_\lambda$, and consequently $\overline{L_\theta}=T\overline{L_\lambda}$ ($\overline{L_\theta}$ being the closure of L_θ in L_τ). In virtue of the properties 1) and 3) of the aggregate L_λ $1\in L_\theta$ and, if $g\in L_\theta$, then $\exp(i\theta)\,g(\theta)\in L_\theta$. Consequently, by Theorem 1 of our note (7) L_θ is not dense in L_τ if and only if

α) $-\infty<\int\limits_{-\pi}^{\pi}\log\tau'(\theta)\,d\theta$

β) there exists a function $G(\theta)\in H_\theta$ such that

$$\int\limits_{-\pi}^{\pi}\frac{|G(\theta)|^2}{\tau'(\theta)}\,d\theta<\infty$$

and $G(\theta)\,g(\theta)\in H_\theta$ for any $g(\theta)\in L_\lambda$.

But if $\overline{L_\theta}\neq L_\tau$, then $\overline{L_\lambda}\neq L^\sigma$, and conversely; on the other hand, in view of (4), the condition a) is equivalent to the condition α), and the condition β), to the condition b), if we observe the transformation $G(\theta)=\Gamma(\lambda)$ $(\lambda=\operatorname{tg}\theta/2)$, the transformation (4) and what has been stated above on the classes H_θ and H_λ.

The theorem is proved.

C o r o l l a r y. In order that the linear envelope L_λ^+ of the set of functions $\exp(it\lambda)\in L_\sigma$ $(t\geqslant 0)$ should not be dense in L_σ, it is necessary and sufficient that the function $\sigma(\lambda)$ should satisfy the condition (1).

2. If the conditions a) and b) of Theorem 1, or, which is the same thing, the conditions α) and β), are satisfied, we may assert in view of Theorem 2 (6) that there exists a convergent Blashke product

$$B(z)=z^p\prod\frac{a_k-z}{1-\bar{a}_k z}\cdot\frac{|a_k|}{a_k} \qquad (p\geqslant 0,\ 0<|a_k|<1,\ k=1,2,\ldots)$$

such that the values of every function $B(e^{i\theta})f(e^{i\theta})$ $(f(e^{i\theta})\in L_\theta)$ coincide almost everywhere with the limit values of a certain meromorphic function in the unit circle of a bounded type. Moreover, if $B(z)$ is chosen in such a way that it possesses no «superfluous» zeroes, and if we put

$$F_\tau(z)=\sqrt{2\pi}\,B(z)\exp\left\{\frac{1}{4\pi}\int\limits_{-\pi}^{\pi}\frac{e^{i\theta}+z}{e^{i\theta}-z}\log\tau'(\theta)\,d\theta\right\} \tag{5}$$

then the function $F_\tau(z)$ will possess the following properties:

I′. $F_\tau(z)\in H_z$, $F_\tau(e^{i\theta})f(e^{i\theta})\in H_\theta$ for any $f(e^{i\theta})\in L_\theta$.

II′. $\tau'(\theta)=\frac{1}{2\pi}|F(e^{i\theta})|^2$ (almost everywhere).

III'. If we put $\Phi_\tau(e^{i\theta}) = F^{-1}(e^{i\theta})$ at the points θ, where $F_\tau(e^{i\theta}) \neq 0$ and where there exists a finite derivative $\tau'(\theta)$, and put $\Phi_\tau(e^{i\theta}) = 0$ at all the remaining points, then $\Phi_\tau(e^{i\theta}) \in \overline{L_\theta}$.

Putting $\sqrt{2}F_\tau(z) = (\zeta + i)D(\zeta)$, $B(z) = b(\zeta)$, $z = \frac{i-\zeta}{i+\zeta}$ $(I_\zeta > 0)$, we obtain from (4) and (5)

$$D(\zeta) = \sqrt{2\pi}\, b(\zeta) \exp\left\{\frac{1}{4\pi i}\int_{-\infty}^{\infty} \frac{1+\lambda\zeta}{\lambda-\zeta}\, \frac{\log \sigma'(\lambda)}{1+\lambda^2}\, d\lambda\right\} \tag{6}$$

In virtue of I' — III' we shall have:

I. $D(\zeta) \in H_\zeta$, $D(\lambda)\varphi(\lambda) \in H_\lambda$ for any $\varphi(\lambda) \in \overline{L_\lambda}$.

II. $\sigma'(\lambda) = \frac{1}{2\pi}|D(\lambda)|^2$ (almost everywhere).

III. If we put $Q(\lambda) = (i+\lambda)^{-1}D^{-1}(\lambda)$ for those λ for which $D^{-1}(\lambda) \neq 0$ and for which there is a finite derivative $\sigma'(\lambda)$, and put $Q(\lambda) = 0$ for the rest of λ, then $Q(\lambda) \in \overline{L_\lambda}$.

Since $D(\zeta) \in H_\zeta$ (cf. [8]),

$$D(\zeta) = \int_0^\infty \exp(i\zeta t)\,\gamma(t)\,dt, \quad D(\lambda) = \underset{A\to\infty}{\mathrm{l.i.m.}} \int_{-A}^{A} \exp(it\lambda)\,\gamma(t)\,dt$$

where

$$\gamma(t) = \underset{A\to\infty}{\mathrm{l.i.m.}}\, \frac{1}{2\pi}\int_{-A}^{A} \exp(-i\lambda t)\,\gamma(t)\,dt, \quad \int_{-\infty}^{\infty} |\gamma(t)|^2\,dt = \frac{1}{2\pi}\int_{-\infty}^{\infty} |D(\lambda)|^2\,d\lambda$$

Theorem 2 *. *The distance* $d(s)$ $(s > 0)$ *of the element* $\exp(is\lambda)$ *from* L *in the metrics of the space* L_σ *is given by the formula*

$$d(s) = \int_0^s |\gamma(t)|^2\,dt \qquad (s > 0)$$

For want of space we omit the proof of Theorem 2.

Remark. From the construction of $B(z)$ and $b(\zeta)$ follows that if L_λ is L_λ^+, *i. e.* the linear envelope of the elements $\exp(it\lambda)$ $(t \geqslant 0)$, then in (6) we have to put $b(s) \equiv 1$. In this case, as may be easily seen, $d(s) > 0$ for all $s > 0$.

The results obtained by us may be interpreted as certain propositions relating to the problem of continuation of semi-infinite unitary lines ξ_t $(-\infty < t \leqslant a)$ (cf. [5]). These propositions may be generalized to the case of semi-infinite helical arcs.

The V. Kuibyshev Industrial Institute.
Kuibyshev.

Received
20.IV.1944.

REFERENCES

[1] A. Kolmogoroff, C. R., **208**, 2043 (1939). [2] А. Н. Колмогоров, Бюлл. МГУ, **2**, 6 (1941). [3] A. N. Kolmogoroff, Bull. Acad. Sci. URSS, sér. math., **5**, No. 1 (1941). [4] M. Krein, C. R. Acad. Sci. URSS, XXVI, No. 1 (1940.) [5] M. Krein, ibid., XLV, No. 4 (1944). [6] V. I. Kryloff Rec. Math., **6** (48), 1, 95 (1939). [7] M. Krein, C. R. Acad. Sci. URSS. XLVI, No. 3 (1945). [8] R. E. A. C. Paley and N. Wiener, Fourier Transforms in the Complex Domain, N. Y., 1934.

* Cp. Theorem 2 of A. Kolmogoroff's note [3].

5

Reprinted with the permission of the publisher from pp. 127–131 of *Selected Translations in Mathematical Statistics and Probability*, Volume 4, American Mathematical Society, 1963, 393 pp.

ON A FUNDAMENTAL APPROXIMATION PROBLEM IN THE THEORY OF EXTRAPOLATION AND FILTRATION OF STATIONARY RANDOM PROCESSES

M. G. KREǏN

Let $\sigma(\lambda)$ $(-\infty < \lambda < \infty)$ be a nondecreasing function which satisfies the condition

$$\int_{-\infty}^{\infty} \frac{d\sigma(\lambda)}{1+\lambda^2} < \infty. \tag{1}$$

Denote by Λ_∞ the Hilbert space of functions $F(\lambda)$ $(-\infty < \lambda < \infty)$ σ-measurable and σ-integrable, together with their squares, with the usual definition of the scalar product (F, G) $(F, G \in \Lambda_\infty)$ as the integral over the whole axis of $F(\lambda)\overline{G(\lambda)}d\sigma(\lambda)$.

Let $I = (a, b)$ $(-\infty \leq a < b < \infty)$ be some interval of the real axis $-\infty < t < \infty$. Denote by Λ_I the closed linear manifold in Λ_∞ of the family of functions of argument λ:

$$\frac{\exp(it_2\lambda) - \exp(it_1\lambda)}{i\lambda} = \int_{t_1}^{t_2} \exp(i\lambda s)\; ds \quad (t_1, t_2 \in I).$$

Note that if $\sigma(\lambda)$ is a function of bounded variation, then Λ_I can be defined more simply, viz. as the closed linear manifold in Λ_∞ of the family of functions $\exp(it\lambda)$ $(t \in I)$. Two questions can be posed:

I. *Under what criteria $\Lambda_I = \Lambda_\infty$?*

II. *If $\Lambda_I \neq \Lambda_\infty$, then how to express analytically the orthogonal projection $P_I F$ of any element $F \in \Lambda_\infty$ on Λ_I?*

For the case of $I = (-\infty, a)$ the first question was already answered by the author in 1944 [1] and (when $\sigma(\lambda)$ is of bounded variation) he also obtained $P_I F$ for $F(\lambda) = \exp(it\lambda)$ $(t > a)$*. For special cases of functions σ and F the question of the effective determination of $P_I F$ for $I = (-\infty, a)$ was studied by N. Wiener [3] who explained its applications in technology (see also [4]). The analytically formulated problems I and II can be reformulated as problems in the theory of prediction (extrapolation) and filtration of stationary random processes from their observation in the time interval I. Naturally problems I and II are of greatest interest when I is a finite interval. However their investigation is particularly difficult in this case. Unaware of the applications which problems I and II have, the author had already obtained in 1940 [5] and then in 1943–1944 [6] a series of essential results directly connected with them. Certain results have been obtained

*These results were obtained by the author in answer to the question raised by A. N. Kolmogorov [1]; K. Karhunen [2] repeated them in 1950.

in recent times which give the complete solution of these problems. It is derived below for an odd function $\sigma(\lambda)$ because the investigation of the solution in the most general case would require us to undertake much preliminary work. Besides in problems of the theory of real stationary random processes only this case occurs.

In considering a finite interval I, it is possible without loss of generality to assume that $I = (-a, a)$ and then it will be convenient to write Λ_a and P_a respectively for Λ_I and P_I.

1. *Solution of problem* I. Thus let $\sigma(-\lambda) = -\sigma(\lambda)$ $(0 < \lambda < \infty;\ \sigma(0) = 0)$. Assume $\tau(\lambda) = 2\sigma(\sqrt{\lambda} - 0)$ $(0 < \lambda < \infty;\ \tau(0) = 0)$. By virtue of (1)

$$\int_0^\infty d\tau(\lambda)/(1+\lambda) < \infty.$$

Consequently [7,8] $\tau(\lambda)$ will be the *principal spectral function* of some string S stretched with unit force between the points $x = 0$ and $x = l\ (\leq \infty)$ where the length l and mass $M(x)$ of any right open interval $[0, x)$ $(0 < x < l;\ M(0) = 0)$ of the string S will be uniquely determined by the function $\tau(\lambda)$.

Theorem 1. *For* $\Lambda_a = \Lambda_\infty$ *it is necessary and sufficient that*

$$t(x) = \int_0^x \sqrt{M'(s)}\,ds \leq a \quad (0 < x \leq l)$$

and that for all $x = x_a$, *for which* $t(x) = a$; $M(x_a) = M(l)$.

2. Let us consider the case $\Lambda_a \neq \Lambda_\infty$.

Denote by x_{a-0} and x_{a+0} respectively the least and greatest of the roots of the equation $t(x) = a$. Thus if $x_{a-0} < x_{a+0}$ then in the interval (x_{a-0}, x_{a+0}) the function $M(x)$ will have a derivative almost everywhere equal to zero but this property no longer holds in any larger interval containing the interval (x_{a-0}, x_{a+0}).

Let us find the solution of the integral equation

$$\phi(x;\lambda) = 1 - \lambda \int_0^x (x-s)\,\phi(s;\lambda)\,dM(s) \quad (0 \leq x < l).$$

Denote by $L_M^{(2)}(0, \xi)$, where $0 < \xi \leq l$, the set of all functions $f(x)$ $(0 \leq x \leq \xi)$ which are M-measurable and which satisfy the condition

$$\int_0^\xi |f(x)|^2\,dM(x) < \infty.$$

Theorem 2. *Entire functions* $F(\lambda)$ *of form*

$$F(\lambda) = \int_0^{x_{a-0}} f(x)\,\phi(x;\lambda^2)\,dM(x) \quad (f \in L_M^{(2)}(0, x_{a-0}))$$

exhaust all even functions of Λ_a.

Entire functions $G(\lambda)$ *of form*

$$G(\lambda) = \frac{1}{\lambda}\int_0^{x_{a-0}} g(x)\,\phi'(x;\lambda^2)\,dx \quad \left(\int_0^{x_{a-0}} |g(x)|^2 dx < \infty;\ \phi' = \frac{d\phi}{dx}\right)$$

exhaust all odd functions of Λ_a.

Theorem 3. *Entire functions* $F(\lambda)$ *of form*

$$F(\lambda) = \int_0^{x_{a+0}} f(x)\,\phi(x;\lambda^2)\,dM(x) \quad (f \in L_M^{(2)}(0, x_{a+0}))$$

exhaust all even entire functions $F(\lambda)$ *satisfying the conditions*

$$\int_{-\infty}^{\infty} |F(\lambda)|^2 d\sigma(\lambda) < \infty, \quad \overline{\lim_{\zeta\to\infty}} \frac{\ln|F(\zeta)|}{|\zeta|} \leq a \quad (\zeta = \lambda + i\mu). \tag{2}$$

Entire functions $G(\lambda)$ *of form*

$$G(\lambda) = \frac{1}{\lambda}\int_0^{x_{a+0}} g(x)\,\phi'(x;\lambda^2)\,dx \quad \left(\int_0^{x_{a+0}} |g(x)|^2 dx < \infty\right)$$

exhaust all odd entire functions $G(\lambda)$ *satisfying condition* (2).

For $d\sigma = d\lambda/\pi$ we will have: $l < \infty$, $dM = dx$, $\phi(x;\lambda^2) = \cos\lambda x$, $\phi'(x;\lambda^2)/\lambda = -\sin\lambda x$, and Theorems 2, 3 transform into the well known Paley–Wiener [9] theorem.

3. *Solution of problem* II. Let $F \in \Lambda_\infty$. Form the functions

$$f(x) = \underset{N\to\infty}{\text{l.i.m.}} \int_{-N}^{N} F(\lambda)\,\phi(x;\lambda^2)\,d\sigma(\lambda), \quad g(x) = \underset{N\to\infty}{\text{l.i.m.}} \int_{-N}^{N} F(\lambda)\,\phi'(x;\lambda^2)\frac{d\sigma(\lambda)}{\lambda};$$

the limit in the first exists in view of convergence in $L_M^{(2)}(0, l)$ and in the second in view of convergence in $L^{(2)}(0, l)$, i.e., convergence in mean square in the interval $(0, l)$. The solution of problem II for $I = (-a, a)$ is given by the formula

$$P_a F(\lambda) = \int_0^{x_{a-0}} f(x)\,\phi(x;\lambda^2)\,dM(x) + \frac{1}{\lambda}\int_0^{x_{a-0}} g(x)\,\phi'(x;\lambda^2)\,dx.$$

Let us also derive the formula for the square of the distance of F from Λ_a:

$$\int_{-\infty}^{\infty} |F(\lambda) - P_a F(\lambda)|^2 d\sigma(\lambda) = \int_{x_{a-0}}^{l} |f(x)|^2 dM(x) + \int_{x_{a-0}}^{l} |g(x)|^2 dx.$$

The family of projective operators P_a ($a \geq 0$, $P_0 = I$) form some spectral decomposition of identity; here the operator P_a is a weakly continuous function, from the left, of the argument a: $P_{a-0} = P_a$ ($a \leq 0$). The point $a \geq 0$ is a point of weak continuity on the right: $P_{a+0} = P_a$ if and only if $M(x_{a+0} + 0) = M(x_{a-0} - 0)$, i.e., when the points x_{a-0} and x_{a+0} are the points of continuity of the function $M(x)$ and between them the function $M(x)$ is of constant value.

4. Let l^* ($\leq l$) be the greatest of the points of growth of the function $M(x)$. If the function $M(x)$ for $x \geq +0$ and $< l^*$ has two absolutely continuous derivatives

where $M'(x) > 0$ for $x \geq +0$ then taking

$$T = \int_0^l \sqrt{M'(x)}dx, \quad t = \int_0^x \sqrt{M'(s)}ds, \quad \Phi(t;\lambda) = \sqrt[4]{M'(x)}\,\phi(x;\lambda)$$

$$(0 \leq x < l),$$

we get $\Phi(t;\lambda)$ as the normalized solution ($\Phi(0;\lambda) = [M'(+0)]^{1/4}$) of the differential system

$$\frac{d^2u}{dt^2} - q(t)u + \lambda u = 0, \quad u'(0) + (m\lambda - h)u(0) = 0 \quad (0 \leq t \leq T), \tag{3}$$

where $q(t) = \Phi''(t;\ 0)/\Phi(t;\ 0)$ $(0 \leq t < T)$, $h = M''(+0)/2[M'(+0)]^{3/2}$ and

$$\sqrt{M'(+0)}\,m = M(+0) = 1 \Big/ \int_{-\infty}^{\infty} d\sigma(\lambda) \ (\geq 0).$$

In this case for any $F \in \Lambda_\infty$ we will have*

$$P_aF(\lambda) = \int_0^a f(t)\Phi(t;\lambda^2)dt + \frac{1}{\lambda}\int_0^a g(t)\Psi(t;\lambda^2)dt + m\int_{-\infty}^{\infty} F(\lambda)d\sigma(\lambda), \tag{4}$$

where

$$\Psi(t,\lambda) = \Phi(t;0)\frac{d}{dt}[\Phi(t;\lambda)/\Phi(t;0)] \quad (0 \leq t < T), \tag{5}$$

$$f(t) = \underset{N\to\infty}{\text{l.i.m.}} \int_{-N}^{N} F(\lambda)\Phi(t;\lambda^2)d\sigma(\lambda), \quad g(t) = \underset{N\to\infty}{\text{l.i.m.}} \int_{-N}^{N} F(\lambda)\Psi(t;\lambda^2)\frac{d\sigma(\lambda)}{\lambda}, \tag{6}$$

where the limits are understood in the sense of convergence in mean square in the interval $(0, T)$. Besides this we have

$$\int_{-\infty}^{\infty} |F(\lambda) - P_aF(\lambda)|^2 d\sigma(\lambda) = \int_a^T |f(t)|^2 dt + \int_a^T |g(t)|^2 dt \quad (0 < a \leq T).$$

Conversely if for some system (3) (with $m \geq 0$, $-\infty < h < \infty$) the spectrum is positive and $\Phi(t;\lambda)$ is the solution of the system (3), normalized, for example, by the condition $\Phi(0;\lambda) = 1$ and $\tau(\lambda)$ $(0 \leq \lambda \leq \infty;\ \tau(0) = 0)$ is, in the corresponding manner, the normalized orthogonal spectral function (any, if it is not uniquely determined) then for the function $\sigma(\lambda) = \operatorname{sign}\lambda\cdot\tau(\lambda^2)$ $(-\infty < \lambda < \infty)$ Λ_a will be different from Λ_∞ only for $a < T$, and for $a < T$ problem II is solved by the formulas (4), (5) and (6).

Thus the effective solution of problems I and II depends on the effective construction of the functions $M(x)$, and $\phi(x;\lambda)$ (and in some cases only on one function $\Phi(t;\lambda)$) corresponding to the sense of the given function $\sigma(\lambda)$ indicated above. The several cases where it is possible to make such effective construction have been mentioned in the preceding note [8]. In a following note we will show that this is possible in all cases where the function σ is analytic and $d\sigma/d\lambda$ is a

*For $m = 0$ the last term on the right side of (4) is assumed to be zero quite independently of whether the integral multiplying m has sense or not.

rational function and also in many other cases (for example when $d\sigma/d\lambda$ can be obtained by dividing unity by a trigonometric polynomial).

BIBLIOGRAPHY

[1] M. G. Kreĭn, *On a problem of extrapolation of A. N. Kolmogoroff*, Dokl. Akad. Nauk SSSR 46 (1945), 306-309. (Russian)

[2] K. Karhunen, *Über die Struktur stationärer zufälliger Funktionen*, Ark. Mat. 1 (1950), 141-160

[3] N. Wiener, *Extrapolation, interpolation and smoothing of stationary time series*, Wiley, New York, 1949.

[4] A. Jaglom, *Introduction to the theory of stationary random functions*, Uspehi Mat. Nauk 7 (1952), no. 5 (51), 3-168. (Russian)

[5] M. G. Kreĭn, Dokl. Akad. Nauk SSSR 24 (1940), no. 1.

[6] ———, ibid. 44 (1944), no. 5; 45 (1944), no. 4; no. 5.

[7] ———, *On a generalization of investigations of Stieltjes*, ibid. 87 (1952), 881-884. (Russian)

[8] ———, ibid. 93 (1953), no. 4.

[9] N. I. Ahiezer, *Lectures on the theory of approximation*, Kharkov, 1940. (Russian)

Translated by:
D. V. Thampuran

6

Reprinted from RM–3091–PR, I. Selin, trans., RAND Corporation, 1962, pp. 1–29

ON THE STRUCTURE OF STATIONARY RANDOM FUNCTIONS

K. Karhunen

In an earlier paper (Karhunen [1]), we endeavored to develop a unified linear theory of random functions, and to consider some applications to stationary random functions. We shall investigate the latter questions somewhat more deeply, and in so doing we shall generalize the results obtained by Wold [1] and Kolmogorov [1,2] for the case of continuous stationary random functions. Hanner [1] has treated a similar problem by another method without the aid of the spectral representation.

Unless stated to the contrary, in the sequel we shall follow the terminology and definitions of our paper cited above.

This work was carried out during our stay as teaching fellow at the Institute for Actuarial Mathematics and Mathematical Statistics of the University of Stockholm. May we express our deeply felt thanks to the head of the Institute, Professor Harald Cramér, for his unflagging cooperation and for his many worthwhile suggestions.

I. SUBORDINATE RANDOM FUNCTIONS

1. Let $x(t)$ be a random function and $L_2(x)$ the corresponding linear space (cf. Karhunen [1], p. 26). Another random function $y(u)$ is called subordinate if $L_2(y) \leqq L_2(x)$. In particular, if $L_2(y) = L_2(x)$, then $x(t)$ and $y(u)$ are linearly coördered.

If $x(t)$ is stationary (cf. Karhunen [1], p. 54) and $y(t)$ and $x(t)$ are defined on the real line, if $y(t)$ is subordinate (coördered) to $x(t)$, and if $x(t)$ and $y(t)$ have a stationary correlation, i.e.,

$$r_{xy}(s-t) = Ex(s)\overline{y(t)} \qquad (1)$$

is a function of s-t alone, then we say that $y(t)$ is stationarily subordinate (coördered) to the random function $x(t)$.

We denote by $L_2(x;t)$ the closed linear hull of the set $\left\{x(s), s \leq t\right\}$. If $y(t)$ is stationarily subordinate (co-ordered) to $x(t)$ and therefore $L_2(y;t) \subseteq L_2(x;t)$ $\left[L_2(y;t) = L_2(x;t)\right]$ for all t, then the subordering (coördering) is called uniform.

2. It is well-known (Cramér [2]) that a continuous stationary random function $x(t)$ may be represented in the form

$$x(t) = \int_{-\infty}^{\infty} e^{i\lambda t} dZ_x(\lambda) \tag{2}$$

where Z_x is a spectral-function coordered with $x(t)$ (cf. Karhunen [1]). For the correlation function of $x(t)$, $r_{xx}(t) = Ex(s+t)\overline{x(s)}$, one has the corresponding representation (Khinchin [1])

$$r_{xx}(t) = \int_{-\infty}^{\infty} e^{i\lambda t} dF_x(\lambda), \tag{3}$$

where F_x is non-decreasing and bounded and the condition $\Delta F_x = ||\Delta Z_x||^2$ is satisfied.

Each element z in $L_2(Z_x) = L_2(x)$ can be represented in the form

$$z = \int_{-\infty}^{\infty} f_z(\lambda) dZ_x(\lambda),$$

where f_z is a complex function which is quadratically integrable with respect to F_x (Karhunen [1], Theorem 9). When we consider a particular random function $y(t)$ which is stationarily subordinate to $x(t)$, there also exists for each t a function of λ, $f(\lambda;t)$, such that

$$y(t) = \int_{-\infty}^{\infty} f(\lambda;t) dZ_x(\lambda). \tag{4}$$

From (1), (2) and (4), for arbitrary s and t, it follows that

$$Ex(s)\overline{y(t)} = \int_{-\infty}^{\infty} e^{i\lambda s}\overline{f(\lambda;t)}dF_x(\lambda) = Ex(s-t)\overline{y(0)} = \int_{-\infty}^{\infty} e^{i\lambda(s-t)}\overline{f(\lambda;0)}dF_x(\lambda) \quad (4')$$

and therefore

$$\int_{-\infty}^{\infty} e^{i\lambda s}\overline{[f(\lambda;t) - e^{i\lambda t}f(\lambda;0]}dF_x(\lambda) \equiv 0. \quad (4'')$$

This is possible only if, for all t, the expression

$$f(\lambda;t) = e^{i\lambda t}f(\lambda;0) \quad (4''')$$

holds almost everywhere with respect to F_x (i.e., for all λ except a set M on which the variation of F_x vanishes).

If we denote $f(\lambda;0) = \gamma_{yx}(\lambda)$, then we obtain

Theorem 1. Let x(t) be a continuous stationary random function; let y(t) be stationarily subordinate to x(t). Then there exists a complex function $\gamma_{yx}(\lambda)$ which is quadratically integrable and, with respect to F_x, almost everywhere unique, such that

$$y(t) = \int_{-\infty}^{\infty} e^{i\lambda t}\gamma_{yx}(\lambda)dZ_x(\lambda). \quad (5)$$

y(t) is also stationary and continuous, and with the definitions corresponding to (2) and (3),

$$Z_y(S) = \int_S \gamma_{yx}(\lambda)dZ_x(\lambda), \quad (6)$$

$$F_y(\lambda) = \int_{-\infty}^{\lambda} |\gamma_{yx}(\lambda)|^2 dF_x(\lambda), \quad (7)$$

$$r_{yx}(t) = \int_{-\infty}^{\infty} e^{i\lambda t}\gamma_{yx}(\lambda)dF_x(\lambda). \quad (8)$$

From (6) it follows that $L_2(Z_x) = L_2(Z_y)$, i.e., $L_2(x) = L_2(y)$, if and only if $\gamma_{yx}(\lambda)$ is different from zero almost everywhere with respect to F_x (Karhunen [1], p.50). Then and only then are x(t) and y(t) stationarily co-ordered.

3. If y(t) is stationary and continuous and is stationarily correlated with x(t), although not necessarily subordinate, then the projection of y(t) on $L_2(x)$,

$$P_{L_2(x)}y(t) = y_x(t), \tag{8'}$$

is stationarily subordinate to x(t) because $Ex(s)\overline{y_x(t)} = Ex(s)\overline{y(t)}$. Since y(t) and $y_x(t)$ are stationary and $y_x(s)$ is orthogonal to $[y(t)-y_x(t)]$, $y(t)-y_x(t)$ is stationary and one obtains

$$y_x(t) = \int_{-\infty}^{\infty} e^{i\lambda t}\gamma_{yx}(\lambda)dZ_x(\lambda), \tag{9}$$

$$y(t)-y_x(t) = \int_{-\infty}^{\infty} e^{i\lambda t}dZ_{y-y_x}(\lambda), \tag{10}$$

where $L_2(Z_{y-y_x})$ is orthogonal to $L_2(Z_x)$. It follows therefore from (7) and (8)

$$r_{xy}(t) = \int_{-\infty}^{\infty} e^{i\lambda t}dF_{xy}(\lambda) \tag{11}$$

and

$$r_{yy}(t) = \int_{-\infty}^{\infty} e^{i\lambda t}dF_y(\lambda) \tag{12}$$

with

$$F_{xy}(\lambda) = \int_{-\infty}^{\lambda} \overline{\gamma_{yx}(\lambda)dF_x(\lambda)} \tag{13}$$

and

$$F_y(\lambda) = \int_{-\infty}^{\lambda} |\gamma_{yx}(\lambda)|^2 dF_x(\lambda) + F_{y-y_x}(\lambda). \tag{14}$$

The Schwartz Inequality holds because $\Delta F_{y-y_x}(\lambda) \geqq 0$

$$|\Delta F_{xy}|^2 = |\int_{\Delta} \gamma_{yx}(\lambda)dF_x(\lambda)|^2 \leqq \int_{\Delta} |\gamma_{yx}(\lambda)|^2 dF_x(\lambda) \int_{\Delta} dF_x(\lambda) \leqq \Delta F_x \Delta F_y. \tag{14'}$$

Thus, F_{xy} is a function of bounded variation, whose differences satisfy the inequality

$$|\Delta F_{xy}|^2 \leqq \Delta F_x \Delta F_y \;. \tag{15}$$

On the other hand, it follows from (9) and (10) that, for arbitrary bounded and non-decreasing real functions F_x and F_y and a complex function F_{xy} which satisfy inequality (15), one can always construct two continuous stationary random functions x(t) and y(t) with a stationary correlation function such that (3), (11) and (12) hold. One need only define

$$\gamma_{yx} = \frac{d\overline{F_{xy}}}{dF_x}, \quad F_{y-y_x} = F_y - \int_{-\infty}^{\lambda} |\gamma_{yx}(\lambda)|^2 dF_x(\lambda). \tag{15'}$$

It follows from (15) that F_{y-y_x} is non-decreasing, so that one can find two orthogonal random spectral functions Z_x and Z_{y-y_x}, such that $||\Delta Z_x||^2 = \Delta F_x$, $||\Delta Z_{y-y_x}||^2 = \Delta F_{y-y_x}$. This result, which can be easily extended to any number of stationary random functions, was proved by Cramér [1] by another method. (For stationary time series, also cf. Kolmogorov [1], § 3.)

4. $y(t)$ is called a stationary component of $x(t)$ if $y(t)$ is stationarily subordinate to $x(t)$ and $y(s)$ is orthogonal to $x(t) - y(t)$ for all s and t.

Theorem 2. $y(t)$ is a stationary component of $x(t)$ if and only if

$$\gamma_{yx}(\lambda) = e_S(\lambda), \tag{16}$$

almost everywhere with respect to F_x. $e_S(\lambda)$ is the characteristic function of a measurable set S (i.e., $e_S(\lambda) = 1$ for $\lambda \in S$, otherwise $e_S(\lambda) = 0$). One can say that any decomposition of a stationary continuous random function into stationary components corresponds to a decomposition of its spectrum into complementary component spectra.

Proof: From (5) and the fact that $x(t)-y(t)$ is orthogonal to $L_2(y)$, it follows that

$$\int_{-\infty}^{\infty} e^{i\lambda t}[1-\gamma_{yx}(\lambda)]\overline{\gamma_{yx}(\lambda)}dF_x(\lambda) = 0 \tag{16'}$$

for all t and therefore

$$[1-\gamma_{yx}(\lambda)]\overline{\gamma_{yx}(\lambda)} = 0, \tag{16"}$$

i.e., either $\gamma_{yx}(\lambda) = 0$ or 1, almost everywhere with respect to F_x. Since $\gamma_{yx}(\lambda)$ is a measurable function, the set S, on which $\gamma_{yx}(\lambda) = 1$, is measurable, whence the assertion follows.

II. DETERMINISTIC AND REGULAR STATIONARY RANDOM FUNCTIONS

5. Let $x(t)$ be a continuous stationary random function. For $s \leq t$ it is obvious that $L_2(x;s) \subseteq L_2(x;t)$. When $t \to -\infty$, $L_2(x;t)$ converges to a sub-space in $L_2(x)$, which we shall call $L_2(x,-\infty)$. If $L_2(x,-\infty) = L_2(x)$,

i.e., if for every t, $L_2(x;t) = L_2(x)$, then every $x(t)$ is an element in $L_2(x;s)$, where s may be chosen arbitrarily. This means that the values of $x(t)$ for $t \leq s$ uniquely determine $x(t)$ for all values of the argument. The random function is then called deterministic. The opposite extreme case occurs when $L_2(x,-\infty)$ contains only the null element. Then one can say that the values of $x(t)$ for $t \leq s$ give arbitrarily little information for $t \geq u$ when $u - s$ is sufficiently large. Then we say that the random function is purely regular.

In general, naturally $0 \subset L_2(x;-\infty) \subset L_2(x)$. One can easily see that $x(t)$ may be uniquely decomposed into two stationary components, one of which is deterministic and one of which is purely regular. In fact, the projection $x_d(t)$ of $x(t)$ on $L_2(x,-\infty)$ is a stationary component of $x(t)$. $x_d(t)$ is subordinate to $x(t)$ since $L_2(x;-\infty) \subseteq L_2(x)$. Then we have

$$E[x_d(s)\overline{x(t)}] = E[P_{L_2(x;-\infty)}x(s)\overline{x(t)}] = E[T_h P_{L_2(x;-\infty)}x(s)\overline{T_h x(t)}] =$$

$$= E[P_{T_h L_2(x;-\infty)}x(s+h)\overline{x(t+h)}], \tag{16'''}$$

where T_h is the linear transformation on $L_2(x)$ which transforms each $x(t)$ into $x(t+h)$ (cf. Karhunen [1], p.55). Now it is obvious that $T_h L_2(x;-\infty) = L_2(x;-\infty)$, and we get

$$E[x_d(s)\overline{x(t)}] = E[x_d(s+h)\overline{x(t+h)}]. \tag{16''''}$$

Therefore the correlation between $x_d(t)$ and $x(t)$ is stationary. Since

$$L_2(x;-\infty) = P_{L_2(x;-\infty)}L_2(x) = L_2(x_d) \text{ and } L_2(x_d;-\infty) = P_{L_2(x;-\infty)}L_2(x_d) = L_2(x_d),$$

we may conclude that $x_d(t)$ is deterministic. $L_2(x;-\infty)\,\theta\,L_2(x_d)^* = 0$, and it follows that $x(t)-x_d(t)$ has no deterministic component.

---

*θ represents the 'exclusive or'

The concept of 'deterministic' and 'purely regular' is due to Wold [1] for stationary time series. He calls the components 'singular' and 'regular' respectively. The term 'deterministic' was introduced by Doob [1]. Wold [1] and Kolmogorov [1] have shown that a decomposition of a stationary time series into a deterministic and a purely regular component is possible.

6. Let h be an arbitrary positive number. We denote

$$x(t;h) = x(t) - P_{L_2(x;t-h)}x(t). \tag{17}$$

It is obvious that $x(t,h)$ is stationarily subordinate to $x(t)$ (although in general not a stationary component of $x(t)$, since $L_2(x;t-h)$ depends on t and therefore $x(t)$ need not be orthogonal to $x(s) - x(s;h)$). If $x(t)$ is deterministic, $x(t;h)$ vanishes identically. Conversely, when $x(t;h)$ vanishes identically for each h, $x(t) = P_{L_2(x;t-h)}x(t)$ for each h, i.e., $x(t) = P_{L_2(x;-\infty)}x(t)$ and therefore $x(t)$ is deterministic.

It follows from Theorem 1 that there exists a function $\gamma(\lambda;h)$ which is quadratically integrable with respect to F_x and satisfies the equation

$$x(t;h) = \int_{-\infty}^{\infty} e^{i\lambda t}\gamma(\lambda;h)dZ_x(\lambda). \tag{18}$$

For $s-t > h$, $Ex(s;h)\overline{x(t;h)} = 0$ because $x(t;h)$ is an element in $L_2(x;t)$, $t < s - h$ is also in $L_2(x;s-h)$, and $x(s;h)$ is orthogonal to $L_2(x;s-h)$ due to (17). From (18) and (7),

$$\int_{-\infty}^{\infty} e^{i\lambda t}|\gamma(\lambda;h)|^2 dF_x(\lambda) = 0 \quad \text{for} \quad |t| > h. \tag{19}$$

Finally we note that the function

$$\int_{-\infty}^{\lambda} |\gamma(\lambda;h)|^2 dF_x(\lambda) \tag{19'}$$

is absolutely continuous and therefore, since the derivative $F'_x(\lambda)$ of the monotone function $F_x(\lambda)$ exists almost everywhere,

$$\int_{-\infty}^{\lambda} |\gamma(\lambda;h)|^2 dF_x(\lambda) = \int_{-\infty}^{\lambda} |\gamma(\lambda;h)|^2 F'_x(\lambda)d\lambda. \tag{20}$$

It is well-known that $F_x(\lambda)$ may be written as the sum of three functions,

$$F_x(\lambda) = F_x^{(1)}(\lambda) + F_x^{(2)}(\lambda) + F_x^{(3)}(\lambda) \tag{21}$$

each of which is non-decreasing. $F_x^{(1)}(\lambda)$ is absolutely continuous and $F'_x(\lambda) = F_x^{(1)'}(\lambda)$ almost everywhere. $F_x^{(2)}$ is a step-function and $F_x^{(3)}(\lambda)$ is continuous. The derivatives of $F_x^{(2)}(\lambda)$ and $F_x^{(3)}(\lambda)$ vanish except on a set M of measure zero. Let M_2 be the countable set of discontinuities of $F_x(\lambda)$, $M_3 = M - M_2$, and $M_1 = R - M$, where R is the whole real axis. Then for every measurable set S

$$\int_S dF_x^{(\nu)}(\lambda) = \int_{S \cdot M_\nu} dF_x(\lambda) \quad (\nu = 1, 2, 3). \tag{21'}$$

We can now define three spectral functions by

$$Z_x^{(\nu)}(S) = \int_{S \cdot M_\nu} dZ_x(\lambda), \quad (\nu = 1, 2, 3) \tag{21''}$$

such that $\|\Delta Z_x^{(\nu)}\|^2 = \Delta F_x^{(\nu)}$, and

$$Z_x(S) = Z_x^{(1)}(S) + Z_x^{(2)}(S) + Z_x^{(3)}(S). \tag{22}$$

$L_2(Z_x^{(1)})$, $L_2(Z_x^{(2)})$, and $L_2(Z_x^{(3)})$ are mutually orthogonal. If we write

$$x_\nu(t) = \int_{-\infty}^{\infty} e^{i\lambda t} dZ_x^{(\nu)}(\lambda), \tag{23}$$

then

$$x(t) = x_1(t) + x_2(t) + x_3(t) \tag{24}$$

is a decomposition of x(t) into three stationary components by Theorem 2. We say (cf. Karhunen [1], p.70) that $x_1(t)$ possesses an absolutely continuous spectrum, $x_2(t)$ has a discrete spectrum, and $x_3(t)$ possesses a singular spectrum.

Since $x_1(t)$, $x_2(t)$ and $x_3(t)$ are mutually orthogonal, it follows from (17) that

$$x(t;h) = x_1(t;h) + x_2(t;h) + x_3(t;h). \tag{25}$$

By (18) and (20), x(t;h) vanishes identically if $F'_x(\lambda)$ vanishes almost everywhere. In this case x(t) is deterministic. It follows that the components $x_2(t)$ and $x_3(t)$ are also deterministic. Therefore we have

$$x(t;h) = x_1(t;h) = \int_{-\infty}^{\infty} e^{i\lambda t}\gamma(\lambda;h) dZ_x^{(1)}(\lambda). \tag{26}$$

There is no essential restriction in assuming in the sequel that x(t) has an absolutely continuous spectrum.

From (19), we have

$$\int_{-\infty}^{\infty} e^{i\lambda t}|\gamma(\lambda;h)|^2 F'_x(\lambda)d\lambda = 0 \qquad \text{for } |t| > h. \tag{27}$$

Furthermore, since $x(s;h)$ is orthogonal to $L_2(x;s-h)$ and $x(t) \in L_2(x;s-h)$ for $t \leqq s - h$, it follows from (8) that

$$\int_{-\infty}^{\infty} e^{i\lambda t}\gamma(\lambda;h)F'_x(\lambda)d\lambda = 0 \qquad \text{for } t > h. \tag{28}$$

The functions $|\gamma(\lambda;h)|^2 F'_x(\lambda)$ and $\gamma(\lambda;h)F'_x(\lambda)$ are both absolutely integrable on $(-\infty, \infty)$ (this follows for $\gamma(\lambda;h)F'_x(\lambda)$ from the absolute integrability of $F'_x(\lambda)$ by the Schwarz inequality). By a theorem of Hille and Tamarkin [1], the Fourier transform of an absolutely integrable function $f(\lambda)$ can vanish on the semi-infinite line only if the integral

$$\int_{-\infty}^{\infty} \frac{|\log|f(\lambda)||}{1 + \lambda^2} d\lambda \tag{28'}$$

converges. It follows from (27) and (28) that

$$\int_{-\infty}^{\infty} \frac{|\log[|\gamma(\lambda;h)|^2 F'_x(\lambda)]|}{1 + \lambda^2} d\lambda < \infty, \quad \int_{-\infty}^{\infty} \frac{|\log[|\gamma(\lambda;h)|F'_x(\lambda)]|}{1 + \lambda^2} d\lambda < \infty. \tag{29}$$

But

$$0 \leqq |\log F'_x| = |2\log|\gamma| + 2\log F'_x - \log|\gamma|^2 F'_x| \leqq 2|\log|\gamma|F'_x|$$
$$+ |\log|\gamma|^2 F'_x|, \tag{29*}$$
$$0 \leqq |\log|\gamma|| = |\log|\gamma|F'_x - \log F'_x| \leqq |\log|\gamma|F'_x| + |\log F'_x|,$$

so that the conditions (29) are equivalent to

$$\int_{-\infty}^{\infty} \frac{|\log F'_x(\lambda)|}{1+\lambda^2}\, d\lambda < \infty, \qquad \int_{-\infty}^{\infty} \frac{|\log|\gamma(\lambda;h)||}{1+\lambda^2}\, d\lambda < \infty. \tag{29'}$$

When the first condition (29') is not fulfilled, no function $\gamma(\lambda;h)$ different from zero can exist. Next we have

Theorem 3. When the integral

$$\int_{-\infty}^{\infty} \frac{|\log F'_x(\lambda)|}{1+\lambda^2}\, d\lambda \tag{30}$$

diverges, the random function $x(t)$ is deterministic. In particular, $x(t)$ is deterministic if $F'_x(\lambda)$ vanishes on a set of positive measure.

Let us note that small values of F'_x can cause the divergence of the integral (30) because $\log F_x < F'_x$.

III. THE CANONICAL EXPANSION OF A REGULAR STATIONARY RANDOM FUNCTION

7. We now assume that the spectrum of $x(t)$ is absolutely continuous and that the integral (30) converges. Then we know that $x(t)$ may be expanded in the form

$$x(t) = \int_0^{\infty} g(u)\, d_u\xi(t-u) = \int_{-\infty}^{t} g(t-u)\, d\xi(u) \tag{31}$$

where $\xi(S)$ is a spectral function defined for all Lebesque-measurable sets on the real line, and which corresponds to the Lebesque-measure m, i.e.,

$$\|\xi(S)\|^2 = m(S) \tag{32}$$

(Karhunen [1], Theorem 16). $x(t)$ and $\xi(t)$ are co-ordered.

A spectral function with the property (32) is called homogeneous. It has properties which closely relate it to stationary random functions. In fact: denote by T_h a translation along the real line, which carries a point t into the point t + h. Then $\xi(T_t S)$ is obviously a stationary random function on t. In particular if S is an interval (a,b), for $\xi(T_t S)$ one gets the spectral representation

$$\xi(T_t S) = \xi(a+t,b+t) = \frac{1}{\sqrt{2\pi}} \int_{-\infty}^{\infty} e^{i\lambda t} \frac{e^{i\lambda b} - e^{i\lambda a}}{i\lambda} d\xi^*(\lambda), \tag{33}$$

where ξ^* is again a homogeneous spectral function (cf. Karhunen [1] p.53). More generally,

$$\xi(S) = \int_{-\infty}^{\infty} \left\{ \frac{1}{\sqrt{2\pi}} \int_S e^{i\lambda t} dt \right\} d\xi^*(\lambda) \tag{34}$$

and conversely

$$\xi^*(S) = \int_{-\infty}^{\infty} \left\{ \frac{1}{\sqrt{2\pi}} \int_S e^{-i\lambda t} d\lambda \right\} d\xi(t). \tag{35}$$

The proof of these formulae is completely analogous to that of (33). ξ^* is the Fourier transform of ξ. From (34) and (35) ξ and ξ^* are obviously co-ordered.

Theorem 4. Let $\xi(S)$ be a homogeneous spectral function and f(t) a quadratically integrable complex function; $\xi^*(S)$ and $f^*(t)$ are their respective Fourier transforms. Then Parseval's theorem holds:

$$\int_{-\infty}^{\infty} f(t)d\xi(t) = \int_{-\infty}^{\infty} f^*(t)d\xi^*(t). \tag{36}$$

Proof: Since ξ and ξ^* are co-ordered, every element

$$z = \int_{-\infty}^{\infty} f(t)d\xi(t) \tag{36'}$$

of $L_2(\xi)$ is also an element of $L_2(\xi^*)$. Therefore there exists a quadratically integrable function f^* such that (36) holds. Now we must show that f^* is the Fourier transform of f. From (34), it is true for every S

$$Ez\overline{\xi(S)} = \int_S f(t)dt = \frac{1}{\sqrt{2\pi}} \int_{-\infty}^{\infty} f^*(\lambda)\left\{\int_S e^{-i\lambda t}dt\right\} d\lambda. \tag{36''}$$

Consequently, almost everywhere

$$f(t) = \frac{d}{dt}\left\{\frac{1}{\sqrt{2\pi}} \int_{-\infty}^{\infty} f^*(\lambda)\frac{e^{-i\lambda t} - 1}{- i\lambda} d\lambda\right\}. \tag{36'''}$$

This is just a well-known expression for the Fourier transform of the quadratically integrable function f^*.

8. We say that a homogeneous spectral function $\xi(S)$ is subordinate to (co-ordered with) a stationary random function $x(t)$ when $\xi(T_t S)$ possesses the corresponding property as a function of t for every S. It is seen that $x(t)$ and $\xi(S)$ in (31) are stationarily co-ordered.

Let $L_2(\xi;t)$ be any subspace of $L_2(\xi)$ spanned by all $\xi(S)$, such that S is a measurable subset of the half-line $(-\infty, t)$. The elements of $L_2(\xi;t)$ can be expressed in the form

$$z = \int_{-\infty}^{t} f(u)d\xi(u) \tag{36''''}$$

where $f(u)$ is quadratically integrable over $(-\infty,t)$. We say that $x(t)$ and $\xi(S)$ are uniformly co-ordered if $L_2(x;t) = L_2(\xi;t)$ for all t.

We ask if we may choose $\xi(S)$ in (31) such that $\xi(S)$ is uniformly co-ordered with $x(t)$. First we show that, assuming that such a $\xi(S)$ exists, it is uniquely determined to within a constant scale factor of absolute magnitude unity.

Let

$$x(t) = \int_{-\infty}^{t} g_1(t-u)d\xi_1(u) = \int_{-\infty}^{t} g_2(t-u)d\xi_2(u) \tag{37}$$

and $L_2(\xi_1;t) = L_2(\xi_2;t) = L_2(x;t)$ for each t. Then for every measurable set S, a function $f(t;S)$ is defined for each $t \in S$, such that

$$\xi_2(S) = \int_S f(t;S)d\xi_1(t). \tag{37'}$$

Because $\xi_2(S_1 + S_2) = \xi_2(S_1) + \xi_2(S_2)$ for $S_1 \cdot S_2 = 0$,

$$\begin{aligned} \xi_2(S_1 + S_2) &= \int_{S_1+S_2} f(t;S_1 + S_2)d\xi_1(t) \\ &= \int_{S_1} f(t;S_1)d\xi_1(t) + \int_{S_2} f(t;S_2)d\xi_1(t), \end{aligned} \tag{37''}$$

and $f(t;S_1) = f(t;S_1 + S_2)$ if $t \in S_1$ for all S_2. Then $f(t;S)$ is independent of S for all t, and we can simply write $f(t)$ instead of $f(t;S)$. (37) becomes, for each t,

$$x(t) = \int_{-\infty}^{t} g_1(t-u)d\xi_1(u) = \int_{-\infty}^{t} g_2(t-u)f(u)d\xi_1(u), \qquad (37''')$$

and since $L_2(x;t) = L_2(\xi_1;t)$

$$g_1(t-u) = g_2(t-u)f(u), \qquad (37'''')$$

from which $f(u)$ = a constant k. Thus, we have

$\xi_2(S) = k\xi_1(S)$. Since $\|\xi_2(S)\|^2 = |k|^2\|\xi_1(S)\|^2 = m(S)$, $|k| = 1$.

It is clear that $x(t)$ is purely regular if there exists a $\xi(S)$ which is uniformly co-ordered with it. In fact: if z is an element in $L_2(x;-\infty)$, it is element in $L_2(\xi;t)$, i.e., expressable in the form

$$z = \int_{-\infty}^{t} f(u)d\xi(u). \qquad (38)$$

This representation, which is unique, must hold for arbitrary t, which is clearly possible only when $f(u)$ vanishes identically.

9. In order to prove the existence of a $\xi(S)$ which is uniformly co-ordered with $x(t)$, we need some analytic lemmas which are contained partially in the work of Paley and Wiener [1] as well as in Hille and Tamarkin [1,2] (cf. also Nevanlinna [1], pp. 174-197).

Lemma 1. Let $f(u)$ be a complex-valued function, defined for real u, and quadratically (absolutely) integrable on $(-\infty, +\infty)$. Let its Fourier transform $f^*(t)$ vanish identically for $t < 0$. Then

$$f(w) = \frac{1}{\sqrt{2\pi}} \int_0^\infty e^{-itw} f^*(t)\,dt \qquad (w = u + iv) \tag{39}$$

defines a regular analytic function for $v < 0$, which satisfies the condition

$$\int_{-\infty}^{\infty} |f(u + iv)|^p du < M(f) \quad \text{for all } v \leqq 0 \qquad [p = 2,(1)] \tag{40}$$

Here $M(f)$ is a constant dependent on the choice of the function f but not on the value v. Conversely, any function $f(w)$ with the latter properties can be expressed in the form (39), where $f^*(t)$ is quadratically integrable (bounded). Furthermore

$$\lim_{v \to 0} f(u + iv) = f(u) \qquad \text{for almost all } u \tag{41}$$

and

$$\lim_{v \to 0} \int_{-\infty}^{\infty} |f(u + iv) - f(u)|^p du = 0 \qquad [p = 2(1)] \tag{42}$$

Lemma 2. A function $f(w)$ which is quadratically (absolutely) integrable on the real line, fulfills the conditions of Lemma 1 if and only if the integral

$$\int_{-\infty}^{\infty} \frac{|\log|f(t)||}{1 + t^2}\,dt \tag{42'}$$

converges and if the representation

$$f(w) = \prod_f(w) e^{-\frac{1}{\pi i}\int_{-\infty}^{\infty} \frac{1+tw}{t-w} \frac{\log|f(t)|}{1+t^2} dt + \frac{1}{\pi i}\int_{-\infty}^{\infty} \frac{1+tw}{t-w} d\chi_f(t) + i\alpha_f - i\beta_f w} \tag{43}$$

is valid for $v < 0$. $\prod_f(w)$ denotes a Blaschke product

$$\prod_f(w) = \prod_{\nu} \frac{w-w_\nu}{w-\bar{w}_\nu} \frac{\bar{w}_\nu - i}{w_\nu + i} \tag{44}$$

with $v_\nu < 0$, and

$$\sum_{\nu} \frac{v_\nu}{1+|w_\nu|^2} < \infty, \tag{45}$$

$\chi_f(t)$ is a bounded non-decreasing function whose derivatives vanish almost everywhere, α_f is a real number and β_f is a positive number. The representation (43) is unique.

The proof of Lemma 1 appears in Paley and Wiener [1] for quadratically integrable f and in Hille and Tamarkin [2] for absolutely integrable f.

Proof of Lemma 2. Let $f(w)$ be defined by (43). Obviously $f(w)$ is analytic and regular for $v < 0$. The function

$$e^{-\frac{1}{\pi i} \int_{-\infty}^{\infty} \frac{1+tw}{t-w} \frac{\log|f(t)|}{1+t^2} dt} \tag{45'}$$

satisfies the condition (40)(Paley and Wiener [1], Theorem XII). The expression

$$\prod_f(w) e^{\frac{1}{\pi i} \int_{-\infty}^{\infty} \frac{1+tw}{t-w} d\chi_f(t) - i\beta_f w + i\alpha_f} \tag{45''}$$

is bounded for $v < 0$ and has limits which are bounded almost everywhere as $v \longrightarrow 0$, so that $f(w)$ fulfills the same condition (40).

Conversely, let $f(w)$ be a function which satisfies the conditions of Lemma 1.

$$w = i\frac{z+1}{z-1} \tag{46}$$

is a substitution which transforms f(w) into a function g(z) which is analytic and regular within the unit circle. Furthermore, from (40) it follows that

$$\int_0^{2\pi} |g(re^{i\phi})|^p d\phi \leq M_f \qquad [p = 2\ (1)] \tag{46'}$$

for $r \leq 1$. (Hille and Tamarkin [2]). Denoting as usual by $\log^+ a$ the larger of the two numbers log a and 0, it follows that

$$\int_0^{2\pi} \log^+|g(re^{i\phi})|d\phi \leq M_f \tag{46''}$$

for $r \leq 1$. Since g(z) has no poles within the unit circle, it is a bounded function for $|z| < 1$ (Nevanlinna [1]). Such a function can be represented in the form

$$g(z) = \prod(z)e^{\frac{1}{2\pi}\int_0^{2\pi} \frac{e^{i\theta}+z}{e^{i\theta}-z} d\psi(\theta)+i\alpha} \tag{47}$$

where $\prod(z)$ is a Blaschke product and ψ a function of bounded variation. $\psi(\theta)$ satisfies the equation

$$\psi(\theta) = \lim_{r\to 1} \int_0^{\theta} \log|g(re^{i\phi})|d\phi\ . \tag{48}$$

When the derivative $\psi'(\theta)$ is finite for some value θ,

$$\psi'(\theta) = \lim_{r\to 1} \log|g(re^{i\theta})|. \tag{49}$$

Since $\psi(\theta)$ as a function of bounded variation, it has a finite derivative almost everywhere, (49) holds almost everywhere and the absolutely continuous component of $\psi(\theta)$ is thereby defined. Let ψ^+ be the positive variation of ψ. From (48) it follows that

$$\psi^+(\theta) = \lim_{r \to 1} \int_0^\theta \log^+|g(re^{i\phi})|\,d\phi. \tag{50}$$

Now, from (42) (cf. Hille and Tamarkin [2])

$$\lim_{r \to 1} \int_0^{2\pi} g(re^{i\phi})\,d\phi = \int_0^{2\pi} g(e^{i\phi})\,d\phi \tag{50'}$$

and $g(e^{i\phi}) = \lim_{r \to 1} g(re^{i\phi})$ almost everywhere. Therefore the integral

$$\int_0^\theta |g(re^{i\phi})|\,d\phi \tag{50''}$$

for $r < 1$ is uniformly absolutely continuous. Since $0 \leq \log^+ |g| \leq |g|$

$$\int_0^\theta \log^+|g(re^{i\phi})|\,d\phi \tag{50'''}$$

for $r < 1$ is also uniformly absolutely continuous and hence by (50), $\psi^+(\theta)$ is absolutely continuous (cf. Nevanlinna [1], p. 194). It follows that the function

$$\psi_0(\theta) = \psi(\theta) - \int_0^\theta \psi'(\phi)\,d\phi, \tag{51}$$

whose derivatives vanish almost everywhere, is non-increasing.

Because $\lim_{r\to 1} \log|g(re^{i\phi})| = \log|g(e^{i\phi})|$ almost everywhere, from (49) and (51)

$$\psi(\theta) = \int_0^\theta \log|g(e^{i\phi})|d\phi + \psi_0(\theta). \tag{52}$$

From (47), (52) and (46) the representation (43) of f(w) follows easily if we write

$$d\theta = \frac{du}{1 + u^2} \tag{52'}$$

[which follows easily from (46)], and denote the saltus of ψ_0 at $\theta = 0$ by β.

Lemma 3. Let $f^*(t)$ be a quadratically integrable complex function defined for $t > 0$. In order that there exist no other quadratically integrable function $h^*(t)$, not identically vanishing for $t > 0$, and satisfying

$$q^*(t) = \int_0^\infty f^*(u)h^*(u + t)du = 0 \qquad \text{for all } t > 0, \tag{53}$$

it suffices that the representation

$$f(w) = e^{-\frac{1}{\pi i}\int_{-\infty}^{\infty} \frac{1 + tw}{t - w}\frac{\log|f(t)|}{1 + t^2}dt + i\alpha_f} \tag{54}$$

of the function f(w) defined by (39) be valid. (We shall later see that this condition is not necessary.)

Proof: Set $h^*(t) = 0$ for $t < 0$, thus defining $q^*(t)$ for all t. Since f^* and h^* are quadratically integrable, $q^*(t)$ is bounded for all t. Similarly, we define $f^*(t) = 0$ for $t < 0$. In addition, we write

$$\begin{cases} h(w) = \dfrac{1}{\sqrt{2\pi}} \displaystyle\int_{-\infty}^{\infty} e^{itw} h^*(t)dt & \text{for } v > 0, \\ q(w) = \dfrac{1}{\sqrt{2\pi}} \displaystyle\int_{-\infty}^{\infty} e^{itw} q^*(t)dt & \text{for } v < 0. \end{cases} \tag{55}$$

From Lemma 1, it follows that almost everywhere

$$\lim_{v \to 0} f(w) = f(u), \qquad \lim_{v \to 0} h(w) = h(u), \tag{55'}$$

where f and h are the Fourier transforms of f* and h*. The Parseval equation and (53) result in

$$q^*(t) = \int_{-\infty}^{\infty} e^{-iut} f(u)h(u)du. \tag{55"}$$

By Lemma 1, since f(t)h(t) is absolutely integrable,

$$q(u) = \lim_{v \to 0} q(u + iv) = f(u)h(u). \tag{56}$$

The function q(w) is analytic for $v < 0$ and can be represented in the form (43). Since h(w) is analytic for $v > 0$, $\overline{h(\overline{w})}$ is also analytic for $v < 0$. $\overline{h(\overline{w})}$ obviously satisfies the conditions of Lemma 1 and therefore can be represented in the form (43). Consequently, when the condition (54) is fulfilled,

$$\Phi(w) = \overline{h(\bar{w})}\frac{q(w)}{f(w)} = \prod_h(w)\prod_q(w) \times$$

$$\exp\left\{-\frac{1}{\pi i}\int_{-\infty}^{\infty}\frac{1+tw}{t-w}\,\frac{\log|h(t)| + \log|q(t)| - \log|f(t)|}{1+t^2}\,dt \right. \tag{56'}$$

$$\left. + \frac{1}{\pi i}\int_{-\infty}^{\infty}\frac{1+tw}{t-w}\,d[\chi_h(t)+\chi_q(t)] - i(\beta_h+\beta_q)w + i(\alpha_h+\alpha_q+\alpha_f)\right\}.$$

$\prod_h(w)\prod_q(w)$ is a Blaschke product, $\chi_h(t) + \chi_q(t)$ is non-decreasing and $\beta_h + \beta_q \geqq 0$. $\Phi(w)$ may be represented in the form (45) and has the limit

$$\Phi(u) = \lim_{v\to 0}\Phi(u+iv) = \overline{h(u)}\,\frac{g(u)}{f(u)} = |h(u)|^2\,. \tag{57}$$

almost everywhere. Since $h(u)$ is quadratically integrable, $\Phi(u)$ is absolutely integrable, and, by Lemmas 1 and 2, has a Fourier transform $\Phi^*(t)$ which vanishes for $t < 0$. From (57), $\Phi(u)$ is real and hence $\Phi^*(-t) = \overline{\Phi^*(t)}$. Hence one may conclude that $\Phi^*(t)$, and with it $\Phi(u)$, $h(u)$, and $h^*(t)$, vanish identically.

10. Let $\xi(S)$ be a homogeneous random spectral function, for which (31) holds. Since $L_2(x;t) \subseteq L_2(\xi;t)$ in all cases, $x(t)$ and $\xi(S)$ are uniformly co-ordered if and only if $L_2(\xi;t)$ contains no element orthogonal to $L_2(x;t)$. Because every element in $L_2(\xi;t)$ can be represented in the form (38),

$$E[x(u)\bar{z}] = \int_{-\infty}^{u} g(u-v)\overline{f(v)}\,dv = \int_{0}^{\infty} g(v)\overline{f(u-v)}\,dv \tag{57'}$$

holds for $u \leq t$. Since z is orthogonal to $L_2(x;t)$, this expression must vanish for all $u \leq t$. Denoting $\overline{f(t-v)}$ by $h(v)$, it follows for all $u > 0$ that

$$\int_0^\infty g(v)h(v+u)dv = 0 \qquad (57")$$

because z is orthogonal to $L_2(x;t)$. $L_2(x;t) = L_2(\xi;t)$ if and only if there exists no $h(v)$ with this property. From Lemma 3, in order that this be so, it is sufficient that for the function

$$G(w) = \frac{1}{\sqrt{2\pi}} \int_0^\infty e^{-itw} g(t)dt \qquad (58)$$

the representation

$$G(w) = e^{-\frac{1}{\pi i} \int_{-\infty}^{\infty} \frac{1+tw}{t-w} \frac{\log|G(t)|}{1+t^2} dt + i\alpha} \qquad (59)$$

hold. $G(u)$ is the Fourier transform of $g(t)$ and because of (31) it satisfies

$$|G(\lambda)|^2 = F'_x(\lambda) \qquad (60)$$

almost everywhere. When we delete the indeterminate factor $e^{i\alpha}$, then

$$G_0(w) = e^{-\frac{1}{2\pi i} \int_{-\infty}^{\infty} \frac{1+\lambda w}{\lambda - w} \frac{\log F'_x(\lambda)}{1+\lambda^2} d\lambda} \qquad (61)$$

is the unique G-function for which (59) holds. Denoting the corresponding g-function by $g_0(t)$, there exists a homogeneous spectral function $\xi_0(S)$ such that

$$x(t) = \int_{-\infty}^{t} g_0(t-u)d\xi_0(u) \qquad (62)$$

(Karhunen [1], p.72), and $\xi_0(S)$ is uniformly co-ordered with $x(t)$. $\xi_0(S)$ and $g_0(t)$ are now uniquely defined to within a constant factor. In particular, the sufficient condition in Lemma 3 is also necessary.*

From Lemma 2 one can obtain all functions $g(t)$ for which (31) holds, if the corresponding G-function is defined by

$$G(w) = G_0(w)\Pi(w)e^{\frac{1}{\pi i}\int_{-\infty}^{\infty}\frac{1+\lambda w}{\lambda - w}d\chi(\lambda) - i\beta w + i\alpha} \tag{63}$$

where $\Pi(w)$, $\chi(\lambda)$, β and α satisfy the conditions of Lemma 2 but are otherwise arbitrary.

11. Finally we wish to construct an expression for the spectral function $\xi(S)$ in (31) and in particular for $\xi_0(S)$.

Due to the demonstrated absolute continuity of F_x, and defining $G(w)$ by (63), it follows from (60) that

$$\xi^*(S) = \int_S \frac{dZ_x(\lambda)}{G(\lambda)} \tag{64}$$

defines a homogeneous spectral function. In fact

$$||\xi^*(S)||^2 = \int_S \frac{dF_x(\lambda)}{|G(\lambda)|^2} = \int_S \frac{dF_x(\lambda)}{F'_x(\lambda)} = m(S) . \tag{64'}$$

$\xi(S)$ is the Fourier transform of $\xi^*(S)$, and from Theorem 4 follows:

---

*One can for instance show this directly by constructing a function $h(w)$ different from zero, which is possible when the latter condition is not met. However this construction is fairly intricate, and we shall not carry it out since we shall not need the results.

$$x(t) = \int_{-\infty}^{\infty} e^{i\lambda t} dZ_x(\lambda) = \int_{-\infty}^{\infty} e^{i\lambda t} G(\lambda) d\xi^*(\lambda)$$

$$= \int_{-\infty}^{\infty} g(t-u) d\xi(u) = \int_{-\infty}^{t} g(t-u) d\xi(u), \qquad (64'')$$

because $g(t)$, the Fourier transform of $G(\lambda)$, vanishes for $t < 0$ when $G(w)$ is defined by (63).

12. From the results of paragraphs 10 and 11 we obtain

Theorem 5. A continuous stationary random variable $x(t)$ is purely regular if and only if its spectrum is absolutely continuous and complete, and the integral

$$\int_{-\infty}^{\infty} \frac{|\log F'_x(\lambda)|}{1+\lambda^2} d\lambda \qquad (64''')$$

converges. This is equivalent to the condition that $x(t)$ may be represented in the form

$$x(t) = \int_{-\infty}^{t} g(t-u) d\xi(u) \qquad (64^{iv})$$

where $\xi(S)$ is a spectral function homogeneously co-ordered with $x(t)$. Such a representation is obtained by setting

$$\left. \begin{aligned} g(t) &= \frac{1}{\sqrt{2\pi}} \int_{-\infty}^{\infty} e^{i\lambda t}\, G(\lambda) d\lambda , \\ \xi(S) &= \frac{1}{\sqrt{2\pi}} \int_{-\infty}^{\infty} \left\{ \int_S e^{i\lambda t} dt \right\} \frac{dZ_x(\lambda)}{G(\lambda)} \end{aligned} \right\} \qquad (64^{v})$$

where

$$G(\lambda) = \lim_{\mu \to 0} G(\lambda+i\mu) \tag{64^{vi}}$$

and $G(\lambda+i\mu) = G(w)$ (for $\mu < 0$) is defined by (61) and (63). $\xi(S)$ can be chosen so that it is uniformly co-ordered with $x(t)$. The representation in question is obtained by taking $G(w) = G_0(w)$. This representation

$$x(t) = \int_{-\infty}^{t} g_0(t-u)d\xi_0(u) \tag{65}$$

is uniquely determined to within a normalizing facter $e^{i\alpha}$ (α real).

<u>Theorem 6</u>: In the decomposition (24) of a continuous stationary random function $x(t)$, the components $x_2(t)$ and $x_3(t)$ are deterministic, and the component $x_1(t)$ is either deterministic or purely regular depending on whether the integral

$$\int_{-\infty}^{\infty} \frac{|\log F'_x(\lambda)|}{1+\lambda^2} d\lambda \tag{65'}$$

diverges or converges. Since $x_1(t)$ is the only stationary component of $x(t)$ which can have a complete and absolutely continuous spectrum, $x_1(t)$ is the only component of $x(t)$ which can be purely regular. In particular, every stationary component of a purely regular random function $x(t)$ is deterministic (if we exclude the trivial component $x(t)$ itself).

IV. LINEAR EXTRAPOLATION OF A STATIONARY RANDOM FUNCTION

13. Let $x(t)$ be continuous and stationary. The problem of linear extrapolation of $x(t)$ is to find an element $z \in L_2(x;s)$ such that $\|x(t) - z\|$, $(t > s)$ is a minimum. It is well known that this minimum can be achieved for $z = P_{L_2(x;s)}x(t)$. From Theorems 5 and 6 it follows that

$$P_{L_2(x;s)}x(t) = x_2(t) + x_3(t) + \int_{-\infty}^{s} g_0(t-u)d\xi_0(u). \tag{66}$$

The extrapolation error is

$$x(t)-P_{L_2(x;s)}x(t) = \int_{s}^{t} g_0(t-u)d\xi_0(u) \tag{67}$$

and its quadratic norm is

$$D_x(t-s) = \left\|\int_{s}^{t} g_0(t-u)d\xi_0(u)\right\|^2 = \int_{0}^{t-s} |g_0(u)|^2du. \tag{68}$$

The non-decreasing function $D_x(t)$ is a measure of the inaccuracy of the extrapolation. When $D_x(t)$ vanishes identically, $x(t)$ is deterministic.

Note that we have not given an explicit solution to the problem of extrapolation, since we use the entire space $L_2(x)$ instead of the sub-space $L_2(x;s)$ in the expression for ξ_0 given above. However it is still possible to construct an explicit solution. We do not wish to explore this question here, but we hope to be able to return to it in another connection.

REFERENCES

Cramér, H.: [1] "On the Theory of Stationary Random Processes," Annals of Mathematics, 41, (1940).

——— : [2] "On Harmonic Analysis in Certain Functional Spaces," Arkiv för Matematik etc., 28 B (1942).

Doob, J. L.: [1] "The Elementary Gaussian Processes," Annals of Mathematical Statistics, 15 (1944).

Hanner, O.: [1] "Deterministic and Non-deterministic Stationary Stochastic Processes," Arkiv för Matematik, 1 (1949).

Hille, E. and Tamarkin, J. D.: [1] "A Theorem of Paley and Wiener," Annals of Mathematics, 34 (1933).

——— : [2] "On the Absolute Integrability of Fourier-Transforms," Fundamenta Mathematicae, 25 (1935).

Karhunen, K.: [1] "Über Lineare Methoden in der Wahrscheinlichkeitsrechnung," Ann. Ac. Sci. Fennicae, A, I, 37 (1947).

Khinchin, A.: [1] "Korrelationstheorie der Stationären Stochastischen Prozesse," Mathematische Annalen, 109 (1934).

Kolmogorov, A.: [1] "Stationary Time Series in Hilbert Space," (Russian), Bulletin de l'Université de Moscou, 2 (1941).

——— : [2] "Interpolation und Extrapolation von Stationären Zufälligen Folgen," Bulletin Acad. Sci. URSS, Sér. Math., 5 (1941).

Nevanlinna, R.: [1] "Eindeutige Analytische Funktionen, Berlin 1936.

Paley, R. and Weiner, N.: [1] Fourier Transforms in the Complex Domain, New York 1934.

Wold, H.: [1] A Study in the Analysis of Stationary Time Series, Uppsala 1938.

Part III

WIENER-HOPF EQUATIONS AND OPTIMUM FILTERS

Editor's Comments on Papers 7 Through 13

7 **YAGLOM**
On a Problem of Linear Interpolation of Stationary Random Sequences and Processes

8 **KREIN**
Introduction from *Integral Equations on a Half-Line with Kernel Depending Upon the Difference of the Arguments*

9 **GOHBERG and FEL'DMAN**
Introduction from *Convolution Equations and Projection Methods for Their Solution*

10 **ZADEH and RAGAZZINI**
An Extension of Wiener's Theory of Prediction

11 **YAGLOM**
Effective Solutions of Linear Approximation Problems for Multivariate Stationary Processes with a Rational Spectrum

12 **ROZANOV**
Spectral Properties of Multivariate Stationary Processes and Boundary Properties of Analytic Matrices

13 **JAFFE and RECHTIN**
Design and Performance of Phase-Lock Circuits Capable of Near-Optimum Performance Over a Wide Range of Input Signal and Noise Levels

Independent of Kolmogorov and Krein, Wiener was led to least-squares problems in an investigation of antiaircraft fire control problems in the early days of World War II [1]. Wiener's original approach was less mathematically comprehensive because he had clearly in mind the possibilities of actually applying his results. Thus Wiener considered the problem of separating (filtering) a signal random process from a noise random process. He was especially interested in explicit formulas for the optimum filter (weighting function) rather than in the more mathematical questions of limiting error, singular prediction, etc., which were thoroughly explored by Kolmogorov. This engineering motivation led to a different and equally significant class of mathematical problems as that described in Part II. Paper 7 provides a good illustration of the point. Yaglom studies the reconstruction of a continuous-time stochastic process from knowledge of its values at the discrete instants, $t = 0, \pm 1, \pm 2, \ldots$. Under the influence of Kolmogorov, he worked out expressions for the minimum error and also the conditions under which the error could be zero. These conditions (see the remark (15) in Paper 7) include the result that perfect reconstruction can be obtained for suitably bandlimited processes. Yaglom stops here. However, he noted in a conversation with the editor that if he had also, following Wiener, tried to determine the optimum reconstructing filter, he would have discovered the widely known sampling theorem for bandlimited processes, a result that was first proved by Balakrishnan [2] and Lloyd [3].

Concern with the optimum weighting function leads immediately to a Wiener-Hopf integral equation. This was a particular spur to Wiener's interest. Though the equation had been known since at least 1894 in the field of radiative transfer, the name actually originated after the appearance in 1931 of the brilliant solution of Wiener and Hopf [4], based on the remarkably fruitful notion of spectral factorization.

By now the Wiener-Hopf technique has been described in many books and papers. A paper by N. Levinson [5] gives a fine heuristic exposition, as does a little monograph by P. Whittle [6], but both of these are readily available. We have, therefore, selected as Paper 8 the introduction from a fundamental survey paper by M. G. Krein, which deserves a much wider reading than it has received so far. Krein gives a good account of the problem and of the results obtained by various investigators. In particular, he discusses the alternative methods of Ambartzumian and Chandrasekhar, developed in the mid-forties in radiative-transfer theory, that first introduced a nonlinear equation into the Wiener-Hopf problem. This work and its extensions were unknown to engineers until recently (see Section VI of Paper 1).

Nevertheless, the central idea of least-squares theory is spectral factorization. Thus we have decided to include here, as Paper 9, the introduction from a recent book by Gohberg and Fel'dman that uses a simple operator calculus to explain the fundamental idea in a general context. This extract also gives a brief outline of certain methods of approximating the solution via "finite sections of Wiener-Hopf equations," a technique originated by Baxter [7] and Hirschman [8]. More will be said about such results and about factorization and other approaches to the Wiener-Hopf equation in the second volume on this topic.

We present as Paper 10 a remarkably insightful and prescient paper by Zadeh and Ragazzini, who rediscovered the innovations method of Kolmogorov at the same time as Bode and Shannon [9], and in fact gave an extended form to it (see the discussion in Paper 1, Section VIII). Moreover, Zadeh and Ragazzini were the first to treat the problems of a finite data interval and to include a nonstationary drift term (a polynomial with random coefficients). Generalizations of these results were considered by many workers, and detailed treatments appeared in various textbooks, for example, those of Solodovnikov [10] and of Laning and Battin [11]. As will be explained in Part IV, these studies were among the factors leading to the use of state-space models in estimation theory.

Following Wiener, the engineering workers formulated the problem in terms of Wiener-Hopf type integral equations, valid for arbitrary stochastic processes, but soon it was found that all the explicit solutions were for the (important) class of processes with rational spectral densities. However, it was pointed out by Yaglom [12] that if this special assumption was introduced into the problem at the very beginning, the explicit solutions of Wiener, Zadeh and Ragazzini, and several others could be obtained by elementary (and rigorous) methods. Yaglom's method, which he notes is related to techniques independently developed in electromagnetic theory by Wainstein and others, exploits in detail the fact that rational spectral densities are the boundary values of analytic functions of a complex variable. A similar approach was developed by Darlington [13]. Paper 11 is a presentation of the method in the more general context of multivariate stationary processes. A detailed treatment of the scalar case can be found in Yaglom [14], which notes, but only in passing, that the methods of Krein (Paper 5) provide an elegant alternative route.

Yaglom's approach circumvented the Wiener-Hopf equation and the technique of spectral factorization. However, as mentioned before, these equations and techniques arise in other problems as well. In particular, Youla [15] reinterpreted some results of Oono and Yasuura [16] in network theory to solve the multivariate spectral factorization problem. It was found that such results were also useful in optimal control theory and in a class of stability problems (the circle and Popov criteria) for certain nonlinear systems. An exposition of these subjects is given by Popov [17], who, in Appendix B of his book, develops several factorization results under weaker conditions than those of Youla. As mentioned by Popov, a very thorough study of the matrix factorization problem (and not only for spectral-density matrices) was made by Gohberg and Krein [18]. A variety of algorithms have been proposed for actually obtaining the spectral factors, and others are still being developed. Here we only refer to the discussion in Paper 1 (page **31**).

Many theoretical questions arise in studying general (i.e., not necessarily with rational spectral densities) multivariate processes, and these have been studied especially by P. Masani and his students. A good survey of this work appears in Masani [19, 20, and 21]. Other studies were made in the USSR by Krein, Gohberg, Rozanov, and others. Some of the problems studied are now arising in an engineering context (see Willems [22] and Anderson [23]) and this was kept in mind in selecting Paper 12.

The needs of certain applications in communications and control made it necessary to extend the Wiener filter to handle certain (power-type) constraints. This is described in Paper 13, in the context of phase-lock circuits. The results of this analysis have found wide application in space communications. Related problems in control and communication theory are described by Newton et al. [24], Chang [25], and Lindsey [26].

REFERENCES

1. N. Wiener, *Extrapolation, Interpolation and Smoothing of Stationary Time Series, with Engineering Applications*, New York: Technology Press and J. Wiley, 1949. (Originally issued in Feb. 1942 as a classified Nat. Defense Res. Council Rept.) available in paperback form as *Time Series*, Cambridge, Mass.: MIT Press, 1964.
2. A. V. Balakrishnan, "A Note on the Sampling Principle for Continuous Signal," *IEEE Trans. Inf. Theory* **IT–3** (2):143–146, June 1957.
3. S. P. Lloyd, "Sampling Theorem for Stationary (Wide-Sense) Stochastic Processes," *Trans. Amer. Math. Soc.* **92**:1–12, 1959.
4. N. Wiener and E. Hopf, "Über eine Klasse Singulärer Integralgleichungen," *S-B Preuss. Akad. Wiss.* 696–706, 1932.
5. N. Levinson, "A Heuristic Exposition of Wiener's Mathematical Theory of Prediction and Filtering," *J. Math. Phys.* 110–119, July 1947; reprinted as an appendix in [1].
6. P. Whittle, *Prediction and Regulation*, New York: Van Nostrand Reinhold, 1963.
7. G. Baxter, "A Norm Inequality for a 'Finite-Section' Wiener-Hopf Equation," *Ill. J. Math.* **7**:97–103, 1963.
8. I. I. Hirschman, Jr., "Finite Section Wiener-Hopf Equations on a Compact Group with Ordered Dual," *Bull. Amer. Math. Soc.* **70**:508–510, 1964.

9. H. W. Bode and C. E. Shannon, "A Simplified Derivation of Linear Least Square Smoothing and Prediction Theory," *Proc. IRE* **38:** 417–425, Apr. 1950.
10. V. V. Solodovnikov, *Introduction to the Statistical Dynamics of Automatic Control System*, ed. J. B. Thomas and L. A. Zadeh, New York: Dover, 1960. (Russian original, 1952).
11. H. Laning and R. Battin, *Random Processes in Automatic Control*, New York: McGraw-Hill, 1958.
12. A. M. Yaglom, "Introduction to the Theory of Stationary Random Functions," *Usp. Mat. Nauk* **7**(5):3–168, 1958. An expanded version appears in an English translation with the same title, trans. R. A. Silverman, Englewood Cliffs, N.J.: Prentice-Hall, 1962.
13. S. Darlington, "Linear Least-Squares Smoothing and Prediction, with Applications," *Bell Syst. Tech. J.* **37**:1221–1294, Sept. 1958.
14. A. M. Yaglom, "Extrapolation, Interpolation and Filtration of Stationary Random Processes with Rational Spectral Density," *Trudy Mosk. Mat. Ob.* **4:** 333–374, 1955; English translation in *Sel. Translations in Probability and Statistics* **4**:345–387, 1964.
15. D. C. Youla, "On the Factorization of Rational Matrices," *IRE Trans. Inf. Theory* **7**:172–189, July 1961.
16. Y. Oono and K. Yasuura, "Synthesis of Finite Passive 2n-Terminal Networks with Prescribed Scattering Matrices," *Mem. Kyushu University* (Engineering), Japan, **14**(2):125–177, 1954.
17. V. M. Popov, *Hyperstability of Control Systems*, Berlin: Springer-Verlag, 1972. (Romanian edition, 1966).
18. I. C. Gohberg and M. G. Krein, "Systems of Integral Equations on a Half-Axis with Kernels Depending on the Difference of the Arguments," *Usp. Mat. Nauk* **13**:3–72, 1958.
19. P. Masani, "Recent Trends in Multivariate Prediction Theory," in P. R. Krishnaiah, ed., *Multivariate Analysis*, New York: Academic Press, 1966, pp. 351–382.
20. P. Masani, "Review of Stationary Random Processes, by Yu. A. Rozanov," *Ann. Math. Statist.* **42**:1463–1467, 1971.
21. P. Masani, "Wiener's Contribution to Generalized Harmonic Analysis, Prediction Theory and Filter Theory," *Bull. Amer. Math. Soc.* **72**(1):73–125, Jan. 1966.
22. J. C. Willems, "Least Squares Stationary Optimal Control and the Algebraic Riccati Equation," *IEEE Trans. Automat. Control* **AC–16** :621–634, Dec. 1971. See also *ibid.* **AC–19**(5):592–593, Oct. 1974.
23. B. D. O. Anderson, "Algebraic Properties of Minimal Degree Spectral Factors," *Automatica* **9**:491–500, 1973.
24. G. C. Newton, L. A. Gould, and J. F. Kaiser, *Analytical Design of Linear Feedback Controls*, New York: J. Wiley, 1957.
25. S. S. L. Chang, *Synthesis of Optimum Control Systems*, New York: McGraw-Hill, 1961.
26. W. C. Lindsey, *Synchronization Systems in Communication and Control*, Englewood Cliffs, N.J.: Prentice-Hall, 1972.

7

Reprinted with the permission of the publisher from pp. 339–344 of *Selected Translations in Mathematical Statistics and Probability*, Volume 4, American Mathematical Society, 1963, 393 pp.

ON A PROBLEM OF LINEAR INTERPOLATION OF STATIONARY RANDOM SEQUENCES AND PROCESSES

A. M. JAGLOM

According to A. Ja. Hinčin [1] a stationary random process is a one-parameter family of real random variables $x(t)$ (t assumes all the possible real values $-\infty < t < \infty$) the square of which has a finite mathematical expectation, such that its mathematical expectations m and $B(\tau)$ do not depend on t, where

$$m = Mx(t) \text{ and } B(\tau) = M[x(t+\tau) - m][x(t) - m]. \tag{1}$$

The definition of a stationary random sequence differs from that of a stationary random process only in that for a sequence the parameter t takes only integral values.

We will always assume in the sequel that $m = Mx(t) = 0$; it is clear that this assumption does not affect the generality of our treatment. Moreover, we will find it convenient to consider the random variable $x(t)$ as complex (with a finite expectation for the square of its modulus) and the quantity

$$B(\tau) = Mx(t+\tau)\overline{x(t)} \tag{2}$$

as independent of t.

In his articles [2, 3] A. N. Kolmogorov has dealt with the following general problem connected with linear interpolation of stationary sequences and processes. Let the set of all values of the argument t (i.e., the set of all real or all integral numbers) be divided into two disjoint subsets K and K' and $s \in K'$. Consider all posible finite sums

$$L_n = \sum_{i=1}^{n} a_i x(t_i), \quad (t_i \in K, \ i = 1, 2, \ldots, n), \tag{3}$$

where $a_1, \cdots, a_n$ are arbitrary complex numbers, and for each such sum define the mean square deviation $\sigma^2(s, L_n)$ of the sum L_n from the $x(s)$ as:

$$\sigma^2(s, L_n) = M|x(s) - L_n|^2. \tag{4}$$

For a given correlation $B(\tau)$ it is necessary to determine the lower bound

$$\sigma^2(s, K) = \inf_{L_n} \sigma^2(s, L_n) \tag{5}$$

of the mean square deviation $\sigma^2(s, L_n)$ for all admissible sums L_n.

A. N. Kolmogorov [2, 3] has considered only stationary sequences $\{x(t)\}$ and solved the general problem in the two following cases, which are of importance in the theory of probability: (a) for the set K composed of all integers less than a fixed number t_0 (the case of extrapolation of stationary sequences) and (b) for the set K containing all integers except s.

In the extrapolation of stationary random processes (the set K now consists of all real numbers, less than a fixed number t_0) this problem has been solved by M. G. Kreĭn [4] (see also [5], Supplement B). In this paper we will give the solution of this problem in two special cases which are of interest in probability theory. That is, we will deal with the interpolation of stationary processes for known values $x(t)$ at integral instants of time (the set K consists of all integers) and interpolation of stationary sequences for a finite number of unknown elements of the sequence (the set K' is finite).

These problems can be easily solved by using elementary results of Hilbert space theory. Following A. N. Kolmogorov [2], let us consider the set of all possible sums

$$\sum_{k=1}^{n} a_k x(t_k), \quad (-\infty < t_k < \infty, \qquad k = 1, 2, \ldots, n)$$

and their limits in mean square as a Hilbert space H_x with scalar product $(x, y) = Mx\bar{y}$. Take the spectral resolution of the correlation function (2) of the set of random variables $\{x(t)\}$

$$B(\tau) = \int e^{i\tau\lambda}\, dF(\lambda) \tag{6}$$

(in all cases where the limits of integration are not stated, it is implied that these are $-\infty$ and $+\infty$ for stationary processes and $-\pi$ and $+\pi$ for stationary sequences). By virtue of (6), the Hilbert space H_x can be isomorphically mapped on the Hilbert space L_x^2 of functions $\phi(\lambda)$ $(-\infty < \lambda < \infty$ or $-\pi \leq \lambda \leq \pi)$ the square of the modulus being integrable with respect to measure $dF(\lambda)$; $x(t)$ will here transform into $e^{it\lambda}$ (cf. [2], Lemma 4). Denote by $L_x^2(K)$ the subspace of the space L_x^2, spanned by the set of functions $e^{it\lambda}$, $t \in K$. It is clear $\sigma^2(s, K)$ will be equal to the square of the norm of the perpendicular drawn from the "point" $e^{is\lambda}$ of the space L_x^2 to the subspace $L_x^2(K)$.

Let $\phi_\delta(\lambda)$ be the perpendicular drawn from $e^{is\lambda}$ to $L_x^2(K)$ and $\phi_\nu(\lambda)$ the projection of $e^{is\lambda}$ on the same subspace. In this case, with respect to $dF(\lambda)$, we have almost everywhere

$$e^{is\lambda} = \varphi_\delta(\lambda) + \varphi_\nu(\lambda). \tag{7}$$

Since $\phi_\delta(\lambda)$ in L_x^2 is orthogonal to $\phi_\nu(\lambda)$, it is clear that $(\phi_\nu(\lambda), \phi_\nu(\lambda)) = (\phi_\nu(\lambda), e^{is\lambda})$ and $(\phi_\delta(\lambda), \phi_\delta(\lambda)) = (\phi_\delta(\lambda), e^{is\lambda})$, i.e.,

$$\int |\varphi_\nu(\lambda)|^2 dF(\lambda) = \int e^{-is\lambda} \varphi_\nu(\lambda) dF(\lambda) \tag{8}$$

and

$$\int |\varphi_\delta(\lambda)|^2 dF(\lambda) = \int e^{-is\lambda} \varphi_\delta(\lambda) dF(\lambda). \tag{8'}$$

Conversely, from (7) and (8) or (7) and (8′) it follows that the vectors $\phi_\nu(\lambda)$ and $\phi_\delta(\lambda)$ of the space L_x^2 are orthogonal to each other.

Now consider the set of functions $\phi_\nu(\lambda)$ belonging to the subspace $L_x^2(K)$ and satisfying (8); the projection of the vector $e^{is\lambda}$ on $L_x^2(K)$ will obviously be the same as that of the function $\phi_\nu(\lambda)$ for which the integral

$$\nu^2 = \int |\varphi_\nu(\lambda)|^2 dF(\lambda) \tag{9}$$

takes the largest value. In exactly the same way, it is easily seen that the perpendicular drawn from $e^{is\lambda}$ to $L_x^2(K)$ is the same as that of the function $\phi_\delta(\lambda)$, orthogonal to $L_x^2(K)$ and satisfying (8′), for which the largest value is the integral

$$\delta^2 = \int |\phi_\delta(\lambda)|^2 dF(\lambda). \tag{9'}$$

These facts enable us to determine easily $\sigma^2(s, K)$ for the two cases indicated above.

At first let t assume all possible real values (the case of interpolation of random processes), and let the set K consist of all integers. In this case $L_x^2(K)$ is the subspace of the space L_x^2 spanned by the set of functions $e^{ik\lambda}$ (k is integrable), i. e., the subspace of the periodic functions of period 2π whose square of the modulus is integrable with respect to $dF(\lambda)$. Introducing the functions of bounded variation on $[-\pi, \pi]$

$$F_0^*(\lambda) = \sum_{n=-\infty}^{+\infty} [F(\lambda + 2n\pi) - F(2n\pi)], \tag{10}$$

$$F_s^*(\lambda) = \sum_{n=-\infty}^{+\infty} [F(\lambda + 2n\pi) - F(2n\pi)] e^{-i2\pi rs}, \tag{11}$$

and denoting by $\phi_\nu^*(\lambda)$ the function

$$\varphi_\nu^*(\lambda) = \frac{\varphi_\nu(\lambda)}{\int_{-\infty}^{\infty} |\varphi_\nu(\lambda)|^2 dF(\lambda)} = \frac{\varphi_\nu(\lambda)}{\int^{\pi} |\varphi_\nu(\lambda)|^2 dF_0^*(\lambda)}, \tag{12}$$

we can express (8) and (9) in the following form:

$$\int_{-\pi}^{\pi} \varphi_{\nu}^{*}(\lambda)\, e^{-is\lambda}\, dF_{s}^{*}(\lambda) = 1, \tag{13}$$

and

$$\frac{1}{\nu^2} = \int_{-\pi}^{\pi} |\varphi_{\nu}^{*}(\lambda)|^2\, dF_{0}^{*}(\lambda). \tag{14}$$

Since $F_s^*(\lambda)$ is absolutely continuous with respect to $F_0^*(\lambda)$, we can express (12) in the form

$$\int_{-\pi}^{\pi} \varphi_{\nu}^{*}(\lambda)\, e^{-is\lambda}\, \frac{dF_{s}^{*}(\lambda)}{dF_{0}^{*}(\lambda)}\, dF_{0}^{*}(\lambda) = 1. \tag{13'}$$

Thus our problem is reduced to the simple isoperimetric one: to find the minimum of the integral (14) subject to the condition (13′); it is interesting that $\sigma^2(s, K)$ is related to $V^2 = \max \nu^2$ by the relation $\sigma^2(s, K) = \int_{-\pi}^{\pi} dF_0^*(\lambda) - V^2$. Finding in the usual way the minimum of the integral (14) under the condition (13′), we get the following theorem:

Theorem 1. *For stationary random processes the quantity $\sigma^2(s, K)$ for the set K of all integers is determined by the formula*

$$\sigma^2(s, K) = \int_{-\pi}^{\pi} \left[1 - \left|\frac{dF_{s}^{*}(\lambda)}{dF_{0}^{*}(\lambda)}\right|^2\right] dF_{0}^{*}(\lambda). \tag{15}$$

From (15) it follows in particular that if there do not exist two points where the function $F(\lambda)$ increases, the difference between these points being a multiple of 2π, then $\sigma^2(s, K) = 0$ for all s. Further, taking $s = 1/2$ in this formula we easily get the following theorem on the interpolation of stationary sequences:

Theorem 1′. *For stationary random sequences the quantity $\sigma^2(s, K)$ for the set K of all odd numbers is determined by the formula*

$$\sigma^2(s, K) = 4\int_{0}^{\pi} \frac{dF(\lambda-\pi)}{d[F(\lambda-\pi)+F(\lambda)]}\, dF(\lambda) = 4\int_{0}^{\pi} \frac{dF(\lambda)}{d[F(\lambda-\pi)+F(\lambda)]}\, dF(\lambda-\pi) \tag{16}$$

(*naturally, in this case $\sigma^2(s, K)$ does not depend on s*).

Let us now assume that t takes only the integral values (interpolation of stationary sequences) and the set K' consists of a finite number of elements

$s, t_1, t_2, \cdots, t_n$. In this case, it is convenient to use the relations (8′) and (9′) and find $\phi_\delta(\lambda)$ at once. The condition that the function $\phi_\delta(\lambda)$ be orthogonal to the subspace $L_x^2(K)$ here takes the following form:

$$\int_{-\pi}^{\pi} e^{-it\lambda}\varphi_\delta(\lambda)\, dF(\lambda) = 0 \tag{17}$$

for each integral t different from $s, t_1, \cdots, t_n$. Since the set of values t in the equality (17) is unbounded, it follows from a theorem of Riesz (see [6], p. 108; this theorem can also be derived from the results of [2, 3]) that

$$\int_{-\pi}^{\lambda} \varphi_\delta(\lambda)\, dF(\lambda) = \int_{-\pi}^{\lambda} \varphi_\delta(\lambda) F'(\lambda)\, d\lambda. \tag{18}$$

Now introducing the function

$$\varphi_\delta^*(\lambda) = \frac{2\pi e^{is\lambda}\varphi_\delta(\lambda)}{\int_{-\pi}^{\pi} |\varphi_\delta(\lambda)|^2\, dF(\lambda)}, \tag{19}$$

we can write (8′) and (9′) by virtue of (18) in the form

$$\int_{-\pi}^{\pi} \varphi_\delta^*(\lambda) F'(\lambda)\, d\lambda = 2\pi, \tag{20}$$

and

$$\frac{4\pi^2}{\delta^2} = \int_{-\pi}^{\pi} |\varphi_\delta^*(\lambda)|^2 F'(\lambda)\, d\lambda. \tag{21}$$

We get from (20), (17) and (18) that

$$\varphi_\delta^*(\lambda) F'(\lambda) = 1 + \alpha_1 e^{i(t_1 - s)\lambda} + \cdots + \alpha_n e^{i(t_n - s)\lambda}. \tag{22}$$

Noting that $4\pi^2/\sigma^2(s, K)$ is the maximum of the integral (21) subject to the conditions (17) and (20), we get the following theorem:

Theorem 2. *If the spectral function $F(\lambda)$ of the stationary random sequence $\{x(t)\}$ is such that there exist complex numbers $\alpha_1, \alpha_2, \cdots, \alpha_n$ for which the integral*

$$R(\alpha_1, \alpha_2, \ldots, \alpha_n) = \int_{-\pi}^{\pi} \frac{\left|1 + \sum_{k=1}^{n} \alpha_k e^{i(t_k - s)\lambda}\right|^2}{F'(\lambda)}\, d\lambda \tag{23}$$

converges, then the quantity $\sigma^2(s, K)$, *when* K' *consists of the finite number of elements* $s, t_1, t_2, \cdots, t_n$, *for this sequence is determined by the formula*

$$\sigma^2(s, K) = \frac{4\pi^2}{\min\limits_{\alpha_k} R(\alpha_1, \ldots, \alpha_n)}. \tag{24}$$

But if for any choice of the complex numbers $\alpha_1, \alpha_2, \cdots, \alpha_n$ *the integral* (23) *diverges, then* $\sigma^2(s, K) = 0$.

Thus the evaluation of the maximum precision of linear interpolation of a stationary sequence when a finite number of elements are not known is reduced to an extremum problem similar to the one whuch arises in the solution of the very same question for a finite number of known elements (in the last case, it is evidently reduced to the finding of the minimum of the integral

$$\int_{-\pi}^{\pi} \left| 1 + \sum_{k=1}^{n} \alpha_k e^{i(t_k - s)\lambda} \right|^2 dF(\lambda))$$

Theorem 2 has a particularly simple form when $n = 0$; this special case has already been treated in detail by A. N. Kolmogorov [2, 3].

BIBLIOGRAPHY

[1] A. Ja. Hinčin, *Korrelationstheorie der stationären stochastischen Prozesse*, Math. Ann. 109 (1934), 604–615.

[2] A. N. Kolmogorov, *Stationary sequences in Hilbert's space*, Bjull. Moskov. Gos. Univ. Matematika 2 (1941), no. 6. (Russian)

[3] ———, *Interpolation and extrapolation of stationary random sequences*, Izv. Akad. Nauk SSSR 5 (1941), 3–14. (Russian)

[4] M. G. Kreĭn, *On a problem of extrapolation of A. N. Kolmogoroff*, Dokl. Akad. Nauk SSSR 46 (1945), 339–342. (Russian)

[5] N. I. Ahiezer, *Lectures on the theory of approximation*, OGIZ, Moscow, 1947. (Russian)

[6] I. I. Privalov, *Boundary properties of one-valued analytic functions*, Moscow 1941. (Russian)

Translated by:
D. V. Thampuran

8

Reprinted with the permission of the publisher from pp. 163–172 of *Am. Math. Soc. Transl.* Ser. 2, **22**:162–288 (1962)

INTEGRAL EQUATIONS ON A HALF-LINE WITH KERNEL DEPENDING UPON THE DIFFERENCE OF THE ARGUMENTS

M. G. KREǏN

Table of contents

INTRODUCTION

1. It has been known for a comparatively long time that the theory of integral Fourier transforms can give a simple procedure for the solution of the integral equation

$$\chi(t) - \int_{-\infty}^{\infty} k(t-s)\,\chi(s)\,ds = f(t) \quad (-\infty < t < \infty), \tag{0.1}$$

where $k(t)$ and $f(t)$ are given functions and $\chi(t)$ is the function to be found.

In fact, if the functions $k(t)$ and $f(t)$ belong to the class $L_1(-\infty, \infty)$, and the function χ is sought within this same class, then, applying the Fourier transformation to both sides of the equality (0.1), we transform it into the algebraic relation

$$X(\lambda)\,[1 - K(\lambda)] = F(\lambda) \quad (-\infty < \lambda < \infty). \tag{0.2}$$

Here $X(\lambda)$, $K(\lambda)$, and $F(\lambda)$ denote, respectively, the Fourier transforms of the functions $\chi(t)$, $k(t)$, and $f(t)$. Thus, for example,

$$K(\lambda) = \int_{-\infty}^{\infty} e^{i\lambda t} k(t)\,dt. \tag{0.3}$$

Having found $X(\lambda)$, we then apply the inverse Fourier transformation to find the desired function $\chi(t)$.

Since the Fourier transform of a function of the class $L_1(-\infty, \infty)$ is always continuous, it is obviously *necessary* for the solvability of the equation (0.1) in this class for an arbitrary function $f \in L_1(-\infty, \infty)$ that

$$1 - K(\lambda) \neq 0 \quad (-\infty < \lambda < \infty). \tag{0.4}$$

It turns out that this condition *is also sufficient* for the equation (0.1) to have a solution $\chi \in L_1(-\infty, \infty)$ for an arbitrary right side $f \in L_1(-\infty, \infty)$. This simply stated fact became clear, however, only after the work of N. Wiener in 1932 ([1] and [2]). In the works cited, Wiener was the first to prove an *exact* theorem stating that for a function $k \in L_1(-\infty, \infty)$, the inequality (0.4) is necessary and sufficient for the existence of a function $k_1 \in L_1(-\infty, \infty)$ such that $1 + K_1(\lambda) = [1 - K(\lambda)]^{-1}(-\infty < \lambda < \infty)$, that is

$$1 + \int_{-\infty}^{\infty} e^{i\lambda t} k_1(t)\,dt = \left[1 - \int_{-\infty}^{\infty} e^{i\lambda t} k(t)\,dt\right]^{-1} \quad (-\infty < \lambda < \infty). \tag{0.5}$$

This assertion shows that if condition (0.4) is satisfied, then the relation (0.2) is equivalent to the relation $X(\lambda) = [1 + K_1(\lambda)]\,F(\lambda)$, and the last equality is equivalent to the equality

$$\chi(t) = f(t) + \int_{-\infty}^{\infty} k_1(t-s)\,f(s)\,ds. \tag{0.6}$$

Since the convolution of two functions of the class $L_1(-\infty, \infty)$ belongs to this class, we come to the following conclusion.

If $k(t) \in L_1(-\infty, \infty)$ and if condition (0.4) is satisfied, (but only in this case), the equation (0.1) will have a solution $\chi(t) \in L_1(-\infty, \infty)$ for an arbitrary right side $f(t) \in L_1(-\infty, \infty)$. This solution is unique and is obtained from formula (0.6), where $k_1(t) \in L_1(-\infty, \infty)$ and is connected with $k(t)$ by the relation (0.5).

Thus if $k(t) \in L_1(-\infty, \infty)$ and satisfies the condition (0.4), the kernel

$k(t-s)$ $(-\infty < t, s < \infty)$ has the resolvent $\gamma(t, s) = k_1(t-s)$ $(-\infty < t, s < \infty)$ of the same type as the original kernel ($k_1(t) \in L_1(-\infty, \infty)$).

The situation is more complicated with the theory of integral equations of the form

$$\chi(t) - \int_0^\infty k(t-s)\,\chi(s)\,ds = f(t) \quad (0 \le t < \infty), \tag{A}$$

to which the present article is devoted.

Many problems of mathematical physics lead to integral equations of type (A). In particular, the theory of equations (A) finds important applications in questions of transfer of radiant energy. Indeed, these questions led a number of investigators, beginning successfully in the year 1921, to study equations (A) (both homogeneous and nonhomogeneous). These investigators placed various special restrictions on the structure of the functions $k(t)$ $(-\infty < t < \infty)$ and $f(t)$ $(0 \le t < \infty)$ and the character of growth of the function $\chi(t)$ $(0 \le t < \infty)$.

It appeared at first that, in contradistinction to the equation (0.1), the discovery of explicit analytic formulas for the solutions of equation (A) was in general impossible. However, this point of view was refuted in 1931 by the brilliant achievement of E. Hopf and N. Wiener [3], who succeeded in developing the theory and technique for analytic solution of the homogeneous equation

$$\phi(t) - \int_0^\infty k(t-s)\,\phi(s)\,ds = 0. \tag{B}$$

In doing this, they assumed that $k(t) = O(e^{-a|t|})$ for some $a > 0$, or, more generally,

$$e^{\alpha|t|}k(t) \in L_2(-\infty, \infty) \text{ for } \alpha < a, \tag{0.7}$$

and sought a solution $\phi(t)$ satisfying the condition

$$\phi(t) = O(e^{\alpha t}) \text{ for some } \alpha < a \tag{0.8}$$

as $t \to \infty$.

The *idea of factorization* of a function $\mathfrak{K}(\lambda)$ holomorphic in the strip $|\mathrm{Im}(\lambda)| < a$ was first used in the work of Wiener and Hopf. This is the idea of the possibility of writing $\mathfrak{K}(\lambda)$ in the form of a product $\mathfrak{K}_+(\lambda)\,\mathfrak{K}_-(\lambda)$, where $\mathfrak{K}_+(\lambda)$ and $\mathfrak{K}_-(\lambda)$ are functions holomorphic respectively in the half-planes $\mathrm{Im}(\lambda) > -a$ and $\mathrm{Im}(\lambda) < a$, and satisfying certain other restrictions.

This idea was also decisive in the construction of the theory of nonhomogeneous integral equations (A) with "kernel" function $k(t)$ satisfying the conditions of exponential decrease described above (or something similar) as $|t| \to \infty$. Furthermore, as will be shown in the present article, the further development of this idea and its systematic application have made it possible to produce a theory of equations (A) of the most general type.

It should be mentioned that in his famous monograph [4] E. Hopf gave a series of important analytic formulas not only for solutions ϕ of the homogeneous equations but also for solutions χ of nonhomogeneous equations (A) with symmetric kernel for special right sides f.

In 1941, E. Reissner showed in the brief communication [5] how it is possible, using the idea of Wiener and Hopf, to obtain a certain procedure for solving the integral equation (A) for an arbitrary $k(t)$ satisfying the condition (0.7), under the assumption that $f(t) = O(e^{\alpha t})$ for some $\alpha < a$ and that χ satisfies the same condition.

V. A. Fok ([6], [7]) in the years 1942–1943 gave independently a significantly more complete investigation of the equation (A) under the assumptions of evenness and exponential decrease of the function $k(t)$.

An essential point in the investigation of V. A. Fok was the fact that he turned the question of finding solutions of the homogeneous and nonhomogeneous equations for right sides of a fixed class into the problem of constructing only one function, connected with the function $k(t)$ (the function $\psi_1(w)$ in his notation, and the function $\mathfrak{G}_+(\lambda)$ in ours).

I. M. Rapoport made a further important step in the development of the theory ([8], [9]). He connected the problem of solving the equations (B) and (A), respectively, with Hilbert's homogeneous and nonhomogeneous boundary problems. This problem at this moment is fairly well worked out, thanks to the work of I. Plemelj, T. Carleman, I. I. Privalov, and F. D. Gahov (see [10]). Thanks to this, I. M. Rapoport was the first to free himself from restrictions of the types (0.7) and (0.8), replacing them by the following conditions:*

1) $k(t) \in L_{1,2}(-\infty, \infty)$;

2) $K(\lambda) \in \mathrm{Lip}\,\alpha\,(\alpha > 0)$ for $-\infty < \lambda < \infty$;

3) $K(\lambda) = O(|\lambda|^{-\beta})$ $(\beta > 0)$ for $|\lambda| \to \infty$.

Rapoport's work was the first to make clear that the integer

$$\nu = -\operatorname{ind}(1 - K(\lambda)) = -\frac{1}{2\pi}\int_{-\infty}^{\infty} d_\lambda \arg(1 - K(\lambda)),$$

which will be called from now on the *index of the equation* (A), must play an essential role in the theory of equations (A) and (B) in the case of a nonsymmetric kernel $k(t - s)$.

I. M. Rapoport showed that the number of linearly independent solutions of the equation (B) in the class $L_2(0, \infty)$ is equal to $\max(0, \nu)$, and hence, the num-

*Comparison of the results of I. M. Rapoport with Theorems I, II, and III below shows in particular that his condition 1) can be replaced by the condition $k(t) \in L_1(-\infty, \infty)$, and his conditions 2) and 3) are inessential and superfluous. These last two conditions appeared in the constructions of I. M. Rapoport as a result of the fact that Hilbert's problem is usually solved only for functions satisfying Hölder's condition (for Hilbert's problem, see Appendix I).

ber of linearly independent solutions in this same class of the transposed homogeneous equation

$$\psi(t) - \int_0^\infty k(s-t)\,\psi(s)\,ds = 0 \quad (0 \le t < \infty) \tag{C}$$

is equal to $\max(0, -\nu)$.

At the same time, Rapoport proved a theorem about normal solvability of the nonhomogeneous equation (A) in the class $L_2(0, \infty)$.

2. As will be shown in this article, a sufficiently general and in the known sense complete theory of the integral equations (A) and (B) can be obtained by enlisting the use, on the one hand, of certain comparatively new results of harmonic analysis (the theorems of Wiener and Wiener-Lévy),* and, on the other hand, of certain general assertions of the theory of operators in a Banach space.

We shall state three principal theorems established in the present article.

For this, we first explain that a certain Banach space figures in the statements of these theorems. This space is one of an entire family of spaces, described and listed in detail in §6. Here we note only that the space E can be in particular any space $L_p(0, \infty)$ $(1 \le p < \infty)$ or the space $M(0, \infty)$ of all bounded measurable functions or the subspace $M_c(0, \infty)$ of all bounded continuous functions, and so on.

The following theorems are valid.

Theorem I. *Let* $k(t) \in L_1(-\infty, \infty)$. *Then the equation* (A) *has exactly one solution* $\chi \in E$ *for an arbitrary* $f \in E$ *if and only if the conditions*

$$1 - K(\lambda) \neq 0 \quad (-\infty < \lambda < \infty), \quad \nu = -\operatorname{ind}(1-K) = 0 \tag{0.9}$$

are satisfied.

If these conditions are satisfied, the solution $\chi \in E$ *of equation* (A) *is obtained by the formula*

$$\chi(t) = f(t) + \int_0^\infty \gamma(t, s)\, f(s)\, ds, \tag{0.10}$$

where the resolvent kernel $\gamma(t, s)$ $(0 \le t, s < \infty)$ *has the special structure*

$$\gamma(t, s) = \gamma(t-s, 0) + \gamma(0, s-t) + \int_0^\infty \gamma(t-r, 0)\,\gamma(0, s-r)\,dr \tag{0.11}$$

$$(\gamma(t, 0) = \gamma(0, t) = 0 \text{ for } t < 0).$$

The functions $\gamma(t, 0)$ *and* $\gamma(0, t)$ *belong to the class* $L_1(0, \infty)$ *and are defined uniquely by the following relations:*

*The theorems of Wiener-Lévy appeared some 2 to 3 years after the work of Wiener-Hopf [3]. It is evident that N. Wiener ceased at this time to occupy himself seriously with equations (B) and (A). This is the only possible explanation for the fact that Wiener failed to think of applying the theorems of Wiener-Lévy to equations (A), since he played such an important role in establishing the Wiener-Lévy theorems.

1) $(1-K(\lambda))^{-1}=\mathfrak{G}_+(\lambda)\,\mathfrak{G}_-(\lambda)\quad(-\infty<\lambda<\infty)$;

2) $\mathfrak{G}_+(\lambda)=1+\int_0^\infty \gamma(t,0)e^{i\lambda t}dt,\quad \mathfrak{G}_-(\lambda)=1+\int_0^\infty \gamma(0,t)e^{-i\lambda t}dt$;

3) $\mathfrak{G}_+(\lambda)\neq 0\quad(\operatorname{Im}\lambda\geq 0),\quad \mathfrak{G}_-(\lambda)\neq 0\quad(\operatorname{Im}\lambda\leq 0)$.

If the function $k(t)$ is even, as it will be in many applications, then $\mathfrak{G}_-(\lambda)=\mathfrak{G}_+(-\lambda)$ $(\operatorname{Im}(\lambda)\geq 0)$ and $\gamma(t,0)=\gamma(0,t)$ $(0\leq t<\infty)$.

In §§2, 7, and 16, we give various general and partial methods for analytic definition of the functions $\mathfrak{G}_\pm(\lambda)$.

Theorem II. *Let $k(t)\in L_1(-\infty,\infty)$ and suppose that condition* (0.4) *is satisfied. Then the homogeneous equation* (B) *has a nontrivial solution in an arbitrary space E if and only if $\nu=-\operatorname{ind}(1-K)>0$.*

When this condition is satisfied, the equation (B) *will have one and the same solution in all of the spaces E. Furthermore, the set of these solutions will have a basis consisting of ν functions $\phi_0(t),\phi_1(t),\cdots,\phi_{\nu-1}(t)$, which are absolutely continuous,* go to zero as $t\to\infty$, and are connected with each other by the relations*

$$\phi_{k+1}(t)=\frac{d\phi_k}{dt},\quad \phi_k(0)=0\quad(k=0,1,\cdots,\nu-2),\quad \phi_{\nu-1}(0)\neq 0.$$

Analytic procedures are given in §9 for constructing the basis just mentioned.

Theorem III. *Let $k(t)\in L_1(-\infty,\infty)$ and satisfy condition* (0.4). *Then, if $\nu>0$, the equation* (A) *has an infinite set of solutions $\chi\in E$ for an arbitrary $f\in E$.*

On the other hand, if $\nu<0$, then for a given $f\in E$, the equation (A) *either has no solution at all in the space E or has a unique solution. The last case occurs if and only if the condition*

$$\int_0^\infty f(t)\,\psi_j(t)\,dt=0\quad(j=0,1,\cdots,|\nu|-1)$$

is satisfied. (*Here $\psi_j(t)$ $(j=0,1,2,\cdots,|\nu|-1)$ is an arbitrary basis for the set of all solutions of the transposed homogeneous equation* (C).)

As a complement to this, it will be shown in §§9 and 10 that in the case $\nu\neq 0$, it is always possible to construct a general resolvent $\gamma_0(t,s)$ having the structure (0.11), with the help of which one can obtain a certain solution of equation (A) by formula (0.10), whenever this equation is solvable.

Chapter II contains proofs of these assertions, except for the proof of the necessity of the first of the conditions (0.9) in Theorem I.

The proof of the necessity of the second of the conditions (0.9) in Theorem I is given in §12 of chapter III. It uses the methods of the abstract theory of operators, briefly set forth in §11 of chapter II. Use of these methods was also used

*Furthermore, the derivatives of these functions belong to $L_1(0,\infty)$.

(§12) to study the spectrum of the operator K, given in E by the equality

$$K\phi(t) = \int_0^\infty k(t-s)\,\phi(s)\,ds \quad (0 \leq t < \infty).$$

In chapter III (§13), we also give "discrete" analogues of Theorems I, II, and III, that is, analogues of these theorems for an infinite system of equations

$$\sum_{s=0}^{\infty} k_{t-s}\chi_s = f_t \qquad (t = 0, 1, 2, \ldots). \tag{0.12}$$

We point out that analogues of this sort were first described by I. M. Rapoport [11]. Subsequently a number of his results were partly repeated, partly extended by other authors [12], [13], [14].

As in the case of integral equations (A), the methods set forth here have permitted the attainment of great completeness also in the theory of systems of equations (0.12).

In chapter IV, we study equations (A) and (B) of Wiener-Hopf type, where $k(t)$ satisfies the additional condition $\exp(h|t|)\,k(t) \in L_1(-\infty, \infty)$ $(h > 0)$.

The results of the first two sections (§§14, 15) of this chapter contain all of the essential facts obtained for the Wiener-Hopf equations by these two authors, and also by V. A. Fok [7] and E. Reissner [5]. In particular, in the second section (§15) of this chapter, one can find some new details regarding the asymptotic distribution of solutions of the homogeneous equation in exponential functions with polynomial coefficients.

In the same chapter (§16), we give some examples of the solution of the Wiener-Hopf equation. At the end of chapter IV (§17), we give examples illustrating the possibility of carrying over an important formula of Hopf-Fok to the Wiener-Hopf equation of the first kind

$$\int_0^\infty k(t-s)\,\phi(s)\,ds = f(t) \quad (0 \leq t < \infty). \tag{0.13}$$

Although we give here references to certain literature dealing with the equation (0.13), which plays a role in many questions of mathematical physics and the theory of random processes (see [15]), nevertheless, it is our opinion that the theory of these equations is far from any sort of completeness.

At the root of all of the proofs of the results of chapters II, III, and IV lies the idea of factorization of functions, which was put forth, as already mentioned, in the work of Wiener-Hopf [3]. The first chapter of this work is devoted to a detailed solution of the problem of factorization in its several aspects. For the sake of completeness, we also consider the problem of factorization of entire functions. The results connected with this problem are not used in the sequel, but they have interest of their own, and among other things they play a role in certain questions of functional analysis [16].

We remark that at the present time the principal results of this article, at any rate the part of them dealing with existence theorems, have been carried over to systems of integral equations

$$\chi_p(t) - \sum_{q=1}^{n} \int_0^{\infty} k_{pq}(t-s)\chi_q(s)\,ds = f_p(t) \qquad (0 \leqslant t < \infty;\ p = 1, 2, \ldots, n), \tag{0.14}$$

which demanded a deeper utilization of new techniques of functional analysis. The basic facts of the theory of systems (0.14) have already been set down in a report to the Third All-Soviet Union Mathematical Conference [17] and thereafter in the article of I. C. Gohberg and the author [18].

As is known, the integro-differential equation for transfer of radiant energy under known assumptions relative to the so-called indicatrix of dispersion (see [19] or [20]) can be turned into an equation or a system of equations of Wiener-Hopf type. On the other hand, starting with certain physical arguments (the so-called "principle of invariance"), V. A. Ambarzumian succeeded in reducing the problem of integrating the integro-differential equation of transfer of radiant energy to the solution of one or more non-linear integral equations. These *non-linear* equations admit simple methods of computational solution, and furthermore, they permitted V. V. Sobolev [19] and S. Chandrasekhar [20] to obtain explicit analytic expressions for a series of characteristic functions of the problem of transfer of radiant energy. Among these are characteristic functions for solutions of the corresponding system of *linear* integral equations.

This connection between linear integral equations of Wiener-Hopf type and the corresponding non-linear equations can be obtained by purely analytic means under certain general hypotheses concerning the kernels of the equations.

For the purpose of explaining this fact for the case considered in this article of one integral equation (A), we suppose that the function $k(t) \in L_1(-\infty, \infty)$ admits a representation

$$k(t) = \int_0^{\infty} e^{-|t|u}\,d\sigma(u) \quad (|t| > 0), \tag{0.15}$$

where $\sigma(u)$ is a certain function with bounded variation on every finite interval. Among other things, by a known theorem of S. N. Bernšteĭn, an even function $k(t)$ will admit a representation (0.15) with a nondecreasing function $\sigma(u)$ if and only if the function $k(t)$ is absolutely monotone for $t < 0$, that is, has non-negative derivatives of all orders.

If the function $k(t)$ of the form (0.15) satisfies condition (0.4) and hence, being even, also the second condition (0.9), the function $\mathfrak{G}_+(\lambda)$ referred to in Theorem I is expressed in terms of its values on the positive imaginary axis by the formula

$$\mathfrak{G}_+^{-1}(\lambda) = 1 - \int_0^\infty \frac{\mathfrak{G}_+(iu)}{u - i\lambda}\, d\sigma(u) \qquad (\lambda \,\bar{\ni}\, (0, -i\infty)).$$

Setting $\lambda = iv$ in this equality, we obtain the following non-linear equation for the function $F(v) = \mathfrak{G}_+(iv)$ $(0 \le v < \infty)$:

$$F^{-1}(v) = 1 - \int_0^\infty \frac{F(u)\, d\sigma(u)}{u+v}. \tag{0.16}$$

If $k(t) = O(\exp[(\epsilon - 1)|t|])$ for an arbitrary $\epsilon > 0$, then in the representation (0.16), the differential $d\sigma(u) = 0$ for $0 \le u \le 1$ and corresponding to this, we apply the substitutions

$$u = \frac{1}{\mu'}, \qquad v = \frac{1}{\mu}, \qquad F\left(\frac{1}{\mu}\right) = H(\mu)$$

to rewrite the equation (0.16) in the following form:

$$H^{-1}(\mu) = 1 - \mu \int_0^1 \frac{H(\mu')\, d\tau(\mu')}{\mu + \mu'} \qquad (0 \leqslant \mu \leqslant 1), \tag{0.17}$$

where

$$\tau(\mu) = \int_{\frac{1}{\mu}}^\infty \frac{d\sigma(u)}{u} \qquad (0 \leqslant \mu \leqslant 1).$$

The equation (0.16) differs from Chandrasekhar's equation (see chapter V in [20]) only in the fact that in the latter $\Psi(u')\, d\mu'$ appears instead of $d(\mu')$, where $\Psi(\mu')$ is a real polynomial satisfying certain additional conditions.

In the case of the Hvol'son-Milne kernel (see the example in §15),

$$k(t) = \frac{\omega}{2} \int_1^\infty \frac{e^{-|t|u} du}{u},$$

where $d\tau(\mu) = d\mu$ $(\Psi \equiv 1)$, the equation (0.17) assumes the form

$$H^{-1}(\mu) = 1 - \frac{\omega\mu}{2} \int_0^1 \frac{H(\mu')\, d\mu'}{\mu + \mu'}. \tag{0.18}$$

This equation is the known non-linear equation of V. A. Ambarzumian, appearing in the simplest problem of radiative equilibrium in a plane-parallel medium, filling up a half-space, under isotropic dispersion.

The author of the present article has worked out a comparatively detailed theory of the non-linear equations (0.16) with their consequences for the linear equations (A). The exposition of this theory, which contains the theory of equations (0.18) as a special case, had to be omitted from the present article for reasons of space.

In §16 (example 3), as an example of the integral equation of Grinberg-Fok (from the theory of boundary refraction of electromagnetic waves), we explain the method of going from linear equations to non-linear ones.

We note that Theorems I, II, and III, along with the principal facts of the theory of the non-linear equations (0.16), were reported by the author at the session of the Scientific Council of the Mathematical Institute of the Academy of Sciences of the Ukrainian Soviet Socialist Republic on 30 October 1954.

[*Editor's Note:* Material has been omitted at this point.]

REFERENCES

[1] N. Wiener, *Tauberian theorems*, Ann. of Math. (2) 33 (1932), 1–100.

[2] ———, *The Fourier integral and certain of its applications*, University Press, Cambridge, 1933.

[3] N. Wiener and E. Hopf, *Über eine Klasse singulärer Integralgleichungen*, S.-B. Preuss. Akad. Wiss. Berlin. Phys.-Math, Kl. 30/32 (1931), 696–706.

[4] E. Hopf, *Mathematical problems of radiative equilibrium*, Cambridge Univ. Press, London, 1933.

[5] E. Reissner, *On a class of singular integral equations*, J. Math. Phys. Mass. Inst. Tech. 20 (1941), 219–233.

[6] V. A. Fok, *On certain integral equations of mathematical physics*, Dokl. Akad. Nauk SSSR 37 (1942), 147. (Russian)

[7] ———, *On certain integral equations of mathematical physics*, Mat. Sb. (N.S.) 14 (56) (1944), 3–50. (Russian)

[8] I. M. Rapoport, *On a class of singular integral equations*, Dokl. Akad. Nauk SSSR 59 (1948), 1403–1406. (Russian)

[9] ———, *On certain "twin" integral and integro-differential equations*, Sb. Trudov Inst. Mat. Akad. Nauk. Ukrain. RSR 12 (1949), 102–118. (Russian)

[10] N. I. Mushelišvili, *Singular integral equations*, OGIZ, Moscow-Leningrad, 1946; English translation, Noordhoff, Groningen, 1953.

[11] I. M. Rapoport, *On a class of infinite systems of algebraic equations*, Dokl. Akad. Nauk Ukrain. RSR 3 (1948), 6–10. (Russian)

[12] I. C. Gohberg, *On the bounds of applicability of theorems of F. Noether*, Uč. Zap. Kišinevsk. Univ. 17 (1955), 35–44. (Russian)

[13] Ya. N. Fel'd, *On infinite systems of linear algebraic equations connected with problems on simi-infinite period structures*, Dokl. Akad. Nauk SSSR 102 (1955), 257–260. (Russian)

[14] V. S. Rogožin, *A class of infinite systems of linear algebraic equations*, Dokl. Akad. Nauk SSSR 114 (1957), 486–489. (Russian)

[15] N. Wiener, *Extrapolation, interpolation and smoothing of stationary time series*; Appendix C by N. Levinson, Wiley, New York, 1949.

[16] B. Ya. Levin, *On a generalization of the Fejér-Riesz theorem*, Dokl. Akad. Nauk SSSR 52 (1946), 291–294. (Russian)

[17] I. C. Gohberg and M. G. Kreĭn, *On some basic propositions of the theory of systems of integral equations on a half-line with kernels depending on the difference of the arguments*, Proceedings of the third all-Union mathematical conference, Moscow, June–July, 1952, Vol. 2, pp. 37–38, Izdat. Akad. Nauk SSSR, Moscow, 1956. (Russian)

[18] ———, *Systems of integral equations on the half-line with kernels depending on the difference of the arguments*, Uspehi Mat. Nauk 13 (1958), no. 2 (80), 3–72. (Russian) = Amer. Math. Soc. Transl. (2) 14 (1960), 217–287.

[19] V. V. Sobolev, *Transfer of radiant energy in the atmosphere of stars and planets*, Gosurdarstv, Izdat. Tehn.-Teor. Lit., Moscow, 1956.

[20] S. Chandrasekhar, *Radiative transfer*, Oxford Univ. Press, New York, 1950; Russian transl., IL, Moscow, 1953.

9

Reprinted with the permission of the publisher from pp. 1–7 of *Convolution Equations and Projection Methods for Their Solution*, Translations of Mathematical Monographs, vol. 41, American Mathematical Society, 1974, 261 pp.

CONVOLUTION EQUATIONS AND PROJECTION METHODS FOR THEIR SOLUTION

I. C. Gohberg and I. A. Fel'dman

I. C. Gohberg and I. A. Fel'dman

Dedicated to
Mark Grigor'evič
KREĬN

PREFACE

The first profound results concerning integral equations on a semiaxis, with kernels depending on the difference of the arguments, were obtained in 1931 by N. Wiener and E. Hopf. After the appearance of their classical paper [1] such equations were given the name Wiener-Hopf equations.

By the present time the theory of Wiener-Hopf equations has been quite fully developed. A great many mathematicians (mainly Soviet) participated in its elaboration. We mention only M. G. Kreĭn's fundamental paper [4]. Beginning with that paper the ideas and methods of functional analysis have had extensive application in the analysis of Wiener-Hopf equations.

The presentation adopted in this book is also based on functional analysis. It has its origins in a distinctive operational calculus.

This route leads naturally to a definite class of convolution equations which contains in particular the original Wiener-Hopf integral equations. Besides that, discrete and difference analogues, the so-called pair equations, singular integral equations on the circle, etc. belong here.

Various projection methods of solving convolution equations make up a significant part of the book. Justification of these methods is also obtained within the framework of the general scheme.

To a considerable extent the presentation is based on papers of the authors which were published in 1963–1967.

The book was conceived in 1964. At that time the authors began reading special courses at Kishinev University. Preliminary publication of the mimeographed brochure "Projection Methods of Solving Wiener-Hopf Equations" (Kishinev, 1967; MR **37** # 1915) helped us substantially in the work on the book.

We assume that the reader is acquainted with the elements of the theory of operators in Hilbert and Banach spaces and the theory of Banach algebra.

The authors sincerely thank M. G. Kreĭn for support and abiding interest in this paper, as well as N. Ja. Krupnik, A. S. Markus, A. A. Semencul and I. B. Simonenko for discussion of various questions and valuable comments.

The authors express sincere gratitude to the book's editor, F. V. Širokov. His assistance contributed significantly to the simplicity and clarity of the presentation.

Kishinev
February 18, 1970

INTRODUCTION

1. Various classes of convolution equations are considered in this book. Among these are integral and discrete Wiener-Hopf equations, integral-difference equations, equations in finite differences, various pair equations, singular integral equations, etc.

The discrete Wiener-Hopf equation

$$\sum_{k=1}^{\infty} a_{j-k}\xi_k = \eta_j \qquad (j = 1, 2, \cdots), \tag{0. 1}$$

where $\xi = \{\xi_j\}_1^\infty$ is the vector sought, and $\eta = \{\eta_j\}_1^\infty$ and $\{a_j\}_{-\infty}^{\infty}$ are given vectors, is an elementary representive of such equations.

We shall consider equation (0.1) in the space l_p $(1 \leqq p < \infty)$. For the sake of simplicity let us assume that the condition

$$\sum_{-\infty}^{\infty} |a_j| < \infty \tag{0. 2}$$

is fulfilled.

Let us denote by A the operator generated in the space l_p by the matrix $\|a_{j-k}\|_1^\infty$, It is easy to see that A expands into the series

$$A = \sum_{-\infty}^{\infty} a_j V_j,$$

which is convergent in the operator norm. Here $V_0 = I$, and the V_j $(j = \pm 1, \cdots)$ are shift operators defined by the equalities

$$V_j\{\xi_1, \xi_2, \cdots\} = \{0, \cdots, 0, \xi_1, \xi_2, \cdots\}, \qquad j = 1, 2, \cdots,$$

and

$$V_{-j}\{\xi_1, \xi_2, \cdots\} = \{\xi_{j+1}, \xi_{j+2}, \cdots\}, \quad j = 1, 2, \cdots.$$

The operator $V = V_1$ is isometric, while the operator $V^{-1} = V_{-1}$ is a left inverse of V. Obviously $V_j = V^j\,(j = 2, 3, \cdots)$ and $V_k = (V^{-1})^{|k|}$ $(k = -2, -3, \cdots)$.

The operator A can be regarded as the value of the function

$$a(\zeta) = \sum_{-\infty}^{\infty} a_j \zeta^j \qquad (|\zeta| = 1)$$

of the operator V, i. e. $A = a(V)$.

This definition differs in some sense from the one usually adopted in functional analysis. The fact is that a fundamental principle of constructing functions of operators is *violated* here: the spectrum of the operator V coincides with the unit disk at the same time that the function $a(\zeta)$ is defined only on the unit circle. As a result, the multiplicativity of the correspondence $a(\zeta) \to a(V)$ is violated, i. e. in general the equality

$$a_1(V)a_2(V) = a(V)$$

does not hold, where $a_1(\zeta)$ and $a_2(\zeta)$ ($|\zeta| = 1$) are functions expanding into absolutely convergent Fourier series, and $a(\zeta) = a_1(\zeta)a_2(\zeta)$. This is immediately evident in the example

$$a_1(\zeta) = \zeta, \qquad a_2(\zeta) = \zeta^{-1}, \qquad a(\zeta) = 1,$$

because $VV^{-1} \neq I$. However, if the function $a_0(\zeta)$ expands into an absolutely convergent Fourier series, if $a_+(\zeta)$ ($a_-(\zeta)$) expands into an absolutely convergent Fourier series in nonnegative (nonpositive) powers of ζ, and if $b(\zeta) = a_-(\zeta)a_0(\zeta)a_+(\zeta)$, then

$$b(V) = a_-(V)a_0(V)a_+(V). \tag{0.3}$$

(The order of multiplying is vital!) Relation (0,3), the *partial multiplicativity* of the correspondence, permits a distinctive operational calculus to be constructed not only for this particular operator V but also for the unilaterally invertible operators of quite a broad class.

The question of the invertibility of functions of unilaterally invertible operators of the form

$$a(V) = \sum_{-\infty}^{\infty} a_j V^j$$

is central here.

Let V be an *arbitrary* isometric operator operating in the Banach space $\mathfrak{B}$, and let it have a left inverse V^{-1} such that $|V^{-1}| = 1$. Then the following theorem is valid.

THEOREM 0.1. *Let the function $a(\zeta)$ ($|\zeta| = 1$) expand into an absolutely convergent Fourier series. For the operator $A = a(V)$ to be invertible, even if only on one side, it is necessary and sufficient that the function $a(\zeta)$ vanish nowhere on the unit circle ($a(\zeta) \neq 0$, $|\zeta| = 1$).*

If this condition is fulfilled, the operator A is invertible on the left, on the right, or on both sides, depending on whether the number

$$\kappa = \operatorname{ind} a(\zeta) = (1/2\pi)[\arg a(e^{i\varphi})]_{\varphi=0}^{2\pi}$$

is positive, negative, or equal to zero.[1)]

[1)]The symbol $[\ \ ]_0^{2\pi}$ denotes the increment of the function on the segment $[0, 2\pi]$.

Let us explain how the appropriate inverse operator of $a(V)$ whose existence is guaranteed by Theorem 0.1 is *constructed.*

If the function $a(\zeta)$ does not vanish on the circle $|\zeta| = 1$, then it admits *factorization,* i. e. is representable in the form

$$a(\zeta) = a_-(\zeta)\zeta^\kappa a_+(\zeta) \qquad (|\zeta| = 1), \tag{0.4}$$

where the functions $a_-(\zeta)$ and $1/a_-(\zeta)$ expand into absolutely convergent series in nonpositive powers of ζ; analogously $a_+(\zeta)$ and $1/a_+(\zeta)$ expand into series in nonnegative powers of ζ.

Using the partial multiplicativity (0.3), we obtain

$$a(V) = a_-(V)V^\kappa a_+(V).$$

Then the operator $[a(V)]^{-1}$, which is inverse to $a(V)$ on the appropriate side, is given by the equality

$$[a(V)]^{-1} = a_+^{-1}(V)V^{-\kappa}a_-^{-1}(V).$$

(Let us recall that V^{-1} is a left inverse of V.)

This is an existence theorem, and this inversion method is pivotal to the entire general theory.

2. Making practical use of the above described method of inverting the operator $a(V)$ is sometimes difficult. Therefore various projection methods for solving convolution equations play an important part in the book. We examine in particular the reduction method, the Galerkin method and some generalizations of the latter.

The most prevalent among the projection methods of solving the *general* infinite system of linear equations

$$\sum_{k=-\infty}^{\infty} a_{jk}\xi_k = \eta_j \qquad (j = 0, \pm 1,\cdots) \tag{0.5}$$

is the *reduction method.* Here an approximate solution is sought in the form

$$\xi^{(n)} = \{\cdots, 0, \xi_{-n}^{(n)},\cdots,\xi_n^{(n)}, 0,\cdots\},$$

where the vector $\{\xi_j^{(n)}\}_{j=-n}^{n}$ is the solution of the finite *truncated* system

$$\sum_{k=-n}^{n} a_{jk}\xi_k = \eta_j \qquad (j = 0, \pm 1,\cdots, \pm n). \tag{0.6}$$

Sometimes a convenient procedure leads to the objective, i. e. system (0.6), beginning with some n, turns out to be uniquely solvable, and the vectors $\xi^{(n)}$ converge to a solution of system (0.5) in the norm of the appropriate space.

This is automatically true if system (0.5), when considered in the Hilbert[2)]

[2)] $\tilde{l}_2$ is the Hilbert space of sequences $\xi = \{\xi_j\}_{-\infty}^{\infty}$ of complex numbers indexed from $-\infty$ to ∞. The symbol $\tilde{l}_p$ is used analogously.

space $\tilde{l}_2$, is solvable for any right side in $\tilde{l}^2$ and if

$$\sum_{j,k=-\infty}^{\infty} |a_{jk} - \delta_{jk}|^2 < \infty.$$

This and other well-known cases of applying the reduction method in $\tilde{l}_2$ are embraced on the whole by the following. The operator A defined by system (0.5) must be bilaterally invertible and representable as a sum of positive definite and completely continuous operators.[3]

Unfortunately this classical method is inapplicable even to the simple system

$$\xi_{j+1} = \eta_j \qquad (j = 0, \pm 1, \cdots), \tag{0.7}$$

because here the matrix of the truncated system (0.6) is degenerate for any n. At the same time the operator A is unitary here. A similar situation also arises in the case of the more general equation

$$\sum_{k=-\infty}^{\infty} a_{j-k}\xi_k = \eta_j \qquad (j = 0, \pm 1, \cdots), \tag{0.8}$$

where $\sum_{-\infty}^{\infty} |a_j| < \infty$.

If the corresponding function

$$a(\zeta) = \sum_{-\infty}^{\infty} a_j\zeta^j \tag{0.9}$$

does not vanish on the unit circle, then by virtue of Wiener's well-known theorem on division by an absolutely convergent Fourier series the operator $A = \|a_{j-k}\|_{j,k=-\infty}^{\infty}$ will be invertible in the space $\tilde{l}_p$ $(p \geqq 1)$, and a solution of equation (0.8) will be obtained by the formula

$$\xi_j = \sum_{k=-\infty}^{\infty} b_{j-k}\eta_k \qquad (j = 0, \pm 1, \cdots),$$

where the b_j are the Fourier coefficients of the function $1/a(\zeta)$.

The question of the applicability of the classical reduction method to equation (0.8) is completely solved by the following proposition.

THEOREM 0.2. *Let the function $a(\zeta)$ expand into an absolutely convergent Fourier series. For the truncated equation*

$$\sum_{k=-n}^{n} a_{j-k}\xi_k = \eta_j \qquad (j = 0, \pm 1, \cdots, \pm n)$$

to have, beginning with some n, the unique solution $\{\xi_j^{(n)}\}_{j=-n}^{n}$ and for the vectors

---

[3] Under these conditions the operator A admits reduction relatively to *any* orthonormal basis (cf. 1°, §2, and Theorem 3.1, Chapter II).

$$\xi^{(n)} = \{\cdots, 0, \xi_{-n}^{(n)}, \cdots, \xi_n^{(n)}, 0, \cdots\}$$

to converge (as $n \to \infty$) in the norm of the space $\tilde{l}_p$ to a solution of the full equation (0.8) *for any right side $\{\eta_j\}_{-\infty}^{\infty} \in \tilde{l}_p$, it is necessary and sufficient that*

a) $a(\zeta) \neq 0$ $(|\zeta| = 1)$,

b) $\kappa = \operatorname{ind} a(\zeta) = 0$.

Condition a) in Theorem 0.2 is natural; it is equivalent to the invertibility of operator A. If *only* condition a) is fulfilled, it turns out that after shifting the "center" of equation (0.8) by the *index* κ, i.e. after writing it in the form

$$\sum_{k=-\infty}^{\infty} a_{j-k+\kappa}\xi_k = \eta_{j+\kappa} \qquad (j = 0, \pm 1, \cdots),$$

we obtain the equation to which we now apply the reduction method. The equation, truncated relative to the new "center"

$$\sum_{k=-n}^{n} a_{j-k+\kappa}\xi_k = \eta_{j+\kappa} \qquad (j = 0, \pm 1, \cdots, \pm n),$$

will have, beginning with some n, a unique solution, and the vectors $\xi^{(n)}$ will converge to a solution of the *full* equation (0.8).

Thus considerations involving applicability of the reduction method distinguish the indicial diagonal $j - k = \kappa$ in the matrix $\|a_{j-k}\|_{j,k=-\infty}^{\infty}$; for the method to be applicable, exactly this diagonal must become principal in the matrices of the truncated systems.

An analogue of Theorem 0.2 holds for the discrete Wiener-Hopf equation

$$\sum_{k=1}^{\infty} a_{j-k}\xi_k = \eta_j \qquad (j = 1, 2, \cdots), \tag{0.10}$$

considered in the space l_p $(p \geq 1)$.

Now, in contrast with the foregoing, conditions a) and b) together serve as the condition for the invertibility of the corresponding operator A. It is amusing that although equation (0.8) is essentially simpler than equation (0.10), proof of the convergence of the reduction method for (0.8) is derived from the corresponding theorem for equation (0.10).

An analogous situation arises in the continuous case. For example, let us cite the theorem for the Wiener-Hopf equation

$$\varphi(t) - \int_0^{\infty} k(t-s)\varphi(s)\, ds = f(t) \qquad (0 < t < \infty). \tag{0.11}$$

Let us assume that the kernel $k(t)$ belongs to $L_1(-\infty, \infty)$. Here the analogue of the function $a(\zeta)$ is the Fourier transform

$$K(\lambda) = \int_{-\infty}^{\infty} k(t)e^{i\lambda t}\, dt.$$

THEOREM 0.3 *For the truncated equation*

$$\varphi(t) - \int_0^\tau k(t-s)\varphi(s)\,ds = f(t) \qquad (0<t<\tau)$$

to have, beginning with some τ, *the unique solution* $\varphi_\tau(t) \in L_p(0, \tau)$, *and for the functions*

$$\tilde{\varphi}_\tau(t) = \begin{cases} \varphi_\tau(t), & 0<t\leqq\tau, \\ 0, & \tau<t<\infty, \end{cases}$$

which are continued as zero, to converge, as $\tau \to \infty$, *in the norm of* $L_p(0, \infty)$ *to a solution of equation* (0.11) *for any right side* $f(t) \in L_p(0, \infty)$, *it is necessary and sufficient that*

a) $1 - K(\lambda) \neq 0 \ (-\infty < \lambda < \infty)$,

b) $\kappa = (1/2\pi)[\arg(1 - K(\lambda))]_{\lambda=-\infty}^{\infty} = 0$.

All of these theorems on the reduction method are comprised in the general propositions concerning the applicability of projection methods to functions of unilaterally invertible operators.

Besides the case of the continuous function $a(\zeta)$ we consider the case of *discontinuous* $a(\zeta)$. For the time being let us confine ourselves to formulating just a few corollaries of the general theorem, passing in the process from the circle to the real axis.

Let A be a bounded linear operator of multiplication by the continuous function $\omega(t)$, let it operate in $L_2(0, 1)$ by the rule

$$(A\varphi)(t) = \omega(t)\varphi(t),$$

and let $\|a_{jk}\|_{-\infty}^{\infty}$ be the matrix of this operator in the orthogonal basis

$$\varphi_0(t) = 1, \quad \varphi_j(t) = \sin 2\pi jt, \quad \varphi_{-j}(t) = \cos 2\pi jt \qquad (j = 1, 2, \cdots).$$

THEOREM 0.4. *For the reduction method to be applicable to the system of equations with matrix* $\|a_{jk}\|_{-\infty}^{\infty}$ *it is necessary and sufficient that*

1) $\omega(t) \neq 0 \ (0 \leq t \leq 1)$,

2) $|[\arg \omega(t)]_{t=0}^{t=1}| < \pi$.

The reduction method described for the equation

$$\omega(t)\varphi(t) = f(t)$$

appearing in the theorem is actually the Galerkin method. Thus the theorem just formulated gives the necessary and sufficient condition for the applicability of the Galerkin method to this elementary equation on a basis of the simplest orthogonal system. It is curious that the "necessary" condition 2) appears even in this, one would think, "trivial" case.

With this we complete our consideration of the "scalar" case.

The next step is considering the "matricial" case, i. e. the case when the a_j ($j = 0, \pm 1, \ldots$) are matrices and $k(t)$ is an nth order matrix-function. For such equations the convergence of one or another approximating method is determined by the *left* and *right indices* of the respective matrix-functions. It appears that now the vanishing of all left and right indices is necessary and sufficient for the *convergence* of the reduction method, whereas only the vanishing of the right indices is necessary and sufficient for the *invertibility* of the corresponding operator.

3. The reader already has some idea of the book's content. Let us pass to a review of it by chapters.

The theory of elementary classes of convolution equations is set forth in the first chapter.

The third chapter is devoted to justifying the reduction method for discrete Wiener-Hopf equations, the Galerkin method for integral Wiener-Hopf equations, etc. The essential role here, as in the subsequent chapters, is played by the general theorems concerning the projection methods for solving linear equations which are set forth in the second chapter.

The applicability of one or another projection method is determined by the index of the corresponding function. An iterative method for calculating the index of a rational function is set forth in § 5 of the third chapter. In the same chapter, in review order, theorems are cited and questions formulated concerning projection methods for solving Wiener-Hopf equations on a "multi-dimensional" half-space. In addition a method for inverting a finite Toeplitz matrix and its continuous analogue is set forth in the same chapter.

In the fourth chapter several results of the first and third chapters are generalized to the case of piecewise continuous functions of isometric operators in Hilbert space.

In the following two chapters the theory and projection methods are set forth for various classes of pair convolution equations. In particular, singular integral equations on the circle are put here.

The seventh chapter is devoted to integral-difference Wiener-Hopf equations.

The eighth contains generalizations to the matricial case of several results of the preceding chapters.

Asymptotic methods of solving homogeneous convolution equations are set forth in the Appendix.

[*Editor's Note:* Material has been omitted at this point.]

10

Reprinted from *J. Appl. Phys.* **21**(7):645–655 (1950)

An Extension of Wiener's Theory of Prediction*

LOTFI A. ZADEH AND JOHN R. RAGAZZINI
Department of Electrical Engineering, Columbia University, New York, New York

(Received November 14, 1949)

The theory of prediction described in this paper is essentially an extension of Wiener's theory. It differs from the latter in the following respects.

1. The signal (message) component of the given time series is assumed to consist of two parts, (a) a non-random function of time which is representable as a polynomial of degree not greater than a specified number n and about which no information other than n is available; and (b) a stationary random function of time which is described statistically by a given correlation function. (In Wiener's theory, the signal may not contain a non-random part except when such a part is a known function of time.)

2. The impulsive response of the predictor or, in other words, the weighting function used in the process of prediction is required to vanish outside of a specified time interval $0 \leq t \leq T$. (In Wiener's theory T is assumed to be infinite.)

The theory developed in this paper is applicable to a broader and more practical class of problems than that covered in Wiener's theory. As in Wiener's theory, the determination of the optimum predictor reduces to the solution of an integral equation which, however, is a modified form of the Wiener-Hopf equation. A simple method of solution of the equation is developed. This method can also be applied with advantage to the solution of the particular case considered by Wiener. The use of the theory is illustrated by several examples of practical interest.

I. INTRODUCTION

PREDICTION—in the broad sense of the term—consists essentially of estimating the values of some function of time on the basis of a time series which may or may not contain random errors. For instance, a typical problem in prediction is as follows. Given a time series $e_1(t)$ which is composed of a signal $s(t)$ and a random disturbance (noise) $N(t)$; provide an estimate of $s(t+\alpha)$, α being a positive constant, as a continuous function of time. More generally, the quantity to be estimated may be a functional of $s(t)$ such as ds/dt, $\int s dt$, etc. In forming such estimates the mathematical operations that may be employed are usually limited by practical considerations. Thus, in most cases the operator furnishing the estimate must be linear and fixed in addition to the obvious requirement of being physically realizable. The physical counterpart of such an operator is what is commonly known as a predictor or an estimator.

It is evident that a function of time cannot be predicted intelligently unless sufficient *a priori* information is available about both the function and the errors. The nature of such information, as well as the characteristics of the signal and noise, can assume a variety of forms. Of these the more common ones have been investigated in recent years with the result that for certain classes of time series it is now possible to design predictors which make optimum use of the *a priori* information concerning the signal and the noise. Thus, when the given time series is stationary and the correlation functions of the signal and noise are known, one can use Wiener's theory[1] to arrive at the specifications of the optimum predictor, that is, one minimizing the mean-square value of the prediction error. On the other hand, when, as is often the case in practice, the given time series is non-stationary, the available theories of prediction, notably Phillips and Weiss' theory,[2] do not lead to the best possible predictor except for a narrow class of time series. It is possible, however, to extend Wiener's theory in many different directions thereby making it applicable to a wider class of problems than is covered by either Wiener's or Phillips and Weiss' theories in their present forms. One such extension is discussed in this paper. A feature of the extension is that the signal (message) is assumed to consist of a stationary component superimposed on a non-random function of time which is known to be representable as a polynomial of degree less than or equal to a specified number n. It will also be shown that the general method developed for treating this problem can be applied with advantage to the solution of many cases of practical

* This work was performed in association with the Special Projects Department of The M. W. Kellogg Company for the Watson Laboratories of the Air Materiel Command.

[1] N. Wiener, "The extrapolation, interpolation, and smoothing of stationary time series," Report of the Services 19, Research Project DIC-6037, M.I.T. (February, 1942). Published in book form by John Wiley and Sons, Inc., New York (1949).

[2] R. S. Phillips and P. R. Weiss, "Theoretical calculation on best smoothing of position data for gunnery prediction," Report 532, Rad. Lab., M.I.T. (February, 1944).

interest as well as the particular case considered by Wiener.

II. FORMULATION

Consider a given time series $e_1(t)$ which is the sum of a function $s(t)$ (signal) and a stationary random disturbance $N(t)$. Let $s^*(t)$ be the quantity to be estimated and let $s^*(t)$ be related to $s(t)$ through a given linear operator $K(p)$, i.e.,

$$s^*(t)=K(p)s(t). \tag{1}$$

$K(p)$ may be thought of as the system function of an *ideal predictor*, i.e., a predictor capable of perfect prediction in the absence of noise. In many cases, particularly those involving actual prediction, the operator $K(p)$ is not physically realizable so that the process of estimation cannot be carried out exactly even in the absence of random disturbances.

Frequently it will be convenient to use a different, though equivalent representation of Eq. (1), i.e.,

$$s^*(t)=\int_{-\infty}^{\infty} k(\tau)s(t-\tau)d\tau, \tag{2}$$

where τ is the variable of integration and $k(t)$ represents the impulsive response of the ideal predictor. $K(p)$ shall be referred to as the ideal prediction operator. As a matter of convenience, the more common of the many possible forms which $K(p)$ and $k(t)$ can assume are given in Table I.

Like all theories of prediction, the theory to be described applies only to a special class of time series. The time series to be considered in the work which follows will be assumed to consist of a signal $s(t)$ and noise $N(t)$, with the signal being composed of a random component $M(t)$ superposed upon a non-random function of time $P(t)$, i.e.,

$$s(t)=M(t)+P(t). \tag{3}$$

The assumptions made concerning the characteristics of $P(t)$, $M(t)$, and $N(t)$, are as follows:

(a) $P(t)$ is assumed to be representable as a polynomial in t of degree not higher than a specified number n.

(b) $M(t)$ and $N(t)$ are stationary functions of time described respectively by their auto-correlation functions $\psi_M(\tau)$ and $\psi_N(\tau)$.

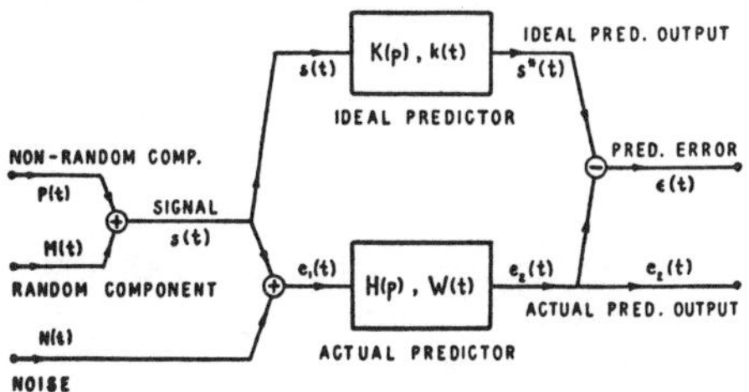

FIG. 1. Flow diagram of prediction process.

TABLE I. Common forms of the prediction operators $K(p)$ and $k(t)$.

Relation between $s^*(t)$ and $s(t)$	Significance of the quantity to be estimated	$K(p)$	$k(t)$
$s^*(t)=s(t)$	Present value of $s(t)$	1	$\delta(t)$
$s^*(t)=\dot{s}(t)$	Present value $\dot{s}(t)$	p	$\delta^{(1)}(t)$
$s^*(t)=\ddot{s}(t)$	Present value of $\ddot{s}(t)$	p^2	$\delta^{(2)}(t)$
$s^*(t)=s(t-\alpha)$	Past or future value of $s(t)$ depending respectively on whether α is a positive or negative constant	$e^{-\alpha p}$	$\delta(t-\alpha)$

Note.—$\delta(t)$ denotes a unit impulse at $t=0$, and $\delta^{(\nu)}(t)$ stands for the νth derivative of $\delta(t)$ with respect to t (time).

(c) $M(t)$ and $N(t)$ have zero mean and are uncorrelated. This assumption is introduced only for the purpose of simplification and is not essential to the analysis. The condition expressed by (c) prevails in most practical cases.

Referring to Fig. 1, these inputs are shown being applied to the actual predictor whose system function is $H(p)$ and whose impulsive response is $W(t)$. The output of the predictor, $e_2(t)$, may be expressed in operational form

$$e_2(t)=H(p)e_1(t) \tag{4}$$

or, alternatively, in the form of a superposition integral

$$e_2(t)=\int_0^{\infty} W(\tau)e_1(t-\tau)d\tau. \tag{5}$$

An important characteristic of the actual predictor is the so-called prediction or estimation error ϵ, which is defined as the difference between the output of the predictor and the quantity to be estimated, $s^*(t)$. Equation-wise this is:

$$\epsilon=e_2(t)-s^*(t). \tag{6}$$

If there were no noise and if $K(p)$ were physically realizable there would be no prediction error and $H(p)$ would be identical with $K(p)$. This, of course, is the trivial case of the prediction problem. In what follows it will be assumed that either because of the presence of noise or physical unrealizability of $K(p)$, or both, $H(p)$ cannot be the same as $K(p)$.

The available *a priori* information about $s(t)$ and $N(t)$ is assumed to consist of n, $\psi_M(\tau)$, and $\psi_N(\tau)$. The problem is to specify the system function or the impulsive response of a predictor that would minimize in a certain sense the prediction error $\epsilon=e_2(t)-s^*(t)$. By analogy with Wiener's theory it will be postulated that the optimum predictor is the one in which: (a) the ensemble mean of ϵ is equal to zero (for all values of t), and (b) the ensemble variance of ϵ is a minimum. Denoting the ensemble average by the symbol $\langle\ \rangle_{Av}$, these conditions read:

(a) $\langle\epsilon\rangle_{Av}=0$ or, equivalently, $\langle e_2(t)\rangle_{Av}\equiv\langle s^*(t)\rangle_{Av}$, (7)

(b) $\sigma^2=\langle\epsilon^2\rangle_{Av}=\text{minimum},$ (8)

where σ^2, the ensemble variance of ϵ, is equal to the mean-square value of the prediction error. In what follows, conditions (a) and (b) will be used as the basis for the determination of the optimum predictor.

III. DETERMINATION OF THE IMPULSIVE RESPONSE OF THE OPTIMUM PREDICTOR

It will be recalled that the output of a predictor may be expressed in the form of a superposition integral

$$e_2(t)=\int_0^\infty W(\tau)e_1(t-\tau)d\tau \qquad (9)$$

where τ is a dummy variable and $W(t)$ represents the impulse response of the predictor. In practice it is usually found necessary to restrict the duration of sampling of the input time series to a finite constant T, meaning in other words that $W(t)$ must be zero outside the interval $0\leq t\leq T$. To place this property in evidence Eq. (9) will be written in the following form:

$$e_2(t)=\int_0^T W(\tau)e_1(t-\tau)d\tau. \qquad (10)$$

In the limiting case where the duration of sampling is infinite ($T\rightarrow\infty$) Eq. (10) becomes identical with Eq. (9).

By hypothesis,

$$e_1(t)=P(t)+M(t)+N(t). \qquad (11)$$

Substituting Eq. (11) into Eq. (10) and making use of the identity

$$P(t-\tau)\equiv P(t)-\tau\dot{P}(t)+\frac{\tau^2}{2!}\ddot{P}(t)+\cdots+(-1)^n\frac{\tau^n}{n!}P^{(n)}(t), \qquad (12)$$

it is found that $e_2(t)$ may be expressed as:

$$e_2(t)=\mu_0P(t)-\mu_1\dot{P}(t)+\frac{\mu_2}{2!}\ddot{P}(t)+\cdots+(-1)^n\frac{\mu_n}{n!}P^{(n)}(t)+\int_0^T W(\tau)M(t-\tau)d\tau+\int_0^T W(\tau)N(t-\tau)d\tau, \qquad (13)$$

where μ_0, μ_1, μ_2, etc., designate the moments of $W(t)$, i.e.,

$$\mu_\nu=\int_0^T \tau^\nu W(\tau)d\tau, \quad \nu=0,1,2,\cdots n. \qquad (14)$$

Since $M(t)$ and $N(t)$ are stationary (with zero mean), it follows that the ensemble means of $e_2(t)$ and $s^*(t)$ depend only on the non-random component of the signal, i.e.,

$$\langle e_2(t)\rangle_{Av}=\int_0^T W(\tau)P(t-\tau)d\tau \qquad (15)$$

or

$$\langle e_2(t)\rangle_{Av}=\mu_0P(t)-\mu_1\dot{P}(t)+\frac{\mu_2}{2!}\ddot{P}(t)+\cdots+(-1)^n\frac{\mu_n}{n!}P^{(n)}(t), \qquad (16)$$

and

$$\langle s^*(t)\rangle_{Av}=\langle K(p)s(t)\rangle_{Av} \qquad (17)$$

or

$$\langle s^*(t)\rangle_{Av}=K(p)P(t). \qquad (18)$$

Comparing Eqs. (16) and (18), condition (a) is reduced to

$$K(p)P(t)\equiv\mu_0P(t)-\mu_1\dot{P}(t)+\frac{\mu_2}{2!}\ddot{P}(t)+\cdots+(-1)^n\frac{\mu_n}{n!}P^{(n)}(t). \qquad (19)$$

Equation (19), being an identity, determines the values of μ_0, μ_1, $\cdots$, and μ_n. In other words, *the ideal prediction operator $K(p)$ determines through Eq. (19) the first $n+1$ moments of the impulsive response of the optimum predictor.*

As an illustration of the foregoing statement consider a case where the quantity to be estimated is the derivative of $s(t)$, i.e., $s^*(t)=\dot{s}(t)$. For this case Eq. (19) reduces to

$$\dot{P}(t)\equiv\mu_0P(t)-\mu_1\dot{P}(t)+\frac{\mu_2}{2!}\ddot{P}(t)+\cdots+(-1)^n\frac{\mu_n}{n!}P^{(n)}(t), \qquad (20)$$

and a term by term comparison of the left-hand and right-hand sides of Eq. (20) yields:

$$\begin{aligned}\mu_0&=\int_0^T W(\tau)d\tau=0\\ \mu_1&=\int_0^T \tau W(\tau)d\tau=-1\\ \mu_2&=\int_0^T \tau^2W(\tau)d\tau=0\\ &\cdots\\ \mu_n&=\int_0^T \tau^nW(\tau)d\tau=0.\end{aligned} \qquad (21)$$

These, therefore, are the $n+1$ constraints which the impulsive response of a derivative estimating network must satisfy.

As the second example consider a case where $K(p)s(t)=s(t-\alpha)$, α being a positive or negative constant. For this case Eq. (19) reads

$$P(t-\alpha)\equiv\mu_0P(t)-\mu_1\dot{P}(t)+\frac{\mu_2}{2!}\ddot{P}(t)+\cdots+(-1)^n\frac{\mu_n}{n!}P^{(n)}(t). \quad (22)$$

Rewriting $P(t-\alpha)$ as

$$P(t-\alpha)\equiv P(t)-\alpha\dot{P}(t)+\frac{\alpha^2}{2!}\ddot{P}(t)+\cdots+(-1)^n\frac{\alpha^n}{n!}P^{(n)}(t), \quad (23)$$

and making in Eq. (22) a term-by-term comparison of the coefficients of $P(t)$, $\dot{P}(t)$, etc., it is easily found that:

$$\begin{aligned}\mu_0&=\int_0^T W(\tau)d\tau=1\\ \mu_1&=\int_0^T \tau W(\tau)d\tau=\alpha\\ &\cdots\\ \mu_n&=\int_0^T \tau^n W(\tau)d\tau=\alpha^n,\end{aligned} \quad (24)$$

which thus represent the constraints imposed upon $W(t)$ in case the quantity to be estimated is $s(t-\alpha)$.

The problem that remains to be solved is that of minimizing σ^2. For this purpose it will be necessary to develop an explicit expression for σ^2 in terms of $W(t)$ and the auto-correlation functions of the signal and noise. Assuming that condition (a) is satisfied, it follows from inspection of Eqs. (6), (13), and (19) that the prediction error is given by the expression

$$\epsilon=\int_0^T W(\tau)[M(t-\tau)+N(t-\tau)]d\tau-K(p)M(t) \quad (25)$$

or equivalently

$$\epsilon=\int_0^T W(\tau)[M(t-\tau)+N(t-\tau)]d\tau-\int_{-\infty}^{\infty} k(\tau)M(t-\tau)d\tau \quad (26)$$

where $k(t)$ is the impulsive response of the ideal predictor. The mean-square value of ϵ may be written as

$$\sigma^2=\langle\epsilon^2\rangle_{Av}=\lim_{L\to\infty}\frac{1}{L}\int_0^L \epsilon^2 dt \quad (27)$$

or

$$\sigma^2=\lim_{L\to\infty}\frac{1}{L}\int_0^L dt\left\{\int_0^T W(\tau)[M(t-\tau)+N(t-\tau)]d\tau-\int_{\infty}^{\infty} k(\tau)M(t-\tau)d\tau\right\}^2. \quad (28)$$

A typical term of Eq. (28) such as

$$\lim_{L\to\infty}\frac{1}{L}\int_0^L dt\left[\int_0^T W(\tau)M(t-\tau)d\tau\right]^2 \quad (29)$$

is expressible in the form of a triple integral

$$\int_0^T\int_0^T d\tau_1 d\tau_2 W(\tau_1)W(\tau_2)\times\lim_{L\to\infty}\frac{1}{L}\int_0^L M(t-\tau_1)M(t-\tau_2)dt \quad (30)$$

which in view of the definition of the auto-correlation function of $M(t)$, i.e.,

$$\psi_M(\tau)=\lim_{L\to\infty}\frac{1}{L}\int_0^L M(t)M(t-\tau)dt \quad (31)$$

may be written as

$$\int_0^T\int_0^T W(\tau_1)W(\tau_2)\psi_M(\tau_1-\tau_2)d\tau_1 d\tau_2. \quad (32)$$

Proceeding similarly in the case of other terms, Eq. (28) reduces finally to the following expression:

$$\begin{aligned}\sigma^2=&\int_0^T\int_0^T W(\tau_1)W(\tau_2)[\psi_M(\tau_1-\tau_2)+\psi_N(\tau_1-\tau_2)]d\tau_1 d\tau_2\\ &-2\int_{\infty}^{\infty}\int_0^T W(\tau_1)k(\tau_2)\psi_M(\tau_1-\tau_2)d\tau_1 d\tau_2\\ &+\int_{-\infty}^{\infty}\int_{-\infty}^{\infty} k(\tau_1)k(\tau_2)\psi_M(\tau_1-\tau_2)d\tau_1 d\tau_2\end{aligned} \quad (33)$$

where, to recapitulate: τ_1, τ_2 = dummy variables; $W(t)$ = impulsive response of the predictor; $\psi_M(\tau)$ = auto-correlation function of $M(t)$ [$M(t)$ is the stationary part of the input signal]; $\psi_N(\tau)$ = auto-correlation function of $N(t)$ [$N(t)$ is the input noise]; $k(t)$ = impulsive response of the ideal predictor.

Returning to the problem of minimization of σ^2 it will be noted first that the last term in Eq. (33) is independent of $W(t)$ and hence, insofar as minimization of σ^2 is concerned, need not be considered. Second, it will be recalled that $W(t)$ is subject to the $n+1$ constraints expressed by Eq. (14); therefore, the problem of minimizing σ^2 with respect to the class of $W(t)$'s satisfying Eq. (14) reduces essentially to an isoperimetric problem in the calculus of variations. Following the standard approach to such problems, one is led to minimizing the following expression:

$$I=\sigma^2-2\lambda_0\mu_0-2\lambda_1\mu_1-\cdots-2\lambda_n\mu_n \tag{34}$$

or, more explicitly

$$I=\int_0^T W(\tau_1)d\tau_1\left\{\int_0^T W(\tau_2)[\psi_M(\tau_1-\tau_2)+\psi_N(\tau_1-\tau_2)]d\tau_2-2\int_{-\infty}^{\infty}k(\tau_2)\psi_M(\tau_1-\tau_2)d\tau_2-2\lambda_0-2\lambda_1\tau_1-\cdots-2\lambda_n\tau_1^n\right\} \tag{35}$$

where the constants $\lambda_0, \lambda_1, \cdots, \lambda_n$, are the Lagrangian multipliers. Proceeding in the usual manner, that is, setting the variation of I equal to zero, it is easily found that I and hence σ^2 is a minimum provided $W(t)$ satisfies the following integral equation:

$$\int_0^T W(\tau)[\psi_M(t-\tau)+\psi_N(t-\tau)]d\tau=\lambda_0+\lambda_1 t+\cdots+\lambda_n t^n+\int_{-\infty}^{\infty}k(\tau)\psi_M(t-\tau)d\tau,\quad 0\le t\le T. \tag{36}$$

This equation together with the $n+1$ constraints expressed by Eq. (14) provides the basis for the determination of the optimum predictor. It will be observed that in the particular case where $n=0$, $T=\infty$, and $k(t)=\delta(t+\alpha)$ [$\delta(t)$ standing, as usual, for a unit impulse at $t=0$], Eq. (36) reduces to

$$\int_0^{\infty} W(\tau)[\psi_M(t-\tau)+\psi_N(t-\tau)]d\tau=\psi_M(t+\alpha),\quad t>0 \tag{37}$$

which is essentially the integral equation of Wiener's theory. On the other hand, in the special case where $M(t)\equiv 0$, Eq. (36) reduces to

$$\int_0^T W(\tau)\psi_N(t-\tau)d\tau=\lambda_0+\lambda_1 t+\cdots+\lambda_n t^n,\quad 0\le t\le T \tag{38}$$

which is the integral equation of Phillips and Weiss' theory. Thus, the integral equations of Wiener's, and Phillips and Weiss' theories are special cases of Eq. (36).

IV. SOLUTION OF THE INTEGRAL EQUATION

In the general case where $\psi_M(\tau)$ and $\psi_N(\tau)$ are prescribed but otherwise arbitrary auto-correlation functions, the complicated nature of the integral equation makes it appear that the solution of Eq. (36) is a formidable problem. In reality, the problem is not as difficult as it may seem, for by using a procedure to be described, the general case can be reduced to a special case which has a simple solution.

Preliminary to the discussion of this procedure it will be expedient to introduce the spectral densities of $M(t)$, $N(t)$, and $M(t)+N(t)$. Denoting these by $S_M(\omega^2)$, $S_N(\omega^2)$, and $S(\omega^2)$, respectively, and recalling that the spectral density of a function is the inverse Fourier transform of its auto-correlation function,[1] it follows that

$$S_M(\omega^2)=\int_{-\infty}^{\infty}\psi_M(\tau)e^{-j\omega\tau}d\tau \tag{39}$$

$$S_N(\omega^2)=\int_{-\infty}^{\infty}\psi_N(\tau)e^{-j\omega\tau}d\tau \tag{40}$$

and

$$S(\omega^2)=S_M(\omega^2)+S_N(\omega^2). \tag{41}$$

Now the spectral density function $S(\omega^2)$ may be factored[1] into the product of two conjugate factors

$$S(\omega^2)=G(j\omega)\cdot G(-j\omega) \tag{42}$$

such that both $G(j\omega)$ and $1/G(j\omega)$ are analytic in the right half of the $j\omega$-plane. Usually $S(\omega^2)$ is assumed to be a rational function of ω^2 of the form

$$S(\omega^2)=[A(\omega^2)/B(\omega^2)], \tag{43}$$

where $A(\omega^2)$ and $B(\omega^2)$ are polynomials in ω^2. For such cases the process of factorization is quite straightforward as can be seen from the following examples:

(a) $S(\omega^2)=\omega^2$; $\quad G(j\omega)=j\omega$.

(b) $S(\omega^2)=\dfrac{1}{\omega^2+\omega_0^2}$; $\quad G(j\omega)=\dfrac{1}{j\omega+\omega_0}$.

(c) $S(\omega^2)=\dfrac{\omega^2+a^2}{\omega^4+b^2\omega^2+c^4}$;

$$G(j\omega)=\frac{j\omega+a}{(j\omega)^2+j\omega(b^2+2c^2)^{\frac{1}{2}}+c^2}.$$

To summarize, a rational spectral density function may be written as

$$S(\omega^2)=|G(p)|^2_{p=j\omega} \tag{44}$$

where $G(p)$ is of the form:

$$G(p)=\frac{Q(p)}{R(p)}=\frac{a_0+a_1p+\cdots+a_mp^m}{b_0+b_1p+\cdots+b_lp^l}, \tag{45}$$

and the polynomials $Q(p)$ and $R(p)$ do not have any zeros in the right half of the p-plane.

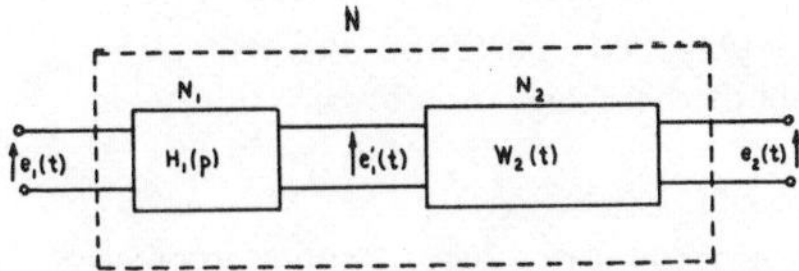

FIG. 2. Division of the predictor into the component networks N_1 and N_2.

The outline of the procedure used[3] for the solution of Eq. (36) can be best explained with reference to Fig. 2. The predictor N is assumed to be composed of two networks N_1 and N_2. The function of N_1 is to suitably modify some of the characteristics of the input time series $e_1(t)$, while that of N_2 is to provide the desired prediction through operating on the time series $e_1'(t)$, which is the output of N_1. It will be seen later that it is possible to choose N_1 in such a manner that the determination of the impulsive response of N_2 becomes an easily solvable problem. Then, once $W_2(t)$ (the impulsive response of N_2) is determined, the impulsive response of N, $W(t)$, can easily be found from the relation

$$W(t)=H_1(p)W_2(t), \tag{46}$$

where $H_1(p)$ is the system function of N_1. The choice of $H_1(p)$ and the problem of determination of $W_2(t)$ are discussed in the sequel.

It is evident that the problem of determination of $W_2(t)$ is similar to that of the determination of $W(t)$, except that the characteristics of the input time series are different for the two problems. An inspection of the integral equation (36) shows that it can be solved rather easily when the input to the predictor consists of a polynomial in t and a stationary component whose spectral density is a polynomial in ω^2. Therefore, in order to make the determination of $W_2(t)$ a simple problem, it is necessary to provide N_2 with an input which has this property. It is not difficult to verify that such a condition will obtain if, and only if, the system function of N_1 is chosen to be

$$H_1(p)=R(p), \tag{47}$$

where $R(p)$ is the denominator of $G(p)$ [cf. Eq. (45)]. With this choice of $H_1(p)$ the input to N_2 will consist of a polynomial in t of the same degree[4] as $P(t)$, and a stationary component $M'(t)+N'(t)$ whose spectral density is

$$S'(\omega^2)=|H_1(j\omega)|^2S(\omega^2) \tag{48}$$

or, in view of Eqs. (43), (44), and (47),

$$S'(\omega^2)=A(\omega^2), \tag{49}$$

where $A(\omega^2)$ is the numerator of $S(\omega^2)$. It will be noted that $A(\omega^2)$ is a polynomial of the form

$$A(\omega^2)=\gamma_0+\gamma_1\omega^2+\cdots+\gamma_m\omega^{2m}, \tag{50}$$

and correspondingly the auto-correlation function of $M'(t)+N'(t)$ is

$$\psi_M'(\tau)+\psi_N'(\tau)=\gamma_0\delta(\tau)-\gamma_1\delta^{(2)}(\tau)+\cdots +(-1)^m\gamma_m\delta^{(2m)}(\tau), \tag{51}$$

where $\delta^{(\nu)}(\tau)$ represents the impulse function of νth order [i.e., the νth derivative of the unit impulse function $\delta(\tau)$].

In addition to $\psi_M'(\tau)+\psi_N'(\tau)$, a number of other quantities associated with the input to N_2 enter the integral equation satisfied by $W_2(t)$. The significance of each of these quantities, as well as their expressions, are as follows:

(a) $S_M'(\omega^2)=$ spectral density of $M'(t)$
$=S_M(\omega^2)|R(j\omega)|^2.$ (52)

(b) $\psi_M'(\tau)=$ auto-correlation function of $M'(t)$

$$=\frac{1}{2\pi}\int_{-\infty}^{\infty}S_M(\omega^2)|R(j\omega)|^2e^{j\omega\tau}d\omega. \tag{53}$$

(c) $k'(t)=$ ideal impulsive response for N_2
$=[1/R(p)]k(t).$ (54)

In terms of these quantities the integral equation satisfied by $W_2(t)$ reads:

$$\int_0^{\infty}W_2(\tau)[\psi_M'(t-\tau)+\psi_N'(t-\tau)]d\tau =\lambda_0'+\lambda_1't+\cdots+\lambda_n't^n +\int_{-\infty}^{\infty}k'(\tau)\psi_M'(t-\tau)d\tau,\quad t\geq 0. \tag{55}$$

It will be noticed that in the case of $W_2(t)$ the upper limit of the integral is infinity, while in the case of $W(t)$ [cf. Eq. (36)] it is T. The explanation for this difference is that $W_2(t)$ need not vanish for $t>T$, even though $W(t)$ is required to do so. Thus in general, $W_2(t)$ will be piecewise analytic in the interval $0<t<\infty$ as is illustrated in Fig. 3. Denoting the parts of $W_2(t)$ extending over the intervals $0\leq t\leq T$ and $T<t<\infty$ by $U(t)$ and $V(t)$, respectively, the relation connecting $W(t)$ and $W_2(t)$ [cf. Eq. (46)] may be rewritten in the following form:

$$W(t)=R(p)U(t) \tag{56a}$$

and

$$0=R(p)V(t). \tag{56b}$$

These relations show that $W(t)$ is completely determined by the part of $W_2(t)$ which extends over the interval $0\leq t\leq T$; the form of $W_2(t)$ outside this interval is irrelevant to the determination of $W(t)$.

[3] The appendix of a report by Bode, Blackman, and Shannon, "Data smoothing and prediction in fire-control systems," Research and Development Board, Washington, D. C. (August, 1948), contains a brief exposition of a method which is similar in certain respects to the method described here.

[4] It is tacitly assumed that $R(p)$ does not have a zero at the origin or, in other words, that $S(\omega^2)$ does not have a pole at zero frequency.

Returning to the integral equation (55), it will be noted that the range of integration $0 \leq \tau < \infty$ may be divided into two parts, $0 \leq \tau \leq T$ and $T < \tau < \infty$, involving $U(t)$ and $V(t)$, respectively. Since $V(t)$ is determined by Eq. (56b) to within a finite number of constants, the integral equation in question degenerates into an integral equation involving only $U(t)$:

$$\int_0^T U(\tau)[\psi_M'(t-\tau)+\psi_N'(t-\tau)]d\tau = \lambda_0' + \lambda_1' t + \cdots$$

$$+\lambda_n' t^n + \int_{-\infty}^{\infty} k'(\tau)\psi_M'(t-\tau)d\tau, \quad 0 \leq t \leq T. \quad (57)$$

Upon substitution of Eqs. (52), (53), and (54), and performing minor simplifications, Eq. (57) reads

$$\int_0^T U(\tau)[\gamma_0\delta(t-\tau) - \gamma_1\delta^{(2)}(t-\tau) + \cdots$$

$$+(-1)^m\gamma_m\delta^{(2m)}(t-\tau)]d\tau = \lambda_0' + \lambda_1' t + \cdots + \lambda_n' t^n$$

$$+\frac{1}{2\pi}\int_{-\infty}^{\infty} S_M(\omega^2)K(j\omega)R(-j\omega)e^{j\omega t}d\omega. \quad (58)$$

Making use of the identity

$$\int_0^T U(\tau)\delta^{(2\nu)}(t-\tau)d\tau \equiv p^{2\nu}U(t), \quad (59)$$

Equation (58) may be rewritten as

$$[\gamma_0 - \gamma_1 p^2 + \cdots + (-1)^m\gamma_m p^{2m}]U(t)$$

$$= \lambda_0' + \lambda_1' t + \cdots + \lambda_n' t^n$$

$$+\frac{1}{2\pi}\int_{-\infty}^{\infty} S_M(\omega^2)K(j\omega)R(-j\omega)e^{j\omega t}d\omega. \quad (60)$$

Since in this equation the left-hand side operator is simply $A(-p^2)$ [cf. Eq. (50)], the integral equation (57) finally reduces to the following differential equation:

$$A(-p^2)U(t) = \lambda_0' + \lambda_1' t + \cdots + \lambda_n' t^n$$

$$+\frac{1}{2\pi}\int_{-\infty}^{\infty} S_M(\omega^2)K(j\omega)R(-j\omega)e^{j\omega t}d\omega. \quad (61)$$

The general solution of this equation is of the form:

$$U(t) = A_0' + A_1' t + \cdots + A_n' t^n + B_1' \exp(\alpha_1 t)$$

$$+ B_2' \exp(\alpha_2 t) + \cdots + B_{2m}' \exp(\alpha_{2m} t)$$

$$+\frac{1}{2\pi}\int_{-\infty}^{\infty} \frac{S_M(\omega^2)}{A(\omega^2)}K(j\omega)R(-j\omega)e^{j\omega t}d\omega,$$

$$0 \leq t \leq T, \quad (62)$$

where $A_0', A_1', \cdots, A_n'$ and $B_1', B_2', \cdots, B_{2m}'$ are as yet undetermined constants, and $\alpha_1, \alpha_2, \cdots, \alpha_{2m}$ are the roots of the characteristic equation

$$A(-p^2) = 0. \quad (63)$$

In brief, Eq. (62) provides an explicit expression for $U(t)$ involving $2m+n+1$ undetermined constants. Availability of such an expression reduces the problem of determination of $W(t)$ to a relatively routine matter which is discussed in the following section.

V. DERIVATION OF AN EXPLICIT EXPRESSION FOR $W(t)$

Recalling that $W(t)$ is related to $U(t)$ through the operational relation

$$W(t) = R(p)U(t), \quad (56a)$$

and substituting $U(t)$ as given by Eq. (62) into Eq. (56a), it is readily found that in the most general case $W(t)$ is given by the following expression:

$$W(t) = [u(t) - u(t-T)]\Big\{A_0 + A_1 t + \cdots + A_n t^n$$

$$+ B_1 \exp(\alpha_1 t) + \cdots + B_{2m} \exp(\alpha_{2m} t)$$

$$+\frac{1}{2\pi}R(p)\int_{-\infty}^{\infty} \frac{S_M(\omega^2)}{A(\omega^2)}K(j\omega)R(-j\omega)e^{j\omega t}d\omega\Big\}$$

$$+ C_1\delta(t) + \cdots + C_{l-m}\delta^{(l-m-1)}(t)$$

$$+ D_1\delta(t-T) + \cdots + D_{l-m}\delta^{(l-m-1)}(t-T) \quad (64)$$

where the A's, B's, C's, and D's are as yet undetermined constants, and the unit step functions $u(t)$ and $u(t-T)$ are used simply to indicate that $W(t)$ is zero outside the interval $0 \leq t \leq T$. The impulse functions contained in the expression for $W(t)$ arise from operation by $R(p)$ on the discontinuities of $U(t)$ at $t=0$ and $t=T$. It will be observed that the order of these impulse functions does not exceed $l-m-1$, which is one-half the order of the zero of $S(\omega^2)$ at infinity minus one. This is due to the fact that the first $m-1$ derivatives of $U(t)$ vanish at $t=0$ and $t=T$. It is not difficult to verify that if this would not have been the case, the mean-square error at the output of N_2 would be infinite.

Having obtained the general expression for $W(t)$ in the form of Eq. (64), there remains the problem of determination of the $2l+n+1$ unknown constants. These can be found in the following manner:

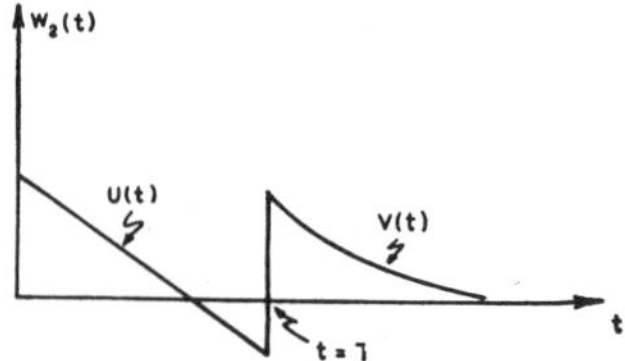

FIG. 3. Form of the impulsive response of N_2.

1. Substituting $W(t)$ as given by Eq. (64) into the integral equation (36) and requiring that the equation be satisfied identically, leads to $2l$ linear homogeneous equations in the A's, B's, C's, and D's.

2. Substituting $W(t)$ as given by Eq. (64) into the $n+1$ moment equations

$$\int_0^T \tau^\nu W(\tau)d\tau=\mu_\nu, \quad \nu=0,1,2,\cdots,n \qquad (65)$$

yields other $n+1$ linear equations. These $n+1$ equations, together with the $2l$ equations obtained in (1), provide a system of $2l+n+1$ linear equations in the unknown constants. Solution of this system gives the values of the A's, B's, C's, and D's and thus completes the process of determination of $W(t)$.

It should be remarked that in some cases it is advantageous to deal with the system function $H(p)$ of the predictor, rather than with its impulsive response $W(t)$. In such cases one can use a transformed form of the integral equation (36) which is as follows:

$$\frac{1}{2\pi}\int_{-\infty}^{\infty} H(j\omega)S(\omega^2)e^{j\omega t}d\omega=\lambda_0+\lambda_1 t+\cdots+\lambda_n t^n$$
$$+\frac{1}{2\pi}\int_{-\infty}^{\infty} S_M(\omega^2)K(j\omega)R(-j\omega)e^{j\omega t}d\omega. \qquad (66)$$

Using Eq. (64), the solution of this equation may be written directly as

$$H(p)=\int_0^T (A_0+A_1t+\cdots+A_nt^n)e^{-pt}dt$$
$$+\frac{B_1}{p+\alpha_1}+\frac{B_2}{p+\alpha_2}+\cdots+\frac{B_{2m}}{p+\alpha_{2m}}$$
$$-\left\{\frac{B_1\exp(\alpha_1T)}{p+\alpha_1}+\frac{B_2\exp(\alpha_2T)}{p+\alpha_2}+\cdots\right.$$
$$\left.+\frac{B_{2m}\exp(\alpha_{2m}T)}{p+\alpha_{2m}}\right\}e^{-pT}+\frac{1}{2\pi}R(p)\int_0^T dte^{-pt}$$
$$\times\int_{-\infty}^{\infty}\frac{S_M(\omega^2)}{A(\omega^2)}K(j\omega)R(-j\omega)e^{j\omega t}d\omega$$
$$+C_1+C_2p+\cdots+C_{l-m}p^{l-m-1}$$
$$+(D_1+\cdots+D_{l-m}p^{l-m-1})e^{-pT}. \qquad (67)$$

The undetermined constants involved in this expression are found in the same manner as in the case of $W(t)$, that is, $H(p)$ as given by Eq. (67) is substituted into the integral equation (66) and the resulting expression is treated as an identity. The $2l$ linear relations between $A_0, A_1, A_2, \cdots$, etc., which are obtained in this manner are adjoined to the $n+1$ relations resulting from Eq. (65); then the system of linear equations in the unknown constants is solved for $A_0, A_1, A_2, \cdots$, etc.

In order to facilitate application of the techniques described in the preceding sections, a summary of the procedure for the determination of $W(t)$ (or $H(p)$) is given in Section VI. Furthermore, actual use of the procedure is illustrated by a few practical examples at the end of the section.

VI. SUMMARY OF THE PROCEDURE FOR DETERMINATION OF $W(t)$ AND $H(p)$

The complete expressions for $W(t)$ (the impulsive response of the optimum predictor) and $H(p)$ (the system function of the optimum predictor) are given by Eqs. (64) and (67). In order to avoid the necessity for reference to preceding sections, the meaning of all symbols appearing in these equations is given:

$u(t)$ = unit step function.
T = duration of sampling (settling time).
$A_0, A_1, \cdots, A_n, B_1, B_2, \cdots, B_{2m}, C_1, C_2, \cdots, C_{l-m}, D_1, D_2, \cdots, D_{l-m}$ = undetermined constants.
n = degree of the polynomial component of the input signal.
$S_M(\omega^2)$ = spectral density of $M(t)$[$M(t)$ is the stationary part of the input signal].
$S(\omega^2)$ = spectral density of $M(t)+N(t)$[$N(t)$ is the input noise].
$A(\omega^2)$ = numerator of $S(\omega^2)$.
$B(\omega^2)$ = denominator of $S(\omega^2)$.
$Q(j\omega)$ = a factor of $A(\omega^2)$ containing all the zeros in right half of the $j\omega$-plane.
$R(j\omega)$ = a factor of $B(\omega^2)$ containing all the zeros in the left half of the $j\omega$-plane.
$2l$ = degree of $B(\omega^2)$.
$2m$ = degree of $A(\omega^2)$.
$\alpha_1, \alpha_2, \cdots, \alpha_{2m}$ = roots of the characteristic equation $A(-p^2)=0$.
$\delta(t)$ = unit impulse function.
$\delta^{(\nu)}(t)$ = νth derivative of $\delta(t)$.

The undetermined constants occurring in the expression for $W(t)$ [and $H(p)$] can be found in the following manner.

1. $W(t)$ as given by Eq. (64) is substituted into the integral equation (36) and the resulting expression is treated as an identity. This gives $2l$ homogeneous linear equations in the unknown constants. Same equations can be obtained by substituting $H(p)$, as given by Eq. (67), into the integral equation (66).

2. $W(t)$ as given by Eq. (64) is substituted into the $n+1$ constraint equations

$$\int_0^T \tau^\nu W(\tau)d\tau=\mu_\nu, \quad \nu=0,1,\cdots,n. \qquad (14)$$

where the μ_ν are determined by the choice of the prediction operator $K(p)$ [cf. Eq. (19)]. The resulting $n+1$ linear equations in the unknown constants are adjoined to the $2l$ equations obtained from (1). The set of $2l+n+1$ linear equations thus obtained is solved for the undetermined constants $A_0, A_1, \cdots, D_{l-m}$. This concludes the process of determining $W(t)$ [or $H(p)$].

VII. ILLUSTRATIVE EXAMPLES

Example 1. Wiener's Theory

Wiener's theory is, in the main, a study of the particular case in which $P(t)\equiv 0$, $T=\infty$ and $K(p)=e^{-\alpha p}$. For this case Eq. (64) gives

$$W(t)=u(t)\left\{\frac{1}{2\pi}R(p)\times\int_{-\infty}^{\infty}\frac{S_M(\omega^2)}{A(\omega^2)}K(j\omega)R(-j\omega)e^{j\omega(t-\alpha)}d\omega+B_1\exp(\alpha_1 t)+\cdots+B_{2m}\exp(\alpha_{2m}t)\right\}. \quad (68)$$

The exponential terms appearing in Eq. (68) may be made to vanish through a slight rearrangement of the factors in the first term of Eq. (68). The resulting expression for $W(t)$ is the same as that obtained by using Wiener's theory, namely,

$$W(t)=u(t)\frac{1}{2\pi}\frac{R(p)}{Q(p)}\times\int_{-\infty}^{\infty}\frac{S_M(\omega^2)}{Q(-j\omega)}R(-j\omega)e^{j\omega(t-\alpha)}d\omega. \quad (69)$$

The rearrangement amounts, essentually, to choosing a particular solution of Eq. (61) which differs from the one chosen before by the exponential terms of Eq. (68). The same result may be achieved directly by choosing $H_1(p)$ [cf. Eq. (47)] as

$$H_1(p)=R(p)/Q(p).$$

With this choice of $H_1(p)$ [in place of the one expressed by Eq. (47)] the various quantities entering Eq. (55) become:

$$S'(\omega^2)=1,$$
$$\psi_M'(\tau)+\psi_N'(\tau)=\delta(\tau),$$
$$S_M'(\omega^2)=S_M(\omega^2)\,|R(j\omega)/Q(j\omega)|^2,$$
$$K'(j\omega)=K(j\omega)Q(j\omega)/R(j\omega),$$

and hence the integral equation (55) reduces to

$$W_2(t)=u(t)\frac{1}{2\pi}\int_{-\infty}^{\infty}\frac{S_M(\omega^2)}{Q(-j\omega)}K(j\omega)R(-j\omega)e^{j\omega t}d\omega; \quad (55a)$$

Eq. (69) then follows immediately from the relation connecting $W(t)$ and $W_2(t)$ [cf. Eq. (46)].

Example 2

The assumptions made here are as follows:

1. $M(t)\equiv 0.$
2. $n=1.$
3. $\psi_N(\tau)=e^{-a|\tau|};\quad S_N(\omega^2)=\dfrac{2a}{\omega^2+a^2}.$

The choice of the prediction operator is left open.

Solution

For this case $A(-p^2)=2a$, and hence $\alpha_1=\cdots=\alpha_{2m}=0$. Also, $l=1$, $M\equiv 0$; hence Eq. (64) gives

$$W(t)=A_0+A_1t+C_1\delta(t)+D_1\delta(t-T), \quad (70)$$

and

$$H(p)=\frac{A_0}{p}(1-e^{-pT})+\frac{A_1}{p}\left(\frac{1}{p}-\frac{e^{-pT}}{p}+Te^{-pT}\right)+C_1+D_1e^{-pT}. \quad (71)$$

Substituting $H(p)$ as given by Eq. (71) into the integral equation

$$\frac{1}{2\pi}\int_{-\infty}^{\infty}\frac{2a}{a^2-p^2}H(p)e^{pt}dp=\lambda_0+\lambda_1t,\quad 0\leq t\leq T, \quad (72)$$

and requiring that this equation be satisfied identically, leads to the following relations:

$$aA_0-A_1-a^2C_1=0, \quad (73)$$

$$aA_0+(aT+1)A_1-a^2D_1=0. \quad (74)$$

Furthermore, substituting $W(t)$ as given by Eq. (70) into the constraint equations

$$\int_0^T W(\tau)d\tau=\mu_0 \quad (75)$$

and

$$\int_0^T \tau W(\tau)d\tau=\mu_1, \quad (76)$$

yields

$$A_0T+A_1(T^2/2)+C_1+D_1=\mu_0, \quad (77)$$

and

$$A_0(T^2/2)+A_1(T^3/3)+D_1T=\mu_1. \quad (78)$$

The unknown constants A_0, A_1, C_1, and D_1 can be readily found from the solution of Eqs. (73), (74), (77), and (78). Thus,

$$A_0=\mu_0\frac{4a(a^2T^2+3aT+3)}{(a^2T^2+6aT+12)(aT+2)}-\mu_1\frac{6a^2}{a^2T^2+6aT+12}, \quad (79)$$

$$A_1=-\mu_0\frac{6a^2}{a^2T^2+6aT+12}+\mu_1\frac{12a^2}{T(a^2T^2+6aT+12)}, \quad (80)$$

$$C_1=\mu_0\frac{2(2a^2T^2+9aT+12)}{(a^2T^2+6aT+12)(aT+2)}-\mu_1\frac{6(aT+2)}{T(a^2T^2+6aT+12)}, \quad (81)$$

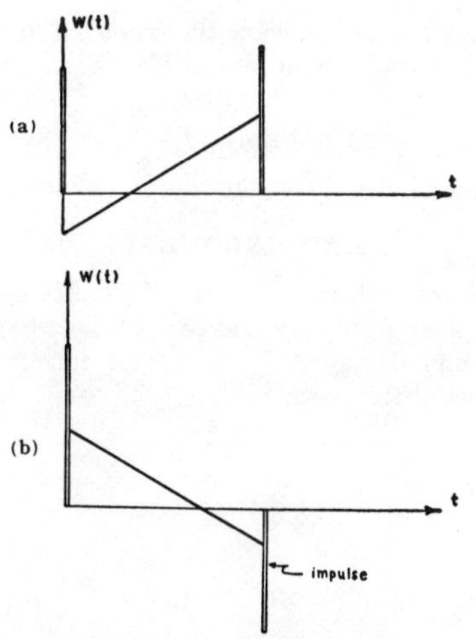

FIG. 4. Shapes of optimum weighting functions. (a) For the best value of $s(t)$. (b) For the best value of $s(t-\alpha)$ (α-negative).

and

$$D_1 = -\mu_0 \frac{2aT(aT+3)}{(a^2T^2+6aT+12)(aT+2)} + \mu_1 \frac{6(aT+2)}{T(a^2T^2+6aT+12)}. \quad (82)$$

Whenever the stationary part of the input signal is zero [i.e., $M(t)\equiv 0$], the mean-square value of the prediction error assumes the simple form

$$\sigma^2 = \mu_0\lambda_0 + \mu_1\lambda_1 + \cdots + \mu_n\lambda_n, \quad (83)$$

which may be readily established by substituting Eq. (36) into the general expression for σ^2 [cf. Eq. (33)]. For the particular case under consideration Eq. (83) gives

$$\sigma^2 = \mu_0^2 \frac{8(a^2T^2+3aT+3)}{(a^2T^2+6aT+12)(aT+2)} + \mu_1^2 \frac{24a}{T(a^2T^2+6aT+12)} - \mu_0\mu_1 \frac{24a}{(a^2T^2+6aT+12)}. \quad (84)$$

The expressions given above are valid for any choice of the prediction operator. For the particular case in which the predictor is called upon to furnish the best possible estimate of the present value of $\dot{s}(t)$, the values of μ_0 and μ_1 are, respectively,

$$\mu_0 = 0 \qquad \text{[cf. Eq. (21)]} \qquad \mu_1 = -1. \quad (85)$$

On the other hand, in the case of the estimation of $s(t+\alpha)$ (i.e., the value of $s(t)$ α-seconds in the future)

$$\mu_0 = 1 \qquad \text{[cf. Eq. (24)]} \qquad \mu_1 = -\alpha. \quad (86)$$

The shapes of $W(t)$ for these two particular cases are illustrated in Fig. 4.

Example 3

The case to be considered here is the same as that treated in Example 2, except that the auto-correlation function of $N(t)$ is assumed to be of the form

$$\psi_N(\tau) = e^{-a|\tau|}\cos\omega_0\tau, \quad (87)$$

with the associated spectral density function being

$$S_N(\omega^2) = \frac{2a(a^2+\omega_0^2+\omega^2)}{\omega^4+2(a^2-\omega_0^2)\omega^2+(a^2+\omega_0^2)^2}. \quad (88)$$

This form of spectral density function is of considerable practical importance since it provides a reasonably good approximation to many of the actual spectra encountered in practice.

Solution

By Eq. (64), the weighting function for this case is of the form

$$W(t) = A_0 + A_1t + B_1e^{bt} + B_2e^{-bt} + C_1\delta(t) + D_1\delta(t-T), \quad 0 \leq t \leq T, \quad (89)$$

where $b = (a^2+\omega_0^2)^{\frac{1}{2}}$. Substituting $W(t)$ as given by Eq. (89) into Eq. (36), and requiring that Eq. (36) be satisfied by $W(t)$ establishes four linear algebraic equations between the six constants A_0, A_1, B_1, B_2, C_1, D_1. These are:

$$\begin{aligned} &-2ab^2A_0 + 2(a^2-\omega_0^2)A_1 - b^3B_1 + b^3B_2 + C_1 = 0, \\ &2\omega_0^2b^2A_0 - 4a\omega_0^2A_1 + b^3(b-a)B_1 + b^3(b+a)B_2 = 0, \\ &-2ab^2A_0 + 2(\omega_0^2 - a^2 - aTb^2)A_1 + \frac{e^{bT}}{2b}B_1 - \frac{e^{-bT}}{2b}B_2 + D_1 = 0, \\ &2\omega_0^2b^2A_0 + 2\omega_0^2(2a+b^2T)A_1 + b^3(b+a)e^{bT}B_1 + b^3(b-a)e^{-bT}B_2 = 0. \end{aligned} \quad (90)$$

The constraints imposed by the prediction operator $K(p)$ are given by Eq. (14); they are:

$$\int_0^T W(\tau)d\tau = \mu_0, \quad (75)$$

and

$$\int_0^T \tau W(\tau)d\tau = \mu_1. \quad (76)$$

The requirement that $W(t)$ must satisfy Eqs. (75) and (76) leads to two additional linear equations. These are:

$$2bTA_0+bT^2A_1+2(e^{bT}-1)B_1-2(e^{-bT}-1)B_2 \\ +2bC_1+2bD_1=\mu_0,$$

and

$$3b^2T^2A_0+2b^2T^3A_1+b[e^{bT}(bT-1)+1]B_1 \\ -6[e^{-bT}(bT+1)-1]B_2+6b^2D_1=\mu_1. \quad (91)$$

The coefficients A_0, A_1, B_1, B_2, C_1, D_1 of Eq. (89) are the solutions of the six linear equations (90) and (91).

The expression for the mean-square value of the prediction error can be conveniently expressed in terms of A_0 and A_1. Thus, making use of Eq. (83) it is readily found that

$$\sigma^2=(2a/b^2)(\mu_0A_0+\mu_1A_1). \quad (92)$$

This completes essentially the solution of the problem.

ACKNOWLEDGMENTS

The authors wish to thank Mr. H. T. Marcy of The M. W. Kellogg Company and Mr. H. Sherman of the Watson Laboratories, AMC, for their encouragement and support of this work.

11

Reprinted with permission from *Theory Probab. Appl.* **5**(3):239–264 (1960)

EFFECTIVE SOLUTIONS OF LINEAR APPROXIMATION PROBLEMS FOR MULTIVARIATE STATIONARY PROCESSES WITH A RATIONAL SPECTRUM

A. M. YAGLOM

(*Translated by Joel R. Lowenthal*)

The problem of the linear extrapolation of multivariate random stationary processes, first considered as long ago as 1941 in Zasukhin's article [1], has in recent years again attracted the attention of a whole series of investigators (see, for example, [2–8] [1]). In spite of all this, up to now we have almost no results that permit us to construct effectively, from the values of all the components of the considered multivariate process in the past, a linear functional which gives the best estimate of the value of some one of its components in the next moment, i.e., which permits us to give the solution of the extrapolation problem in a form allowing practical utilization in applied problems of a different type. The only exception in this respect is the well-known monograph of N. Wiener [10], Chapter IV of which contains a method for the practical solution of extrapolation and filtration problems (for data on the half-line) for the same special class of multivariate stationary processes which will be considered in the present article. However, Wiener's account is not entirely clear, and the method proposed by him is so unwieldy that its application in all cases except the very simplest is rather inconvenient. In the present article we shall also consider the extrapolation problem and other approximation problems related to it (extrapolation for data on the finite interval, filtration, interpolation) for multivariate stationary processes (with continuous time) from a purely practical point of view; in this connection we shall not strive for maximum generality, but shall limit ourselves to one particular (but sufficiently wide from the point of view of application) class of multivariate processes, which we shall call *non-degenerate processes with a rational spectrum* (an exact definition is given in the beginning of § 1). As will be evident later on, for processes of this class the effective solution of the relatively wide class of linear approximation problems may be reduced to a calculation of the roots of some algebraic equa-

[1] The complete proof of the results stated in 1 was first given in [3] and [5] (the last of which is composed partly of material from lectures given in 1955–56 at Moscow University by M. G. Krein). We note further that only processes with discrete time are considered in [1–7]; however, as was shown in [8], for the extrapolation of multivariate stationary processes the transition from discrete to continuous time may be accomplished the same way as in the one-dimensional case (cf. [9], Ch. XII, § 5). In the case of other linear approximation problems considered in §§ 2–4 of this article (with the exception of the filtration problem for data on the half-line), such a transition proves to be impossible

tion and to the subsequent solution of a simple system of linear algebraic equations; the general method which allows us to carry out this reduction is a natural generalization of the method applied in the author's works [11] and [12] to the solution of analogous problems for one-dimensional random processes. In the final § 5 of this article, some further applications of this method will also be briefly pointed out. [2]

1. The Simplest Extrapolation Problem

Let $\xi(t) = \{\xi_1(t), \cdots, \xi_n(t)\}$, $-\infty < t < \infty$, be a multivariate stationary random process. Without any loss of generality we can assume that the mean value of the process $\xi(t)$ is identically equal to zero:

$$\mathbf{M}\xi_k(t) \equiv 0. \tag{1.1}$$

In addition, the fundamental statistical characteristic of the process $\xi(t)$ is the correlation matrix

$$||B_{jk}(\tau)|| = ||\mathbf{M}\xi_j(t+\tau)\overline{\xi_k(t)}||. \tag{1.2}$$

We restrict ourselves to the case where all the functions $B_{jk}(\tau)$ decrease sufficiently rapidly for large values, so that we can represent them by the Fourier integral:

$$B_{jk}(\tau) = \int_{-\infty}^{\infty} e^{i\tau\lambda} f_{jk}(\lambda) d\lambda. \tag{1.3}$$

Then, according to H. Cramer's well-known theorem,

$$||f_{jk}(\lambda)|| \tag{1.4}$$

is a Hermitian, non-negative definite matrix, which is called a matrix of the spectral densities (spectral matrix, for short) of the process $\xi(t)$. Moreover, we shall assume further that all functions $f_{jk}(\lambda)$ are rational functions of the variable λ:

$$f_{jk}(\lambda) = \frac{Q_{jk}(\lambda)}{P_{jk}(\lambda)}, \tag{1.5}$$

where $Q_{jk}(\lambda)$ and $P_{jk}(\lambda)$ are some polynomials. Multivariate stationary processes which satisfy conditions (1.3) and (1.5) we shall call processes with a rational spectrum.

For processes with a rational spectrum, the determinant of a spectral matrix is also rational:

$$D(\lambda) = \det ||f_{jk}(\lambda)|| = \frac{Q(\lambda)}{P(\lambda)}. \tag{1.6}$$

[2] After the completion of the work on this article, the author became acquainted with the work of S. Darlington [13], who discusses from a different point of view the same method of solution for the problems of extrapolation and filtration of stationary processes with a rational spectral density that was proposed in [11] and [12] (the latter have, apparently, remained unknown to S. Darlington). Sections **3.4** and **4.4** of [**13**] also contain a discussion of the application of the method to the solution of some special cases of the filtration problem (for data on the half-line) for multivariate stationary processes with a rational spectrum.

Since $||f_{jk}(\lambda)||$ is non-negative, $D(\lambda) \geqq 0$ here for all real λ. Henceforth we shall always assume that the function $D(\lambda)$ is not identically equal to zero. According to Zasukhin's fundamental result [1] (see also [8]), the process $\xi(t)$ in this case will certainly be regular to the highest rank, i.e., will possess simple extrapolation properties. We shall by the way never use this fact: the elementary solution for the extrapolation problem described in what follows requires no information from the general theory.

The condition that $D(\lambda)$ not be identically equal to zero is a basic condition for the non-degeneracy of a process with a rational spectrum. However, it will be convenient for us to interpret non-degeneracy somewhat more narrowly. If the determinant $D(\lambda)$ is not equal to zero for all λ, it can still vanish at discrete points on the real axis — at those points which coincide with real roots of the polynomial $Q(\lambda)$. In the main part of the present article we shall consider only processes for which $D(\lambda)$ is not equal to zero anywhere on the real axis, i.e., the polynomial $Q(\lambda)$ has no real zeros (nor a zero at the point $\lambda = \infty$, cf. footnote on p. 244). It is exactly these kinds of processes $\xi(t)$ with a rational spectrum that we shall call non-degenerate. As a matter of fact, the solution cited below for the extrapolation problem for non-degenerate (in the sense indicated here) stationary processes with a rational spectrum is carried over almost without change to those processes for which $D(\lambda)$ has a finite number of real zeros. However, the obvious meaning of the obtained extrapolation formulas in the last case is noticeably more complicated than in the case of the processes which we call non-degenerate (see the remark on p. 247). We note further that in practical problems, the degeneration consisting of the vanishing of the non-negative polynomial $Q(\lambda)$ at discrete points on the real axis occurs only in some exceptional cases, where it is usually difficult to determine accurately the presence of such degeneration.

In what follows, besides the spectral decomposition (1.3) of the correlation functions of the process $\xi(t)$, we shall also need a spectral decomposition of the process itself:

$$\xi_k(t) = \int_{-\infty}^{\infty} e^{it\lambda}\, dZ_k(\lambda), \qquad k = 1, 2, \cdots, n. \tag{1.7}$$

Here $Z_k(\lambda)$, $k = 1, \cdots, n$, are random processes with uncorrelated increments such that for $\lambda_1 < \lambda_2$ and $\lambda_3 < \lambda_4$,

$$\mathbf{M}[Z_j(\lambda_2) - Z_j(\lambda_1)]\overline{[Z_k(\lambda_4) - Z_k(\lambda_3)]} = \int_{[\lambda_1, \lambda_2] \cap [\lambda_3, \lambda_4]} f_{jk}(\lambda)\, d\lambda. \tag{1.8}$$

The simplest extrapolation problem for the process $\xi(t)$ consists in finding, from the values $\xi_k(t')$, $k = 1, \cdots, n$, $t' \leqq t$, the linear functional $\tilde{\xi}_1(t, \tau)$ (where t is arbitrary, and $\tau > 0$) which gives the best estimate of the random variable $\xi_1(t+\tau)$. Here the words "best estimate" are considered in the sense of the classical method of least squares, i.e., in the sense of reducing to a minimum the mean square of the error of extrapolation

$$\sigma_1^2(\tau) = \mathbf{M}\overline{|\xi_1(t+\tau) - \tilde{\xi}_1(t, \tau)|^2}. \tag{1.9}$$

If we consider the set of all the random variables of bounded variance to be a

Hilbert space H (with a scalar product $(\xi, \eta = \mathbf{M}\xi\bar{\eta})$, then geometrically $\tilde{\xi}_1(t, \tau)$ will coincide with a projection of the vector $\xi_1(t+\tau)$ onto the linear subspace

$$H_\xi(t) = [\xi_k(t'), k = 1, \cdots, n, t' \leqq t], \tag{1.10}$$

spanned by the vectors $\xi_k(t')$, $k = 1, \cdots, n, t' \leqq t$. From this it is clear that $\tilde{\xi}_1(t, \tau)$ is given uniquely (as an element of H, i.e., correct to the random variables which are almost certainly equal to zero) and is the only element in H which satisfies the conditions

$$\text{(a)} \qquad \tilde{\xi}_1(t, \tau) \in H_\xi(t), \tag{1.11}$$

and

$$\text{(b)} \qquad (\xi_1(t+\tau)-\tilde{\xi}_1(t, \tau), \xi_k(t')) \equiv \mathbf{M}[\xi_1(t+\tau)-\tilde{\xi}_1(t, \tau)]\overline{\xi_k(t')} = 0 \tag{1.12}$$

for all $k = 1, \cdots, n$ and $t' \leqq t$. Thus to solve the extrapolation problem one needs only select, in whatever way one likes, a random variable $\tilde{\xi}_1(t, \tau)$ which satisfies these conditions.

Restricting ourselves for the time being to the case of the non-degenerate process $\boldsymbol{\xi}(t)$ with a rational spectrum, we shall select a random variable $\tilde{\xi}_1(t, \tau)$ from the variables of the form

$$\tilde{\xi}_1(t, \tau) = \sum_{k=1}^{n} \int_{-\infty}^{\infty} e^{it\lambda} \Phi_k(\lambda) dZ_k(\lambda), \tag{1.13}$$

where all integrals are considered in the mean square.[3] To solve the problem it suffices to find one system of functions $\Phi_k(\lambda)$, $k = 1, \cdots, n$, for which (1.13) satisfies conditions (a) and (b). Then we still have to rewrite the corresponding variable $\tilde{\xi}_1(t, \tau)$ in the form of a linear functional of values in the past (for $t' \leqq t$) of components of the process $\boldsymbol{\xi}(t)$; however, as we shall see, for the class of processes $\boldsymbol{\xi}(t)$ considered here it is very easy to do this (see p. 246).

In order that all integrals in (1.13) might exist as integrals in the mean square, the functions $\Phi_k(\lambda)$ must satisfy the conditions

1)

$$\int_{-\infty}^{\infty} |\Phi_k(\lambda)|^2 f_{kk}(\lambda) d\lambda < \infty, \qquad k = 1, \cdots, n. \tag{1.14}$$

In order to satisfy conditions (a) and (b), it is sufficient to impose some natural restrictions on the behaviour of the functions $\Phi_k(\lambda)$ for complex values of the argument λ. Indeed, condition (b) (see (1.12)) is obviously equivalent to the following equality (where $t' = t-s$, i.e., $s = t-t' \geqq 0$):

$$\int_{-\infty}^{\infty} e^{is\lambda}\{[e^{i\tau\lambda}-\Phi_1(\lambda)]f_{1k}(\lambda)-\sum_{j=2}^{k}\Phi_j(\lambda)f_{jk}(\lambda)\}d\lambda = 0, \quad k = 1, \cdots, n, s \geqq 0. \tag{1.15}$$

But because of Cauchy's theorem on integration in a complex region, this equality will certainly hold if the following condition is satisfied:

---

[3] We note that, in the case of degenerate processes with a rational spectrum, the selection of the variable $\tilde{\xi}_1(t+\tau)$ in such a form may prove to be completely impossible (see the remark on p. 247).

2) *The functions*

$$\Psi_k(\lambda) = [e^{i\tau\lambda} - \Phi_1(\lambda)]f_{1k}(\lambda) - \sum_{j=2}^{n} \Phi_j(\lambda) f_{jk}(\lambda), \qquad k = 1, \cdots, n, \tag{1.16}$$

can all be extended analytically into the upper half-plane $\Lambda_1 = \{\lambda; \operatorname{Im} \lambda \geq 0\}$ *in such a way that they have no singularities in the whole half-plane and decrease for* $|\lambda| \to \infty$ *not slower than* $|\lambda|^{-1-\varepsilon}$, $\varepsilon > 0$ (cf. [11]). Condition (a) is certainly satisfied, if all the integrals $\int_{-\infty}^{\infty} e^{it\lambda} \Phi_k(\lambda) dZ_k(\lambda)$ depend only on the values $\xi_k(t')$ for $t' \leq t$; we can achieve the latter by demanding that the following condition be satisfied:

3) *The functions* $\Phi_k(\lambda)$, $k = 1, \cdots, n$, *can all be extended analytically into the lower half-plane* $\Lambda_2 = \{\lambda; \operatorname{Im} \lambda \leq 0\}$ *in such a way that they have no singularities in the whole half-plane and increase for* $|\lambda| \to \infty$ *not faster than some finite power of* $|\lambda|$ (cf. [11]). Thus to solve the extrapolation problem for the process $\xi(t)$ it is sufficient to select functions $\Phi_k(\lambda)$ which satisfy conditions 1, 2, and 3.

We now show that it is not difficult to satisfy all these conditions, even if we limit ourselves to only the rational functions $\Phi_k(\lambda)$. Indeed, let us assume that

$$\Phi_k(\lambda) = \frac{\omega_k(\lambda)}{(\lambda - \theta_1)(\lambda - \theta_2) \cdots (\lambda - \theta_L)}, \qquad k = 1, 2, \cdots, n, \tag{1.17}$$

where $\omega_k(\lambda)$ are polynomials with respect to λ, and θ_i, $i = 1, \cdots, L$, are complex numbers. In such a case, in order to satisfy condition 3 we need only restrict ourselves to the numbers θ_i with $\operatorname{Im} \theta_i > 0$. In order that condition 2 be satisfied at the same time, all the θ_i must not be poles of any of the functions $\Psi_k(\lambda)$. Thus the following equalities must be valid:

$$\sum_{j=1}^{n} \omega_j(\theta_i) f_{jk}(\theta_i) = 0, \qquad k = 1, \cdots, n,\ i = 1, \cdots, L, \tag{1.18}$$

(other restrictions which follow from condition 2 will be indicated below). From this it is obvious that all the poles θ_i of the functions $\Phi_k(\lambda)$ must coincide with the zeros of the determinant $\|f_{jk}(\lambda)\|$ — the function (1.6).

We shall now attempt to calculate the number of zeros of the function (1.6), i.e., the degree of the polynomial $Q(\lambda)$. In (1.5) let the denominator $P_{jk}(\lambda)$ have the form

$$P_{jk}(\lambda) = \prod_{m=1}^{N_{jk}} (\lambda - \alpha_{jk}^{(m)}) \prod_{m=1}^{N_{kj}} (\lambda - \overline{\alpha_{kj}^{(m)}}), \tag{1.19}$$

where separate $\alpha_{jk}^{(m)}$ may coincide and

$$\begin{aligned} &\operatorname{Im} \alpha_{jk}^{(m)} > 0, && m = 1, \cdots, N_{jk}, \\ &\operatorname{Im} \overline{\alpha_{kj}^{(m)}} < 0, && m = 1, \cdots, N_{kj}. \end{aligned} \tag{1.20}$$

In this case

$$P(\lambda) = \left| \prod_{j,k=1}^{n} \prod_{m=1}^{N'_{jk}} (\lambda - \alpha_{jk}^{(m)}) \right|^2, \tag{1.21}$$

where $\alpha_{jk}^{(m)}$, $m = 1, \cdots, N'_{jk}$, are those zeros of the polynomial $P_{jk}(\lambda)$ in the upper half-plane which are not at the same time either zeros of some $P_{lk}(\lambda)$ with $l < j$ or zeros of some $P_{ji}(\lambda)$ with $i < k$, but not of $P_{li}(\lambda)$, $l \neq j$, and not of

$P_{lk}(\lambda)$, $l > j$.[4] Therefore the degree $2K$ of the polynomial $P(\lambda)$ is given by the equality

$$2K = 2\sum_{j,k=1}^{n} N'_{jk}. \tag{1.22}$$

Let the degree of the polynomial $Q_{jj}(\lambda)$ equal $2(N_{jj}-m_j)$ (we recall that N_{jj} is the degree of $P_{jj}(\lambda)$). In this case the degree of the denominator of the term $f_{11}(\lambda)f_{22}(\lambda)\cdots f_{nn}(\lambda)$ of the determinant $D(\lambda)$ exceeds the degree of the numerator by $2(m_1+m_2+\cdots+m_n)$. And for all the other terms of the determinant $D(\lambda)$ the difference between the degrees of the denominator and the numerator is certainly not less than this (because $|f_{jk}(\lambda)| \leq \sqrt{f_{jj}(\lambda)f_{kk}(\lambda)}$ for all j and k). Therefore after reducing to a common denominator all the terms of $D(\lambda)$, the degree of the denominator exceeds the degree of the numerator by $2(m_1+m_2+\cdots+m_n)$.[5] Consequently, the degree $2L$ of the polynomial $Q(\lambda)$ equals

$$2L = 2K - 2\sum_{j=1}^{n} m_j = 2\left(\sum_{j,k=1}^{n} N'_{jk} - \sum_{j=1}^{n} m_j\right). \tag{1.23}$$

In the non-degenerate case all the zeros of $Q(\lambda)$ are complex: they are broken up naturally into complex conjugate pairs and may be designated by

$$\theta_1, \theta_2, \cdots, \theta_L, \bar{\theta}_1, \bar{\theta}_2, \cdots, \bar{\theta}_L$$

(the roots θ_i lie in the upper half-plane, the roots $\bar{\theta}_i$ in the lower). We shall now seek $\Phi_k(\lambda)$ in the form of (1.17), where $\theta_1, \cdots, \theta_L$ are all L roots of the polynomial $Q(\lambda)$ in the upper half-plane. Moreover, condition 1 (i.e., (1.14)) is satisfied if a polynomial of $(L+m_k-1)$-th (but not higher!) degree is chosen as $\omega_k(\lambda)$. Such a polynomial obviously has only $L+m_k$ coefficients. Therefore by demanding that all $\Phi_k(\lambda)$ satisfy (1.14), we have at our disposal $nL+m_1+\cdots+m_n$ additional arbitrary coefficients which we must choose in such a way that condition 2 is also satisfied.

Thus it remains for us to find values of the polynomials $\omega_k(\lambda)$ for which all the functions $\Psi_k(\lambda)$ do not have poles in the upper half-plane. In other words, the coefficients of the polynomials $\omega_k(\lambda)$ must be selected so that after substituting (1.17) into (1.16) and reducing all the rational functions on the right side to a

[4] The polynomial $P(\lambda)$ is obviously the least common multiple of the polynomials $P_{j_1k_1}(\lambda)$ $P_{j_2k_2}(\lambda)\cdots P_{j_nk_n}(\lambda)$, where $j_1 j_2 \cdots j_n$ and $k_1 k_2 \cdots k_n$ are permutations of the numbers 1, 2, $\cdots$, n. In order not to make the discussion cumbersome, we have deliberately permitted an inaccuracy here by simplifying somewhat the general rule for the construction of this least multiple. For example, if some root of the polynomial $P_{jk}(\lambda)$ is at the same time also a root of $P_{ji}(\lambda)$ and $P_{jh}(\lambda)$ (where $i < k$, $h < k$) and of $P_{li}(\lambda)$ and $P_{lh}(\lambda)$ (with $l \neq j$), but not of $P_{mi}(\lambda)$ or $P_{mh}(\lambda)$ with $m \neq l$, $m \neq j$, then it is not necessary to include it in the number of roots of $\alpha_{jk}^{(m)}$, $m = 1, \cdots, N'_{jk}$. It is not difficult to verify, however, that in this case even the number (1.26) of those equations which will be discussed on p. 245 must be reduced by one. A strict proof of the fact that the total number of all these equations is always exactly 2 times less than the degree of the polynomial $P(\lambda)$ may be given by means of the method of mathematical induction.

[5] Generally speaking, in reducing all the terms of $D(\lambda)$ to a common denominator, the major terms of the numerator may be mutually reduced, and then the difference between the degrees of the denominator and numerator proves to be greater than $2(m_1+m_2+\cdots+m_n)$. However, in this case we say that the polynomial $Q(\lambda)$ has a root $\lambda = \infty$ and we include the corresponding process $\xi(t)$ among the degenerate ones.

common denominator, all the zeros $\alpha_{jk}^{(m)}$, $m = 1, 2, \cdots, N_{jk}$, $j = 1, 2, \cdots, n$, and all the zeros θ_i, $i = 1, 2, \cdots, L$, in the denominator might be cancelled with zeros of no less multiplicity in the numerator. We begin with the zeros $\alpha_{jk}^{(m)}$. In order that all the zeros $\alpha_{j1}^{(m)}$, $m = 1, \cdots, N_{j1}$, $j = 1, \cdots, n$, in the denominator of the function $\Psi_1(\lambda)$ might be cancelled (i.e., that the function $\Psi_1(\lambda)$ have no poles for $\lambda = \alpha_{j1}^{(m)}$), we must demand that after reducing to a common denominator all the rational functions $f_{j1}(\lambda)$, $j = 1, \cdots, n$, on the right side of (1.16) with $k = 1$, the numerator should vanish at all zeros of the denominator in the upper half-plane. Since this numerator is a linear combination of the polynomials $\omega_j(\lambda)$, every condition of its vanishing gives one linear equation with respect to the coefficients of the indicated polynomials. The total number of equations obtained in this way is obviously equal to the degree of the least common denominator of the functions $f_{11}(\lambda), \cdots, f_{n1}(\lambda)$, i.e., equal to

$$N'_{11}+N'_{21}+\cdots+N'_{n1}, \tag{1.24}$$

where N'_{j1} is the number of zeros of the polynomial $P_{j1}(\lambda)$ in the upper half-plane which do not coincide with any zero of $P_{l1}(\lambda)$, $l < j$. Similar reasoning is also applicable to the function $\Psi_2(\lambda)$ with the only difference that if the zero $\alpha_{j2}^{(m)}$ coincides with some $\alpha_{j1}^{(r)}$ but does not coincide with any $\alpha_{l1}^{(s)}$, $l \neq j$, nor with any $\alpha_{l2}^{(t)}$, $l > j$, then the linear equation corresponding to this zero (having the form $\omega_j(\alpha_{j2}^{(m)}) = 0$) will already have been counted among the equations for the function $\Psi_1(\lambda)$. Therefore the total number of new equations here is equal to

$$N'_{j2}+N'_{22}+\cdots+N'_{n2}, \tag{1.25}$$

where N'_{j2} has the same meaning as in (1.21) and (1.22). Similarly it is shown that the condition that the function $\Psi_k(\lambda)$ does not have poles at the points $\alpha_{jk}^{(m)}$, provided that all $\Psi_l(\lambda)$, $l < k$, do not have poles at $\alpha_{lk}^{(m)}$, gives us

$$N'_{1k}+N'_{2k}+\cdots+N'_{nk} \tag{1.26}$$

new linear equations with respect to the coefficients of the polynomials $\omega_j(\lambda)$, where all N'_{jk} have the same meaning as in (1.21) and (1.22). Consequently, the condition that all functions $\Psi_k(\lambda)$, $k = 1, \cdots, n$, do not have poles at all points $\alpha_{jk}^{(m)}$, $j = 1, \cdots, n$, $m = 1, \cdots, N_{jk}$, is equivalent to the system $\sum_{j,k} N'_{jk} = K$ of linear equations with respect to the coefficients of the polynomials $\omega_j(\lambda)$.

Let us now examine what is implied by the condition that the $\Psi_k(\lambda)$ have no poles at the points θ_i. These conditions are obviously equivalent to the system of equations (1.18). We shall consider at first the general case, when all the roots θ_i, $i = 1, \cdots, L$, are different. In this case the rank of the spectral matrix $\|f_{jk}(\lambda)\|$ for $\lambda = \theta_i$ is equal to $n-1$, i.e., out of n equations of (1.18) with fixed index i, exactly $n-1$ will be linearly independent. Taking into consideration that i runs over L values, we can conclude that the condition that $\Psi_k(\lambda)$ has no poles at the points θ_i is equivalent to the system of $(n-1)L$ linear equations with respect to the coefficients of the polynomials $\omega_j(\lambda)$. Consequently, to satisfy condition 2, we must impose on the coefficients of the polynomials $\omega_k(\lambda)$ in (1.17)

$$K+(n-1)L = nL+m_1+m_2+\cdots+m_n \tag{1.27}$$

linear relations. But this number coincides exactly with the maximum number of all coefficients of the polynomials $\omega_k(\lambda)$, $k = 1, \cdots, n$, allowed by condition 1. Thus the final calculation of all the functions $\Phi_k(\lambda)$, $k = 1, \cdots, n$, reduces to a solution of the system of $K+(n-1)L$ linear equations with $K+(n-1)L$ unknowns, which determines uniquely all coefficients of the polynomials $\omega_k(\lambda)$. [6]

If some roots θ_i coincide, then the indicated system of equations may change its form somewhat. It is possible, for example, that for some θ_i only $n-2$ of the equations of (1.18) will be linearly independent, but we must still require that the derivatives of all the functions $\omega_1(\lambda)f_{1k}(\lambda)+ \cdots +\omega_n(\lambda)f_{nk}(\lambda)$ vanish at the point $\lambda = \theta_i$, which gives us n (and not $n-1$) more equations. However, in this instance the system obtained defines uniquely all the coefficients of $\omega_k(\lambda)$; actually we can always first rearrange the parameters of the spectral matrix somewhat, so that the roots θ_i become different, then solve the system of equations for a new spectral matrix, and finally equate the coefficients in $\omega_k(\lambda)$ to the limits of the obtained solutions, with the rearrangement converging to zero.

An obvious expression for the functional $\tilde{\xi}_1(t, \tau)$ is also obtained very simply from (1.17) and (1.13). Indeed, we can represent each of the functions $\Phi_k(\lambda)$ in the form

$$\Phi_k(\lambda) = \sum_{j=0}^{m_k-1} W_k^{(j)}(i\lambda)^j + \int_0^\infty e^{-is\lambda} w_k(s)\, ds, \tag{1.28}$$

where $\Phi_k(\lambda) - \sum W_k^{(j)}(i\lambda)^j$ is a function of L_2, and $w_k(s)$ is its Fourier transform. Substituting (1.28) into (1.13), we find that

$$\tilde{\xi}_1(t, \tau) = \sum_{k=1}^{n} \left[\sum_{j=0}^{m_k-1} W_k^{(j)} \xi_k^{(j)}(t) + \int_0^\infty w_k(s)\, \xi_k(t-s)\, ds \right]. \tag{1.29}$$

Thus we have the following

Theorem 1. *The solution $\tilde{\xi}_1(t, \tau)$ of the extrapolation problem for a non-degenerate stationary process with a rational spectrum is given by* (1.29). *The constants $W_k^{(j)}$ and the function $w_k(s)$ in this formula are selected uniquely by the condition that all functions of* (1.28) *have the form* (1.17), *where $\theta_1, \cdots, \theta_L$ are zeros of $D(\lambda)$ in the upper half-plane, and $w_k(\lambda)$, $k = 1, \cdots, n$, are polynomials of degree no higher than $(L+m_k-1)$, respectively, with coefficients determined by a system of $nL+m_1+ \cdots +m_n$ linear equations, which follows from the condition that all the functions* (1.16) *are analytic in the upper half-plane.*

After determining all the $\Phi_k(q)$, it is very easy to estimate also the mean square of the error of extrapolation $\sigma_1^2(\tau)$ (see (1.9)): because of (1.13) and (1.12) the square of the error here is equal to

[6] The uniqueness of these coefficients follows at once from the uniqueness of the solution $\tilde{\xi}_1(t, \tau)$ for the extrapolation problem in the space H. For a proof of the existence of the solution, owing to the properties of linear equations, it is sufficient to show that a solution (unique) exists for at least one τ; but for $\tau = 0$ such a solution will obviously be

$$\omega_1(\lambda) = (\lambda-\theta_1) \cdots (\lambda-\theta_L),\ \omega_j(\lambda) = 0 \quad \text{for} \quad j \neq 1.$$

$$\sigma_1^2(\tau) = \int_{-\infty}^{\infty} f_{11}(\lambda)d\lambda - \int_{-\infty}^{\infty} \left\{ \sum_{k,j=1}^{n} \Phi_k(\lambda)\overline{\Phi_j(\lambda)} f_{kj}(\lambda) \right\} d\lambda \tag{1.30}$$

$$= B_{11}(0) - \int_{-\infty}^{\infty} \left\{ \sum_{k,j=1}^{n} \Phi_k(\lambda)\overline{\Phi_j(\lambda)} f_{kj}(\lambda) \right\} d\lambda.$$

REMARK. Now let $\xi(t)$ be a degenerate stationary process with a rational spectrum such that $D(\lambda)$ vanishes at some points of the real axis. Since the function $D(\lambda)$ is non-negative, all of its real zeros have an even order. We also include among the zeros $\theta_i, \cdots, \theta_L$ exactly half of the real roots of the equation $D(\lambda) = 0$. In this case we shall again be able to determine uniquely all the coefficients of the polynomials $\omega_k(\lambda)$ in (1.17) from the condition that the function does not increase for large values faster than $|\lambda|^{m_k-1}$ and that the functions $\Psi_k(\lambda)$ have no singularities in the whole upper half-plane (including the real axis). The obtained functions $\Phi_k(\lambda)$ are obviously limits of some sequence $\Phi_k^{(n)}(\lambda)$, $n = 1, 2, \cdots$, which gives the solution of the extrapolation problem for a sequence of non-degenerate processes with a rational spectral density. But do the functions yield a solution of the extrapolation problem for our degenerate process?

Substituting our $\Phi_k(\lambda)$ directly into (1.13), we run the risk of obtaining a meaningless expression; if, for example, the real zeros of $D(\lambda)$ do not coincide with the zeros of some $f_{kk}(\lambda)$, then the integrals on the right side of (1.13) do not exist as integrals in the mean square (since condition (1.14) is not satisfied here). It is not difficult to show, however, that the integral

$$\tilde{\xi}_1(t, \tau) = \int_{-\infty}^{\infty} e^{it\lambda} \left\{ \sum_{k=1}^{n} \Phi_k(\lambda) dZ_k(\lambda) \right\} \tag{1.31}$$

(considered as the limit in the mean square of the Stieltjes integral sums for the whole expression under the integral sign) has meaning. Indeed, the integral (1.31) exists if only the following condition is satisfied:

$$\int_{-\infty}^{\infty} \left\{ \sum_{k,j=1}^{n} \Phi_k(\lambda)\overline{\Phi_j(\lambda)} f_{kj}(\lambda) \right\} d\lambda < \infty, \tag{1.32}$$

and in (1.32) the integrand no longer has singularities on the real axis because of (1.18). It is easy to see that (1.31) is also exactly the solution of the extrapolation problem for our degenerate stationary process with a rational spectrum.

In order that we can consider our extrapolation problem effectively solved, it is still necessary for us to represent expression (1.31) in the form of a functional of values of all the processes $\xi_k(t')$ in the past (for $t' \leq t$). It is not difficult to realize, however, that in the given case, $\tilde{\xi}_1(t, \tau)$ cannot at all be represented in the form (1.29). The only thing that we can accomplish with our formulas is to represent $\tilde{\xi}_1(t, \tau)$ in the form of a limit (in the mean square) of a sequence of functionals of the form on the right side of (1.29) (of course, such a representation is far from simple).

It is the same thing in the case of degeneration, which consists of the major terms of the polynomial in the numerator of $D(\lambda)$ being mutually cancelled. In this case, condition (1.32) allows us to choose as our $\Phi_k(\lambda)$ functions which increase for large values more rapidly than $|\lambda|^{m_k-1}$. For such a selection the number of unknown coefficients which $\omega_k(\lambda)$ has again proves to be equal to $nL+m_1+\cdots+m_n$, where L is the same as before (in spite of the fact that the degree of the denominator of $D(\lambda)$ here is less than L—$D(\lambda)$ "has not enough zeros"). The subsequent solution is completely analogous to the solution for the non-degenerate case with the only difference that here again $\tilde{\xi}_1(t, \tau)$ is given by (1.31) and cannot be represented in the form (1.29) (but only in the form of the limit of a sequence of similar functionals).

2. Extrapolation of a Process Given on a Finite Interval

The problem of the linear extrapolation of the process $\xi(t)$, which is given on a finite interval $[t-T, t]$ of the time axis, consists in finding a linear functional $\tilde{\xi}_1(t, \tau, T)$ from the values $\xi_k(t')$, $k = 1, \cdots, n$, $t-T \leqq t' \leqq t$, which gives the best (in the sense of the method of least squares) estimate of $\xi_1(t+\tau)$, $\tau > 0$. It is obvious that in the given case $\tilde{\xi}_1(t, \tau, T)$ coincides with a (unique) perpendicular which is dropped in H from the point $\xi_1(t+\tau)$ onto the subspace

$$H_\xi(t, T) = [\xi_k(t'), k = 1, \cdots, n, t-T \leqq t' \leqq t], \tag{2.1}$$

defined by the vectors indicated in the square brackets. Therefore the element $\tilde{\xi}_1(t, \tau, T)$ is defined uniquely by the following two conditions:

$$\text{(a)} \quad \tilde{\xi}_1(t, \tau, T) \in H_\xi(t, T), \tag{2.2}$$

$$\text{(b)} \quad \mathbf{M}[\xi_1(t+\tau)-\tilde{\xi}_1(t, \tau, T)]\overline{\xi_k(t')} \quad \text{for } k=1,\cdots,n \text{ and } t-T \leqq t' \leqq t. \tag{2.3}$$

In the case of non-degenerate processes $\xi(t)$ with a rational spectrum $\tilde{\xi}_1(t, \tau, T)$ we can choose from among variables of the form (1.13)

$$\tilde{\xi}_1(t, \tau, T) = \sum_{k=1}^{n} \int_{-\infty}^{\infty} e^{it\lambda} \Phi_k(\lambda) dZ_k(\lambda), \tag{2.4}$$

where all integrals on the right are considered in the mean square, so that
1)

$$\int_{-\infty}^{\infty} |\Phi_k(\lambda)|^2 f_{kk}(\lambda) d\lambda < \infty, \qquad k = 1, \cdots, n; \tag{2.5}$$

condition (b) is then equivalent to the equality

$$\int_{-\infty}^{\infty} e^{is\lambda} \Psi_k(\lambda) d\lambda = 0, \qquad k = 1, \cdots, n, 0 \leqq s \leqq T, \tag{2.6}$$

where $\Psi_k(\lambda)$ is given by (1.16). It is easy to see that for (2.6) to be correct it is sufficient that the following condition be satisfied:

2) *All functions $\Psi_k(\lambda)$ may be represented in the form*

$$\Psi_k(\lambda) = \Psi_k^{(1)}(\lambda) + e^{-iT\lambda} \Psi_k^{(2)}(\lambda), \qquad k = 1, \cdots, n, \tag{2.7}$$

where $\Psi_k^{(1)}(\lambda)$ are functions of a complex variable which are analytic on the upper half-plane, $\Psi_k^{(2)}(\lambda)$ are functions which are analytic on the lower half-plane, and all $\Psi_k^{(1)}(\lambda)$, $\Psi_k^{(2)}(\lambda)$ for $|\lambda| \to \infty$ in the corresponding half-plane decrease not slower than $|\lambda|^{-1-\varepsilon}$, $\varepsilon > 0$ (see [11]).

As far as condition (a) is concerned, it is certainly satisfied if all the integrals $\int_{-\infty}^{\infty} e^{it\lambda}\Phi_k(\lambda)dZ_k(\lambda)$ depend only on values of $\xi_k(t')$ for $t-T \leq t' \leq t$. We can obtain the latter by demanding that the following condition be satisfied:

3) *The functions $\Phi_k(\lambda)$, $k = 1, \cdots, n$, are all analytic (integral) functions of the complex variable λ on the whole plane, and can be represented in the form*

$$\Phi_k(\lambda) = \Phi_k^{(1)}(\lambda) + e^{-iT\lambda}\Phi_k^{(2)}(\lambda), \tag{2.8}$$

where $\Phi_k^{(1)}(\lambda)$ and $\Phi_k^{(2)}(\lambda)$ are rational functions which increase for $|\lambda| \to \infty$ no more rapidly than a finite power of $|\lambda|$ (see [11]).

It remains for us only to select functions $\Phi_k(\lambda)$, $k = 1, \cdots, n$, which satisfy these conditions 1, 2, and 3.

Substituting (2.8) into (1.16) and comparing with (2.7), we find that for $\tau > 0$ we must assume

$$\Psi_k^{(1)}(\lambda) = [e^{i\tau\lambda} - \Phi_1^{(1)}(\lambda)]f_{1k}(\lambda) - \sum_{j=2}^{n} \Phi_j^{(1)}(\lambda)f_{jk}(\lambda), \tag{2.9}$$

$$\Psi_k^{(2)}(\lambda) = -\sum_{j=1}^{n} \Phi_j^{(2)}(\lambda)f_{jk}(\lambda). \tag{2.10}$$

Since the function $\Phi_k(\lambda)$ must be integral, all the poles of the function $\Phi_k^{(1)}(\lambda)$ must coincide with the poles of $\Phi_k^{(2)}(\lambda)$. Further from (2.9) and (2.10) it is easy to conclude that in order that some pole of $\Phi_k^{(1)}(\lambda)$ and $\Phi_k^{(2)}(\lambda)$ not be simultaneously a pole of the function $\Psi_k^{(1)}(\lambda)$ (if it is located in the upper half-plane) or a pole of $\Psi_k^{(2)}(\lambda)$ (if it is located in the lower half-plane), it must coincide with a zero of the determinant $D(\lambda)$ of the spectral matrix of the process $\xi(t)$. Thus in the given case we must assume

$$\Phi_k^{(i)}(\lambda) = \frac{\omega_k^{(i)}(\lambda)}{(\lambda-\theta_1)\cdots(\lambda-\theta_L)(\lambda-\bar\theta_1)\cdots(\lambda-\bar\theta_L)}, \quad k = 1, \cdots, n,\ i = 1, 2, \tag{2.11}$$

where $\theta_1, \cdots, \theta_L, \bar\theta_1, \cdots, \bar\theta_L$ are all possible roots of the equation $D(\lambda) = 0$. In order that condition 1 be satisfied, the degrees of the polynomials $\omega_k^{(1)}(\lambda)$ and $\omega_k^{(2)}(\lambda)$ must not exceed $2L + m_k - 1$. Thus, by using (2.11), we have at our disposal $4nL + 2(m_1 + \cdots + m_n)$ arbitrary coefficients. In order that all the functions (2.8) be integral functions, the following equalities must be satisfied:

$$\omega_k^{(1)}(\lambda) + e^{iT\lambda}\omega_k^{(2)}(\lambda) = 0, \qquad k = 1, \cdots, n;\ \lambda = \theta_1, \cdots, \theta_L, \bar\theta_1, \cdots, \bar\theta_L; \tag{2.12}$$

this gives us $2nL$ linear equations with respect to the coefficients of the polynomials $\omega_k^{(i)}(\lambda)$ (if some of the roots θ_i coincide, then only the form of the equations (2.12) is changed, but not their number). After this one must only make sure that condition 2 is satisfied.

Further reasoning is very similar to that which we used in § 1. In order that all the functions $\Psi_k^{(1)}(\lambda)$, $k = 1, \cdots, n$, have no poles at the points $\lambda = \alpha_{jk}^{(m)}$, $m = 1, \cdots, N_{jk}$; $j = 1, \cdots, n$, the coefficients of the polynomials $\omega_k^{(1)}(\lambda)$ must satisfy $K = L + \sum m_k$ linear equations. Similarly, the requirement that all functions $\Psi_k^{(2)}(\lambda)$, $k = 1, \cdots, n$, have no singularities for $\lambda = \alpha_{kj}^{(m)}$, $m = 1, \cdots, N_{kj}$; $j = 1, \cdots, n$, gives us K linear equations which the coefficients of the polynomials $\omega_k^{(2)}(\lambda)$ must satisfy. Finally, in order that the points $\lambda = \theta_i$,

$i = 1, \cdots, L$, not be poles of the functions $\Psi_k^{(1)}(\lambda)$, the following relationships must be satisfied:

$$\sum_{j=1}^{n} \omega_j^{(1)}(\theta_i) f_{jk}(\theta_i) = 0, \qquad k = 1, \cdots, n; \tag{2.13}$$

and in order that the points $\lambda = \bar{\theta}_i$ not be poles of the functions $\Psi_k^{(2)}(\lambda)$, the following relationships must be satisfied:

$$\sum_{j=1}^{n} \omega_j^{(2)}(\bar{\theta}_i) f_{jk}(\bar{\theta}_i) = 0, \quad k = 1, \cdots, n; \; i = 1, \cdots, L. \tag{2.14}$$

If all the roots θ_i are different, then equations (2.13) are equivalent to $(n-1)L$ linear equations with respect to the coefficients of the polynomials $\omega_k^{(1)}(\lambda)$, and equations (2.14) are equivalent to $(n-1)L$ linear equations with respect to the coefficients of $\omega_k^{(2)}(\lambda)$. Finally, we obtain in this way the system

$$2nL + 2K + 2(n-1)L = 4nL + 2(m_1 + \cdots + m_n) \tag{2.15}$$

of linear equations, which defines uniquely $4nL + 2(m_1 + \cdots + m_n)$ unknown coefficients of the polynomials $\omega_k^{(1)}(\lambda)$ and $\omega_k^{(2)}(\lambda)$ (cf. footnote 6, which is completely applicable in the given case). If some roots θ_i coincide, the system of equations may be changed somewhat, but the final outcome remains the same.

In order to obtain an obvious form for the functional $\tilde{\xi}_1(t, \tau, T)$, it is convenient to separate out the polynomial terms of $\Phi_k^{(1)}(\lambda)$ and $\Phi_k^{(2)}(\lambda)$, i.e., to represent $\Phi_k(\lambda)$ in the form

$$\Phi_k(\lambda) = \sum_{j=0}^{m_k-1} [W_k^{(j)} + W_{k,1}^{(j)} e^{-iT\lambda}](i\lambda)^j + \varphi_k(\lambda), \tag{2.16}$$

where $\varphi_k(\lambda)$ is a function of L_2. Taking further advantage of the analytic properties of $\Phi_k(\lambda)$, it is not difficult to show that the Fourier transform $w_k(s)$ of the function $\varphi_k(\lambda)$ vanishes for $s > 0$ and for $s < -T$, so that we can also rewrite (2.16) in the form

$$\Phi_k(\lambda) = \sum_{j=0}^{m_k-1} [W_k^{(j)} + W_{k,1}^{(j)} e^{-iT\lambda}](i\lambda)^j + \int_0^T e^{-is\lambda} w_k(s) ds. \tag{2.17}$$

Substituting the last formula into (2.4), we find

$$\xi(t, \tau, T) = \sum_{k=1}^{n} \left\{ \sum_{j=0}^{m_k-1} [W_k^{(j)} \xi_k^{(j)}(t) + W_{k,1}^{(j)} \xi_k^{(j)}(t-T)] + \int_0^T w_k(s) \xi_k(t-s) ds \right\}. \tag{2.18}$$

The results obtained may be stated in the form of the following theorem:

Theorem 2. *The solution $\tilde{\xi}_1(t, \tau, T)$ of the problem of the linear extrapolation of the non-degenerate stationary process with a rational spectrum, given on the finite interval $[t-T, t]$, is provided by (2.18). The constants $W_k^{(j)}$, $W_{k,1}^{(j)}$ and the functions $w_k(s)$ in this formula are defined uniquely by the condition that all functions (2.17) have the form (2.8), where $\Phi_k^{(1)}(\lambda)$ and $\Phi_k^{(2)}(\lambda)$ are the rational functions (2.11), $\theta_1, \cdots, \theta_L$ and $\bar{\theta}_1, \cdots, \bar{\theta}_L$ are all possible zeros of the determinant of the spectral matrix of our process, and $\omega_k^{(1)}(\lambda)$ and $\omega_k^{(2)}(\lambda)$ are the polynomials of respectively no higher degree than $(2L+m_k-1)$ which satisfy (2.12),*

(2.13), *and* (2.14) *and 2K more linear equations which result from the condition that* (2.9) *and* (2.10) *must be analytic functions in the upper and lower half-planes, respectively.*

The mean square of the error of extrapolation

$$\sigma_1^2(\tau) = \mathbf{M}|\xi_1(t+\tau)-\tilde{\xi}_1(t, \tau, T)|^2 \tag{2.19}$$

here is given by the same formula (1.30) as in the case of extrapolation from the half-axis $t' \leqq t$.

Just as in § 1, most of the results we have obtained may be applied automatically to the case of degenerate processes $\xi(t)$ with a rational spectrum such that $D(\lambda) = 0$ at isolated points. The only exceptions are the results about the obvious form (2.18) of the functional $\tilde{\xi}_1(t, \tau, T)$ and also the interpretation of (2.4) (in its application to degenerate processes we must in many cases interpret it similarly to (1.31)). We note further that in the case of extrapolation from a finite interval (in contrast to extrapolation from the half-axis $t' \leqq t$), it is not difficult to select examples of degenerate processes to which all the results of Theorem 2 may be applied, without any changes. Such is the case, for example, if degeneracy consists only of the fact that the spectral densities $f_{kk}(\lambda)$ vanish at discrete points on the real axis (then at these points, of course, all $f_{kj}(\lambda)$ with $j \neq k$ and $D(\lambda)$ will also be equal to zero).

3. Linear Filtration of Multivariate Processes

The problem of linear filtration of a multivariate stationary process $\xi(t)$ (given on the half-line $t' \leqq t$ or on the finite interval $t-T \leqq t' \leqq t$) is formulated in the following manner. Together with the process $\xi(t)$ let us have some random variable Ξ which is related to this process by a correlation. The correlation relation between Ξ and $\xi(t)$ is characterized naturally by the functions

$$B_{\Xi k}(t) = \mathbf{M}\Xi\overline{\xi_k(t)} \tag{3.1}$$

(for simplicity we assume that $\mathbf{M}\Xi = 0$; there is no loss of generality with this assumption). One is required to construct, according to the known correlation matrix of the process $\xi(t)$ and according to (3.1), a linear functional $\tilde{\Xi}$ from values of all the components of the process $\xi(t')$ for $t' \leqq t$ or for $t-T \leqq t' \leqq t$, which gives the best (in the sense of the method of least squares) approximation to Ξ.

Taking advantage of the spectral decomposition (1.7) of the process $\xi_k(t)$, it is not difficult to show that the function $B_{\Xi k}(t)$ can always be represented in the form

$$B_{\Xi k}(t) = \int_{-\infty}^{\infty} e^{-it\lambda} f_{\Xi k}(\lambda)d\lambda, \tag{3.2}$$

where $f_{\Xi k}(\lambda)$ is a complex function such that

$$\int_{-\infty}^{\infty} \frac{|f_{\Xi k}(\lambda)|^2}{f_{kk}(\lambda)} d\lambda < \infty \tag{3.3}$$

(cf., for example, the derivation of (3.51) in [12]). The methods developed in §§ 1–2 of this article allow us to solve effectively the problem of filtration, provided that the process $\xi(t)$ has a rational spectrum (and is non-degenerate) and that all the functions $f_{\Xi k}(\lambda)$ have the form

$$f_{\Xi k}(\lambda) = \sum_{r=1}^{R} q_{rk}(\lambda) e^{it_r \lambda}, \tag{3.4}$$

where $t_1, \cdots, t_R$ are real numbers, and $q_{1k}(\lambda), \cdots, q_{Rk}(\lambda)$ are rational functions which do not have real poles (some of these functions may vanish).

As a matter of fact, the desired best estimate $\tilde{\Xi}$ obviously coincides with the origin of the perpendicular which is dropped in H from the point Ξ onto the subspace $H_\xi(t)$ (in the case of filtration of the process given on the half-line) or onto $H_\xi(t, T)$ (in the case where $\xi(t')$ is given only on the interval $t-T \leqq t' \leqq t$). Thus $\tilde{\Xi}$ is defined uniquely by the conditions:

(a) $\tilde{\Xi} \in H_\xi(t)$ or, respectively, $\tilde{\Xi} \in H_\xi(t, T)$,

(b) $\mathbf{M}[\Xi - \tilde{\Xi}]\overline{\xi_k(t')} = 0$ for all $k = 1, \cdots, n$ and $t' \leqq t$ or $t-T \leqq t' \leqq t$, respectively. If we seek $\tilde{\Xi}$ in the form

$$\tilde{\Xi} = \sum_{k=1}^{n} \int_{-\infty}^{\infty} e^{it\lambda} \Phi_k(\lambda) dZ_k(\lambda), \tag{3.5}$$

then in order for the integrals to exist (in the sense of the mean square), we must demand that the following inequality be satisfied:

1)

$$\int_{-\infty}^{\infty} |\Phi_k(\lambda)|^2 f_{kk}(\lambda) d\lambda < \infty, \qquad k = 1, \cdots, n, \tag{3.6}$$

and we can satisfy conditions (a) and (b) by imposing certain restrictions on the behaviour of the functions $\Phi_k(\lambda)$ for complex values of the argument λ.

We note, first of all, that condition (b) is equivalent to the relationship

$$\int_{-\infty}^{\infty} e^{is\lambda} \left\{ \sum_{r=1}^{R} e^{i(t_r - t)\lambda} q_{rk}(\lambda) - \sum_{j=1}^{n} \Phi_j(\lambda) f_{jk}(\lambda) \right\} d\lambda = 0, \tag{3.7}$$

$k = 1, \cdots, n$; $s \geqq 0$ or $0 \leqq s \leqq T$, respectively. Further we shall consider separately the problem of filtration of the process given on the half-axis $t' \leqq t$ and the problem of filtration of the process given on the interval $t-T \leqq t' \leqq t$.

In the case where we must find the best approximation $\tilde{\Xi}$ which is linearly dependent on all $\xi_k(t')$, $t' \leqq t$, conditions (a) and (b) will certainly hold, provided:

2) *The functions*

$$\Psi_k(\lambda) = \sum_{r=1}^{R} e^{i(t_r - t)\lambda} q_{rk}(\lambda) - \sum_{j=1}^{n} \Phi_j(\lambda) f_{jk}(\lambda), \qquad k = 1, \cdots, n, \tag{3.8}$$

can all be continued analytically (without singularities) into the upper half-plane, in such a way that for $|\lambda| \to \infty$ *they decrease not slower than* $|\lambda|^{-1-\varepsilon}$, $\varepsilon > 0$, *in this half-plane.*

3) *The functions* $\Phi_k(\lambda)$, $k = 1, \cdots, n$, *can all be continued analytically into the lower half-plane, in such a way that for* $|\lambda| \to \infty$ *they do not increase faster than exponentially in this half-plane.*

In selecting the functions $\Phi_k(\lambda)$ which satisfy all the conditions 1, 2, and 3, it is convenient to represent all numbers t_r in the form $t_r = t+\tau_r$ and to write the sum (3.4) in the following form:

$$f_{\Xi k}(\lambda) = e^{it\lambda}\left\{\sum_{r=1}^{R_1} q_{rk}(\lambda)e^{i\tau_r\lambda}+\sum_{r=R_1+1}^{R} q_{rk}(\lambda)e^{i\tau_r\lambda}\right\}, \tag{3.9}$$

where $\tau_r < 0$ for $r = 1, \cdots, R_1$ and $\tau_r \geqq 0$ for $r = R_1+1, \cdots, R$. We note now that because of conditions 2 and 3 the functions $\Phi_k(\lambda)$ can contain only exponential terms $e^{i\tau\lambda}$ with negative exponents τ, and the functions $\Psi_k(\lambda)$ only exponential terms with positive exponents. Therefore it is natural to seek the functions $\Phi_k(\lambda)$ in the form

$$\Phi_k(\lambda) = \sum_{r=1}^{R_1} \varphi_{rk}(\lambda)e^{i\tau_r\lambda}+\Phi_k^*(\lambda), \qquad k = 1, \cdots, n, \tag{3.10}$$

where $\varphi_{rk}(\lambda)$, $r = 1, \cdots, R_1$, and $\Phi_k^*(\lambda)$ are rational functions. Here the functions $\varphi_{rk}(\lambda)$ are determined uniquely by the condition that in (3.8) all exponential functions with positive indices vanish, i.e., by the condition

$$\sum_{j=1}^{n} \varphi_{rj}(\lambda)f_{jk}(\lambda) = q_{rk}(\lambda), \qquad k = 1, \cdots, n;\ r = 1, \cdots, R_1. \tag{3.11}$$

For fixed r, (3.11) gives us a linear system of n equations with n unknowns $\varphi_{rj}(\lambda)$ whose determinant equals $D(\lambda) > 0$; hence explicit expressions for all rational functions $\varphi_{rj}(\lambda)$ are found at once. At the same time, because of condition (3.3), the functions $|\varphi_{rk}(\lambda)|^2 f_{kk}(\lambda)$ for all r and k are integrable; therefore, in order that condition 1 might be satisfied, it is necessary only that the degree of the numerator of the rational function $\Phi_k^*(\lambda)$ exceed the degree of its denominator by no more than m_k-1. The last condition also guarantees at the same time the necessary order of increase of the functions $\Psi_k(\lambda)$ for $|\lambda| \to \infty$ (in the corresponding half-plane).

We now reduce to a common denominator all rational functions $\varphi_{rk}(\lambda)$, $r = 1, \cdots, R'$, $k = 1, \cdots, n$; because of the very definition of these functions, generally speaking, all the zeros of the determinant $D(\lambda)$ are among the zeros of the indicated common denominator. As a result, we can write the sum on the right side in the form

$$\sum_{r=1}^{R_1} \varphi_{rk}(\lambda)e^{i\tau_r\lambda} = \frac{A_k(\lambda)}{Q(\lambda)C(\lambda)C^{(1)}(\lambda)}, \qquad k = 1, \cdots, n, \tag{3.12}$$

where $Q(\lambda)$ is the same as in (1.6), $C(\lambda)$ is some polynomial (independent of k) with zeros in the upper half-plane, $C^{(1)}(\lambda)$ is a polynomial (of degree ν) all of whose zeros are located in the lower half-plane, and $A_k(\lambda)$ is a linear combination of the functions $e^{i\tau_r\lambda}$, $r = 1, \cdots, R_1$, with polynomial coefficients ((3.12) obviously cannot have real poles). Similarly we represent the second sum on the right side of (3.9) in the form

$$\sum_{r=R_1+1}^{R} q_{rk}(\lambda)e^{i\tau_r\lambda} = \frac{B_k(\lambda)}{E(\lambda)E^{(1)}(\lambda)}, \qquad k = 1, \cdots, n, \tag{3.13}$$

where $E(\lambda)$ is some polynomial (independent of k) of degree μ, all of whose zeros are located in the upper half-plane, and $E^{(1)}(\lambda)$ is a polynomial, all of whose zeros are located in the lower half-plane. It is not difficult to realize that we should seek the rational functions $\Phi_k^*(\lambda)$ of interest to us in the form

$$\Phi_k^*(\lambda) = \frac{\omega_k(\lambda)}{Q(\lambda)C^{(1)}(\lambda)E(\lambda)}, \tag{3.14}$$

where $\omega_k(\lambda)$ is a polynomial of degree not higher than $(2L+\nu+\mu+m_k-1)$.

Assuming (3.14), we have at our disposal

$$n(2L+\nu+\mu)+\sum_{k=1}^{n} m_k \tag{3.15}$$

as yet undetermined coefficients of the polynomials $\omega_k(\lambda)$, $k=1,\cdots,n$. In order that condition 3 be satisfied, the zeros $\bar{\theta}_1,\cdots,\bar{\theta}_L$ of the polynomial $Q(\lambda)$ and the ν zeros of the polynomial $C^{(1)}(\lambda)$ in the denominator on the right side of (3.10) must be cancelled with corresponding zeros of the numerator, i.e., the relation

$$\frac{A_k(\lambda)\,\varepsilon(\lambda)-\omega_k(\lambda)C(\lambda)}{(\lambda-\bar{\theta}_1)\cdots(\lambda-\bar{\theta}_L)C^{(1)}(\lambda)} \tag{3.16}$$

for all $k=1,\cdots,n$ must be an integral function of λ. This condition is obviously equivalent to $n(L+\nu)$ linear equations with respect to the coefficients of the polynomials $\omega_k(\lambda)$. Further, because of (3.8)—(3.14),

$$\Psi_k(\lambda) = \frac{B_k(\lambda)}{E(\lambda)E^{(1)}(\lambda)} - \sum_{j=1}^{n} \frac{\omega_j(\lambda)}{Q(\lambda)C^{(1)}(\lambda)E(\lambda)} f_{jk}(\lambda); \tag{3.17}$$

therefore for condition 2 to be correct, the functions of (3.17) must not have poles in the upper half-plane. As is usually the case (cf. § 1), the requirement that, for any k, the poles $\alpha_{jk}^{(m)}$ of the functions $f_{jk}(\lambda)$ may be cancelled in (3.17) with the zeros of the numerator gives us $K=L+m_1+\cdots+m_n$ more linear equations with respect to the coefficients of the polynomials $\omega_k(\lambda)$. A similar requirement concerning the zeros of the polynomial $E(\lambda)$ in the expression for $\Psi_k(\lambda)$ gives us μ more linear equations; so for all possible k we obtain $n\mu$ equations. Finally the conditions that $\Psi_k(\lambda)$ has no poles at the points $\theta_1,\cdots,\theta_L$ is equivalent to the relations

$$\sum_{j=1}^{n} \frac{\omega_j(\theta_i)}{C^{(1)}(\theta_i)E(\theta_i)} f_{jk}(\theta_j) = 0, \quad k=1,\cdots,n; = 1,\cdots,L, \tag{3.18}$$

i.e., they give us $(n-1)L$ more linear equations (for simplicity we consider here that all the roots of the polynomials $E(\lambda)$, $D(\lambda)$, and $P_{jk}(\lambda)$ are different; the changes which arise in the case where some of these roots coincide are obvious). Altogether we obtain

$$n(L+\nu)+L+\sum_{k=1}^{n} m_k+n\mu+(n-1)L$$

linear equations — just as many as we had unknowns.

An explicit formula for the functional $\tilde{\Xi}$ here has the form

(3.19) $$\tilde{\Xi}=\sum_{k=1}^{n}\left\{\sum_{j=0}^{m_k-1}\sum_{r=0}^{R_1}W_{k,r}^{(j)}\xi_k^{(j)}(t_r)+\int_0^\infty w_k(s)\xi_k(t-s)ds\right\},$$

where $t_0 = t$, so that $\tau_0 = 0$, and the constants $W_{k,r}^{(j)}$ and the functions $w_k(s)$ are defined by the condition

(3.20) $$\sum_{j=0}^{m_k-1}\sum_{r=0}^{R_1}W_{k,r}^{(j)}e^{i\tau_r\lambda}(i\lambda)^j+\int_0^\infty e^{is\lambda}w_k(s)ds=\sum_{r=1}^{R_1}\varphi_{rk}(\lambda)e^{i\tau_r\lambda}+\frac{\omega_k(\lambda)}{Q(\lambda)C^{(1)}(\lambda)E(\lambda)}.$$

Thus we have the following

Theorem 3. *The solution $\tilde{\Xi}$ of the problem of linear filtration of a multivariate non-degenerate stationary process with a rational spectrum, in the case where the mutual correlation of the variable Ξ and of the components of the process is given by* (3.2) *and* (3.4), *has the form* (3.19), *where $t_0 = t$, and t_r, $r = 1, \cdots, R_1$, are all less than zero. The constants $W_{k,r}^{(j)}$ and the functions $w_k(s)$ in* (3.19) *are defined uniquely by condition* (3.20), *where $\varphi_{rk}(\lambda)$, $C^{(1)}(\lambda)$, $E(\lambda)$, and $Q(\lambda)$ are found from* (3.11)–(3.13) *and* (1.6), *and the $\omega_k(\lambda)$, $k = 1, \cdots, n$, are polynomials of degree no higher than $(2L+\nu+\mu+m_k-1)$ whose coefficients are solutions of the system of linear equations which follows from conditions* 2 *and* 3.

In order to calculate the mean square

(3.21) $$\sigma^2 = \mathbf{M}|\Xi-\tilde{\Xi}|^2$$

of the error of filtration we must also know the variable

(3.22) $$B = \mathbf{M}|\Xi|^2.$$

The following formula holds:

(3.23) $$\sigma^2 = B-\int_{-\infty}^{\infty}\left\{\sum_{j,k=1}^{n}\Phi_j(\lambda)\overline{\Phi_j(\lambda)}f_{jk}(\lambda)\right\}d\lambda$$

(cf. (1.30)).

We shall not linger here (and in subsequent sections) to extend some of the indicated results to degenerate stationary processes with a rational spectrum; in this respect the problem of filtration does not differ from the problem of extrapolation.

We now turn to the problem of linear filtration of a non-degenerate process $\xi(t')$ given on a finite interval $t-T \leqq t' \leqq t$. Reasoning in the same way as in § 2, we can show that the quantity (3.5) certainly gives a solution for this problem if the functions $\Phi_k(\lambda)$ satisfy condition 1 (i.e., (3.6)) and the following two conditions:

2) *The functions $\Psi_k(\lambda)$, $k = 1, \cdots, n$, given by* (3.8), *may all be represented in the form*

(3.24) $$\Psi_k(\lambda) = \sum_{s=1}^{S_1}\psi_k^{(s)}(\lambda)e^{i\tau_s\lambda}+\sum_{s=S_1+1}^{S}\psi_k^{(s)}(\lambda)e^{i\tau_s\lambda},$$

where $\tau_s \leqq -T$, $s = 1, \cdots, S_1$; $\tau_s \geqq 0$, $s = S_1+1, \cdots, S$; the first sum on the right side of (3.24) *is the function $\Psi_k^{(1)}(\lambda)$, which is analytic on the lower half-plane, and the second sum is the function $\Psi_k^{(2)}(\lambda)$, which is analytic on the upper half-plane, with both of these functions decreasing on their respective half-planes not slower than $|\lambda|^{-1-\varepsilon}$, $\varepsilon > 0$.*

3) *The functions $\Phi_k(\lambda)$, $k = 1, \cdots, n$, are all integral functions of the complex variable λ, which can be written in the form*

$$(3.25) \qquad \Phi_k(\lambda) = \sum_{u=1}^{U} \Phi_k^{(u)}(\lambda) e^{-i\tau_u \lambda},$$

where $0 \leqq \tau_u \leqq T$, $u = 1, \cdots, U$, and all the functions $\Phi_k^{(u)}$, $u = 1, \cdots, U$, are rational (cf. [11]).

We now write the function (3.4) in the form

$$(3.26) \quad f_{\Xi k}(\lambda) = e^{it\lambda}\left\{\sum_{r=1}^{R'} q_{rk}(\lambda)e^{i\tau_r\lambda} + \sum_{r=R'+1}^{R''} q_{rk}(\lambda)e^{i\tau_r\lambda} + \sum_{r=R''+1}^{R} q_{rk}(\lambda)e^{i\tau_r\lambda}\right\},$$

where $\tau_r = t_r - t$ and $-T < \tau_r < 0$ for $r = 1, \cdots, R'$, $\tau_r \leqq -T$ for $r = R'+1, \cdots, R''$, and finally $\tau_r > 0$ for $r = R''+1, \cdots, R$. Proceeding from conditions 2 and 3, it is natural to select the functions $\Phi_k(\lambda)$ in the form

$$(3.27) \qquad \Phi_k(\lambda) = \sum_{r=1}^{R'} \varphi_{rk}(\lambda)e^{i\tau_r\lambda} + \Phi_k^{(1)}(\lambda)e^{-iT\lambda} + \Phi_k^{(2)}(\lambda),$$

where all the functions $\varphi_{rk}(\lambda)$, $r = 1, \cdots, R'$, $\Phi_k^{(1)}(\lambda)$, and $\Phi_k^{(2)}(\lambda)$ are rational. In addition, the functions $q_{rk}(\lambda)$ are given uniquely by the equalities

$$(3.28) \qquad \sum_{j=1}^{n} \varphi_{rj}(\lambda) f_{jk}(\lambda) = q_{rk}(\lambda), \qquad k = 1, \cdots, n;\ r = 1, \cdots, R',$$

which have the same meaning as the equalities (3.11) used above. For $\Psi_k(\lambda)$ we thus obtain the expression

$$(3.29) \ \Psi_k(\lambda) = \sum_{r=R'+1}^{R''} q_{rk}(\lambda)e^{i\tau_r\lambda} + \sum_{r=R''+1}^{R} q_{rk}(\lambda)e^{i\tau_r\lambda} - \sum_{j=1}^{n} [\Phi_j^{(1)}(\lambda)e^{-iT\lambda} + \Phi_j^{(2)}(\lambda)] f_{jk}(\lambda),$$

which agrees completely with (3.24). We note further that for all the conditions concerning the order of increase of the functions $\Phi_k(\lambda)$, $\Psi_k^{(1)}(\lambda)$, and $\Psi_k^{(2)}(\lambda)$ to hold, we must require only that, for any k, the degree of the numerator of the rational functions $\Phi_k^{(1)}(\lambda)$ and $\Phi_k^{(2)}(\lambda)$ exceeds the degree of the denominator by not more than $m_k - 1$.

Analogously, to (3.12) and (3.13), we set

$$(3.30) \qquad \sum_{r=1}^{R'} \varphi_{rk}(\lambda)e^{i\tau_r\lambda} = \frac{A_k(\lambda)}{Q(\lambda)C(\lambda)C^{(1)}(\lambda)},$$

$$(3.31) \quad \sum_{r=R'+1}^{R''} q_{rk}(\lambda)e^{i\tau_r\lambda} = \frac{B_k(\lambda)}{E(\lambda)E^{(1)}(\lambda)}, \quad \sum_{r=R''+1}^{R} q_{rk}(\lambda)e^{i\tau_r\lambda} = \frac{G_k(\lambda)}{H(\lambda)H^{(1)}(\lambda)},$$

where $Q(\lambda)$ has the same meaning as in (1.6), $C(\lambda)$, $E(\lambda)$, and $H(\lambda)$ are certain polynomials all of whose zeros are located in the upper half-plane, and $C^{(1)}(\lambda)$, $E^{(1)}(\lambda)$, and $H^{(1)}(\lambda)$ are polynomials all of whose zeros are located in the lower half-plane. We seek the rational functions $\Phi_k^{(1)}(\lambda)$ and $\Phi_k^{(2)}(\lambda)$ among functions of the form

$$(3.32) \quad \Phi_k^{(1)}(\lambda) = \frac{\omega_k^{(1)}(\lambda)}{Q(\lambda)C(\lambda)E^{(1)}(\lambda)H(\lambda)}, \quad \Phi_k^{(2)}(\lambda) = \frac{\omega_k^{(2)}(\lambda)}{Q(\lambda)C^{(1)}(\lambda)E^{(1)}(\lambda)H(\lambda)}.$$

Here $\omega_k^{(1)}(\lambda)$ and $\omega_k^{(2)}(\lambda)$ are polynomials of degree $2L+\nu+\mu^{(1)}+\mu+m_k-1$ and $2L+\nu^{(1)}+\mu^{(1)}+\mu+m_k-1$, respectively, where the degrees of the polynomials $C(\lambda)$, $C^{(1)}(\lambda)$, $E^{(1)}(\lambda)$, and $H(\lambda)$ are denoted by ν, $\nu^{(1)}$, $\mu^{(1)}$, and μ. Thus formulas (3.32) contain

$$2n(2L+\mu+\mu^{(1)})+n(\nu+\nu^{(1)})+2\sum_{k=1}^{n} m_k \tag{3.33}$$

arbitrary coefficients which we can deal with according to our own judgement.

In order that condition 3 be satisfied, all the zeros of the denominator of the function (3.27) must be cancelled against zeros of not lower degree in the numerator. In other words, the ratios

$$\frac{A_k(\lambda)E^{(1)}(\lambda)H(\lambda)+C^{(1)}(\lambda)\omega_k^{(1)}(\lambda)e^{-iT\lambda}+C(\lambda)\omega_k^{(2)}(\lambda)}{Q(\lambda)C(\lambda)C^{(1)}(\lambda)}, \tag{3.34}$$
$$\frac{C^{(1)}(\lambda)\omega_k^{(1)}(\lambda)e^{-iT\lambda}+C(\lambda)\omega_k^{(2)}(\lambda)}{E^{(1)}(\lambda)H(\lambda)}$$

must both be integral functions of λ. Considering that k can take the values $1,\cdots,n$, this gives us $n(2L+\nu+\nu^{(1)}+\mu+\mu^{(1)})$ equations which relate the coefficients of all the polynomials $\omega_k^{(1)}(\lambda)$ and $\omega_k^{(2)}(\lambda)$.

As regards condition 2, in order that it might be satisfied, all the functions

$$\Psi_k^{(1)}(\lambda)=\frac{B_k(\lambda)}{E(\lambda)E^{(1)}(\lambda)}+\sum_{j=1}^{n}\frac{\omega_j^{(1)}(\lambda)e^{-iT\lambda}}{Q(\lambda)C(\lambda)E^{(1)}(\lambda)H(\lambda)}f_{jk}(\lambda) \tag{3.35}$$

must have no poles in the lower half-plane, and all the functions

$$\Psi_k^{(2)}(\lambda)=\frac{G_k(\lambda)}{H(\lambda)H^{(1)}(\lambda)}+\sum_{j=1}^{n}\frac{\omega_j^{(2)}(\lambda)}{Q(\lambda)C^{(1)}(\lambda)E^{(1)}(\lambda)H(\lambda)}f_{jk}(\lambda) \tag{3.36}$$

must have no poles in the upper half-plane. It is not difficult to calculate that the first of these conditions gives us

$$K+n\mu^{(1)}+(n-1)L=n(L+\mu^{(1)})+\sum_{k=1}^{n} m_k \tag{3.37}$$

linear relations between the coefficients of the polynomials $\omega_k^{(1)}(\lambda)$, $\omega_k^{(2)}(\lambda)$, and the second gives

$$K+n\mu+(n-1)L=n(L+\mu)+\sum_{k=1}^{n} m_k \tag{3.38}$$

linear relations between the coefficients of the same polynomials. Finally we obtain exactly as many linear equations as there are unknowns, and we can determine these coefficients uniquely.

The explicit formula for the functional $\tilde{\Xi}$ evidently has the form

$$\tilde{\Xi}=\sum_{k=1}^{n}\left\{\sum_{j=0}^{m_k-1}\sum_{r=-1}^{K'} W_{k,r}^{(j)}\xi_k^{(j)}(t_r)+\int_0^T w_k(s)\xi_k(t-s)\,ds\right\}, \tag{3.39}$$

where $t_0=t$, $t_{-1}=t-T$, and the constants $W_{k,r}^{j}$ and the functions $w_k(s)$ are determined from the relation

(3.40)
$$\sum_{j=0}^{m_k-1}\sum_{r=-1}^{R'} W_{k,r}^{(j)} e^{i\tau_r\lambda}(i\lambda)^j + \int_0^T e^{-is\lambda} w_k(s)ds = \Phi_k(\lambda),$$

where $\tau_0 = 0$, $\tau_{-1} = -T$.

We state the result obtained in the form of a theorem:

Theorem 4. *If $\xi(t)$ is a non-degenerate stationary process with a rational spectrum, given for $t-T \leqq t' \leqq t$, and Ξ is a random variable whose correlation relation with the process $\xi(t)$ is given by* (3.2) *and* (3.4), *then the solution of the filtration problem for the process $\xi(t)$ has the form* (3.39), *where $t_0 = t$, $t_{-1} = t-T$, and t_r, $r = 1, \cdots, R'$, are those values t_r in* (3.4) *which satisfy the condition $t-T < t_r < t$. In addition, the constants $W_{k,r}^{(j)}$ and the functions $w_k(s)$ in* (3.39) *are defined uniquely by condition* (3.40), *where the functions $\Phi_k(\lambda)$ are found from* (3.27)–(3.28) *and* (3.30)–(3.32), *and $\omega_k^{(1)}(\lambda)$ and $\omega_k^{(2)}(\lambda)$ are polynomials of degree no higher than $(2L+\nu+\mu^{(1)}+\mu+m_k-1)$ and $(2L+\nu^{(1)}+\mu^{(1)}+\mu+m_k-1)$, respectively, whose coefficients are solutions of a system of linear equations which follows from conditions* 2 *and* 3 *on pp.* 255–256.

The mean square σ^2 of the error of filtration of the process $\xi(t)$, which is given on a finite interval, may be calculated by means of the same formula (3.23) which we used in the case of filtration of the process given on the half-axis $t' \leqq t$. Of course, for the functions $\Phi_k(\lambda)$ we must now take the functions (3.27) and not the functions (3.10) which we employed for the derivation of (3.23).

4. Linear Interpolation of Multivariate Processes

The problem of linear interpolation of the process $\xi(t')$ given on the two half-axes $t' \leqq t$ and $t' \geqq t+\tau$ consists of finding a linear functional $\tilde{\xi}_1^*(t, \tau)$, $0 < \tau < T$, from all values $\xi_k(t')$ for $t' \leqq t$ or for $t' \geqq t+T$, which gives the best approximation of the unknown value $\xi_1(t+\tau)$. This problem arises in those cases where for some reason observation of the process $\xi(t)$ was discontinuous on the time interval between t and $t+T$, and it is necessary to interpolate one of the omitted values of the process from the available data.

In the given case $\xi_1(t+\tau)-\tilde{\xi}_1^*(t, \tau)$ will represent the perpendicular which is dropped in H from the point $\xi_1(t+\tau)$ onto the subspace

(4.1)
$$H^*(t, T) = [\xi_k(t'),\quad k = 1, \cdots, n,\ t' \leqq t \quad \text{or} \quad t' \geqq t+T].$$

Thus the basic conditions which define $\tilde{\xi}_1^*(t, \tau)$ uniquely have here the form

$$\text{(a)}\quad \tilde{\xi}_1^*(t, \tau) \in H^*(t, T),$$

$$\text{(b)}\quad \mathbf{M}[\xi_1(t+\tau)-\tilde{\xi}_1^*(t, \tau)]\overline{\xi_k(t')} = 0,\quad k = 1, \cdots, n,\ t' \leqq t \ \text{or}\ t' \geqq t+T.$$

We shall, as always, seek the best approximation in the form

(4.2)
$$\tilde{\xi}_1^*(t, \tau) = \sum_{k=1}^{n}\int_{-\infty}^{\infty} e^{it\lambda}\Phi_k(\lambda)dZ_k(\lambda),$$

where

(4.3)
$$\int_{-\infty}^{\infty} |\Phi_k(\lambda)|^2 f_{kk}(\lambda)d\lambda < \infty, \qquad k = 1, \cdots, n.$$

Condition (b) reduces to the equality

$$(4.4)\qquad \int_{-\infty}^{\infty} e^{is\lambda}\left\{[e^{iT\lambda}-\Phi_1(\lambda)]f_{1k}(\lambda)-\sum_{j=2}^{n}\Phi_j(\lambda)f_{jk}(\lambda)\right\}d\lambda=0$$

for $k=1,\cdots,n$ and $s\geqq 0$ or $s\leqq -T$. As regards condition (a), it is satisfied if $\int_{-\infty}^{\infty}e^{it\lambda}\Phi_k(\lambda)dZ_k(\lambda)$ depends only on the values $\xi_k(t')$ for $t'\leqq t$ and for $t'\geqq t+T$ (but not for $t<t'<t+T$).

Using Cauchy's theorem, it is not difficult to verify that both these conditions are satisfied, provided we can continue the functions $\Phi_k(\lambda)$ into the complex plane so that the following conditions are satisfied:

2) *The functions*

$$(4.5)\qquad \Psi_k(\lambda)=[e^{iT\lambda}-\Phi_1(\lambda)]f_{1k}(\lambda)-\sum_{j=2}^{n}\Phi_j(\lambda)f_{jk}(\lambda),\qquad k=1,\cdots,n,$$

are all everywhere analytic (integral) functions of the complex variable λ, *which can be represented in the form*

$$(4.6)\qquad \Psi_k(\lambda)=\Psi_k^{(1)}(\lambda)+e^{iT\lambda}\Psi_k^{(2)}(\lambda),$$

where $\Psi_k^{(1)}(\lambda)$ *and* $\Psi_k^{(2)}(\lambda)$ *are rational functions which decrease for* $|\lambda|\to\infty$ *not slower than* $|\lambda|^{-1-\varepsilon}$, $\varepsilon>0$ *(and thus not slower than* $|\lambda|^{-2}$*).*

3) *Each of the functions* $\Phi_k(\lambda)$ *can be written in the form*

$$(4.7)\qquad \Phi_k(\lambda)=\Phi_k^{(1)}(\lambda)+e^{iT\lambda}\Phi_k^{(2)}(\lambda),$$

where the function $\Phi_k^{(1)}(\lambda)$ *is analytic on the lower half-plane, the function* $\Phi_k^{(2)}(\lambda)$ *is analytic on the upper half-plane, and they both decrease for* $|\lambda|\to\infty$ *on their respective half-planes not faster than a finite power of* $|\lambda|$ (cf. [11]).

Thus the interpolation problem may be reduced to a selection of functions $\Phi_k(\lambda)$ which satisfy the conditions formulated here.

In the case where $\xi(t)$ is a non-degenerate multivariate process with a rational spectrum, the choice of the functions $\Phi_k(\lambda)$ is made very simply. We seek all functions $\Phi_k(\lambda)$ in the form (4.7), where $\Phi_k^{(1)}(\lambda)$ and $\Phi_k^{(2)}(\lambda)$ are rational functions in which the degree of the numerator exceeds the degree of the denominator by not more than m_k-1. In addition, the functions $\Phi_k^{(1)}(\lambda)$ and $\Phi_k^{(2)}(\lambda)$ obviously cannot have common poles. Therefore, in order that the poles of $\Phi_k^{(1)}(\lambda)$ and $\Phi_k^{(2)}(\lambda)$ not be at the same time poles of the functions $\Psi_k(\lambda)$, the sums

$$(4.8)\qquad \sum_{j=1}^{n}\omega_j^{(i)}(\lambda)f_{jk}(\lambda),\qquad k=1,\cdots,n,$$

where i takes values 1 and 2 and $\omega_j^{(1)}(\lambda)$ and $\omega_j^{(2)}(\lambda)$ are the numerators of the rational functions $\Phi_j^{(1)}(\lambda)$ and $\Phi_j^{(2)}(\lambda)$, must all vanish at every pole of the functions $\Phi_j^{(i)}(\lambda)$. Consequently, we must seek $\Phi_k^{(1)}(\lambda)$ and $\Phi_k^{(2)}(\lambda)$ in the form

$$(4.9)\qquad \Phi_k^{(1)}(\lambda)=\frac{\omega_k^{(1)}(\lambda)}{(\lambda-\theta_1)\cdots(\lambda-\theta_L)},\qquad \Phi_k^{(2)}(\lambda)=\frac{\omega_k^{(2)}(\lambda)}{(\lambda-\bar\theta_1)\cdots(\lambda-\bar\theta_L)},$$

where $\theta_1,\cdots,\theta_L$ are zeros of the determinant $D(\lambda)$ on the upper half-plane, and $\bar\theta_1,\cdots,\bar\theta_L$ are the corresponding zeros on the lower half-plane. In addition, the $2nL+2\sum_{k=1}^{n}m_k$ unknown coefficients of the polynomials $\omega_k^{(i)}(\lambda)$, $i=1,2,$

$k = 1, \cdots, n$, are determined uniquely by the condition that $\Psi_k(\lambda)$ has no poles at the points $\lambda = \alpha_{jk}^{(m)}$ or $\lambda = \overline{\alpha_{jk}^{(m)}}$ (this gives us $2K = 2L + 2\sum_{k=1}^{n} m_k$ linear equations) and by the condition that the sums (4.8) vanish for $\lambda = \theta_i$ or $\lambda = \bar{\theta}_i$ ($2(n-1)L$ more linear equations).

We now separate out from the functions $\Phi_k^{(1)}(\lambda)$ and $\Phi_k^{(2)}(\lambda)$ the polynomial components:

$$(4.10)\quad \Phi_k^{(1)}(\lambda) = \sum_{j=0}^{m_k-1} W_{k,1}^{(j)}(i\lambda)^j + \varphi_k^{(1)}(\lambda), \quad \Phi_k^{(2)}(\lambda) = \sum_{j=0}^{m_k-1} W_{k,2}^{(j)}(i\lambda)^j + \varphi_k^{(2)}(\lambda),$$

where $\varphi_k^{(1)}(\lambda)$ and $\varphi_k^{(2)}(\lambda)$ are functions of L_2. Because of the location of the poles of these functions on the half-plane defined, the Fourier transform $w_k^{(1)}(s)$ of the function $\varphi_k^{(1)}(s)$ vanishes on the negative half-axis, and the Fourier transform $w_k^{(2)}(s)$ of the function $\varphi_k^{(2)}(s)$ vanishes on the positive half-axis. Thus

$$(4.11)\quad \begin{aligned} \Phi_k^{(1)}(\lambda) &= \sum_{j=0}^{m_k-1} W_{k,1}^{(j)}(i\lambda)^j + \int_0^\infty e^{-is\lambda} w_k^{(1)}(s)ds, \\ \Phi_k^{(2)}(\lambda) &= \sum_{l=0}^{m_k-1} W_{k,2}^{(j)}(i\lambda)^j + \int_0^\infty e^{is\lambda} w_k^{(2)}(s)ds. \end{aligned}$$

From this it follows immediately that

$$(4.12)\quad \begin{aligned} \tilde{\xi}_1^*(t, \tau) = \sum_{k=1}^{n} \Bigg\{ &\sum_{j=0}^{m_k-1} [W_{k,1}^{(j)} \xi_k^{(j)}(t) + W_{k,2}^{(j)} \xi_k^{(j)}(t+T)] \\ &+ \int_0^\infty [w_k^{(1)}(s)\xi_k(t-s) + w_k^{(2)}(s)\xi_k(t+T+s)]ds \Bigg\}. \end{aligned}$$

And so we have

Theorem 5. *The solution* $\tilde{\xi}_1^*(t, \tau)$ *of the problem of interpolation of a non-degenerate multivariate stationary process* $\xi(t)$ *with a rational spectrum is given by* (4.12). *The constants* $W_{k,1}^{(j)}$, $W_{k,2}^{(j)}$ *and the functions* $w_k^{(1)}(s)$, $w_k^{(2)}(s)$ *in this formula are calculated from* (4.11), *where* $\Phi_k^{(1)}(\lambda)$ *and* $\Phi_k^{(2)}(\lambda)$ *are the functions* (4.9), θ_u *and* θ_v *are zeros of the determinant* $D(\lambda)$ *of the spectral matrix of the process, and* $\omega_k^{(1)}(\lambda)$ *and* $\omega_k^{(2)}(\lambda)$ *are polynomials of degree not higher than* $(L+m_k-1)$, *given uniquely by the condition that the functions* (4.5) *have no poles in the whole complex plane.*

5. Further Generalizations

In the preceding paragraphs we discussed in detail the problems of extrapolation, interpolation, and filtration of non-degenerate multivariate stationary processes with a rational spectrum. In conclusion, we shall enumerate some further linear approximation problems which also permit an effective solution by means of the method developed here.

I. We note first of all that all the problems considered in the present article are particular cases of the following general problem. On the axis $-\infty < t < \infty$, let us single out a finite number p of non-intersecting intervals Δ_j, $j = 1, \cdots, p$, one or two of which may also be semi-infinite intervals of the type $(-\infty, t_0)$ or $[t_0, +\infty)$. It is assumed that the values of all the components of the process $\xi(t)$ on all intervals Δ_j are unknown to us; it is required to obtain from the data the best linear approximation of a certain random variable Ξ which is related by correlation to the process $\xi(t)$.

We shall assume that the process $\xi(t)$ has a rational spectrum (and is non-degenerate) and that the relation of Ξ with $\xi(t)$ is given by (3.1)–(3.4). We shall again seek the best linear approximation $\tilde{\Xi}$ of the variable Ξ in the form (3.5), where the functions $\Phi_k(\lambda)$ satisfy (3.6). Let $\Psi_k(\lambda)$ be the function (3.8), and let $\bar{\Delta}_l, l = 1, \cdots, g$, (where q equals $p-1$, p, or $p+1$) be intervals which are complementary to the intervals Δ_j (so that the sum of all the intervals $\Delta_j, j = 1, \cdots, p$, and $\bar{\Delta}_l$, $l = 1, \cdots, q$, covers the whole line). We shall assume further that all intervals Δ_j and $\bar{\Delta}_l$ are closed, i.e., the boundary points belong to two intervals at once. In such a case it is not difficult to see that (3.5) is a solution of our problem, provided the functions $\Phi_k(\lambda)$, $k = 1, \cdots, n$, and $\Psi_k(\lambda)$, $k = 1, \cdots, n$, satisfy the following conditions:

1) *All the functions $\Phi_k(\lambda)$ can be written in the form*

$$\Phi_k(\lambda) = \sum_{j=1}^{p} \Phi_k^{(j)}(\lambda), \tag{5.1}$$

where

$$\Phi_k^{(j)}(\lambda) = \sum_l \Phi_{k,l}^{(j)}(\lambda) e^{it_l^{(j)}\lambda}, \qquad t_l^{(j)} \in \Delta_j \quad \textit{for all } l, \tag{5.2}$$

and $\Phi_{k,l}^{(j)}$ are rational functions. In addition, the functions $\Phi_k^{(j)}(\lambda)$, which correspond to the finite intervals Δ_j, must all be analytic on the whole plane (integral), the functions $\Phi_k^{(j)}(\lambda)$, which correspond to an infinite interval Δ_j of the type $(-\infty, t_0)$ (if there is such an interval among the Δ_j), must be analytic on the lower half-plane, and the functions $\Phi_k^{(j)}(\lambda)$, which correspond to an interval Δ_j of the type $[t_0, \infty)$ (if such exists among the Δ_j), must be analytic on the upper half-plane.

2) *All the functions $\Psi_k(\lambda)$ can be written in the form*

$$\Psi_k(\lambda) = \sum_{j=1}^{q} \Psi_k^{(j)}(\lambda), \tag{5.3}$$

where

$$\Psi_k^{(j)}(\lambda) = \sum_l \Psi_{k,l}^{(j)}(\lambda) e^{is_l^{(j)}\lambda}, \qquad s_l^{(j)} \in \bar{\Delta}_j \quad \textit{for all } l, \tag{5.4}$$

and $\Psi_{k,l}^{(j)}(\lambda)$ are rational functions which decrease for $|\lambda| \to \infty$ not slower than $|\lambda|^{-2}$. In addition, the functions $\Psi_k^{(j)}(\lambda)$, which correspond to the finite intervals $\bar{\Delta}_j$, must also be integral, the functions $\Psi_k^{(j)}(\lambda)$, which correspond to the interval $\bar{\Delta}_j$ of the type $(-\infty, t_0)$ (if such an interval exists among the $\bar{\Delta}_j$), must be analytic on the lower half-plane, and the functions $\Psi_k^{(j)}(\lambda)$, which correspond to the interval $\bar{\Delta}_j$ of the type $[t_0, \infty)$ (if such exists among the $\bar{\Delta}_j$), must be analytic on the upper half-plane.

The fact that these two conditions (together with condition (3.6)) are actually sufficient for the variable $\tilde{\Xi}$ to be the desired best approximation is proved exactly in the same way as was done in [11] for particular cases of this problem (concerning the one-dimensional processes $\xi(t)$). The selection of the functions $\Phi_k(\lambda)$ which satisfy the required conditions is carried out in the general case similarly to the way it was done in a particular case of the filtration problem (see § 3); in all cases the choice finally boils down to a solution of some system of linear equations. The explicit form for the functional $\tilde{\Xi}$ here includes both the sum of the linear combinations of values of all the components of the process $\boldsymbol{\xi}(t)$ and of all their existing derivatives at both ends of every interval Δ_j and at those points t_r (see (3.4)) which belong to any of the Δ_j, and also the sum of the integrals of all the components of the process $\boldsymbol{\xi}(t)$ (with continuous weight functions), which are spread over every interval Δ_j, $j = 1, \cdots, p$.

II. In the preceding account we have always assumed that all the components of the process $\boldsymbol{\xi}(t)$ have a zero mean value. This assumption does not restrict the generality, compared to the assumption that all the mean values $\mathbf{M}\xi_k(t)$ are arbitrary numbers known to us. However, we can also assume that the mean values of the components of $\xi_k(t)$ are unknown and seek, for example, the best *non-displaced* linear estimate of the value of the first component of the process at some moment at which there was no detailed observation of the process. In this case the functions $\Phi_k(\lambda)$ must satisfy the additional conditions: $\Phi_1(0) = 1$, $\Phi_k(0) = 0$ for $k = 2, \cdots, n$; at the same time the basic condition of orthogonality ("condition (b)"; cf. pp. 242 or 248) is also changed somewhat, so that the functions $\Psi_k(\lambda)$ include one more arbitrary parameter (cf. [11], §§ 4—5). For non-degenerate processes with a rational spectrum also in this case the solution of the problem can be finally reduced to the solution of some system of linear equations with a non-zero determinant.

Together with the problem of the best approximation of an unknown value of the process $\boldsymbol{\xi}(t)$, the problem of the best linear estimate for the unknown mean values of the components of the process is also of interest. Here also we can seek the best estimate in the form of a functional of the type of the right side of (1.13); moreover, we can choose the functions $\Phi_k(\lambda)$ that we need by imposing natural restrictions, which follow from the nature of the problem, on the behavior of these functions for complex values λ.

III. If we assume that the process $\boldsymbol{\xi}(t)$ is stationary, its mean value can only be a constant vector. In some cases, however, it is natural to assume that all the components $\xi_k(t)$ represent the sum of a time-dependent ordinary (non-random) function $x_k(t)$ and the random stationary oscillations which are superimposed on this function (with a zero mean value). In addition, there arise naturally problems of non-displaced extrapolation or filtration for such a process and of the optimal linear estimate for the unknown mean values $x_k(t)$ (in this respect, we usually assume beforehand that they have a definite functional form). In the more usual statement of problems of this type, all mean values $x_k(t)$ are assumed to be written in the form of linear combinations (with unknown coefficients) of a finite number of given functions $g_j(t)$, $j = 1, \cdots, s$; it is required to give the best linear estimate of these unknown coefficients and to indicate the best method of linear extrapolation or filtration of the original processes $\xi_k(t)$ (cf. [14]—[16]). If, in addition, the process $\boldsymbol{\xi}(t) - \boldsymbol{x}(t)$ has a rational spectrum, and all the $g_j(t)$ are either step functions of t^j or trigonometric functions (in other words, functions of the form $e^{i\lambda_j t}$), an effective solution of the indicated problems may be obtained very simply by means of a slight modification in the reasoning of §§ 2–3. In this case all $\Phi_k(\lambda)$ will have either fixed values for the function itself and some of its derivatives for $\lambda = 0$ (if $g_j(t) = t^j$) or fixed values at the points $\lambda = \lambda_j$ (if $g_j(t) = e^{i\lambda_j t}$); on the other hand, the functions $\Psi_k(\lambda)$ will include s additional parameters. Therefore the number of obtained linear equations here too remains equal to the number of unknowns.

We also employ an analogous method in the case where the coefficients of the polynomials or of the trigonometric polynomials $x_k(t)$ are random variables with given moments of the first two orders. In this case, the conditions imposed on the functions $\Phi_k(\lambda)$ are not changed, but to the functions $\Psi_k(\lambda)$ are added additional terms which are equal to values of these same functions $\Phi_k(\lambda)$ or their derivatives at fixed points with coefficients which are known beforehand.

IV. We note, finally, that in all the considerations of this article it is possible to assume that the process $\xi(t)$ is not stationary but only has stationary increments of some order (cf. [12], where solutions of corresponding problems for one-dimensional processes $\xi(t)$ of this type are discussed in detail). One special example of a similar filtration problem (in which we are dealing with one-dimensional processes with stationary increments of the first order) is also analyzed by means of an altogether different method in [17]. In the case of processes with stationary increments, rational spectral densities may have a pole at the point $\lambda = 0$; however, this does not produce substantial changes in the solutions of all possible linear approximation problems. Therefore we need not dwell at length on the application of the results of this article to the case of processes with stationary increments.

Received by the editors
July 15, 1959

REFERENCES

[1] V. N. Zasukhin, *On the theory of multivariate stationary random processes*, Dokl. Akad. Nauk SSSR, **33**, 1941, pp. 435–437. (In Russian.)

[2] M. Rosenblatt, *A multi-dimensional prediction problem*, Ark. Mat., **3**, 1957, pp. 407–424.

[3] N. Wiener and P. Massani, *The prediction theory of multivariate stochastic processes*, Acta Math., **98**, 1–2, 1957, pp. 111–150; **99**, 1–2, 1958, pp. 93–137.

[4] P. Massani, *Sur les processus vectoriels minimaux de rang maximal*, C. R. Acad. Sci. Paris, **246**, 15, 1958, pp. 2212–2217.

[5] Yu. A. Rozanov, *The spectral theory of multivariate stationary random processes with discrete time*, Uspekhi Mat. Nauk, **13**, 2 (80), 1958, pp. 93–142. (In Russian.)

[6] Yu. A. Rozanov, *The linear extrapolation of multivariate stationary processes of rank* 1 *with discrete time*, Dokl. Akad. Nauk SSSR, **125**, 2, 1959, pp. 277–280. (In Russian.)

[7] R. F. Matveev, *On the regularity of multivariate stationary random processes with discrete time*, Dokl. Akad. Nauk SSSR, **126**, 4, 1959, pp. 713–715. (In Russian.)

[8] E. G. Gladyshev, *On multi-dimensional stationary random processes*, Theory Prob. Applications, **3**, 1958, pp. 425–428. (English translation.)

[9] J. L. Doob, *Stochastic Processes*, Wiley, New York, 1953. (Russian translation, I.L., 1956.)

[10] N. Wiener, *Extrapolation, Interpolation, and Smoothing of Stationary Time Series*, *Technology Press of MIT, Cambridge*, 1949.

[11] A. M. Yaglom, *Extrapolation, interpolation, and filtration of stationary random processes with a rational spectral plane*, Trudy Moskov. Mat. Obsc., **4**, 1955, pp. **333**–374. (In Russian.)

[12] A. M. Yaglom, *Correlation theory of processes with random stationary n-th increments*, Mat. Sb., **37** (79), 1, 1955, pp. 141–196. (In Russian.)

[13] S. Darlington, *Linear least-squares smoothing and prediction with applications*, Bell System Tech. J., **37**, 5, 1958, pp. 1221–1294.

[14] L. A. ZADEH and R. RAGAZZINI, *Extension of Wiener's theory of prediction*, J. Appl. Phys., 21, 7, 1950, pp. 645–655.

[15] J. H. LANING and R. H. BATTIN, *Random Processes in Automatic Control*, McGraw-Hill, New York, 1956. I. L., Moscow, 1958. (Russian translation.)

[16] U. GRENANDER and M. ROSENBLATT, *Statistical Analysis of Stationary Time Series*, Wiley, New York, 1957.

[17] C. L. DOLPH and M. A. WOODBURY, *On the relation between Green's functions and covariances of certain stochastic processes and its application to unbiased linear prediction*, Trans. Amer. Math. Soc., 72, 3, 1952, pp. 519–550.

EFFECTIVE SOLUTIONS OF LINEAR APPROXIMATION PROBLEMS FOR MULTIVARIATE STATIONARY PROCESSES WITH A RATIONAL SPECTRUM

A. M. YAGLOM (*MOSCOW*)

(*Summary*)

We consider a class of multivariate stationary random processes $\boldsymbol{\xi}(t) = \{\xi_1(t), \cdots, \xi_k(t)\}$ having the nonsingular spectral density matrix $||f_{jk}(\lambda)||$, where all $f_{jk}(\lambda)$ are rational functions of λ. The following linear approximation problems for the processes are studied: 1) the simplest extrapolation problem of determining a linear least-square estimate of $\xi_k(t+\tau)$, $\tau > 0$, by known values of $\xi_j(t')$, $j = 1, \cdots, n$, $t' \leqq t$; 2) the finite interval extrapolation problem of a linear least-square estimation of $\xi_k(t+\tau)$ by $\xi_j(t')$, $j = 1, \cdots, n$, $t-T \leqq t' \leqq t$; 3) the interpolation problem of a least-square estimation of $\xi_k(t+\tau)$, $0 < \tau < T$ by $\xi_j(t')$, $j = 1, \cdots, n$, $t' \leqq t$ or $t' \geqq t+T$; 4) the filtration problem of a least-square estimation of the value of some random variable Ξ (such that the functions $f_{\Xi k}(\lambda)$, $k = 1, \cdots, n$, from equations (3.1)–(3.2) have the form (3.4), where all $q_{rk}(\lambda)$ are rational) by the values of $\xi_j(t')$, $j = 1, \cdots, n$, $t' \leqq t$ or $t-T \leqq t' \leqq t$.

In all cases the method used in previous papers [11] and [12] enables the explicit extrapolation, interpolation or filtration formulae to be derived by merely solving the algebraical equation $D(\lambda) = \det ||f_{jk}(\lambda)|| = 0$ and afterwards a simple system of linear algebraical equations. The same method can also be applied to the case when we wish to find a least-square estimate of $\xi_k(t+\tau)$ or Ξ by the values of $\xi_j(t')$, $j = 1, \cdots, n$, on any set of closed intervals on the time axis. Some other generalizations of extrapolation, interpolation and filtration problems may be solved by the same method; they are given in the last section of the paper.

12

Reprinted with permission from *Theory Probab. Appl.* **5**(4):362–376 (1960)

SPECTRAL PROPERTIES OF MULTIVARIATE STATIONARY PROCESSES AND BOUNDARY PROPERTIES OF ANALYTIC MATRICES

YU. A. ROZANOV

(*Translated by Eizo Nishiura*)

The development of the spectral theory of stationary processes [1]—[9] has revealed a deep connection between this theory and the theory of functions of a complex variable, in particular, between the boundary properties of analytic functions (in the one variable case) and analytic matrices (in many variables). While the boundary properties of analytic functions have been well studied (see, for example, the monograph by I. I. Privalov [12]), the corresponding properties of analytic matrices are poorly represented in the mathematical literature.

The present work gives a survey (of both previously published and unpublished) results relating to the spectral theory of stationary processes and boundary properties of analytic matrices.

1. Regularity of Stationary Processes and Factorization of Matrix-Valued Functions

By an n-dimensional stationary process, we shall understand a column vector

$$x(t) = \begin{pmatrix} x_1(t) \\ \cdot \\ \cdot \\ \cdot \\ x_n(t) \end{pmatrix},$$

the components $x_k(t)$ of which are stationarily connected functions in some Hilbert space H, i.e. such that the correlation matrix

$$B(\tau) = \{B_{kl}(\tau)\}_{k,l=1,\cdots,n}; \quad B_{kl}(\tau) = (x_k(t), x_l(t-\tau))$$

is independent of t (the parameter t takes integer values). The matrix $B(\tau)$ may be represented in the form

$$B(\tau) = \int_{-\pi}^{\pi} e^{i\lambda\tau} dF(\lambda), \tag{1}$$

where $F(\lambda)$ is a monotone (i.e. $F(\lambda') - F(\lambda'') \geqq 0$ for $\lambda' \geqq \lambda''$), bounded, matrix function, the spectral function of the process $x(t)$. We shall consider $F(\lambda)$ to be absolutely continuous; $f(\lambda) = F'(\lambda)$ is called the spectral density.

The matrix $f(\lambda)$ is positive definite (we note that for each integrable, positive definite matrix $f(\lambda)$ there exists a process $x(t)$ with such a spectral density [1]).

We shall denote by $H_x^-(\tau)$ the linear closure of the variables $x_k(t)$, $t \leq \tau$, $k = 1, \cdots, n$; the stationary process $x(t)$ is called *regular* if

$$\bigcap_\tau H_x^-(\tau) = 0.$$

We shall say that the matrix function $f(\lambda)$ admits a factorization if there exists a matrix $A(z)$, analytic in the unit circle and having boundary values $a(\lambda) = \lim_{\rho\uparrow 1} A(\rho e^{-i\lambda})$ such that

$$a(\lambda)a^*(\lambda) = 2\pi f(\lambda), \tag{2}$$

where a^* is the matrix conjugate to a.

In fact, the following theorem is proved in [3].

Theorem 1. *The stationary process $x(t)$ is regular if and only if its spectral function $F(\lambda)$ is absolutely continuous and the spectral density $f(\lambda)$ admits a factorization.*

In all of the following, we shall encounter functions $\Gamma(z)$ which are analytic in the unit circle and belong to the so-called class H_δ (see [12]), i.e. such that

$$\lim_{\rho\uparrow 1} \int_{-\pi}^{\pi} |\Gamma(\rho e^{-i\lambda})|^\delta d\lambda = H_\delta(\Gamma) < \infty.$$

We shall say that the function $\Gamma(z)$ belongs to the class N_δ if it may be represented in the form of a quotient $\Gamma(z) = \Gamma_1(z)/\Gamma_2(z)$, where $\Gamma_1(z)$ and $\Gamma_2(z)$ belong to the class H_δ.

We note that the function $\Gamma(z) \in N_\delta$ has a boundary limit $\gamma(\lambda) = \lim_{\rho\uparrow 1}\Gamma(\rho e^{-i\lambda})$ for almost all λ, moreover it is uniquely determined [12].

Theorem 2. *A positive definite, integrable matrix function $f(\lambda)$ admits a factorization if and only if it satisfies the following conditions:*

1. *The rank of $f(\lambda)$ is equal to some m for almost all λ.*
2. *There exists a principal minor $M(\lambda) = \det \{f_{i_p i_q}(\lambda)\}_{p,q=1,\cdots,m}$ of order m, for which*

$$\int_{-\pi}^{\pi} \log M(\lambda) d\lambda > -\infty.$$

3. *The functions $\gamma_{i_p j}(\lambda) = M_{i_p j}(\lambda)/M(\lambda)$ (where $M_{i_p j}(\lambda)$ is obtained from $M(\lambda)$ by replacing the j-th row by the i_p-th row, $p = 1, \cdots, m$; $j \neq i_p$) are boundary values of functions $\Gamma_{i_p j}(z)$ of class H_δ for some $\delta > 0$.*

(We note that if the conditions enumerated above are satisfied for some minor of $M(\lambda)$, then they are automatically satisfied for all principal minors of order m, which are not identically equal to zero.) This theorem (see [7]) is the result of a series of articles [2]—[6].

2. Sliding Summation of Stationary Processes and Analytic Matrices

As is well known (see, for example, [3], p. 118), an n-dimensional, regular, stationary process $x(t)$ may be represented, with the help of sliding summation, in the form

$$x(t) = \sum_{\tau=0}^{\infty} a(\tau)u(t-\tau), \tag{3}$$

where the stationary process

$$u(t) = \begin{pmatrix} u_1(t) \\ \cdot \\ \cdot \\ \cdot \\ u_q(t) \end{pmatrix}$$

has orthonormal values ($\|u_j(t)\| = 1$ and $(u_i(t), u_j(t')) = 0$ for $i \neq j$ or $t \neq t'$) and is stationarily connected with the process $x(t)$; the coefficients $a(\tau)$ are the matrices $a(\tau) = \{a_{kj}(\tau)\}_{k=1,\cdots,n;j=1,\cdots,q}$. (We note that only regular processes may be represented in the form (3).)

Let us consider the matrix $A(z)$,

$$A(z) = \sum_{\tau=0}^{\infty} a(\tau)z\tau.$$

In so far as [1]

$$\sum_{\tau=0}^{\infty} \|a(\tau)\|^2 = \sum_{\tau=0}^{\infty}\sum_{k=1}^{n}\sum_{j=1}^{q} |a_{kj}(\tau)|^2 = \sum_{k=1}^{n} \|x_k(0)\|^2 < \infty,$$

the matrix $A(z) = \{A_{kj}(z)\}_{k=1,\cdots,n;j=1,\cdots,q}$ is analytic in the unit circle $|z| < 1$ (each of its components $A_{kj}(z) = \sum_{\tau=0}^{\infty} a_{kj}(\tau)z^\tau$ is an analytic function of class H_2).

Let us set

$$a(\lambda) = \lim_{\rho \uparrow 1} A(\rho e^{-i\lambda}) = \sum_{\tau=0}^{\infty} a(\tau)e^{-i\lambda\tau}.$$

The boundary value $a(\lambda)$ satisfies the condition (2). Indeed, from the representation (3) of the process $x(t)$, we get that the correlation matrix $B(\tau)$ has the form

$$B(\tau) = \sum_{s=0}^{\infty} a(s)a^*(\tau-s) = \frac{1}{2\pi}\int_{-\pi}^{\pi} e^{i\lambda\tau}a(\lambda)a^*(\lambda)d\lambda;$$

(2) follows from this and (1).

It should be said that the least dimension of the process $u(t)$, with orthonormal values, appearing in the representation (3) of the process $x(t)$ is equal to the rank m of the matrix $f(\lambda)$, the spectral density; the number m is called the rank of the process $x(t)$.

If the process $u(t)$ in (3) has dimension m $(q = m)$, then, as was shown in [3], p. 116, the variables $u_j(t)$, the components of the process $u(t)$, belong to the linear closure H_x of the variables $x_k(t)$, $k = 1, \cdots, n$, $-\infty < t < \infty$.

Theorem 3. *Let $x(t)$ be an n-dimensional, stationary, regular process of rank m with spectral density $f(\lambda)$. To each analytic matrix $A(z) = \{A_{kj}(z)\}_{k=1,\cdots,n;j=1,\cdots,m}$,*

$$A(z) = \sum_{\tau=0}^{\infty} a(z)z^\tau,$$

[1] The norm $\|a\|$ of the matrix $a = \{a_{kj}\}$ is equal to $(\sum_{k,j}|a_{kj}|^2)^{1/2}$.

the boundary value $a(\lambda) = \lim_{\rho\uparrow 1} A(\rho e^{-i\lambda})$ *of which satisfies condition* (2), *corresponds an n-dimensional, stationary process* $u(t)$ *with orthonormal values, such that*

$$x(t) = \sum_{\tau=0}^{\infty} a(\tau)u(t-\tau);$$

moreover, if

$$\Phi(d\lambda) = \begin{pmatrix} \Phi_1(d\lambda) \\ \vdots \\ \Phi_n(d\lambda) \end{pmatrix}$$

is the spectral family for the process $x(t)$,

$$x(t) = \int_{-\pi}^{\pi} e^{i\lambda t}\Phi(d\lambda),$$

then [2]

$$u(t) = \int_{-\pi}^{\pi} e^{i\lambda t}\varphi(\lambda)\Phi(d\lambda), \tag{4}$$

where the matrix $\varphi(\lambda) = \{\varphi_{jk}(\lambda)\}_{j=1,\cdots,m;k=1,\cdots,n}$ *is defined by the relations:*

$$\begin{aligned} \varphi(\lambda)f(\lambda) &= \frac{1}{2\pi}a^*(\lambda), \\ \varphi(\lambda)f(\lambda)\varphi^*(\lambda) &= \frac{1}{2\pi}I. \end{aligned} \tag{5}$$

PROOF. From formulas (4) and (5) it is seen that the joint correlation matrix $B^{ux}(\tau) = \{B^{ux}_{jk}(\tau)\}_{j=1,\cdots,m;k=1,\cdots,n}$, $B^{ux}_{jk}(\tau) = (u_j(\tau), x_k(0))$, has the form

$$B^{ux}(\tau) = \int_{-\pi}^{\pi} e^{i\lambda\tau}\varphi(\lambda)f(\lambda)d\lambda = \frac{1}{2\pi}\int_{-\pi}^{\pi} e^{i\lambda\tau}a^*(\lambda)d\lambda,$$

whence

$$B^{xu}(\tau) = \frac{1}{2\pi}\int_{-\pi}^{\pi} e^{i\lambda\tau}a(\lambda)d\lambda,$$

while

$$B^{xu}(\tau) = \frac{1}{2\pi}\int_{-\pi}^{\pi} e^{i\lambda\tau}f^{xu}(\lambda)d\lambda,$$

where $f^{xu}(\lambda) = \{f^{xu}_{kj}(\lambda)\}_{k=1,\cdots,n;j=1,\cdots,m}$ is the joint spectral density of the processes $x(t)$ and $u(t)$. Comparing these equations, we get that

$$f^{xu}(\lambda) = \frac{1}{2\pi}a(\lambda).$$

On the basis of Lemma 2 of [3], this relation together with (2) allows us to conclude that

---

[2] More precisely, equation (4) means that

$$u_j(t) = \int_{-\pi}^{\pi} e^{i\lambda t}\Psi_j(d\lambda), \quad \text{where} \quad \Psi_j(\Delta) = \int_{\Delta}\sum_{k=1}^{n}\varphi_{jk}(\lambda)\Phi_k(d\lambda).$$

$$x(t) = \int_{-\pi}^{\pi} e^{i\lambda t} a(\lambda)\Psi(d\lambda),$$

where $\Psi(d\lambda)$ is the spectral family of the process $u(t)$. The theorem easily follows from here.

Further, let us introduce the variables $u_j(z)$, $j = 1, \cdots, m$, by the equations

$$u_j(z) = \int_{-\pi}^{\pi} \frac{1}{1-\bar{z}\theta} \sum_{k=1}^{n} \varphi_{jk}(\lambda)\Phi_k(d\lambda), \tag{6}$$

where $\theta = e^{-i\lambda}$ and the matrix $\varphi(\lambda) = \{\varphi_{jk}(\lambda)\}_{j=1,\cdots,m;\,k=1,\cdots,n}$ is defined by relation (5).

Lemma 1. *For the elements $A_{kj}(z)$, $k = 1, \cdots, n$; $j = 1, \cdots, m$, of the analytic matrix $A(z)$ appearing in Theorem 3, we have the following representation*:

$$A_{kj}(z) = (x_k(0), u_j(z)). \tag{7}$$

Indeed,

$$\{(x_k(0), u_j(z))\}_{k=1,\cdots,n;\,j=1,\cdots,m} = \int_{-\pi}^{\pi} \frac{1}{1-z\bar{\theta}} f(\lambda)\varphi^*(\lambda)d\lambda$$

$$= \frac{1}{2\pi}\int_{-\pi}^{\pi} \frac{\theta}{\theta-z} a(\lambda)d\lambda = \frac{1}{2\pi i}\int_{|\theta|=1} \frac{A(\theta)d\theta}{\theta-z} = A(z).$$

As is well known, (see, for example, [3], p. 95), to each stationary process $x(t)$ corresponds a unitary operator U in H_x:

$$Ux(t) = x(t+1).$$

Let us return to the process $u(t)$ with orthonormal values which corresponds to the analytic matrix $A(z)$. Consider the space $D_u(z) = H_u^-(0) \ominus H_u^-(z)$, the orthogonal complement in $H_u^-(0)$ to the subspace $H_u^-(z) = (U^{-1}-zI)H_u^-(0)$. The variables $u_j(z)$, $j = 1, \cdots, m$, form an orthogonal basis in the space $D_u(z)$.

Indeed, $u_j(z) = \sum_{t\geqq 0} z^{-t}u_j(-t) \in H_u^-(0)$, and for arbitrary $t \leqq 0$,

$$(u_i(t-1)-zu_i(t), u_j(z)) = \int_{-\pi}^{\pi} \theta^t(\theta-z)\frac{1}{1-z\bar{\theta}} \sum_{k,l=1}^{n} \varphi_{ik}(\lambda)\overline{\varphi_{jl}(\lambda)}f_{kl}(\lambda)d\lambda$$

$$= \frac{1}{2\pi}\int_{-\pi}^{\pi} \theta^t(\theta-z)\frac{\theta}{\theta-z}d\lambda = \frac{1}{2\pi}\int_{-\pi}^{\pi} \theta^{-t+1}d\lambda = 0;$$

besides, as is not hard to see, the dimension of the space $D_u(z)$ does not exceed m, and

$$\begin{aligned}(u_i(z), u_j(z)) &= \int_{-\pi}^{\pi} \frac{\theta}{(1-\bar{z}\theta)(\theta-z)} \sum_{k,l} \varphi_{ik}(\lambda)\overline{\varphi_{jl}(\lambda)}f_{kl}(\lambda)d\lambda \\ &= \frac{\delta_{ij}}{2\pi}\int_{-\pi}^{\pi} \frac{\theta d\lambda}{(1-\bar{z}\theta)(\theta-z)} = \delta_{ij}\frac{1}{1-|z|^2},\end{aligned} \tag{8}$$

where δ_{ij} is the Kronecker symbol.

Lemma 2. *Let $y = \sum_{k=1}^{n} \alpha_k x_k(0)$. The perpendicular h dropped from y to* *the space* $H_u^-(z)$ *has the form*

$$h_u = (1-|z|^2) \sum_{j=1}^{m} \left(\sum_{k=1}^{n} \alpha_k x_k(0),\ u_j(z) \right) u_j(z), \tag{9}$$

$$\|h_u\|^2 = (1-|z|^2)^2 \sum_{k,l=1}^{n} \alpha_k \bar{\alpha}_l \sum_{j=1}^{m} A_{kj}(z)\, A_{lj}^*(z).$$

Equations (9) follow immediately from Lemma 1 and relations (8).

Further, the process $u(t)$ with orthonormal values appearing in the representation (3) will be called *fundamental* for the process $x(t)$ if

$$H_u^-(\tau) = H_x^-(\tau) \tag{10}$$

for all τ. It is not hard to show (see [3], p. 117) that a fundamental process is defined uniquely "up to unitary transformations", i.e. if $u_1(t)$ and $u_2(t)$ are any two fundamental processes, then

$$u_1(t) = Tu_2(t),$$

where T is a unitary matrix of order m (if $u(t)$ is a fundamental process, then evidently, the process $Tu(t)$, where T is an arbitrary unitary matrix, will also be fundamental).

M. G. Krein proved the following

Theorem 4. *The process* $u(t)$ *corresponding to the analytic matrix* $A(z)$ *is fundamental if and only if the matrix is maximal, i.e. if*

$$A(z)A^*(z) - B(z)B^*(z) \geqq 0 \tag{11}$$

(positive definiteness) for all z, $|z| < 1$, *for any analytic matrix* $B(z)$ *the boundary value* $b(\lambda)$ *of which satisfies condition* (2) *satisfied by the boundary value* $a(\lambda)$ *of the matrix* $A(z)$. *All maximal matrices are uniquely determined up to a constant factor — the unitary matrix* T.

PROOF. Let $v(t)$ be a process with orthonormal values, corresponding to the matrix $B(z)$.

In so far as $H_x^-(\tau) \subseteqq H_v^-(\tau)$, the fact that the process $u(t)$ is fundamental implies $H_u^-(\tau) \subseteqq H_v^-(\tau)$. From this it follows that

$$H_u^-(z) \subseteqq H_v^-(z),$$

and therefore, the perpendiculars h_u and h_v dropped from $y = \sum_{k=1}^{n} \alpha_k x_k(0)$ to the subspaces $H_u^-(z)$ and $H_v^-(z)$ satisfy the inequality (see Lemma 2)

$$\|h_u\|^2 = (1-|z|^2)^2 \sum_{k,l=1}^{n} \alpha_k \bar{\alpha}_l \sum_{j=1}^{m} A_{kj}(z)\bar{A}_{lj}(z)$$

$$\geqq (1-|z|^2)^2 \sum_{k,l=1}^{n} \alpha_k \bar{\alpha}_l \sum_{j=1}^{m} B_{kj}(z)\bar{B}_{lj}(z) = \|h_v\|^2,$$

whence follows inequality (11), in view of the arbitrariness of the numbers α_k, $k = 1, \cdots, n$.

One should note that if the difference (11) vanishes ($BB^*(z) = AA^*(z)$) even for a single z, $|z| < 1$, then the matrix is also maximal.

Theorem 5. *Let* $A(z) = \{A_{kj}(z)\}_{k=1,\cdots,n;j=1,\cdots,m}$ *be a maximal matrix satisfying boundary condition* (4).

Any analytic matrix $B(z) = \{B_{kj}(z)\}_{k=1,\cdots,n;j=1,\cdots,m}$ *satisfying the same boundary conditions as the matrix* $A(z)$ *has the form*

$$B(z) = A(z)C(z), \tag{12}$$

where the analytic matrix $C(z) = \{C_{ij}(z)\}_{i,j=1\cdots,m}$ *which is unitary on the boundary* $z = e^{-i\lambda}$, *is contractive, i.e.*[3]

$$I - C(z)C^*(z) \geqq 0, \tag{13}$$

for all z, $|z| < 1$.

Proof. Let $u(t)$ be a process with orthonormal values, corresponding to a maximal matrix $A(z)$.

The process $u(t)$ is fundamental, therefore

$$H_u^-(\tau) = H_x^-(\tau) \subseteqq H_v^-(\tau),$$

where $v(t)$ is a stationary process corresponding to the matrix $B(z)$.

From this relation it follows that it is possible to represent the process $u(t)$ with the help of sliding summation by means of the process $v(t)$,

$$u(t) = \sum_{\tau=0}^{\infty} c(\tau)v(t-\tau).$$

Insofar as the spectral density of the process $u(t)$ is the matrix $(1/2\pi)I$, the maximal analytic matrix corresponding to it will evidently be the identity. By the preceding theorem, the matrix $C(z) = \sum_{\tau=0}^{\infty} c(\tau)z^\tau$ satisfies inequality (13).

Further,

$$x(t) = \sum_{\tau=0}^{\infty} a(\tau)u(t-\tau) = \sum_{\tau=0}^{\infty} a(\tau) \sum_{s=0}^{\infty} c(s)v(t-\tau-s)$$
$$= \sum_{\tau=0}^{\infty} \sum_{s=0}^{\tau} a(s)c(\tau-s)v(t-\tau) = \sum_{\tau=0}^{\infty} b(\tau)U(t-\tau),$$

whence follows equation (12).

From theorem (5) we may easily obtain

Theorem 6. *Among all analytic matrices satisfying the boundary condition* (2), *the maximal matrix* $A(z) = \{A_{kj}(z)\}_{k=1,\cdots,n;j=1,\cdots,m}$ *is determined up to a unitary factor* $T = \{T_{ij}\}_{i,j=1,\cdots,m}$ *by the modulus of the value* $M_A(z_0)$ *of its minor* $M_A(z)$ *of order* m, *not identically zero at any point* z, $|z_0| < 1$.

3. Reconstruction of a Maximal Matrix from the "Square Modulus" of its Boundary Value

An extremely important problem is that of reconstructing a maximal matrix $A(z)$ from the "square modulus" of its boundary value $a(\lambda)$, i.e., from the spectral density matrix $f(\lambda)$ of the stationary process $x(t)$.

In the one-dimensional case, the answer is simple [1]:

[3] See [12].

$$(14) \qquad A(z) = \sqrt{2\pi} \exp\left\{\frac{1}{4\pi}\int_{-\pi}^{\pi} \log f(\lambda) \frac{e^{-i\lambda}+z}{e^{-i\lambda}-z} d\lambda\right\}.$$

In the multi-dimensional case, in view of the non-commutativity of multiplication of matrices, the solution of this problem raises great difficulties. These can be avoided in the case (see [6]) of stationary processes of rank 1, obtaining explicit formulas for the elements $A_k(z)$ of the maximal matrix $A(z)$ (which consists of only one column):

$$(15) \qquad \begin{aligned} A_k(z) &= \sqrt{2\pi} \exp\left\{\frac{1}{4\pi}\int_{-\pi}^{\pi} \log f_{kk}(\lambda) \frac{e^{-i\lambda}+z}{e^{-i\lambda}+z} d\lambda\right\} \\ &\times z^{r_k} \prod_p \frac{z-\alpha_p^{(k)}}{1-\bar{\alpha}_p^{(k)} z} \cdot \frac{|\alpha_p^{(k)}|}{\alpha_p^{(k)}} \exp\left\{\int_{-\pi}^{\pi} \frac{e^{-i\lambda}+z}{e^{-i\lambda}-z} \sigma_k(d\lambda)\right\}, \quad k = 1, \cdots, n, \end{aligned}$$

(appearing in this formula are variables defined by the spectral density $f(\lambda)$ by a method indicated in the more general Theorem 8 of the present work).

An interesting method of finding the boundary value $a(\lambda)$ of a maximal matrix $A(z)$ by means of the spectral density $f(\lambda)$ in the case where

$$0 < mI \leqq \|f(\lambda)\| \leqq MI < \infty$$

is indicated in [9]; indeed, without loss of generality, we may consider that

$$f(\lambda) = I + g(\lambda), \quad \|g(\lambda)\| \leqq g < 1$$

and then

$$(16) \qquad a(\lambda) = b^{-1}(\lambda)\sqrt{C},$$

where

$$(17) \qquad b(\lambda) = I - g^- + (g^- g)^- - [(g^- g)^- g]^- + \cdots$$

(the matrix $h^-(\lambda)$ is obtained from the matrix $h(\lambda) = \sum_{-\infty}^{\infty} h_k e^{i\lambda k}$ by omitting the positive powers of $e^{i\lambda k}$, $h^-(\lambda) = \sum_{-\infty}^{0} h_k e^{i\lambda k}$), and the matrix C is defined by the equation

$$C = b(\lambda) f(\lambda) b^*(\lambda).$$

It is comparatively simple to reconstruct the maximal matrix in the case where the elements of the spectral density are rational with respect to $e^{i\lambda}$; here, naturally, it turns out that the elements of the maximal matrix are rational with respect to z.

Lemma 3.[4] *A positive definite matrix function $f(\lambda) = \{f_{kl}(\lambda)\}_{k,l=1,\cdots,n}$ of rank m, the elements $f_{kl}(\lambda)$ of which are rational with respect to $e^{i\lambda}$, may be represented in the form*

$$(18) \qquad f(\lambda) = b(\lambda) b^*(\lambda),$$

where the matrix $b(\lambda) = B(e^{-i\lambda}) = \{B_{kj}(e^{-i\lambda})\}_{k=1,\cdots,n;\, j=1,\cdots,m}$ is the boundary value of a matrix $B(z)$ which is analytic in the unit circle and rational with respect to z.

Fundamental to the proof of this lemma is the fact that a linear transfor-

[4] This lemma was proved by D. G. Polyak [8] on suggestion of the author.

mation, "rational" with respect to the matrix elements, $f_{kl}(\lambda)$, may be reduced to a diagonal form.

For the matrix $f(\lambda)$, we may find a principal minor $M(\lambda)$ of order m which vanishes only at a finite number of points; let, for definiteness, this minor stand in the upper left-hand corner of the matrix $f(\lambda)$.

Let us denote by M_j, $j = 1, \cdots, m$, the sequence of principal minors of order j, standing in the upper left-hand corner (we note that by virtue of the positive definiteness of the matrix $f(\lambda)$, $M_j(\lambda) > 0$ at those points λ at which $M(\lambda) > 0$).

If to the k-th row, $k = 2, \cdots, n$, of the matrix f we add the first row multiplied by $-f_{k1}/f_{11}$, and then to the l-th column we add the first column multiplied by $-f_{1l}/f_{11}$, we obtain the matrix

$$f^{(2)} = \begin{pmatrix} f_{11} & 0 \\ 0 & g^{(2)} \end{pmatrix},$$

where the elements of the matrix $g^{(2)} = \{g_{kl}^{(2)}\}_{k,l=2\cdots,n}$ have the form:

$$g_{kl}^{(2)} = g_{kl}^{(1)} - \frac{g_{k1}^{(1)} g_{1l}^{(1)}}{g_{11}^{(1)}}, \quad g_{kl}^{(1)} = f_{kl}.$$

The element $g_{22}^{(2)} = M_2/M_1 > 0$, and with the matrix $g^{(2)}$,, we may go through the same procedure as with the matrix $g^{(1)} = f$. Continuing this process $m-1$ times, we reduce the matrix f to the diagonal form

$$f_{(i)}^{(m)} = \mathrm{diag}\{g_{11}^{(1)}, \cdots, g_{mm}^{(m)}, 0, \cdots, 0\},$$

where $g_{ii}^{(i)}$ is the element standing in the upper left-hand corner of the matrix $g^{(i)} = \{g_{kl}^{(i)}\}_{k,l=i\cdots,n}$ the elements of which are connected with the elements of the matrix $g^{(i-1)}$ by the following recurrence relations:

$$g_{kl}^{(i)} = g_{kl}^{(i-1)} - \frac{g_{k,i-1}^{(i-1)} g_{i-1,l}^{(i-1)}}{g_{i-1,i-1}^{(i-1)}}, \quad g_{kl}^{(1)} = f_{kl}.$$

The diagonal elements $g_{(ii)}^{(i)} = d_i$ are defined as

$$d_i = \frac{M_i(\lambda)}{M_{i-1}(\lambda)}.$$

It is not hard to verify that the matrix $f(\lambda)$ can be represented in the form:

$$f = gdg^*,$$

where

$$\begin{gathered} g = \{g_{kj}\}_{k=1,\cdots,n;\,j=1,\cdots,m}, \quad g_{kj} = 0 \quad \text{for} \quad j > k, \\ g_{jj} = 1, \quad g_{kj} = g_{kj}/d_j \quad \text{for} \quad j < k; \\ d = \{d_{ij}\}_{i,j=1,\cdots,m}, \quad d_{ij} = 0 \quad \text{for} \quad i \neq j, \quad d_{ii} = d_i. \end{gathered}$$

We note that all the elements of the matrices $g(\lambda)$ and $d(\lambda)$ are rational with respect to $e^{-i\lambda}$,

$$g_{kj}(\lambda) = \frac{P_{kj}(e^{-i\lambda})}{Q_{kj}(e^{-i\lambda})}.$$

Let $\alpha_p^{(j)}$, $p = 1, 2, \cdots$, be the zeros of the polynomial $Q_{kj}(z)$, $k = 1, \cdots, n$, lying in the unit circle, counted as many times as their maximal multiplicity (we assume that the polynomials P_{kj} and Q_{kj} have no common zeros).

Let us set

$$C_j(z) = \prod_p (z-\alpha_p^{(j)}),$$

$$D_j(e^{-i\lambda}) = d_j(\lambda)/|C_j(e^{-i\lambda})|^2.$$

The non-negative function $D_j(e^{-i\lambda})$, rational with respect to $e^{-i\lambda}$, may be represented in the form (see, for example, [11], p. 450)

$$D_j(e^{-i\lambda}) = \left|\frac{P_j(e^{-i\lambda})}{Q_j(e^{-i\lambda})}\right|^2,$$

the polynomials $P_j(z)$ and $Q_j(z)$ have no zeros in the unit circle $|z| < 1$.

Let us consider the matrix

$$B(e^{-i\lambda}) = \{B_{kj}(e^{-i\lambda})\}_{k=1,\cdots,n;j=1,\cdots,m},$$

where

$$B_{kj}(e^{-i\lambda}) = g_{kj}(\lambda)C_j(e^{-i\lambda})\frac{P_j(e^{-i\lambda})}{Q_j(e^{-i\lambda})}.$$

The functions $B_{kj}(z)$ are rational with respect to z and analytic in the unit circle.

It is not hard to verify that

$$f(\lambda) = B(e^{-i\lambda})B^*(e^{-i\lambda}),$$

which proves the theorem.

Theorem 7. *Let $f(\lambda) = \{f_{kl}(\lambda)\}_{k,l=1,\cdots,n}$ be a positive definite matrix function of rank m, the elements $f_{kl}(\lambda)$ of which are rational with respect to $e^{i\lambda}$.*

The analytic matrix $A(z) = \{A_{kj}(z)\}_{k=1,\cdots,n;j=1,\cdots,m}$ satisfying the boundary condition (2), is maximal if and only if its rank is equal to m for all z, $|z| < 1$, and all its elements $A_{kj}(z)$ are rational with respect to z.

Proof. By Lemma 3, there is a matrix $B(z) = \{B_{kj}(z)\}_{k=1,\cdots,n;j=1,\cdots,m}$ with rational elements, satisfying the boundary condition (2).

If $A(z)$ is a maximal matrix, then there exists a matrix $C(z) = \{C_{ij}(z)\}_{i,j=1,\cdots,m}$ such that $B(z) = A(z)C(z)$. By virtue of the rationality of the matrix $B(z)$, the function $\det C(z)$ of the analytic class $H_{2/m}$ has only a finite number of zeros, and therefore (see [13]) the matrix $C(z)$ is representable in the form $C(z) = b(z)D(z)$ where the analytic matrix $b(z) = \{b_{ij}(z)\}_{i,j=1,\cdots,m}$ is rational with respect to z, and the matrix $D(z)$ is contractive, unitary on the boundary, and moreover $\det D(z)$ does not vanish in the interior of the unit circle $|z| < 1$.

Let $M_B(z)$ be a minor of order m of the matrix $B(z)$, not identically equal to zero. From the equations obtained above, it follows that

$$M_B(z) = M_A(z) \det b(z) \det D(z),$$

where $M_A(z)$, the corresponding minor of the matrix $A(z)$, is an analytic function of class $H_{2/m}$.

Because of the parametric representation of functions of class H_δ (see, for example [12]), and the rationality of $M_B(z)$ it is not hard to realize that

$$\det D(z) \equiv e^{i\lambda_0}.$$

Because the matrix $D(z)$ is contractive, then from the last equation, it follows immediately that $D(z) = D$ is a constant unitary matrix, and the maximal matrix $A(z) = B(z)D^{-1}b^{-1}(z)$ is rational with respect to z and has rank m for all z, $|z| < 1$ (see Lemma 4 below).

We note that in the case of a non-degenerate matrix function, the maximal matrix $A(z) = \{A_{kj}(z)\}_{k,j=1,\cdots,n}$ is not hard to find, by "chipping off" zeros (see [13]) of the matrix $B(z) = \{B_{kj}(z)\}_{k,j=1,\cdots,n}$ which is constructed as indicated in the proof of Lemma 3.

Indeed, let $z = \alpha$ be a zero of the function det $B(z)$. We represent the matrix $B(\alpha)$ in the form

$$B(\alpha) = U\begin{pmatrix} \lambda_1 & & 0 \\ & \ddots & \\ 0 & & \lambda_{p_0} \end{pmatrix}V,$$

where V is a unitary matrix which reduces the matrix $BB^*(\alpha)$ to diagonal form and $U = V_1V^*$, V_1 being a unitary matrix, figuring in the polynomial decomposition of the matrix $B(\alpha) = V_1H$, $H = \sqrt{BB^*(\alpha)}$ (see, for example, [14], p. 225).

If we multiply the matrix $B(z)$ by $b(z)$,

$$b(z) = V^{-1}\begin{pmatrix} I_p & 0 \\ 0 & \dfrac{1-\bar{\alpha}z}{z-\alpha}\dfrac{\alpha}{|\alpha|}I_{n-p} \end{pmatrix}$$

then the matrix $B_1(z) = B(z)b(z)$ will evidently be analytic and satisfy the same boundary condition (2) as $B(z)$; in addition, the function det $B_1(z)$ will have $n-p$ less zeros than det $B(z)$. Continuing this process, we finally arrive at a matrix $A(z)$ for which the determinant does not vanish in the interior of the unit circle (as was shown in the above theorem, it will be maximal).

We consider further an arbitrary matrix function $f(\lambda) = \{f_{kl}(\lambda)\}_{k,l=1,\cdots,n}$ of rank m, admitting a factorization.

Lemma 4. *The rank of a maximal matrix $A(z)$ is equal to the rank m of the matrix $f(\lambda)$ for all z, $|z| < 1$.* (For the proof, see the paper [3], p. 121).

Let $M(\lambda)$ be a principal minor of order m of the matrix $f(\lambda)$, differing from zero almost everywhere (such a minor exists by virtue of Theorem 3); for definiteness, let this be the principal minor standing in the upper left-hand corner.

Let α_p, $p = 1, 2, \cdots$, be the poles of the functions $\Gamma_{kl}(z)$ of class N_δ, $k = 1, \cdots, m$, $l = m+1, \cdots, n$, the boundary values $\gamma_{kl}(\lambda)$ of which figure in the conditions of Theorem 2, moreover each pole is counted as many times as its maximal multiplicity.

Lemma 5. *The numbers α_p, $p = 1, 2, \cdots$, and only these, are the zeros of* *the upper minor $M_A(z)$ of order m of the maximal matrix $A(z)$ satisfying boundary condition* (2).

PROOF. As is not hard to see (see [7]), the elements $A_{kj}(z)$ of the matrix are connected by means of the following relations

$$A_{lj}(z) = \sum_{k=1}^{m} \Gamma_{kl}(z) A_{kj}(z), \quad l = m+1, \cdots, n, \quad j = 1, \cdots, m. \tag{19}$$

Suppose α is not a pole of any of the functions $\Gamma_{kl}(z)$; then the rows of the matrix $A(\alpha)$ may be linearly expressed in terms of the first m rows. If $z = \alpha$ should be a zero of the minor $M_A(z)$, then the rank of the matrix $A(\alpha)$ is less than m, which is impossible by virtue of Lemma 4. Consequently the zeros of the minor $M_A(z)$ may only be among the poles α_p. But, as is seen from (19), the functions $\Gamma_{kl}(z)$ have the form

$$\Gamma_{kl}(z) = \frac{M_{kl}(z)}{M_A(z)},$$

where the minors $M_{kl}(z)$ are obtained from the $M_A(z)$ by replacing the k-th row, $k = 1, \cdots, m$, by the l-th row, $l = m+1, \cdots, n$. Insofar as the functions $M_A(z)$ and $M_{kl}(z)$ are analytic, the poles α_p, $p = 1, 2, \cdots$, may only be among the zeros of the minor $M_A(z)$. Q.E.D.

We turn again to the matrix function $f(\lambda) = \{f_{kl}(\lambda)\}_{k,l=1,\cdots,n}$ of rank m with elements which are rational with respect to $e^{i\lambda}$.

Let $A(z) = \{A_{kj}(z)\}_{k=1,\cdots,n;j=1,\cdots,m}$ be a maximal matrix, satisfying boundary condition (2) and $M_A(z)$ be a minor of order m of the matrix $A(z)$, not identically equal to zero (let, for definiteness, this be the upper minor).

The principal upper minor $M(\lambda)$ of order m of the matrix $f(\lambda)$ is rational with respect to $e^{-i\lambda}$,

$$M(\lambda) = \frac{P(e^{-i\lambda})}{Q(e^{-i\lambda})}.$$

Let β_q, $q = 1, 2, \cdots$, and γ_s, $s = 1, 2, \cdots$, be the zeros of the polynomials P and Q not lying in the interior of the unit circle, and counted as many times as their multiplicity. Being positive, the minor $M(\lambda)$ is representable in the form (see [11], p. 450)

$$M(\lambda) = c^2 \left| \frac{\prod_q (e^{-i\lambda} - \beta_q)}{\prod_s (e^{-i\lambda} - \gamma_s)} \right|^2.$$

The functions $\Gamma_{kl}(e^{-i\lambda}) = M_{kl}(\lambda)/M(\lambda)$ (see Theorem 2) have the form

$$\Gamma_{kl}(e^{-i\lambda}) = \frac{P_{kl}(e^{-i\lambda})}{Q_{kl}(e^{-i\lambda})}.$$

Let α_p, $\alpha_p \neq 0$, $p = 1, 2, \cdots$, be the zeros of the polynomials $Q_{kl}(z)$, $k = 1, \cdots, m$; $l = m+1, \cdots, n$, lying in the unit circle and counted as many times as their maximal multiplicity; let τ be the maximal multiplicity of the zero $z = 0$.

Let us denote by $D(z)$ the function

$$D(z) = cz^{\tau}\prod_{p}\frac{z-\alpha_p}{1-\bar{\alpha}_p z}\frac{|\alpha_p|}{\alpha_p}\frac{\prod_{q}(z-\beta_q)}{\prod_{s}(z-\gamma_s)}. \tag{20}$$

Since $|M_A(e^{-i\lambda})|^2 = (2\pi)^m M(\lambda)$, the theorem on the parametric representation of functions of class H_δ and the rationality of $M_A(z)$ with respect to z imply

$$M_A(z) = e^{i\lambda_0}(2\pi)^{m/2}D(z). \tag{21}$$

We shall formulate in the form of a hypothesis (confirmed in a series of special cases [5]) the following theorem (see [9], p. 145).

Theorem 8. *Let $f(\lambda) = \{f_{kl}(\lambda)\}_{k,l=1,\cdots,n}$ be a matrix function of rank m, admitting a factorization, and $A(z) = \{A_{kj}(z)\}_{k=1,\cdots,n;j=1,\cdots,m}$ its corresponding maximal analytic matrix.*

A minor $M_A(z)$ of order m of the matrix $A(z)$, not identically zero (for definiteness, let it be the upper minor), has the form

$$M_A(z) = z^{\tau}\prod_{p}\frac{z-\alpha_p}{1-\bar{\alpha}_p z}\frac{|\alpha_p|}{\alpha}\cdot e^{i\lambda_0}(2\pi)^{m/2}\exp\left\{\frac{1}{4\pi}\int_{-\pi}^{\pi}\log M(\lambda)\frac{e^{i\lambda}+z}{e^{-i\lambda}-z}\,d\lambda\right\} \times \exp\left\{\int_{-\pi}^{\pi}\frac{e^{-i\lambda}+z}{e^{-i\lambda}-z}\,\sigma(d\lambda)\right\}, \tag{22}$$

where τ is the maximal multiplicity of the pole $z = 0$ of the functions $\Gamma_{kl}(z)$, $k = 1,\cdots,m$, $l = m+1,\cdots,n$, figuring in Theorem 2, α_p, $p = 1, 2, \cdots$, the remaining poles, taken as many times as their maximal multiplicity. $M(\lambda)$ is the principal minor of the matrix $f(\lambda)$, and $\sigma(d\lambda)$ a positive measure defined by the following form:

$$\sigma(\Delta) = \int_{\Delta} p(\lambda)\tilde{\sigma}(d\lambda),$$

where

$$p(\lambda) = \max_{k,l}\{p_{kl}(\lambda)\},\quad p_{kl}(\lambda) = \frac{\sigma_{kl}^{+}(d\lambda)}{\tilde{\sigma}(d\lambda)},$$

$$\tilde{\sigma}(\Delta) = \sum_{k,l}\sigma_{kl}^{+}(\Delta),$$

the measure $\sigma_{kl}^{+}(d\lambda)$ is the positive part (in the sense of the Hahn-Jordan decomposition) of the measure $\sigma_{kl}(d\lambda)$, being defined on the interval $\Delta = (\lambda_1, \lambda_2)$, $\sigma_{kl}(\lambda_1) = \sigma_{kl}(\lambda_2) = 0$, by the relation

$$\sigma_{kl}(\Delta) = \lim_{\rho\uparrow 1}\int_{\lambda_1}^{\lambda_2}\log|\delta_{kl}(\rho e^{-i\lambda})|\,d\lambda,$$

in which the functions $\delta_{kl}(z)$ have the form

---

[5] Theorem 8 is proved, for example, in the cases $m = 1$ and $m = n$, and also if $f(\lambda)$ is rational with respect to $e^{i\lambda}$.

$$\delta_{kl}(z) = \Gamma_{kl}(z) \exp\left\{ -\frac{1}{4\pi} \int_{-\pi}^{\pi} \log \frac{M_{kl}(\lambda)}{M(\lambda)} \cdot \frac{e^{-i\lambda}+z}{e^{-i\lambda}-z} d\lambda \right\}$$

$$\times z^{\tau_{kl}} \prod_p \frac{\alpha_{kl}^{(p)}-z}{1-\bar{\alpha}_{kl}^{(p)}z} \frac{|\alpha_{kl}^{(p)}|}{\alpha_{kl}^{(p)}} \prod_q \frac{1-\beta_{kl}^{(q)}(z)}{\beta_{kl}^{(d)}-z} \cdot \frac{\beta_{kl}^{(q)}}{|\beta_{kl}^{(q)}|},$$

where τ_{kl} *is the multiplicity of the pole* $z = 0$ *of the function* $\Gamma_{kl}(z)$ *(in case* $z = 0$ *is not a pole,* $-\tau_{kl}$ *is the multiplicity of the zero* $z = 0$ *of the function* $\Gamma_{kl}(z) \not\equiv 0$*),* $\alpha_{kl}^{(p)}$ *are its other poles and* $\beta_{kl}^{(p)}$ *are its other zeros, counted as many times as their multiplicities;* $M_{kl}(\lambda)$ *is the principal upper minor of order* m *of the matrix obtained from* $f(\lambda)$ *by replacing the k-th row and k-th column,* $k = 1, \cdots, m$ *by the l-th row and l-th column,* $l = m+1, \cdots, n$*; in case* $M_{kl}(\lambda) = 0$*, then* $\sigma_{kl}(\Delta) = 0$.

4. Extrapolation of Multidimensional Stationary Processes

The problem of extrapolation of a multidimensional stationary process $x(t)$ consists in finding variables $\hat{x}_k(t, \tau)$, $t > \tau$, satisfying the condition

$$\sigma_k(t, \tau) = \min_{h \in H_x(\tau)} \|x_k(t)-h\| = \|x_k(t)-\hat{x}_k(t, \tau)\|.$$

These variables are easily obtained from the representation (3) of the process $x(t)$ with the help of sliding summation with respect to the fundamental process $u(t)$:

$$\hat{x}_k(t, \tau) = \sum_{s=t-\tau}^{\infty} a(s)u(t-s),$$

where $a(s)$ is the coefficient of z^s in the expansion of the series for the maximal matrix $A(z) = \sum_{s=0}^{\infty} a(s)z^s$ corresponding to the process $u(t)$; the process $u(t)$ is easily determined by the maximal matrix $A(z)$ (see Theorem 3).

Thus the problem of finding the variables $\hat{x}_k(t, \tau)$ reduces to finding the maximal matrix $A(z)$ corresponding to the spectral density $f(\lambda)$ of the stationary process $x(t)$.

In conclusion we mention one theorem which establishes a formula for the minimal error in the extrapolation of the process $x(t)$ in a single step (see [10]).

Let the process $x(t)$ have rank m and let $x_{i_1}(t), \cdots, x_{i_m}(t)$ be any m of its components. We consider the variables

$$y_p = \sum_{q=1}^{m} c_{pq} x_{i_q}(0), \qquad p = 1, \cdots, m,$$

where the matrix $C = \{c_{pq}\}_{p,q=1,\cdots,m}$ has determinant of modulus equal to 1. Let us set

$$\rho(C) = \frac{1}{m} \sum_{p=1}^{m} \inf_{h \in H_x^-(-1)} \|y_p-h\|^2.$$

Theorem 10. *The following relation holds:*

$$\inf_C \rho(C) = |M_A(0)|^{2/m}, \tag{23}$$

where $M_A(z) = \det \{A_{i_p i_j}(z)\}_{p,j=1,\cdots,m}$ *is the minor of the maximal matrix* $A(z)$*, corresponding to the chosen components* $x_{i_1}(t), \cdots, x_{i_m}(t)$.

PROOF. Let $u(t)$ be the fundamental process, corresponding to the maximal matrix $A(z) = \{A_{kj}(z)\}_{k=1,\cdots,n;1,\cdots,m}$.

Since $H_u^-(\tau) = H_x^-(\tau)$, it follows from formula (9) that

$$\min_{h\in H_x(-1)} \|y_p - h\|^2 = \sum_{k,l=1}^{m} c_{pk}\bar{c}_{pl} \sum_{j=1} A_{i_kj}(0)\bar{A}_{i_lj}(0)$$

and

$$\rho(C) = \frac{1}{m}\operatorname{tr} C\tilde{A}\tilde{A}^*C^*$$

is the normalized trace of the matrix $B = C\tilde{A}\tilde{A}^*C^*$, where $\tilde{A} = \{A_{i_kj}(0)\}_{k,j=1,\cdots,m}$. Thanks to the inequality between the arithmetic and the geometric means,

$$\frac{1}{m}\operatorname{tr} B \geqq (\det B)^{1/m} = |\det \tilde{A}|^{2/m} = |M_A(0)|^{2/m}.$$

Let $M_A(0) \neq 0$ (in the case where $M_A(0) = 0$, the assertion of the theorem is evident). If V is a unitary matrix which reduces the matrix $\tilde{A}\tilde{A}^*$ to the diagonal form $\operatorname{diag}(\lambda_1^2, \cdots, \lambda_m^2)$, and $D = M_A(0)^{1/m}\operatorname{diag}(\lambda_1^{-1}, \cdots, \lambda_m^{-1})$, then the matrix $C = DV$ possesses the property that

$$\frac{1}{m}\operatorname{tr} C\tilde{A}\tilde{A}^*C^* = |M_A(0)|^{2/m}.$$

The theorem is proved.

Received by the editors November 18, 1959

REFERENCES

[1] A. N. KOLMOGOROV, *Stationary sequences in Hilbert space*, Byull. Moskov. Gos. Univ., 2, 6, 1941, pp. 1–40. (In Russian.)
[2] V. N. ZASUKHIN, *On the theory of multivariate stationary random processes*, Dokl. Akad. Nauk SSSR, 33, 1941, pp. 435–437. (In Russian.)
[3] YU. A. ROZANOV, *Spectral theory of multivariate stationary random processes with discrete time*, Uspekhi Mat. Nauk, XIII, 2, 80, 1958, p. 93–142. (In Russian.)
[4] N. WIENER, *On the factorization of matrices*, Comm. Mat. Helv., 29, 1955, pp. 97–111.
[5] P. MASANI and N. WIENER, *On bivariate stationary processes and the factorization of matrix-valued functions*, Theory Prob. Applications, 4, 1959, pp. 300–308. (English translation.)
[6] YU. A. ROZANOV, *Linear extrapolation of multivariate stationary processes of rank 1 with discrete time*, Dokl. Akad. Nauk, SSSR, 125, 2, 1959, pp. 277–280. (In Russian.)
[7] R. F. MATVEEV, *On the regularity of multivariate stationary processes*, Dokl. Akad. Nauk SSSR, 126, 5, 1959. (In Russian.)
[8] D. G. POLYAK, Diploma work, Moskov. Gos. Univ., 1959.
[9] N. WIENER and P. MASANI, *The prediction theory of multivariate stochastic processes*, I, Acta Math., 98, 1957, pp. 111–150; II, Acta Math., 99, 1958, pp. 93–137.
[10] H. HELSON and D. LOWDENSLAGER, *Prediction theory and Fourier series in several variables*, Acta. Math., 99, 1958, pp. 165–202.
[11] J. L. DOOB, *Stochastic Processes*, John Wiley and Sons New York, 1953. (Translated into Russian.)
[12] I. I. PRIVALOV, *Boundary Properties of Analytic Functions*, Mosk., 1950. (In Russian.)
[13] V. P. POTAPOV, *Multiplicative structure of J-contractive matrix functions*, Trudy Moskov. Mat. Obshch., 5, 1955. (In Russian.)
[14] F. R. GANTMAKHER, *Theory of Matrices*, Mosk., 1953. (In Russian.)

SPECTRAL PROPERTIES OF MULTIVARIATE STATIONARY PROCESSES AND BOUNDARY PROPERTIES OF ANALYTIC MATRICES

YU. A. ROZANOV (MOSCOW)

(*Summary*)

A survey is presented of some recent results obtained in the spectral theory of multivariate time series closely connected with the boundary properties of analytic matrices.

13

Reprinted from *IRE Trans. Inf. Theory* **IT-1**:66–76 (1955)

Design and Performance of Phase-Lock Circuits Capable of Near-Optimum Performance Over a Wide Range of Input Signal and Noise Levels*

R. JAFFE† AND E. RECHTIN†

INTRODUCTION

PHASE-LOCK LOOPS provide an efficient method for detection and tracking of narrow-band signals in the presence of wide-band noise. This paper explains how minimum-rms-error loops may be designed if the input-signal level, input-noise level, and a specification for transient performance are given. However, the system performance of such loops departs rapidly from the best obtainable performance if either the signal or the noise levels are different from the design levels, and if no compensating changes are made in the loop. A marked improvement results if the total input power is held constant, regardless of signal or noise levels. It will be demonstrated that a fixed-component loop preceded by a bandpass limiter yields near-optimum performance over a wide range of input signal and noise levels. The following topics will be discussed:

1. An outline of the theoretical design of minimum-rms-error, phase-lock loops when input-signal level, input-noise level, and a specification for transient error are given.

2. The effects of different input levels of signal and noise:
   a. On a system having a fixed-component loop that is optimum only for an original set of levels.
   b. On a system in which loop components maintain optimum performance when the new levels are given.

3. Characteristics of a bandpass limiter.

4. A comparison of the effect of different signal and noise levels:
   a. On a loop using a fixed filter preceded by an automatic-gain-control (AGC) system that holds the signal level constant.
   b. On a fixed-filter loop preceded by a bandpass limiter.
   c. On a variable-filter loop continually adjusted to be optimum.

5. Experimental verification of the fixed-component loop preceded by a bandpass limiter.

* Presented at the Western Electronics Show and Convention, Los Angeles, Calif., August 25 27, 1954. This paper presents the results of one phase of research carried out at the Jet Propulsion Laboratory, California Institute of Technology, under Contract No. DA-04-495-Ord 18, sponsored by the Department of the Army, Ordnance Corps.

† Jet Propulsion Laboratory, California Institute of Technology, Pasadena, Calif.

DESCRIPTION OF THE LOOP

The elements of a typical phase-locked loop are shown in Fig. 1. Signal input to such a loop may be assumed to be $\sqrt{2A} \sin [\omega t + \theta_1(t)]$ where A is the rms signal amplitude, ω is the signal-center radian frequency, and $\theta_1(t)$ represents the information content of the signal. It is assumed that the noise input to the loop is narrowbanded about the signal-center frequency and has an essentially flat spectrum over the band; the noise input is represented by $N(t) \sin \omega t$. Output of the voltage-controlled oscillator (VCO) is assumed to have an rms amplitude of C, a center frequency identical to that of the signal, and a phase equal to $\theta_2(t)$.

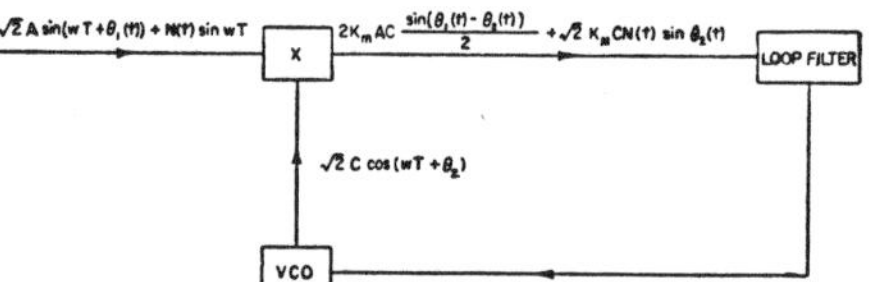

Fig. 1—Typical phase-lock loop.

Briefly, the operating principles of such a loop are as follows: The multiplier beats the signal input and the VCO output together, giving a low-frequency output proportional to the sine of $\theta_1 - \theta_2$. The loop filter accepts only this low-frequency term which is applied as a control voltage to the voltage-controlled oscillator, thereby forcing the VCO output phase θ_2 to be equal to the signal phase θ_1.

It is possible to make a phase-locked loop such that the loop need have a bandwidth only large enough to pass the *difference* between the signal frequency and the VCO estimate of the signal frequency. Since this difference frequency has considerably less variation than the actual signal frequency, the loop does not need nearly as large a bandwidth as would be needed if the loop were merely a tuned circuit placed between the system input and output, which would have to pass all frequencies over which the signal was expected to vary.

Since the bandwidth of the tracking loop is much smaller than that of a comparable nontracking filter, the amount of noise passed on to the output is proportionally smaller, and the loop accordingly picks up a greater resistance to interference.

Linearization of the Loop

To obtain a mathematical description of the loop suitable for subsequent analysis, it is desirable to approximate linearly the nonlinear operation of the actual loop.[1]

The low-frequency multiplier output is

$$2K_M AC\frac{\sin(\theta_1 - \theta_2)}{2} + N(t)\sqrt{2}\,C\sin\theta_2(t),$$

where K_M is the multiplier constant relating the multiplier voltage output both to the amplitude of its two inputs and to their phase difference in radians. Since the filter following the multiplier is low-pass, the high-frequency term in $2\omega t$ is disregarded. Under the assumption that the phase error $\theta_1 - \theta_2$ is small and that the input noise is uncorrelated with sine θ_2,

$$\sin(\theta_1 - \theta_2) \cong \theta_1 - \theta_2$$

and

$$N(t)\sin\theta_2 \cong \frac{N(t)}{\sqrt{2}}.$$

The multiplier output may therefore be represented approximately as

$$K_M C[A(\theta_1 - \theta_2) + N(t)].$$

The nonlinear operation of multiplication may now be linearized to subtraction, as shown in Fig. 2, together with associated changes of (1) making the phase input to the loop proportional to $\theta_1 + (N/A)$, (2) inserting an amplifier of gain A after the subtractor so that the voltage feeding the filter will be the same as in the nonlinearized configuration, and (3) representing the VCO as an integrator which relates its phase output to the integral of its input voltage by the VCO constant K_v. The phases around the loop and the transfer functions of the loop elements have been expressed in terms of their Laplace Transforms rather than as functions of time.

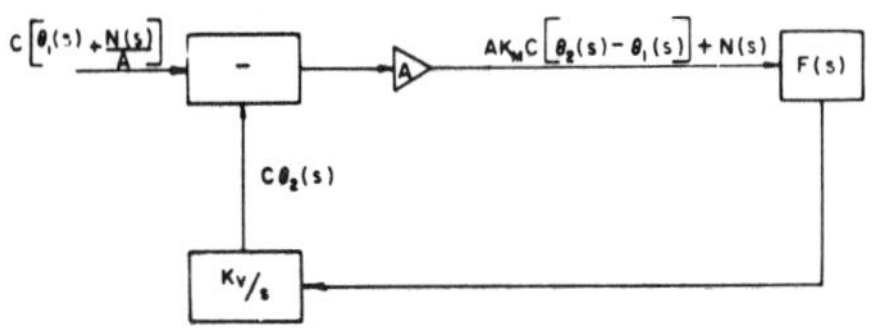

Fig. 2—Partial linearization of phase-lock loop.

Finally, the various gain coefficients may be combined into a single amplifier of gain K, as shown in Fig. 3, where K is evaluated in both radians and degrees. The product AK is defined to be the loop gain.

Criteria for Filter Design

The loop filter is important in insuring good loop operation and should be designed so that it performs two distinct functions:

1. It should minimize VCO phase-noise jitter due to noise interference.

2. It should maintain, at a specified amount, transient error in the VCO phase due to specified changes in signal phase.

Transient error is defined as the infinite-time integral of $(\theta_1 - \theta_2)^2$ and is caused solely by signal phase variations. Otherwise expressed,

$$E_T^2 = \int_0^\infty [\theta_1(t) - \theta_2(t)]^2\,dt.$$

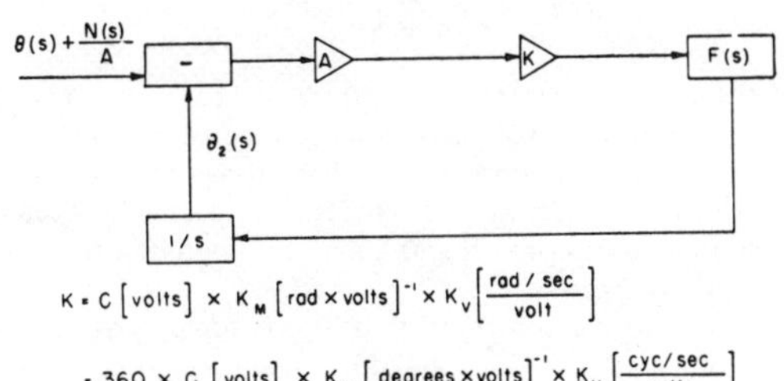

Fig. 3—Complete linearization of phase-lock loop.

Theory of Filter Design

If the rms phase jitter due to noise interference is denoted by σ_N, and if the transient error, as previously defined, is represented by E_T^2, the design criterion may be stated as follows:

$$\sigma_N^2 + \lambda^2 E_T^2 = \Sigma^2 = \text{minimum},$$

where λ is an undetermined constant, a Lagrangian multiplier used in the standard calculus-of-variations procedure. Theoretically, when the filter form has been computed (Appendix I), the value of λ may be determined from the requirement that the transient error to a specified signal must equal a specified value. Practically, λ is evaluated from the loop-bandwidth parameter B_0, which will be described in a subsequent paragraph.

Filters may be designed for many forms of signal phase inputs. It has been found that filters designed to have zero steady-state error to signal *frequency* changes (zero velocity-error loops) are relatively easy to mechanize and work well for a variety of inputs. This form of filter will be the only one discussed in the body of the paper, although the results of similar analyses concerning loops designed to follow phase steps and frequency ramps are discussed in Appendix II.

Derivation of the optimum filter is given in Appendix I. It is shown that such a filter for zero velocity-error loops has the transform

$$F(s) = \frac{B_0^2 + \sqrt{2}\,B_0 s}{A_0 K s},$$

where

$$B_0^2 \equiv \frac{\lambda(\Delta\omega)}{(N_0/A_0)}\sqrt{2\Delta f},$$

[1] The linearization appears quite legitimate, based on experimental evidence.

and where K represents the loop constant defined in Fig. 3, $\Delta\omega$ is the signal radian-frequency step, N_0 is the expected rms noise input of bandwidth Δf, and A_0 is the expected rms signal strength.

The output phase jitter and transient error of such an optimum loop, when the input noise and signal levels are those expected, are, respectively,

$$\sigma_N^2 = \frac{3}{2\sqrt{2}} B_0 \left(\frac{N_0^2}{A_0^2 2\Delta f} \right) \text{rad}^2,$$

and

$$E_T^2 = \frac{(\Delta\omega)^2}{2\sqrt{2}\, B_0^3} \text{rad}^2 \text{ sec}.$$

It may be shown that B_0 is proportional to the expected loop bandwidth; B_0 is approximately equal to two times the loop noise bandwidth and three times the loop 3-db bandwidth. If the loop bandwidth is specified, B_0 may therefore be determined and λ may be evaluated from the definition of B_0.

It is important to note that the parameter B_0 varies with the expected input signal-to-noise ratio and that the value of the optimum filter varies both with B_0 and with the expected signal amplitude A_0.

This filter form is therefore optimum only for a particular input-signal level and a particular input-noise level. If the actual levels differ from their expected values, the filter is no longer optimum.

Performance of Loops Having Levels Other Than Design Levels

It may be shown, by integrating the actual loop transfer function, that the actual output phase jitter and transient error in a fixed-filter loop fed with signal and noise levels different from those for which it was designed are, respectively,

$$\sigma_N^2 = B_0 \left(\frac{N^2}{A^2 2\Delta f} \right) \left[\frac{2(A/A_0) + 1}{2\sqrt{2}} \right] \text{rad}^2 \qquad (1)$$

and

$$E_T^2 = \frac{(\Delta\omega)^2}{2\sqrt{2}\, B_0^3} \frac{1}{(A/A_0)^2} \text{rad}^2 \text{ sec}, \qquad (2)$$

where A is actual signal level, N, actual noise level.

It also may be shown that an optimum variable filter with values continually adjusted with slowly changing signal and noise levels would have the transfer function

$$F(s) = \frac{B_0^2 \left[\frac{(A/A_0)}{(N/N_0)} \right] + \sqrt{2}\, B_0 \left[\frac{(A/A_0)}{(N/N_0)} \right]^{1/2} s}{KAs}; \qquad (3)$$

and noise jitter and transient error, respectively, of

$$\sigma_N^2 = \frac{3B_0}{2\sqrt{2}} \frac{\sqrt{(N_0/A_0)}\,(N/A)^{3/2}}{2\Delta f} \text{rad}^2$$

and

$$E_T^2 = \frac{(\Delta\omega)^2}{2\sqrt{2}\, B_0^3} \frac{(N/A)^{3/2}}{(N_0/A_0)^{3/2}} \text{rad}^2.$$

Thus far, a filter has been derived that is optimum for fixed input levels, and its phase jitter and transient error have been obtained when the input levels matched and mismatched the design levels.

A variable filter has also been derived that was assumed to adjust itself to be optimum for different signal and noise inputs. Expressions for the phase jitter and transient error in this loop have been obtained.

The problem remaining is how to obtain optimum or near-optimum performance over a wide range of input signal and noise levels, yet achieve such performance using a filter than can easily be mechanized—instead of resorting to auxiliary servo loops that continually readjust the filter to keep it optimum.

One possible solution is to use an AGC based on the signal only. The AGC voltage may be obtained if the signal is multiplied by a 90-degree, phase-shifted version of the VCO output; the inputs to the multiplier will then be in phase, and the dc multiplier output will be proportional to the signal level and may be used for the AGC. The AGC method has several disadvantages: (1) It introduces additional components. (2) It introduces additional time constants which may cause two-loop oscillation. (3) It usually results in systems with less dynamic range.

Characteristics of a Bandpass Limiter

An alternate solution, which appeared promising, was to use a limiter preceding the loop. If the noise input to the system increased, the signal strength at the limiter output would decrease because of the limiter property of holding total output power constant. However, the noise bandwidth of the loop is directly dependent on the signal amplitude at the limiter output. Therefore, the increase in input noise, by forcing down the signal amplitude at the limiter output, would reduce the loop bandwidth and require that the phase jitter at the VCO output be a smaller percentage of the input noise. The system would therefore appear to be self-compensating.

The type of limiter to be considered performs perfect snap-action limiting, i.e., its output is assumed to be $+1$ for inputs greater than 0 and -1 for inputs less than 0; such a representation closely approximates an actual limiter fed with input signal and noise levels much greater than its limiting level. The limiter is followed by a filter which restricts the output to the zone centered about the input frequency and excludes the harmonics generated in the limiting action.

Davenport[2] proved that signal-to-noise ratio is essentially preserved in passing through a bandpass limiter. This fact has been experimentally verified at the Jet Propulsion Laboratory. Youla of this laboratory then proved that the total power output of a limiter in a given zone is constant. Mathematically stated:

$$N^2 + A^2 = L^2 = N_0^2 + A_0^2$$

and

$$\frac{N'}{A'} \cong \frac{N}{A},$$

[2] W. B. Davenport, "Signal-to-noise ratios in band-pass limiters," *Jour. of Appl. Phys.*, vol. 24, pp. 720-727; June, 1953.

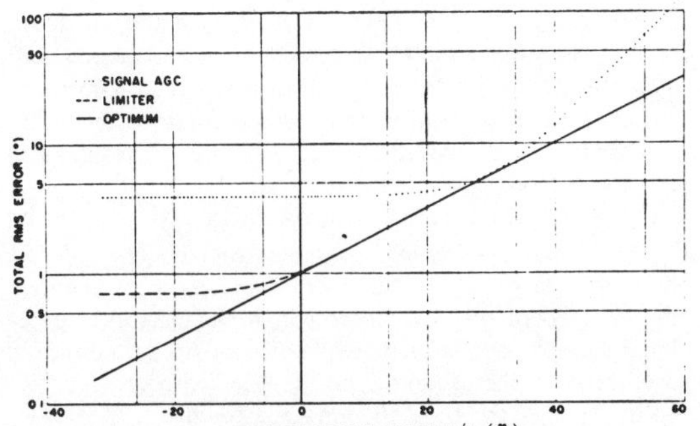

Fig. 4—Total phase error for various types of first-order loops.

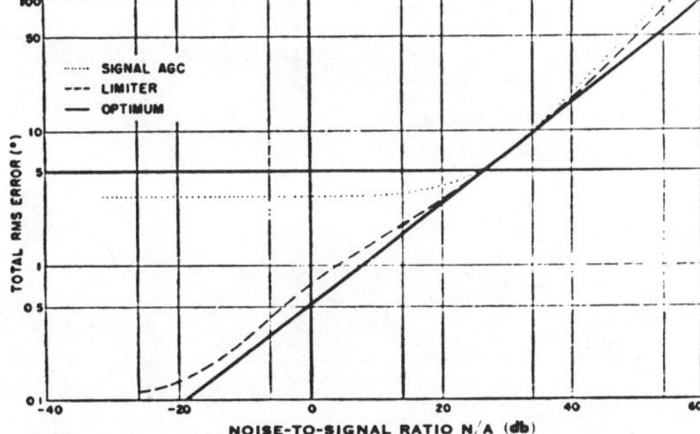

Fig. 5—Total phase error for various types of second-order loops.

where

N' and A' = limiter input noise and signal levels, respectively,

N and A = limiter output noise and signal levels, respectively,

L = total power output in a given zone (constant), and

N_0 , A_0 = output levels of the limiter for which the loop was designed.

These results may be combined to give the ratio both of actual signal level to expected signal level and of actual noise level to expected noise level at the limiter output:

$$\left(\frac{A}{A_0}\right)^2 = \frac{1 + (N_0/A_0)^2}{1 + (N'/A')^2} \qquad \left(\frac{N}{N_0}\right)^2 = \frac{1 + (N_0/A_0)^2}{1 + (A'/N')^2}. \qquad (4)$$

This relationship may in turn be applied to the expressions for the phase jitter and transient error [see (1) and (2)] in a fixed loop, giving the phase jitter and transient error in a fixed loop preceded by a limiter.

Comparison of AGC, Limiter, and Variable-Parameter Optimum Loops

It is now desirable to compare the behavior of the various loops discussed. The basis for comparison will be the relative phase jitter, transient error, and total error in the respective loops.

Total rms error is defined as

$$\Sigma = \sqrt{\sigma_N^2 + \lambda^2 E_T^2} \text{ rad},$$

and was the quantity minimized by the Wiener calculus-of-variations approach used in designing the filter.

Curves have been plotted for a loop using a fixed filter preceded by an AGC system that holds the signal level constant, for a fixed loop preceded by a limiter and for a variable-filter loop continually readjusted to be optimum. Expressions for loops designed to track input phase steps, frequency steps, and frequency ramps are derived in Appendix II and plotted in Figs. 4, 5, and 6.

Other theoretical curves for a zero velocity-error loop will now be discussed. For these plots the various parameters have been chosen to agree with the experimental

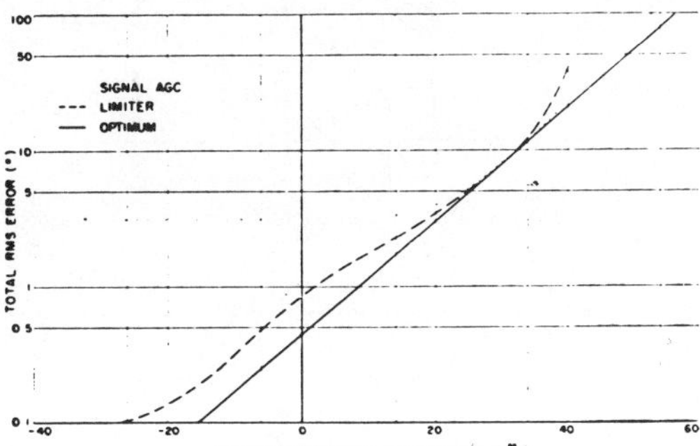

Fig. 6—Total phase error for various types of third-order loops.

and configuration to be described in a later section. Specifically, the input bandwidth of the noise was 900 cycles per second, and the loop noise bandwidth was $12\frac{1}{2}$ cycles per second (i.e., $B_0 = 25$); the loop was designed for an input noise-to-signal ratio of unity.

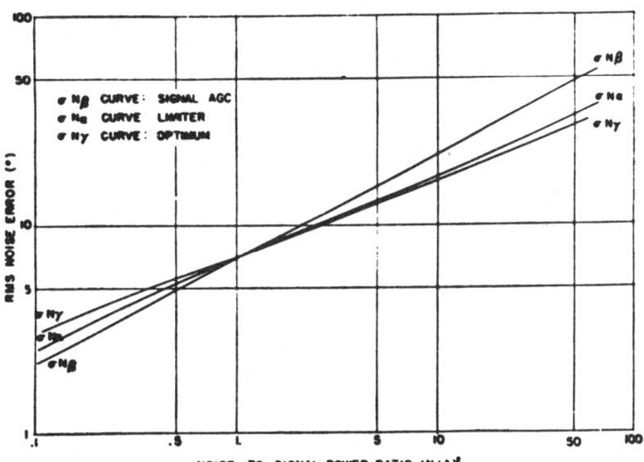

Fig. 7—Comparison of theoretical noise error in various loops.

Fig. 7 shows the rms phase-noise jitter to be expected in the different loops. All the curves coincide at the design point of unity input noise-to-signal ratio but diverge at other input ratios. The phase-noise jitter in a loop employing an optimum variable filter is not necessarily less

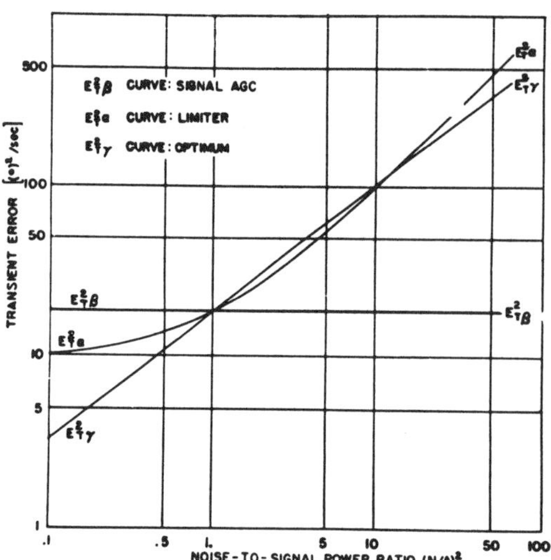

Fig. 8—Comparison of theoretical transient error in various loops.

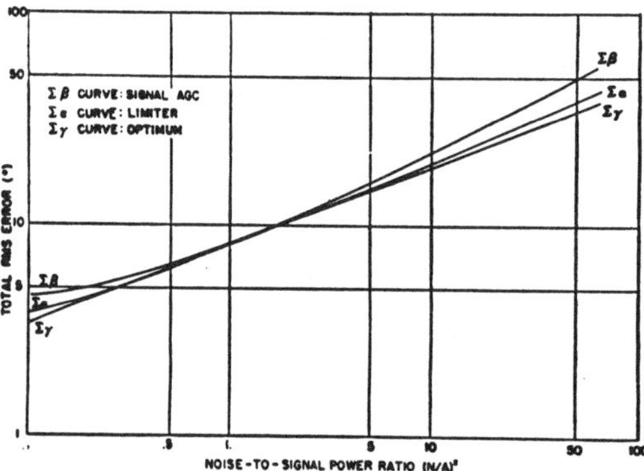

Fig. 9—Comparison of theoretical total error in various loops.

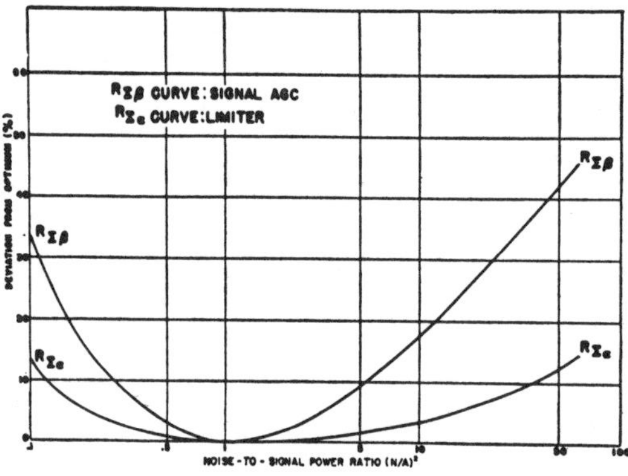

Fig. 10—Total error in signal-AGC and limiter-loop percentage deviation from optimum.

than the jitter in the other loops because the optimum loop is continually minimizing the total error rather than only the phase jitter or the transient error. Optimum performance may require that a slight increase in phase-noise jitter be allowed in order to obtain a considerable decrease in transient error.

Fig. 8 shows the transient error to be expected in the different loops. Since the transient error depends only on the loop gain in a fixed-filter loop tracking a given input frequency step, the AGC loop, which has constant input-signal amplitude, has a fixed loop gain and therefore a fixed amount of transient error.

Fig. 9 shows the total theoretically predicted rms error. However, it is easier to see the significant features of the loops if a plot is made showing the percentage by which the total error in the AGC and limiter loops deviates from the total error in the optimum loop (Fig. 10.) It is seen that the loop preceded by a limiter has a total error not more than 15 per cent greater than optimum, whereas the AGC loop has a total error exceeding that of an optimum loop by 45 per cent.

Experimental Configuration

It has been shown that a limiter loop is theoretically superior to an AGC loop and is nearly as good as a loop continually adjusted to be optimum by auxilliary servos. It remains to be shown that the theoretically derived results for such a limiter loop are experimentally verifiable.

Experimental results were obtained using equipment previously designed.[3] The experimental configuration is shown in Fig. 11. The noise was bandpassed to a 900-cycle

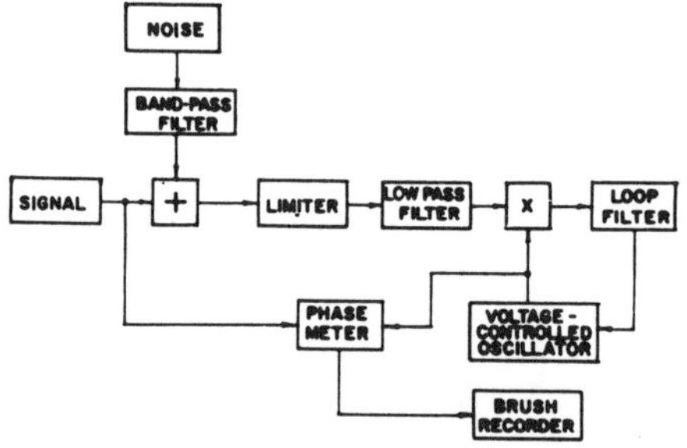

Fig. 11—Experimental configuration for limiter loop.

bandwidth centered at 5 kilocycles by a steep-sided filter and was then added to the 5-kilocycle signal. The sum was fed to a limiter consisting of a four-stage pentode amplifier having back-to-back silicon diodes across the input of each stage. The limiter output was passed through a steep-sided, 7-kilocycle, low-pass filter which removed all but the first zone of the limiter output. The output of this filter then served as the input to the phase-locked loop.

The loop itself consisted of (1) a multiplier using silicon diodes in a bridge circuit, (2) a multivibrator-type VCO

[3] Equipment used was designed by MacMillan of this laboratory.

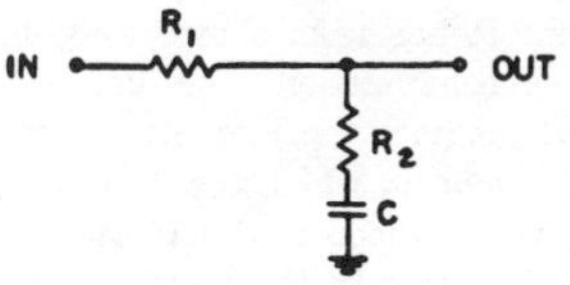

IN — R_1 — OUT; R_2, C to ground

$$\tau_1 = R_1 C \qquad \tau_2 = R_2 C$$

$$\text{DESIRED } F(s) = \frac{B_0^2 + \sqrt{2}\, B_0 s}{K A_0} \qquad \text{ACTUAL } F(s) = \frac{\tau_2 s + 1}{(\tau_1 + \tau_2) s + 1}$$

$$\therefore \tau_1 = \frac{K A_0}{B_0^2} \qquad \cong \frac{\tau_2 s + 1}{\tau_1 s}$$

$$\tau_2 = \frac{\sqrt{2}}{B_0}$$

Fig. 12—Loop filter form.

having a low-pass filter on its output, which passed only the fundamental component of the multivibrator square-wave output, and (3) a loop filter as shown in Fig. 12.

The clean signal was compared with the VCO output in a phase meter, and the recorder output of the phase meter was connected to an oscillograph which was zeroed at the steady no-noise phase difference between the signal and the VCO. Any error in VCO tracking was then plotted directly on the oscillograph.

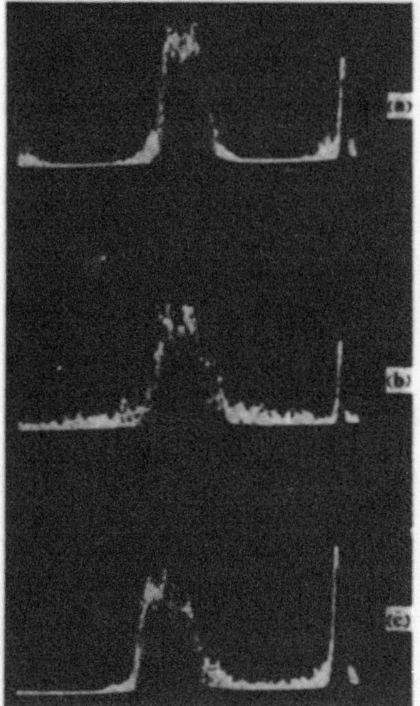

Fig. 13—Noise spectra at various points in limiter loop.

Experimental Results

Before discussing the data obtained from this experiment, it is interesting to look at the spectrum of the noise present in various parts of the loop. Figs. 13(a), 13(b), and 13(c), are, respectively, photographs of the noise spectrum at input to the limiter, at output of the limiter, and at output of the low-pass filter following the limiter.

The spectra are displayed on a logarithmic ordinate, with zero frequency at the right of the photograph. The input noise is centered at 5 kilocycles, with 900 cycles per second bandwidth. The first zone of the limiter output looks quite similar to the input noise except for the presence of some very low-level (considering the logarithmic ordinate) background noise generated in the limiter; also, the limiter input noise has been spread out in a manner similar to that predicted theoretically.[4] The last photograph shows the effect of the 7-kilocycle, low-pass filter in attenuating the limiter background noise past 7 kilocycles, implying that the higher-order noise zones generated in the limiter are cut off.

Experimental data on the noise error were obtained simply by adding a known amount of noise to the signal, and recording the VCO phase jitter on the recorder as shown in Figs. 14(a), 14(b), 14(c), for input noise-to-signal amplitude ratios of 1, 2, and 4, respectively. It is interesting to note how the frequency of the output noise jitter decreases as the input noise-to-signal amplitude ratio increases. This phenomenon implies that the loop bandwidth is decreasing with increasing input noise-to-signal ratio, a result theoretically predicted above.

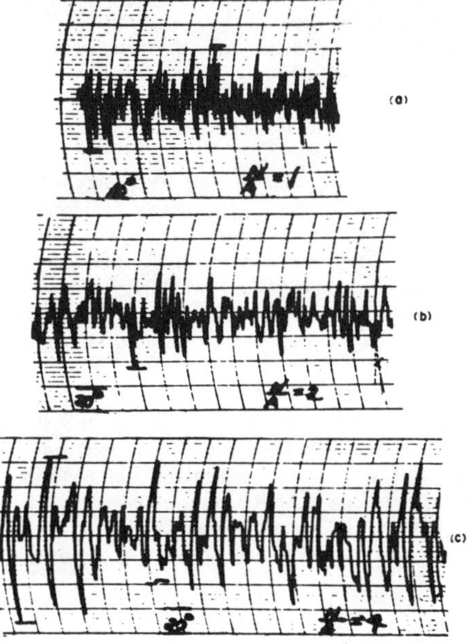

Fig. 14—Experimental phase jitter in limiter loop.

[4] J. L. Lawson and G. E. Uhlenbeck, "Threshold Signals," Radiation Laboratory Series, vol. 24:59, Mass. Inst. Tech., 1950.

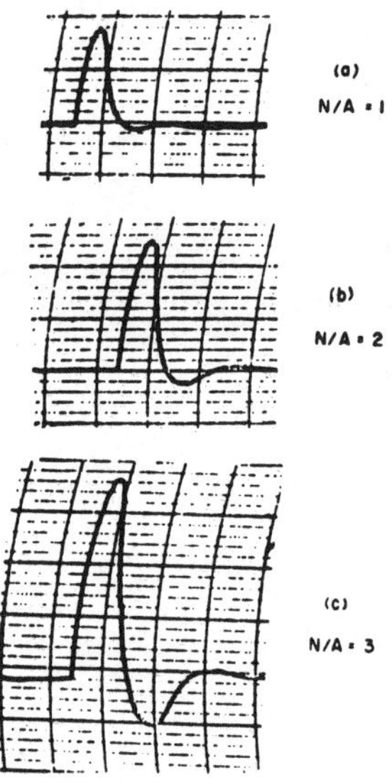

Fig. 15—Experimental transient error in limiter loop.

Experimental data on transient error were more difficult to obtain. In order to determine only the transient performance of the system, it was necessary to disconnect the noise source and reduce the signal output of the limiter in accordance with the limiter relationship already discussed (4). The transient error of the loop was then obtained by introducing a 2.5 cycle per second frequency step into the VCO and recording the loop phase error on the oscillograph. The step could have been introduced into either the signal source or the VCO; the VCO was chosen for convenience.

Figs. 15(a), 15(b), and 15(c) are recordings of the experimental transient error corresponding to input noise-to-signal ratios of 1, 2, and 3, respectively. The ordinate scale is 10 degrees for every heavy line. At the design level of unity signal-to-noise ratio, the Wiener theory requires about 5 per cent overshoot; as was expected, both the initial pip and the overshoot increase as the loop gain and bandwidth decrease.

Finally, the experimental results were combined to plot in Fig. 16, and in Figs. 17 and 18 (opposite) the experimental vs the theoretical values of noise error, transient error, and total error, respectively, in a phase-locked loop preceded by a limiter and employing a fixed filter that satisfied the Wiener optimum criterion for a design level of unity signal-to-noise ratio. The results are commensurate.

Conclusion

It has therefore been theoretically proved that a phase-lock loop preceded by a bandpass limiter approximates, over a wide range of input signal and noise levels, the optimum performance obtainable only with a variable filter that is continually readjusted to be optimum by an auxilliary servo system. It has been shown that, in addition to facility of mechanization, the limiter loop approximates optimum behavior considerably better than does an AGC-controlled loop in which the AGC is based upon the signal. Finally, the theoretical derivations leading to the expressions for error in the limiter loop have been experimentally verified and the effectiveness of a limiter phase-lock loop has been confirmed.

Appendix I

Derivation of Optimum Filter[5]

Let $Y(s)$ be the loop transfer function. By definition

$$\theta_2(s) = Y(s)\theta_1(s) \tag{5}$$

and

$$Y(s) = \frac{AKF(s)}{s + AKF(s)}. \tag{6}$$

The total noise power at the VCO output is

$$\sigma_N^2 = \frac{1}{2\pi j}\int_{-j\infty}^{+j\infty} |Y(s)|^2 \Phi_N(s)\, ds,$$

where $\Phi_N(s)$ is the spectral density of the input phase noise. Assuming this spectral density to be flat, Figs. 2 and 3 show that

$$\Phi_N(s) = \Phi_N(0) = \frac{N^2}{A^2 2\Delta f}$$

where Δf is the input-noise bandwidth.

The transient error, previously defined as

$$E_T^2 = \int_0^\infty [\theta_2(t) - \theta_1(t)]^2\, dt$$

upon application of (5), may be written

$$E_T^2 = \frac{1}{2\pi j}\int_{-j\infty}^{+j\infty} |Y(s) - 1|^2\, |\theta_1(s)|^2\, ds.$$

Since the optimum filter is to be designed so that σ_N^2 is minimized under the constraint that E_T^2 have a specified value, it is desired that

$$\Sigma^2 = \sigma_N^2 + \lambda^2 E_T^2 = \text{minimum}. \tag{7}$$

The optimum filter is to be physically realizable, which for the purpose of this paper means that the loop transfer function has no poles in the right half of the s plane. Therefore, although the standard variational techniques are to be applied to minimize Σ^2, the integral expression for Σ^2 must be set up in such a manner that it yields a realizable filter.

Expansion of (7) gives

$$\Sigma^2 = \frac{1}{2\pi j}\int_{-j\infty}^{+j\infty} ds\, Y(s)Y(-s)\Phi_N(0) + \lambda^2[Y(s) - 1][Y(-s) - 1]\, |\theta_1(s)|^2. \tag{8}$$

[5] This derivation follows the approach to the Wiener filter theory given in unpublished notes by Rechtin of this laboratory.

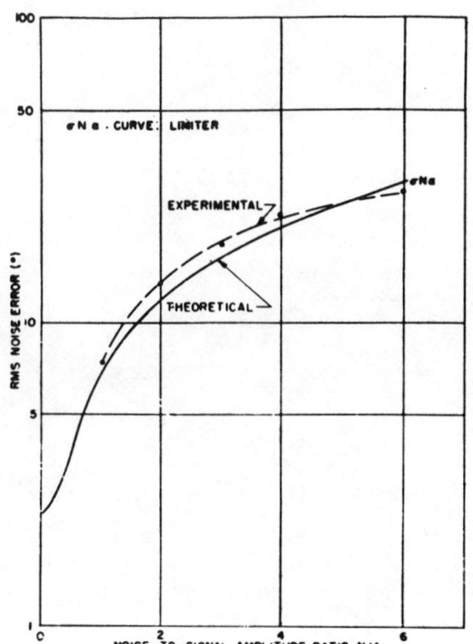

Fig. 16—Experimental vs theoretical noise error in limiter loop.

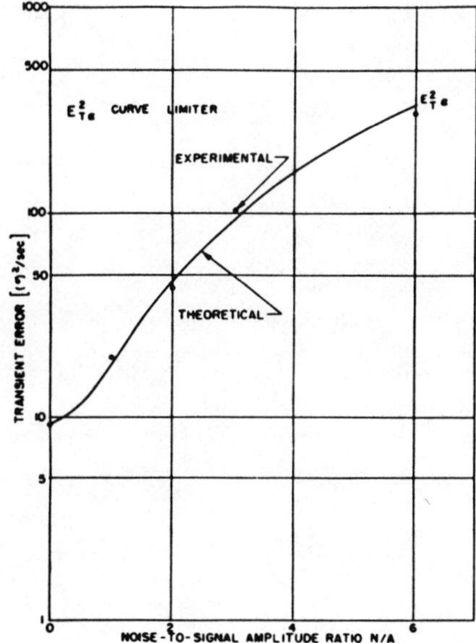

Fig. 17—Experimental vs theoretical transient error for limiter loop.

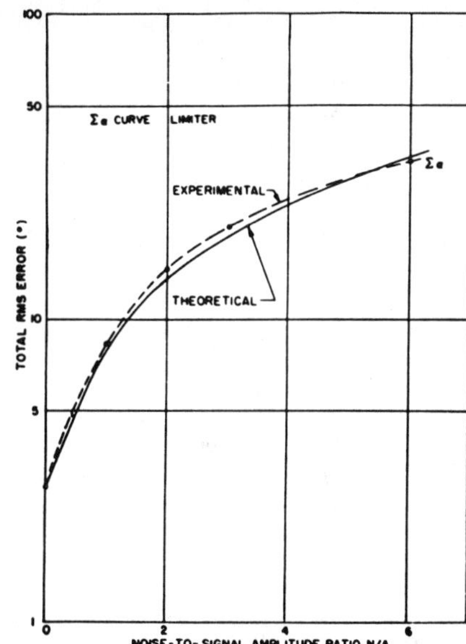

Fig. 18—Experimental vs theoretical total error in limiter loop.

Applying the standard variational procedure to $Y(s)$, let

$$Y^*(s) = Y(s) + \epsilon\eta(s),$$

where $\epsilon\eta(s)$ is the variation to be minimized on the true optimum $Y(s)$, and $Y^*(s)$ is the estimate of $Y(s)$.

When $Y^*(s)$ is split into factors having positive and negative poles and substituted into (8),

$$\Sigma^2[Y(s) + \epsilon\eta(s)]$$
$$= \frac{1}{2\pi j}\int_{-j\infty}^{+j\infty} ds\,[Y(s) + \epsilon\eta(s)][Y(-s) + \epsilon\eta(-s)]\Phi_N(0)$$
$$+ \frac{1}{2\pi j}\int_{-j\infty}^{+j\infty} ds\,\lambda^2[1 - Y(s) - \epsilon\eta(s)]$$
$$\cdot [1 - Y(-s) - \epsilon\eta(-s)]\,|\theta_1(s)|^2.$$

Setting the variation of Σ^2 to zero at ϵ equals zero completes the standard variational procedure.

$$\frac{\partial\Sigma^2}{\partial\epsilon}[Y(s) + \epsilon\eta(s)]\Big|_{\epsilon=0} = \frac{1}{2\pi j}\int_{-j\infty}^{+j\infty} ds\,\eta(s)$$
$$\cdot \{(Y(-s))(\lambda^2|\theta_1(s)|^2 + \Phi_N(0)) - \lambda^2|\theta_1(s)|^2\}$$
$$+ \frac{1}{2\pi j}\int_{-j\infty}^{+j\infty} ds\,\eta(-s)$$
$$\cdot \{(Y(s))(\lambda^2|\theta_1(s)^2| + \Phi_N(0)) - \lambda^2|\theta_1(s)|^2\} \quad (9)$$

In order to keep all terms in (9) split into factors having poles in either, but not both, the right or left half-plane, it is convenient to define

$$\lambda^2|\theta_1(s)|^2 + \Phi_N(0) \equiv |\psi(s)|^2 = \psi(s)\psi(-s). \quad (10)$$

It is now necessary to substitute (10) into (9) in such a way that the latter will be satisfied for realizable $Y(s)$, whereas the remainder of the integral will reduce to zero.

This can be effected by noting that

$$\int_{-j\infty}^{+j\infty} Z(s)W(s)\,ds = 0 \quad (11)$$

if $Z(s)$ and $W(s)$ are algebraic polynomials having poles only in the same half of the s plane.

Substitution of (10) into (9) and keeping together terms having similar poles

$$\int_{-j\infty}^{+j\infty} ds\ \eta(s)\psi(s)[Y(-s)\psi(-s) - \lambda^2 |\theta_1(s)|^2/\psi(s)]$$

$$+ \eta(-s)\psi(-s)[Y(s)\psi(s) - \lambda^2 |\theta_1(s)|^2/\psi(-s)] = 0. \quad (12)$$

It is now advisable to split

$$\frac{|\theta_1(s)|^2}{\psi(s)} = \left[\frac{|\theta_1(s)|^2}{\psi(s)}\right]_+ + \left[\frac{|\theta_1(s)|^2}{\psi(s)}\right]_-, \quad (13)$$

where the right-hand side of the equation is the partial fraction expansion of the left-hand side, having denominator terms in $+s$ and $-s$, i.e., having poles in the left and right halves of the s plane, respectively.

Substitution of (13) into (12) gives

$$\int_{-j\infty}^{+j\infty} ds\ \eta(s)\psi(s)\{Y(-s)\psi(-s) - \lambda^2[(|\theta_1(s)|^2/\psi(s))_+$$

$$+ (|\theta_1(s)|^2/\psi(s))_-]\}$$

$$+ \int_{-j\infty}^{+j\infty} ds\ \eta(-s)\psi(-s)\{Y(s)\psi(s) - \lambda^2[(|\theta_1(s)|^2/\psi(-s))_+$$

$$- (|\theta_1(s)|^2/\psi(-s))_-]\} = 0.$$

Because of (11), the terms having similar poles drop out, leaving

$$\int_{-j\infty}^{+j\infty} ds\ \eta(s)\psi(s)\{Y(-s)\psi(-s) + \lambda^2(|\theta_1(s)|^2/\psi(s))_-\}$$

$$+ \int_{-j\infty}^{+j\infty} ds\ \eta(-s)\psi(-s)$$

$$\cdot\{Y(s)\psi(s) - \lambda^2(|\theta_1(s)|^2/\psi(s))_+\} = 0. \quad (14)$$

These integrals are identical except that one has an integrand in $+s$ and one, in $-s$. Therefore, the integrals are merely the negatives of each other, so that equating one of them to zero satisfies (14).

Setting the second integral to zero yields

$$Y(s) = \frac{\lambda^2}{\psi(s)}\left[\frac{|\theta_1(s)|^2}{\psi(s)}\right]_+, \quad (15)$$

which is the optimum loop transfer function composed of realizable terms. Note that all terms have poles in the left half-plane, thereby satisfying the requirement of physical realizability.

In this paper, (15) will be solved only for input frequency steps, although the results for other type inputs will be given.

Assume the input to be a frequency step

$$\theta_1(s) = \frac{\Delta\omega}{s^2}$$

where $\Delta\omega$ is the magnitude of the step in radians per second.

Therefore

$$|\theta_1(s)|^2 = \frac{(\Delta\omega)^2}{s^4}$$

and

$$|\psi(s)|^2 = \lambda^2 |\theta_1(s)|^2 + \Phi_N(0)$$

$$= \frac{\lambda^2(\Delta\omega)^2}{s^4} + \Phi_N(0) = \frac{B_0^4 + s^4}{s^4}\Phi_N(0),$$

where

$$B_0^4 \equiv \frac{\lambda^2\omega^2}{\Phi_N(0)}. \quad (16)$$

Factoring

$$|\psi(s)|^2 = \psi(s)\psi(-s)$$

$$= \Phi_N(0)\left[\frac{B_0^2 + \sqrt{2}B_0s + s^2}{s^2}\right]\left[\frac{B_0^2 - \sqrt{2}B_0s + s^2}{s^2}\right]$$

gives

$$\psi(s) = \sqrt{\Phi_N(0)}\left[\frac{B_0^2 + \sqrt{2}B_0s + s^2}{s^2}\right]. \quad (17)$$

Substitution of (17) into the optimum-filter equation gives

$$Y(s) = \frac{\lambda^2 s^2(\Delta\omega)^2}{\Phi_N(0)[B_0^2 + \sqrt{2}B_0s + s^2]}\left[\frac{1}{s^2[B_0^2 - 2B_0s - s^2]}\right]_+. \quad (18)$$

Separating the last term of (18) into partial fractions,

$$\frac{1}{s^2[B_0^2 + \sqrt{2}B_0s + s^2]} = \frac{1/B_0^2}{s^2} + \frac{\sqrt{2}/B_0^3}{s}$$

$$+ \frac{c}{s - B_0/\sqrt{2}(-1+j)} + \frac{d}{s - B_0/\sqrt{2}(-1-j)}.$$

The coefficients c and d are of no interest since the terms containing them have poles in the right half-plane and are not to be considered because of the definition of $[\ \ ]_+$ in (13).

The first two terms may be considered as having poles in the left half-plane, since they represent the limiting case of $1/(s - \epsilon)$ where ϵ is a small number representing the reciprocal of the dc gain. Since the pole approaches the origin from the left, in such a limiting process, the first two terms are included in the $[\ \ ]_+$ terms.

Therefore

$$Y(s) = \frac{\lambda^2(\Delta\omega)^2 s^2}{\Phi_N(0)[B_0^2 + \sqrt{2}B_0s + s^2]}\left[\frac{1}{B_0^2s^2} + \frac{\sqrt{2}}{B_0^3s}\right],$$

which, with the aid of (16), reduces to

$$Y(s) = \frac{B_0^2 + \sqrt{2}B_0s}{B_0^2 + \sqrt{2}B_0s + s^2}.$$

As a check, it is interesting to note that $Y(s)$ is quasi-distortionless to a step-frequency change, i.e.,

$$\text{error}(s) \equiv \theta_1(s) - \theta_2(s) = \theta_1(s)[1 - Y(s)]$$

$$= \theta_1(s)\frac{s^2}{B_0^2 + \sqrt{2}B_0s + s^2},$$

so that, if $\theta_1(s) = \Delta\omega/(s^2)$, representing a step-frequency input, the infinite-time error is zero; i.e., the loop is quasi-distortionless to a step-frequency change:

$$\text{error}(s) = \frac{\Delta\omega}{s^2}\,\frac{s^2}{B_0^2 + \sqrt{2}\,B_0 s + s^2} = \frac{\Delta\omega}{B_0^2 + \sqrt{2}\,B_0 s + s^2}$$

$$\lim_{t\to\infty} \text{error}(t) = \lim_{s\to\infty} s \times \text{error}(s) = 0$$

Since the optimum loop transfer function is

$$Y(s) = \frac{B_0^2 + \sqrt{2}\,B_0 s}{B_0^2 + \sqrt{2}\,B_0 s + s^2}$$

and

$$Y(s) = \frac{AKF(s)}{s + AKF(s)}.$$

It follows that

$$F(s) = \frac{B_0^2 + \sqrt{2}\,B_0 s}{A_0 K s},$$

which was to be proved.

The noise error and transient error in such an optimum loop may be calculated as follows:

$$\sigma_N^2 = \frac{1}{2\pi j}\int_{-j\infty}^{+j\infty} |Y(s)|^2\, \Phi_N(0)\, ds$$

$$E_T^2 = \frac{1}{2\pi j}\int_{-j\infty}^{+j\infty} |\theta_1(s)|^2\, [1 - Y(s)]^2\, ds.$$

These integrals may be evaluated by involved contour integration and have been tabulated conveniently;[6] the evaluation of these integrals led to all explicit expressions for σ_N^2 and E_T^2 given in this paper.

Appendix II

Filters for Different Forms of Signal Inputs

Filter forms for fixed-component loops, fixed-component loops preceded by bandpass limiters, and variable-component loops will be given in this appendix, together with their associated transient responses and phase-noise jitters, for two other forms of signal inputs besides those discussed in the text. Derivation of these filters are similar to the derivation in Appendix I and are not repeated. Loops designed to track phase steps will be called first-order loops; loops designed to track frequency ramps will be called third-order loops. Second-order loops designed to track frequency steps are discussed in the text and will not be discussed in this Appendix.

(1) *First-order-loop*—If the signal input to the loop is assumed to be a phase step, i.e.,

$$\theta_1(t) = \Delta\theta \qquad \theta_1(s) = \frac{\Delta\theta}{s},$$

then the filter for the first-order, fixed-component loop has the transform

$$F(s) = \frac{B_1}{KA_0}$$

[6] "Servomechanisms," Radiation Laboratory Series, No. 25, pp. 369–370, Mass. Inst. Tech.; 1947.

and has phase jitter and transient error, respectively, of

$$\sigma_N^2 = B_1\left(\frac{N^2}{2\Delta f A^2}\right)\left(\frac{A}{A_0}\right)$$

and (19)

$$E_T^2 = \frac{(\Delta\theta)^2}{2B_1(A/A_0)},$$

where B_1 is a quantity similar to B_0 in the previous discussion of the loop for tracking frequency steps, i.e.,

$$B_1 \equiv \frac{\lambda(\Delta\theta)}{N_0\ A_0}\sqrt{2\Delta f}.$$

Substitution of (4) into (19) gives a total error in a first-order loop preceded by a bandpass limiter of

$$\Sigma^2 \equiv \sigma_N^2 + \lambda^2 E_T^2 = \left(\frac{B_1}{2}\right)\left(\frac{N^2}{A^2 2\Delta f}\right)\left(\frac{A}{A_0}\right) + \left(\frac{B_1}{2}\right)\left(\frac{N_0^2}{A_0^2 2\Delta f}\right)\left(\frac{A_0}{A}\right)$$

$$= \frac{B_1}{2}\left(\frac{N^2}{A^2 2\Delta f}\right)\left[\sqrt{\frac{1 + (N_0/A_0)^2}{1 + (N/A)^2}} + \frac{N_0^2}{A_0^2}\sqrt{\frac{1 + (N/A)^2}{1 + (N_0/A_0)^2}}\right].$$

It may be shown, using the same techniques employed in Appendix III for second-order loops, that the optimum variable-component filter for the first-order loop has the transform

$$F(s) = \frac{B_1}{KA_0}\,\frac{N_0}{N},$$

and that the total error in such a loop is

$$\Sigma^2 \equiv \sigma_N^2 + \lambda^2 E_T^2 = \left(\frac{B_1}{2}\right)\left(\frac{A\ \ A_0}{N\ \ N_0}\right)\left(\frac{N^2\ \ A^2}{2\Delta f}\right) + \left[\frac{\lambda^2(\Delta\theta)^2}{2B_1}\right]\frac{N\ \ N_0}{A\ \ A_0}$$

$$= B_1\left(\frac{N_0}{A_0}\right)\left(\frac{N\ \ A}{2\Delta f}\right).$$

Plots for the total error in first-order loops with (1) fixed components preceded by a signal-based AGC, (2) fixed components preceded by a bandpass limiter and (3) variable components are given in Fig. 4. Design parameters are

$$\frac{N_0}{A_0} = 1$$

$$B_0 = 10$$

$$\Delta f = 1 \text{ megacycle per second.}$$

The input-signal change was assumed to be a 1-radian phase step.

For comparison purposes, the total error in each of the three types of second-order loops is given in Fig. 5. The input-signal change was assumed to be a 6 radians per second (∼ 1 cycle per second) frequency step.

(2) *Third-order-loop*—If the signal input to the loop is assumed to be a frequency ramp, ie.,

$$\theta_1(t) = \frac{(\Delta\alpha)t^2}{2} \qquad \theta_1(s) = \frac{\Delta\alpha}{s^3},$$

then the third-order, fixed-component loop filter has the transform

$$F(s) = \frac{B_2^3 + 2B_2^2 s + 2B_2 s^2}{KA_0 s^2}$$

and has phase jitter and transient error, respectively, of

$$\sigma_N^2 = \left(\frac{N^2}{A^2}\right)\left(\frac{B_2}{2\Delta f}\right)\left[\frac{4(A/A_0)+1}{4(A/A_0)-1}\right]$$

and

$$E_T^2 = \frac{(\Delta\alpha)^2}{(A/A_0)[4(A/A_0)-1]B_2^3}, \tag{20}$$

where

$$B_2^3 \equiv \left[\frac{\lambda(\Delta\alpha)}{N_0/A_0}\right]\sqrt{2\Delta f}.$$

Substitution of (4) into (20) gives a total error in a third-order loop preceded by a band-pass limiter of

$$\Sigma^2 \equiv \sigma_N^2 + \lambda^2 E_T^2 = \left(\frac{B_2}{2}\right)\left[\frac{N^2}{A^2 2\Delta f}\right]\left[\frac{4(A/A_0)+1}{4(A/A_0)-1}\right] + \left(\frac{B_2}{2}\right)\left[\frac{N_0^2}{A_0^2 2\Delta f}\right]\frac{(A/A_0)}{[4(A/A_0)-1]}$$

$$= B_2\left[\frac{N^2}{A^2 2\Delta f}\right]\left\{\left[\frac{4\sqrt{\frac{1+(N_0/A_0)^2}{1+(N/A)^2}}+1}{4\sqrt{\frac{1+(N_0/A_0)^2}{1+(N/A)^2}}-1}\right]\left(\frac{A}{A_0}\right) + \frac{(A/A_0)^2}{(N/N_0)^2}\left[\frac{1}{\sqrt{\frac{1+(N_0/A_0)^2}{1+(N/A)^2}}\left[4\sqrt{\frac{1+(N_0/A_0)^2}{1+(N/A)^2}}-1\right]}\right]\right\}.$$

It should be noted that the total error becomes infinite when the noise has reduced the signal level to one-fourth of the design level.

It may be shown that the optimum third-order, variable-component loop has the transform

$$F(s) = \frac{B_2^3\left[\frac{(A/A_0)}{(N/N_0)}\right] + 2B_2^2\left[\frac{(A/A_0)}{(N/N_0)}\right]^{2/3} s + 2B_2\left[\frac{(A/A_0)}{(N/N_0)}\right]^{1/3} s^2}{As^2},$$

and that the total error in such a loop is

$$\Sigma^2 = \sigma_N^2 + \lambda^2 E_T^2 = \left(\frac{5}{3}\right)\left(\frac{N}{A}\right)^{5/3}\left(\frac{N_0}{A_0}\right)^{1/3}\left(\frac{B_2}{2\Delta f}\right) + \frac{(\Delta\alpha)^2}{3B_2^5\left[\frac{A/A_0}{N/N_0}\right]^{5/3}}$$

$$= 2B_0\left(\frac{N^2}{A^2 2\Delta f}\right)\left[\frac{(N_0/N)}{(A_0/A)}\right]^{1/3}.$$

Plots for the total error in the three types of third-order loops are given in Fig. 6. The input-signal change was assumed to be a 1 radian per second^2 frequency ramp.

Appendix III

Optimum-Filter Design

The optimum filter is specified by

$$F(s) = \frac{B^2 + \sqrt{2}\,Bs}{KAs},$$

where

$$B^2 \equiv \frac{\lambda(\Delta\omega)}{(N/A)}\sqrt{2\Delta f}.$$

B varies in accordance with the signal- and noise-input levels and may be expressed in terms of B_0, the fixed-filter bandwidth parameter, as follows:

$$B^2 = B_0^2\left[\frac{(A/A_0)}{(N/N_0)}\right].$$

Therefore

$$F(s) = \frac{B_0^2\left[\frac{(A/A_0)}{(N/N_0)}\right] + \sqrt{2}\,B_0\left[\frac{(A/A_0)}{(N/N_0)}\right]s}{KAs}, \tag{21}$$

giving transfer function for optimum variable filter (3).

All results described in this paper as pertaining to second-order, variable-filter loops were obtained by substituting (21) into the expression of $Y(s)$ in (6) and evaluating that which was required.

Part IV

STATE-SPACE MODELS AND RECURSIVE FILTERS

Editor's Comments on Papers 14 Through 20

The papers presented up to this point have essentially dealt only with stationary processes. However, Paper 10 by Zadeh and Ragazzini and Paper 11 by Yaglom treated the problem of a finite data-interval, which almost always introduces transient nonstationarities. Therefore, special attention has to be given to the "initial conditions" $\{y(0), \dot{y}(0), \ldots, y^{(n-1)}(0)\}$. There is no difficulty in principle with handling these initial conditions, but as Papers 10 and 11 showed, there are considerable algebraic and notational complexities. It is well known in linear system theory that state-space methods, with their combination of first-order equations and matrix methods, are useful in organizing the algebra of initial conditions. And it is interesting to see how, for example, Laning and Battin [1] were driven almost inexorably to use such descriptions. As they explain it, "In effect, the essential difficulties associated with the point $t = 0$ are absorbed in the calculations needed to reduce the N-th order equation to N first-order equations" (page 340). However, the state-space models of Laning and Battin did not attract much attention until a few years later, when they were found to be well suited to the introduction of recursive estimation techniques. Such methods had first been devised by Gauss, around 1800, for calculating the trajectories of heavenly objects. One hundred and fifty years later, the advent of Sputnik again brought the need to revive such methods. Paper 14 by Swerling is a detailed presentation of the basic formulas needed for several important calculations in satellite tracking and, since its appearance in 1958, these results have been widely used by groups like those at the Bell Laboratories [2].

For different reasons, Stratonovich in the USSR was also led to recursive methods. Stratonovich was interested in handling non-Gaussian processes, and the only tractable general class of such processes were the Markov processes. In this case, the Fokker-Planck equation for the transition density has a natural recursive structure, which is exploited in Paper 15 in several different problems. However, in retrospect, perhaps the most important of Stratonovich's results are those in Section 5, where he treats the "Gaussian approximation" and obtains (see Eqs. (56) and (59)) the differential equations of the now celebrated Kalman-Bucy filter.

Kalman was independently led to recursive algorithms through his interest in state-space descriptions of linear systems. The continuous-time version of these results was published in a joint paper with Bucy [3], which appears in a companion Benchmark volume [4] and is therefore omitted here.

On the other hand, we present here the work of some of Bucy's colleagues at Johns Hopkins, who, while studying adaptive filters, also set up (in a special case) state-space models and a recursive solution of the Kalman-Bucy type (see Section 5 of Paper 17). According to Bucy [5], these results were a stimulus for his own work on such solutions.

For several reasons, as noted in Section IV of survey Paper 1, the Kalman-Bucy filter has been the center of attraction in the estimation literature in the ten to fifteen years since its appearance. There has been a host of papers providing alternative derivations, pointing out new connections, exploring variations and extensions, and so on. Thus Genin in Paper 18 exhibits how the Kalman-Bucy results could be derived by a proper reinterpretation of classical least-squares theory (see Rao [6] and Rosenbrock [7]) and especially the results of Gauss [8]. Of course, Swerling's Paper 14 is also closely related to such things and in his expository paper [9] he has elaborated on the connections to classical least-squares and to the Kalman-Bucy filter.

In Paper 16, Kalman notes a duality between the least-squares estimation problem and a certain quadratic regulator problem. This observation has led to a very useful interplay with control-theoretical concepts and techniques. Bryson and Frazier in Paper 19 used this link to obtain the first recursive solution for smoothed estimates, a topic that has grown extensively since then [10, 11]. This approach also provides a new derivation of the Kalman filter by classical calculus of variations techniques.

This brings us to the last paper in this volume. In a sense, this paper closes the circle of ideas described above by rederiving the results on filtering and smoothing by using an extension to continuous-time and nonstationary processes of the original innovations approach of Kolmogorov (Paper 2). In this connection, it may be noted that the open problem on nonlinear innovations mentioned on page 306 has been substantially resolved by Benes [12]; it turns out that the conjecture holds for the physically important case of signals generated by a linear or nonlinear dynamical system.

REFERENCES

1. H. Laning and R. Battin, *Random Processes in Automatic Control*, New York: McGraw-Hill, 1958.
2. R. B. Blackman, "Methods of Orbit Refinement," *Bell Syst. Tech. J.* **43**:885–909, May 1964.
3. R. E. Kalman and R. S. Bucy, "New Results in Linear Filtering and Prediction Theory," *Trans. ASME*, Ser. D., *J. Basic Eng.* **83**:95–107, Dec. 1961.
4. A. Ephremides and J. B. Thomas, eds., *Random Processes—Multiplicity Theory*, vol. 1, Benchmark Papers in Electrical Engineering and Computer Science, Stroudsburg, Pa.: Dowden, Hutchinson & Ross, 1973.
5. R. S. Bucy and P. D. Joseph, *Filtering for Stochastic Processes with Applications to Guidance,* New York: J. Wiley, 1968.
6. C. R. Rao, *Linear Statistical Inference*, New York: J. Wiley, 1974.
7. H. H. Rosenbrock, "On the Connection Between Discrete Linear Filters and Some Formulae of Gauss," in *Identification, Optimisation et Stabilité des Systèmes Automatiques*, Paris: Dunod, 1967, pp. 239–249.
8. C. F. Gauss, *Theoria Motus Corporum Coelestium in Sectionibus Conicis Solem Ambientum*, Hamburg, 1809. (Translation, New York: Dover, 1963.)

9. P. Swerling, "Modern State Estimation Methods from the Viewpoint of the Method of Least Squares," *IEEE Trans. Automat. Control* **AC–16** 707–720, Dec. 1971.
10. J. Meditch, "A Survey of Data Smoothing for Linear and Nonlinear Dynamic Systems," *Automatica* **9**:151–162, 1973. See also *ibid*, **11**:109–111, 1975.
11. L. Ljung and T. Kailath, "A Unified Approach to Smoothing Formulas," *Automatica* **12**:147–157, 1976.
12. V. E. Benes, "On Kailath's Innovations Conjecture," *Bell Syst. Tech. J.* **55**:981–1001, Nov. 1976.

14

Reprinted from *J. Astronaut. Sci.* **6**:46–52 (1959)

First Order Error Propagation in a Stagewise Smoothing Procedure for Satellite Observations

Peter Swerling*

Abstract

A practical method of smoothing satellite data by evaluating a finite number of parameters, or elements, is presented.

Introduction

The subject discussed is that of smoothing observational data in cases where the observations, in the absence of observational error, would all be determined by the time of observation plus a finite number of parameters, called elements. The objective is to estimate the elements.

The immediate motivation for this arises in studies of estimation of earth-satellite orbits from observational data. In this case, if, for example, the force field were known exactly, one could regard the elements as the position and velocity components at a particular instant t_0; if the field were a central inverse square field, the elements could alternatively be the six elements of a Keplerian elliptic orbit. (It should be mentioned, however, that satellite orbit prediction is only one of the possible applications of the results.)

In satellite tracking and prediction, it is desired to produce ephemerides—i.e., predictions of the future position as a function of time—as well as to make various other types of decisions and predictions. As new observations become available, one can improve the accuracy of these predictions. Ideally, one would like to store all previous observations of the object, and combine these in some optimum way to yield the desired predictions or decisions. The optimum method for processing the available data would be based on analysis of the error statistics for the individual observations, and on the functional dependence of future predicted quantities on the previous observations.

In satellite tracking one is dealing with a situation in which there may be a large number of observations of varying degrees of accuracy, as well as large numbers of tracked objects. Also, methods of orbit prediction (even in the absence of observation errors) are subject to various sources of error, such as

(a) uncertainty in the forces acting on the body (earth's gravitational and magnetic fields, atmospheric resistance);

(b) cumulative errors in solving the equations of motion.

* The RAND Corporation, Santa Monica, California.

Two features would be desirable in a tracking and prediction method:

(1) The data processing load per tracked object should not exceed a certain maximum, regardless of how many observations are available to be processed. On the other hand, the prediction method should not throw available observations away.

(2) The method should be adaptable to situations in which the underlying prediction functions are subject to the above-mentioned uncertainties.

The stagewise procedures described below are motivated by these considerations.

The particular methods of data smoothing to be discussed are variations of the classical method of minimizing a quadratic form in the residuals (in practise this is usually a weighted sum of squares of residuals). After defining this method, the first order dependence of the errors in the resulting element estimates on the observation errors is established.

We then go on to describe a stagewise procedure for processing the observational data, in which the element estimates at each stage are smoothed in a particular way with some additional observations. This is, in essence, a type of differential correction procedure. It is shown that the errors in the resulting element estimates, after stagewise smoothing of a given set of observations, have the same first order dependence on the observation errors as would the errors in the estimates obtained by simultaneous processing of the same total set of observations.

Some statistical properties of errors in the element estimates are derived for the case in which the observation errors are regarded as statistical variables and in which the matrix of the quadratic form to be minimized is the inverse covariance matrix of the observation errors. (For Gaussian error statistics, this would result in a maximum-likelihood method of estimation.)

The elements are at first regarded as constants, and later the treatment is extended to the case in which the elements are regarded as time-dependent.

We assume there are N observed quantities F^μ, $\mu = 1, \cdots, N$, each F^μ being a real scalar. We also assume that n real constants x_i, $i = 1, \cdots, n$ exist such that in the absence of observation error, all N observed quantities would satisfy the relations

$$F^\mu = f^\mu(x_1, \cdots, x_n, t_\mu) \quad (1)$$

The vector $x = (x_1, \cdots, x_n)$ will be called the "element vector," and its components the "elements"; t_μ is the time at which the μth observation is taken.

When observation errors are present, the F^μ are given by

$$F^\mu = f^\mu(x_1, \cdots, x_n, t_\mu) + \varepsilon^\mu. \tag{2}$$

For purposes of illustration, consider the observation of a satellite following a Keplerian orbit. The elements might then be taken to be the eccentricity, semi-major axis, inclination, etc.; the quantities F^μ might be observations of such things as range, azimuth, elevation, or range rate from particular observation sites; the f^μ would be the functions describing the dependence of these observed quantities on the elements and time; and the ε^μ would be the observation errors.

The times t_μ need not be in any particular temporal order, nor are they necessarily all distinct. They are assumed to be measured without error in all cases in which the functions f^μ depend explicitly on t_μ. There is no loss of generality in this assumption, since timing errors may always be reduced to equivalent observation errors; this would of course modify the statistics of the resulting equivalent observation errors (for example, this might introduce an additional systematic component into the equivalent observation errors).

Many of the formulas below will involve the functions f^μ and their partial derivatives $\partial f^\mu/\partial x_i$. Practical application of these formulas would be possible both for cases where analytic expressions are known for the f^μ and their partial derivatives and for cases where these functions must be evaluated by numerical integration, as well as for cases where some of the functions have known analytic expressions and others must be determined by numerical methods.

Henceforth it will be supposed that $N \geqq n$. The problem to be considered is the estimation of the elements by means of smoothing, in some sense, of the observations.

A classical smoothing method is as follows: writing $P = (P_1, \cdots, P_n)$ for an estimate of the element vector, and writing $f(x, t)$ for $f(x_1, \cdots, x_n, t)$, $f(P, t)$ for $f(P_1, \cdots, P_n, t)$, and so forth, the method consists of minimizing with respect to $P_1, \cdots, P_n$ the quadratic form

$$Q = \sum_{\mu,\nu=1}^{N} \eta_{\mu\nu}[F^\mu - f^\mu(P, t_\mu)][F^\nu - f^\nu(P, t_\nu)] \tag{3}$$

where $(\eta_{\mu\nu})$ is a symmetric, positive definite matrix. Thus, the method consists in minimizing a positive definite quadratic form in the residuals. Differentiating Q with respect to P_i and setting the results equal to zero, we find that the minimizing estimates P_i must satisfy

$$\sum_{\mu,\nu=1}^{N} \frac{\partial f^\nu}{\partial x_i}(P, t_\nu)[F^\mu - f^\mu(P, t_\mu)]\eta_{\mu\nu} = 0 \qquad (i = 1, \cdots, n) \tag{4}$$

It is clear that if $\varepsilon^\mu = 0$, all μ, then $P = x$ is a solution of (4). The first question to be investigated is that of the first order propagation of errors—i.e., the first order dependence of $P - x$ on the errors ε^μ. It will be assumed that the functions f^μ are sufficiently well behaved for the following operations to be valid. We may write

$$f^\mu(P, t_\mu) = f^\mu(x, t_\mu) + \sum_{j=1}^{n} \frac{\partial f^\mu}{\partial x_j}(x, t_\mu)(P_j - x_j) + \cdots \tag{5}$$

$$\frac{\partial f^\mu}{\partial x_i}(P, t_\mu) = \frac{\partial f^\mu}{\partial x_i}(x, t_\mu) + \sum_{j=1}^{n} \frac{\partial^2 f^\mu}{\partial x_i\,\partial x_j}(x, t_\mu)(P_j - x_j) + \cdots \tag{6}$$

Neglecting all terms of higher order in $P - x$, and substituting (5) and (6) into (4), we obtain

$$\sum_{\mu,\nu=1}^{N} \eta_{\mu\nu}\left[\frac{\partial f^\nu}{\partial x_i}(x, t_\nu) + \sum_{j=1}^{n} \frac{\partial^2 f^\nu}{\partial x_i\,\partial x_j}(x, t_\nu)(P_j - x_j)\right] \times \left[F^\mu - f^\mu(x, t_\mu) - \sum_{j=1}^{n} \frac{\partial f^\mu}{\partial x_j}(x, t_\mu)(P_j - x_j)\right] = 0 \tag{7}$$

It is also clear that for sufficiently small ε^μ and $P_j - x_j$, the term involving second derivatives of f^ν may be neglected. The result may conveniently be expressed as follows: let

$$a_i^\mu(x, t_\mu) = \frac{\partial f^\mu}{\partial x_i}(x, t_\mu) \tag{8}$$

$$\rho_i(x) = \sum_{\mu,\nu=1}^{N} \eta_{\mu\nu} a_i^\nu(x, t_\nu)\,[F^\mu - f^\mu(x, t_\mu)] \tag{9}$$

$$\rho(x) = \{\rho_1(x), \cdots, \rho_n(x)\} \tag{10}$$

$$B_{ij}(x) = \sum_{\mu,\nu=1}^{N} \eta_{\mu\nu} a_i^\nu(x, t_\nu) a_j^\mu(x, t_\mu) \tag{11}$$

$$B(x) = \{B_{ij}(x)\} \tag{12}$$

Then, assuming $B(x)$ to be non-singular,

$$P - x = [B(x)]^{-1}\rho(x) \tag{13}$$

Eq. (13) expresses the first order dependence of $P - x$ on the errors ε^μ. This can be expressed equivalently:

$$P_i - x_i = \sum_{\mu=1}^{N} \Gamma_i^\mu(x)\,[F^\mu - f^\mu(x, t_\mu)] \tag{14}$$

$$\Gamma_i^\mu(x) = \sum_{j=1}^{n} \sum_{\nu=1}^{N} [B(x)]_{ij}^{-1} \eta_{\mu\nu} a_j^\nu(x, t_\nu) \tag{15}$$

A special case of this is as follows: suppose one has already an estimated element vector $p = (p_1, \cdots, p_n)$ together with K new observations. One can form a new estimate vector P by smoothing the original estimate vector p with the new observations in the following manner: P is determined by minimizing, with respect to $P_1, \cdots, P_n$, the quadratic form

$$Q = \sum_{\mu,\nu=1}^{n} \eta_{\mu\nu}(p_\mu - P_\mu)(p_\nu - P_\nu) + \sum_{\mu,\nu=n+1}^{n+K} \eta_{\mu\nu}[F^\mu - f^\mu(P, t_\mu)]\,[F^\nu - f^\nu(P, t_\nu)] \tag{16}$$

In this case,

$$f^\mu(x, t_\mu) = x_\mu, \qquad \mu = 1, \cdots, n$$
$$a_i^\mu(x, t_\mu) = \delta_{i\mu} \qquad i = 1, \cdots, n \qquad (17)$$
$$\mu = 1, \cdots, n$$
$$F^\mu = p_\mu \qquad \mu = 1, \cdots, n$$

Also

$$\rho_i(x) = \sum_{j=1}^{n} \eta_{ij}(p_j - x_j) + \sum_{\mu,\nu=n+1}^{n+K} \eta_{\mu\nu}[F^\mu - f^\mu(x, t_\mu]a_i^\nu(x, t_\nu) \qquad (i = 1, \cdots, n) \qquad (18)$$

and

$$B_{ij}(x) = \eta_{ij} + \sum_{\mu,\nu=n+1}^{n+K} \eta_{\mu\nu}a_i^\nu(x, t_\nu)a_j^\mu(x, t_\mu) \qquad (i, j = 1, \cdots, n) \qquad (19)$$

If we also define $r^\mu(p, t_\mu)$ for $u = n + 1, \cdots, n + K$ by

$$r^\mu(p, t_\mu) = F^\mu - f^\mu(p, t_\mu) \qquad \mu = n + 1, \cdots, n + K \qquad (20)$$

then $\rho_i(x)$ becomes, to first order,

$$\rho_i(x) = \sum_{j=1}^{n} B_{ij}(x)(p_j - x_j) + \sum_{\mu,\nu=n+1}^{n+K} \eta_{\mu\nu}a_i^\nu(x, t_\nu)r^\mu(p, t_\mu) \qquad (21)$$

Consequently, (13) reduces to

$$P - x = p - x + [B(x)]^{-1}\rho^*(x) \qquad (22)$$

where

$$\rho_i^*(x) = \sum_{\mu,\nu=n+1}^{n+K} \eta_{\mu\nu}a_i^\nu(x, t_\nu)r^\mu(p, t_\mu) \qquad (23)$$

Eq. (22) may be used as the basis for a first order differential correction to p, given the additional observations F^μ, $\mu = n + 1, \cdots, n + K$. First rewrite (22) as

$$P - p = [B(x)]^{-1}\rho^*(x) \qquad (24)$$

Then, to first order, one can also write

$$P - p = [B(p)]^{-1}\rho^*(p) \qquad (25)$$

Since all the quantities on the right of (25) are known, (25) gives the required first order correction to p.

A Stagewise Smoothing Procedure

Suppose the matrix $(\eta_{\mu\nu})$ can be written as a diagonal array of matrices

$$\eta = \begin{pmatrix} \eta^{(1)} & & \\ & \ddots & \\ & & \eta^{(S)} \end{pmatrix} \qquad (26)$$

where $\eta^{(s)}$, $s = 1, \cdots, S$, are $N_s \times N_s$ matrices. For future notational convenience, we will regard the indices of the components $\eta_{\mu\nu}^{(s)}$ of $\eta^{(s)}$ as running over $\mu, \nu = M_{s-1} + 1, \cdots, M_s$ where

$$M_s = \sum_{r=1}^{s} N_r$$
$$M_0 = 0 \qquad (27)$$
$$M_1 = N_1 \geqq n$$
$$M_S = N$$

We define a stagewise smoothing procedure as follows: Suppose initial element estimates $\{P_i^{(1)}\}$ are obtained by minimization with respect to $P^{(1)}$ of the quadratic form

$$Q^{(1)} = \sum_{\mu,\nu=1}^{N_1} \eta_{\mu\nu}^{(1)}[F^\mu - f^\mu(P^{(1)}, t_\mu)] \cdot [F^\nu - f^\nu(P^{(1)}, t_\nu)] \qquad (28)$$

Now define sequences of matrices $B^{(s)}$ and $B^{(s)}(x)$ as follows:*

$$B_{ij}^{(s)} = B_{ij}^{(s-1)} + \sum_{\mu,\nu=M_{s-1}+1}^{M_s} \eta_{\mu\nu}^{(s)}a_i^\nu(P^{(s)}, t_\nu)a_j^\mu(P^{(s)}, t_\mu)$$
$$B_{ij}^{(0)} = 0 \qquad (29)$$
$$B_{ij}^{(s)}(x) = B_{ij}^{(s-1)}(x) + \sum_{\mu,\nu=M_{s-1}+1}^{M_s} \eta_{\mu\nu}^{(s)}a_i^\nu(x, t_\nu)a_j^\mu(x, t_\mu)$$
$$B_{ij}^{(0)}(x) = 0$$

For $s > 1$, the sth element estimates $\{P_i^{(s)}\}$ are obtained by minimizing, with respect to $P^{(s)}$, the quadratic form

$$Q^{(s)} = \sum_{i,j=1}^{n} B_{ij}^{(s-1)}[P_i^{(s-1)} - P_i^{(s)}][P_j^{(s-1)} - P_j^{(s)}] + \sum_{\mu,\nu=M_{s-1}+1}^{M_s} \eta_{\mu\nu}^{(s)}[F^\mu - f^\mu(P^{(s)}, t_\mu)] \cdot [F^\nu - f^\nu(P^{(s)}, t^\nu)] \qquad (30)$$

Let us also define $\rho_i^{(s)}(x)$, $s = 1, 2, \cdots, S$, by

$$\rho_i^{(s)}(x) = \sum_{j=1}^{n} B_{ij}^{(s-1)}(x)[P_j^{(s-1)} - x_j] + \sum_{\mu,\nu=M_{s-1}+1}^{M_s} \eta_{\mu\nu}^{(s)}a_i^\nu(x, t_\nu)[F^\mu - f^\mu(x, t_\mu)] \qquad (31)$$

Then, using (13), (18), and (19), it can be shown that to first order

$$P^{(s)} - x = [B^{(s)}(x)]^{-1}\rho^{(s)}(x) \qquad (32)$$

* Alternative definitions for the matrices $B^{(s)}$ are possible; see the remark at the end of this Section.

It is also not hard to show by an inductive argument that

$$P^{(S)} - x = [B^{(S)}(x)]^{-1}\rho^{(S)}(x) = [B(x)]^{-1}\rho(x) \quad (33)$$

where $B(x)$ and $\rho(x)$ are defined as in (9), (10), (11), (12). This is proved by showing that $B^{(S)}(x) = B(x)$, and $\rho^{(S)}(x) = \rho(x)$.

Thus, the first order dependence of the errors in the estimates $\{P_i^{(S)}\}$ is the same as that for the $\{P_i\}$ obtained by processing all N observations at once by minimizing Q as defined by (3). Another way of stating this is to say that, to first order, the estimate obtained by processing the N observations by the stagewise procedure just described is the same as that obtained by minimizing Q as defined by (3). (However, the range of magnitudes of $\mathcal{E}^\mu$ for which the first order expressions give good approximations to the estimation errors is not necessarily the same for the stagewise method as for the non-stagewise method.)

A stagewise smoothing procedure may be advantageous in certain situations. For example, suppose that observations are coming in at some average rate; in the stagewise procedure, it is not necessary to store all previous observations and $N \times N$ matrices ($\eta_{\mu\nu}$) with N increasing. It is at any time necessary to store only the 'current' element estimates $\{P_i^{(s)}\}$ and the matrix $B^{(s)}$, that is, $\frac{n}{2}(n + 3)$ quantities (taking into account the symmetry of $B^{(s)}$).

It can be seen that the matrices $B^{(s)}$ play the role of estimates of the matrices $B^{(s)}(x)$. Thus, the particular method of defining the sequence $\{B^{(s)}\}$ above is not the only one possible. For example, one could define

$$\begin{aligned} \bar{B}_{ij}^{(s)} &= B_{ij}^{(s)}, \qquad s = 1 \\ &= \bar{B}_{ij}^{(s-1)} + \sum_{\mu,\nu=M_{s-1}+1}^{M_s} \eta_{\mu\nu}^{(s)} a_i^{\nu}(P^{(s-1)}, t_\nu) \qquad (29') \\ &\quad \cdot a_j^{\mu}(P^{(s-1)}, t_\mu), \qquad s > 1 \end{aligned}$$

and define the stagewise process using the matrices $\bar{B}^{(s)}$ instead of $B^{(s)}$. Comparing with (29), we see that the main difference is that $\bar{B}^{(s)}$ can be computed, for $s > 1$, before one computes $P^{(s)}$.

This lends itself conveniently to first order determination, for $s > 1$, of $P^{(s)} - P^{(s-1)}$ by means of Eq's. (20), (23), and (24). One would write $p = P^{(s-1)}$; $P = P^{(s)}$; and $P^{(s)} - P^{(s-1)} = [\bar{B}^{(s)}]^{-1}\rho^{*(s)}(P^{(s-1)})$. (The vector $\rho^{*(s)}$ would be defined by obvious modifications of (20) and (23).)

Application when the Elements are Functions of Time

Suppose the elements $x_1, \cdots, x_n$ are functions of time: $x_i = x_i(t)$. Also suppose that error-free observations are given by

$$F^\mu = g^\mu[x(t_\mu), t_\mu] \quad (34)$$

Suppose the manner in which $x(t)$ depends on its values at some t_0 is known:

$$x_i(t) = \mathfrak{F}_i[x(t_0), t_0, t] \quad (35)$$

Then we may obtain the case considered in the previous sections by defining the elements x_i in the formulas of those sections to be $x_i = x_i(t_0)$ and by defining

$$f^\mu[x, t_\mu] = g^\mu\{\mathfrak{F}_1[x, t_0, t_\mu], \cdots, \mathfrak{F}_n[x, t_0, t_\mu], t_\mu\} \quad (36)$$

(where $x = x(t_0)$).

Modification when the Functions f^μ are Imperfectly Known

Suppose the observations are given by

$$F^\mu = h^\mu[x, t_\mu] + \mathcal{E}^\mu \quad (37)$$

(where $\mathcal{E}^\mu$ are observation errors),

but that the estimated elements $\{P_i\}$ are obtained by minimizing Q as defined by (3), with the functions f^μ (which differ from h^μ).

So long as $F^\mu - f^\mu$ are sufficiently small, equations (9)–(13) still describe the dependence of $P - x$ on $F^\mu - f^\mu$. This dependence was expressed

$$P_i - x_i = \sum_{\mu=1}^{N} \Gamma_i^\mu(x)[F^\mu - f^\mu(x, t_\mu)] \quad (14)$$

where:

$$\Gamma_i^\mu(x) = \sum_{j=1}^{n}\sum_{\nu=1}^{N} [B(x)]_{ij}^{-1}\eta_{\mu\nu}a_j^\nu(x, t_\nu) \quad (15)$$

The observation errors $\mathcal{E}^\mu$ are now given by

$$\mathcal{E}^\mu = F^\mu - h^\mu(x, t_\mu). \quad (38)$$

Therefore

$$F^\mu - f^\mu(x, t_\mu) = \mathcal{E}^\mu + h^\mu(x, t_\mu) - f^\mu(x, t_\mu) \quad (39)$$

Thus, (14) becomes

$$P_i - x_i = \sum_{\mu=1}^{N} \Gamma_i^\mu(x)\mathcal{E}^\mu + \sum_{\mu=1}^{N} \Gamma_i^\mu(x)[h^\mu(x, t_\mu) - f_\mu(x, t_\mu)]. \quad (40)$$

Statistics of Propagated Errors

Eq. (13) may be used in an obvious manner to determine the means and covariance matrix of $\{P_i - x_i\}$ as functions of the means and covariance matrix of $\{\mathcal{E}^\mu\}$.

We shall deal here with a special case, namely, one in which the ensemble means and covariance matrix of $\{\mathcal{E}^\mu\}$ are known and in which the matrix $\{\eta_{\mu\nu}\}$ of the quadratic form Q has a special relation to the covariance matrix of $\{\mathcal{E}^\mu\}$.

Since the ensemble means of $\{\mathcal{E}^\mu\}$ are assumed known, it is no loss of generality to assume they are zero. In this case the covariance matrix is (denoting ensemble means by $E(\ )$)

$$\phi_{\mu\nu} = E(\mathcal{E}^\mu\mathcal{E}^\nu) \quad (41)$$

It will now be assumed that the matrix $(\eta_{\mu\nu})$ in (3) is

$$(\eta_{\mu\nu}) = (\phi_{\mu\nu})^{-1} \quad \text{(matrix inverse)} \tag{42}$$

If $\{\mathcal{E}^\mu\}$ were to have a Gaussian probability distribution (and if $f^\mu \equiv h^\mu$, all μ) then the resulting method of obtaining P would be the maximum likelihood method.

The covariance matrix of $\{P_i - x_i\}$ assumes a particularly simple form in this case. For generality, we will deal with the case described in the last section, in which f^μ, the functions used in the quadratic form Q, may differ from h^μ.

The means of $\{P_i - x_i\}$ are

$$E(P_i - x_i) = \sum_{\mu=1}^{N} \Gamma_i^\mu(x)[h^\mu(x, t_\mu) - f^\mu(x, t_\mu)] \tag{43}$$

The covariance matrix of $\{P_i - x_i\}$ is readily established to be

$$E\{P_i - x_i - E(P_i - x_i)\} \cdot \{P_j - x_j - E(P_j - x_j)\} = [B(x)]_{ij}^{-1} \tag{44}$$

This holds whether the observations are processed all together or by a stagewise procedure as described. In the latter case, of course, it must be assumed that the covariance matrix of $\{\mathcal{E}^\mu\}$ can be broken up into a diagonal array of covariance matrices corresponding to the different stages. It is, in fact, quite easy to establish that $B^{(s)}(x)$ is the inverse covariance matrix of $\{P_i^{(s)} - x_i\}$:

$$E\{P_i^{(s)} - x_i - E(P_i^{(s)} - x_i)\} \cdot \{P_j^{(s)} - x_j - E(P_j^{(s)} - x_j)\} = [B^{(s)}(x)]_{ij}^{-1} \tag{45}$$

This throws some further light on the stagewise method of Section II. The matrices $B^{(s-1)}$ occurring in the quadratic forms $Q^{(s)}$ are seen to be estimates of the inverse covariance matrices $B^{(s-1)}(x)$ of the element estimates $P^{(s-1)}$. Thus, if the error statistics were Gaussian, this procedure would consist at each stage of a maximum likelihood smoothing of the previous element estimates with the new observations.

The above formulas may be used to determine the rate at which the covariance matrix of the errors $P_i^{(s)} - x_i$ decreases as additional observations are processed. In fact, this information is contained in (44) and (45).

As a special case, consider the case where all observation errors are mutually uncorrelated. (If this is not true originally, it can be made true by means of linear transformations.) In this case, the matrices ϕ and η are diagonal.

If we now regard $P^{(s)}$ as the element estimate vector resulting from the processing of the first s observations, we have the inverse covariance matrix for $\{P_i^{(s)} - x_i\}$ given by

$$B_{ij}^{(s)}(x) = \sum_{\mu=1}^{s} \eta_{\mu\mu} a_i^\mu(x, t_\mu) a_j^\mu(x, t_\mu) \tag{46}$$

(This holds whether the s observations were processed all together or stagewise.)

It is also quite easy to verify that, for $s \geqq n$,

$$[B^{(s)}(x)]_{ij}^{-1} = [B^{(s-1)}(x)]_{ij}^{-1} - d_i^{(s)}(x)\, d_j^{(s)}(x) \tag{47}$$

where

$$d_i^{(s)}(x) = \frac{\sqrt{\eta_{ss}} \sum_{k=1}^{n} a_k^s(x, t_s)\, [B^{(s-1)}(x)]_{ik}^{-1}}{\sqrt{1 + \eta_{ss} \sum_{j,k=1}^{n} [B^{(s-1)}(x)]_{jk}^{-1} a_k^s(x, t_s)\, a_j^s(x, t_s)}} \tag{48}$$

We might close this section by briefly discussing the subject of systematic errors, which can be defined as errors which are highly correlated over a certain group of observations. There are several possible ways of handling systematic errors. They could be regarded as introducing off-diagonal elements into the correlation matrix $\phi_{\mu\nu}$ and the smoothing matrix $\eta_{\mu\nu}$, the magnitude of the off-diagonal elements being chosen according to the approximate magnitude of the systematic errors. Or, alternatively, a systematic error might in some cases profitably be regarded as an additional parameter to be estimated from the data—i.e., as an additional element. This is one of the choices that would have to be made in actual implementation of these results.

Modified Stagewise Procedure for Time-Varying Elements

In this section the elements will be considered functions of time, $x_i = x_i(t)$. It will be assumed that the elements vary much more slowly with respect to time than do the functions describing observations.

We suppose that the matrix $(\eta_{\mu\nu})$ can be written (as in Eq. (26)) as a diagonal array of matrices $\eta^{(s)}, s = 1, 2, \cdots$; the further assumption is made that the times t_μ, $\mu = M_{s-1} + 1, \cdots, M_s$, are sufficiently close together that they may be regarded as equal insofar as variation of the elements is concerned. This will be expressed

$$x_i(t_\mu) = x_i(T_s) \qquad \mu = M_{s-1} + 1, \cdots, M_s \tag{49}$$

Henceforth, the time parameter occurring in the argument of an element or element estimate will be written T.

It will be supposed that error-free observations are given by

$$F^\mu = g^\mu[x(T_s), t_\mu] \qquad \mu = M_{s-1} + 1, \cdots, M_s \qquad s = 1, 2, \cdots \tag{50}$$

and that the variation of the elements with respect to time is given by

$$x_i(T_s) = \mathfrak{F}_i[x(T_r), T_r, T_s]. \tag{51}$$

Now consider the following stagewise procedure for processing the observations (the main reason for which is its adaptability to cases in which the prediction functions f^μ or $\mathfrak{F}_i$ are imperfectly known):

The element estimate vector $P^+(T_1)$ is obtained by minimizing with respect to $P^+(T_1)$ the quadratic form

$$Q^{(1)} = \sum_{\mu,\nu=1}^{N_1} \eta_{\mu\nu}^{(1)}\{F^\mu - g^\mu[P^+(T_1), t_\mu]\} \cdot \{F^\nu - g^\nu[P^+(T_1), t_\nu]\}. \quad (52)$$

For $T_s \leqq T < T_{s+1}$, $s = 1, 2, \cdots$,

$$P_i(T) = \mathfrak{F}_i[P^+(T_s), T_s, T] \quad (53)$$

Also,

$$P_i^-(T_{s+1}) = \mathfrak{F}_i[P^+(T_s), T_s, T_{s+1}] \quad (54)$$

The quantities $P^+(T_s)$ will be defined for $s > 1$ below. To do this, we define sequences of matrices $B^{(s)}$, $B_+^{s)}$, $\psi^{(s)}$, $\psi_+^{s)}$ as follows:

$$B_{+ij}^{(1)} = \sum_{\mu,\nu=1}^{N_1} \eta_{\mu\nu}^{(1)} b_i^\nu[P^+(T_1), t_\nu] b_j^\mu[P^+(T_1), t_\mu] \quad (55)$$

$$b_i^\nu[x,t_\nu] = \frac{\partial g^\nu}{\partial x_i}[x,t_\nu] \quad (56)$$

$$\psi_+^{(s)} = [B_+^{(s)}]^{-1}, \qquad s = 1, 2, \cdots \quad (57)$$

$$\psi^{(s+1)} = \sum_{k,l=1}^{n} \mathfrak{A}_{ik}[P^+(T_s), T_s, T_{s+1}] \cdot \mathfrak{A}_{jl}[P^+(T_s), T_s, T_{s+1}]\psi_{+kl}^{(s)} \quad (58)$$

$$\mathfrak{A}_{ij}[x, T_s, T] = \frac{\partial \mathfrak{F}_i}{\partial x_j}[x, T_s, T] \quad (59)$$

$$\begin{aligned} B^{(s)} &= 0, \qquad s = 1 \\ &= [\psi^{(s)}]^{-1}, \qquad s > 1 \end{aligned} \quad (60)$$

$$B_{+ij}^{(s)} = B_{ij}^{(s)} + \sum_{\mu,\nu=M_{s-1}+1}^{M_s} \eta_{\mu\nu}^{(s)} b_i^\nu[P^+(T_s), t_\nu] b_j^\mu[P^+(T_s), t_\mu] \quad (61)$$

Then $P^+(T_s)$ is obtained for $s > 1$ by minimizing, with respect to $P^+(T_s)$, the quadratic form

$$\begin{aligned} Q^{(s)} = {} & \sum_{i,j=1}^{n} B_{ij}^{(s)}\{P_i^-(T_s) - P_i^+(T_s)\} \cdot \{P_j^-(T_s) - P_j^+(T_s)\} \\ & + \sum_{\mu,\nu=M_{s-1}+1}^{M_s} \eta_{\mu\nu}^{(s)}\{F^\mu - g^\mu[P^+(T_s), t_\mu]\} \cdot \{F^\nu - g^\nu[P^+(T_s), t_\nu]\} \end{aligned} \quad (62)$$

Scrutiny of Eqs. (52)–(62) reveals that a stagewise smoothing procedure has been completely defined. Now, in order to give the first order error equations for the resulting estimates, define the following matrices:

$$B_{+ij}[T_1] = \sum_{\mu,\nu=1}^{N_1} \eta_{\mu\nu}^{(1)} b_i^\nu[x(T_1), t_\nu] b_j^\mu[x(T_1), t_\mu] \quad (63)$$

(where $b_i^\nu(x, t_\nu)$ is defined as in (56));

$$\psi_+[T_s] = \{B_+[T_s]\}^{-1} \quad (64)$$

for $T_s < T \leqq T_{s+1}$,

$$\psi_{ij}[T] = \sum_{k,l=1}^{n} \mathfrak{A}_{ik}[x(T_s), T_s, T] \cdot \mathfrak{A}_{jl}[x(T_s), T_s, T]\psi_{+kl}[T_s] \quad (65)$$

(where $\mathfrak{A}_{ij}(x, T_s, T)$ is defined as in (59));

$$B[T] = \{\psi[T]\}^{-1}, \; T_s < T \leqq T_{s+1} \quad (66)$$

and

$$B_{+ij}[T_s] = B_{ij}[T_s] + \sum_{\mu,\nu=M_{s-1}+1}^{M_s} \eta_{\mu\nu}^{(s)} b_i^\nu[x(T_s), t_\nu] b_j^\mu[x(T_s), t_\mu]. \quad (67)$$

Also define

$$\rho_i[T_s] = \sum_{j=1}^{n} B_{ij}[T_s][P_j^-(T_s) - x_j(T_s)] + \sum_{\mu,\nu=M_{s-1}+1}^{M_s} \eta_{\mu\nu}^{(s)} b_i^\nu[x(T_s), t_\nu]\{F^\mu - g^\mu[x(T_s), t_\mu]\}. \quad (68)$$

Then, the first order dependence of $P^+(T_s) - x(T_s)$ on the observation errors is:

$$P^+(T_s) - x(T_s) = \{B_+[T_s]\}^{-1}\rho[T_s] \quad (69)$$

Also, for $T_s < T < T_{s+1}$, the first order dependence of $P(T) - x(T)$ is

$$P_i(T) - x_i(T) = \sum_{j=1}^{n} \mathfrak{A}_{ij}[x(T_s), T_s, T][P_j^+(T_s) - x_j(T_s)] \quad (70)$$

Further, suppose that $(\eta_{\mu\nu})$ represents the inverse covariance matrix of $\{F^\mu - g^\mu[x(t_\mu), t_\mu]\}$. Then, $B_+[T_s]$ is the inverse covariance matrix of $\{P_i^+(T_s) - x_i(T_s)\}$, and $B[T]$ is the inverse covariance matrix of $\{P_i(T) - x_i(T)\}$, $T_s < T < T_{s+1}$.

In the third section an alternative method was described for dealing with time-varying elements. That method involved a reduction to the case of constant elements.

Choose any time T, $T_s \leqq T < T_{s+1}$, and regard T as now fixed. Define a constant element vector x by: $x = x(T)$. Let $P^*(T)$ represent the estimate of $x = x(T)$ obtained by the method described in the third section, based on the first M_s observations.

Then, it can be verified that, to first order, $P^*(T) = P(T)$, where $P(T)$ is defined by (53). That is, the stagewise procedure described in (52)–(62) yields an estimate $P(T)$ for $x(T)$ having the same first order dependence on $F^\mu - g^\mu[x(t_\mu), t_\mu]$, $\mu = 1, \cdots, M_s$, as the estimate derived by the method of Section III.

The stagewise procedure described above actually goes through the following steps: $P^+(T_s)$ represents an estimate of $x(T_s)$, based on the first M_s observations;

for $T_s < T < T_{s+1}$, $P(T)$ is obtained by (53), that is, simply by prediction from $P^+(T_s)$ according to the functional relation by which the elements are known to vary; $P^-(T_{s+1})$ represents the estimate of $x(T_{s+1})$ just before the observations F^μ, $\mu = M_s + 1, \cdots, M_{s+1}$ are processed.

The particular method chosen above to define the matrices $B_+^{(s)}$—i.e., by (61)—is not the only one possible. One could, for example, define matrices $\tilde{B}_+^{(s)}$ for $s > 1$ by using (61) with $P^-(T_s)$ instead of $P^+(T_s)$ in the arguments of b_i^ν and b_j^μ. Then one could compute $\tilde{B}_+^{(s)}$, for $s > 1$, before computing $P^+(T_s)$.

This would be convenient for purposes of using (20), (23), and (24) for a first order determination of $P^+(T_s) - P^-(T_s)$: one would put $p = P^-(T_s)$; $P = P^+(T_s)$; $P^+(T_s) - P^-(T_s) = [\tilde{B}_+^{(s)}]^{-1}\rho^{*(s)}[P^-(T)]$, with $\rho^{*(s)}$ defined by the appropriate modification of (20) and (23).

In practical applications, for example, in satellite observations where perturbation forces are imperfectly known, the functions $\mathfrak{F}_i$ may not be known exactly. Furthermore, the inaccuracy in knowledge of $\mathfrak{F}_i[x(T), T, T']$ will depend on the time difference $T' - T$—i.e., on how far ahead you are predicting the elements.

Suppose, for example, that the prediction functions used in the stagewise procedure are g^μ, $\mathfrak{F}_i$, but that the correct functions should be h^μ, $\mathfrak{F}_i^*$. Let $\mathcal{E}^\mu$ still represent the observation errors—i.e., $\mathcal{E}^\mu = F^\mu - h^\mu$—and let, for $T \geqq t_\mu$,

$$\begin{aligned}\delta^\mu(T) = {} & h^\mu\{\mathfrak{F}_1^*[x(T), T, t_\mu], \cdots, \\ & \mathfrak{F}_n^*[x(T), T, t_\mu], t_\mu\} \\ & - g^\mu\{\mathfrak{F}_1[x(T), T, t_\mu], \cdots, \\ & \mathfrak{F}_n[x(T), T, t_\mu], t_\mu\}\end{aligned} \tag{71}$$

Suppose that, in the case $g^\mu = h^\mu$, $\mathfrak{F}_i = \mathfrak{F}_i^*$, the first order error in $P(T)$, $T_s \leqq T < T_{s+1}$, could be expressed

$$P_i(T) - x_i(T) = \sum_{\mu=1}^{M_s} \Gamma_i^\mu(T)\mathcal{E}^\mu \tag{72}$$

Then, in the case $g^\mu \neq h^\mu$, $\mathfrak{F}_i \neq \mathfrak{F}_i^*$, one would have

$$P_i(T) - x_i(T) = \sum_{\mu=1}^{M_s} \Gamma_i^\mu(T)[\mathcal{E}^\mu + \delta^\mu(T)] \tag{73}$$

provided $\delta^\mu(T)$ and $\mathcal{E}^\mu$ are sufficiently small.

The errors $\delta^\mu(T)$ would, in general, grow as T increases. In such cases, it would be desirable to modify the smoothing procedure so as to give more recent observations greater weight, and continuously to diminish the weight given to past observations. There are a number of possibilities for accomplishing this; one way, for instance, would be to define the matrices $B^{(s)}$ in (60)–(62) by

$$B_{ij}^{(s)} = [\psi^{(s)}]_{ij}^{-1}\lambda_i^{(s)}\lambda_j^{(s)} \tag{60*}$$

where $\lambda_i^{(s)}$ and $\lambda_j^{(s)}$ are $\leqq 1$.

This would mean that the effective smoothing matrix for the observations would not be the matrix (η) appearing in (52)–(62); in fact, there would really be no fixed smoothing matrix. Also, the first order error equations (63)–(70) would have to be modified appropriately.

There is an analogy between this class of smoothing procedures and the process of filtering of signals. One might regard the smoothing procedures as filters with input $x(T)$ and output $P(T)$. The case where the f^μ and $\mathfrak{F}_i$ are perfectly known is equivalent to a constant signal input to the filter, in which case the filter should have infinite memory to provide maximum smoothing of observation errors. If, say, $\mathfrak{F}_i$ are not exactly known, this corresponds to the existence of an unpredictable time varying component of the filter input; in this case the filter memory must be reduced (its 'bandwidth' increased); the best 'time constant' is related to the rate of variation of the unpredictable part of $x(T)$. One could also, in effect, make the time constant different for the different elements.

15

Reprinted from *Radio Eng. Electron. Phys.* **1**:1–19 (1960)

APPLICATION OF THE THEORY OF MARKOFF PROCESSES IN OPTIMAL SIGNAL DISCRIMINATION*

R. L. Stratonovich

(Physics Faculty of the Moscow State University)

An apparatus based on Markoff conditional processes is used in the design and construction of optimal non-linear systems discriminating signal from noise. The general theory is illustrated by the solution of two concrete problems. In the first of these the optimal system found contains a Woodward - Davies optimal receiver.

Introduction

A signal usually arrives at the receiver input in a form distorted by noise. If the statistical properties of signal and disturbance are previously known, it is possible to construct an optimal receiver to reduce the effect of the disturbance to a minimum.

After receipt of the distorted signal, we have at our disposal more exact, *a posteriori*, knowledge concerning the true signal. An optimal receiver will be so constructed as to automatically extract and process *a posteriori* information on the statistical properties of signal and noise.

When signal and noise are gaussian, and minimum of the mean-square error is taken as the criterion of optimality, the optimal transformation is linear, and can be calculated from the Kolmogorov - Wiener theory [1 - 2]. In the case of non-gaussian signal or noise, the optimal transformation is non-linear. This transformation has been found for the case of separate pulse signals combined with white noise by the aid of the inverse probability method, in the well-known work of Woodward and Davies [3].

In the more complicated cases, use of the inverse probability method leads

---

* *Radiotekhnika i elektronika* 5, No. 11, 1751 - 1763, 1960.

to considerable difficulties in calculation. We may conveniently use to eliminate these difficulties a Markoff process apparatus, involving Markoff conditional processes and their corresponding equations [4-5]. Methods of calculation connected with Markoff conditional processes are applicable to very different problems, and may be various in structure. Markoff processes may be continuous or discontinuous, may have one or many components, may correspond to continuous time or form discrete sequences. Accordingly, the fundamental equations of optimal filtration and the means of modelling them are different; furthermore, the equations can be modelled by various means, as is shown below. The general ideas applied in these cases can be clarified by the consideration of different concrete problems.

The optimal discrimination of a signal with constant parameters from non-gaussian noise has been examined in paper [6]. In the present paper, we shall investigate the more complicated case when the parameters of the signal are not constant. For simplicity, we shall take the noise $n(t)$ to be delta-correlated; $\overline{nn_\tau} = N\delta(\tau)$.

In order to show more clearly the derivation of the present method from the results obtained by Woodward and Davies, we shall begin with the case when the useful signal is a square pulse,

$$s(t) = s_0(t - x) = \begin{cases} A_0 & \text{when } x < t < x + \tau_0; \\ 0 & \text{when } t < x \quad \text{or} \quad t > x + \tau_0, \end{cases} \tag{1}$$

and it is required to determine the moment x of its appearance on the section $0 < x < T$. We shall assume the height of the pulse A_0 and its duration τ_0 to be known.

The signal received, $y(t) = s(t) + n(t)$ is the sum of useful signal and white noise $n(t)$, which can be written as a function probability [6]

$$W[n(t)] = \exp\left\{-\frac{1}{2N}\int n^2(t)\,dt\right\} \tag{2}$$

(the integral is understood in the sense of

$$\Sigma n_k^2\Delta \left(n_k = \frac{1}{\Delta}\int_{t_k}^{t_k+\Delta} n\,dt\right)$$

when Δ is small). Thus the *a priori* joint distribution density for x and $y(t)$ has the form

$$W\,[y\,(t),\,x] = \exp\left\{-\frac{1}{2N}\int [y\,(t) - s\,(t)]^2\,dt\right\} w_{pr}\,(x)$$

(w_{pr} being the *a priori* distribution for x). Hence from the formula of inverse probability we obtain

$$\begin{aligned} w_{ps}\,(x) &= C'\exp\left\{-\frac{1}{2N}\int_0^T [y\,(t) - s_0\,(t - x)]^2\,dt\right\} w_{pr}\,(x) = \\ &= C''\exp\left\{\frac{1}{N}\int_0^{T'} y\,(t)\,s_0\,(t - x)\,dt\right\} w_{pr}\,(x) \end{aligned} \tag{3}$$

(w_{ps} being the *a posteriori* probability density).

At the locus of the real position of the pulse, the probability density has, as a rule, a maximum, the mean relative height of which is determined by the parameter $a = A_0^2\tau_0/N$.

The greater this parameter, the more precise and reliable is the determination of x. Precision improves with repeated observations of the pulse. Let there arrive at intervals T (T being the scanning period) signals $y(t) = s_0$ $(t - kT + T - x_k) + n(t)$, where x_k is the position of the pulse in the kth period $(k-1)\,T < t < kT$. Let the *a priori* statistical data on $x_1, \ldots, x_m$ be written as the probability density $w_{pr}\,(x_1, \ldots, x_m)$. Then, by analogy with (3), we obtain the *a posteriori* distribution density

$$w_{ps}\,(x_1, \ldots, x_m) = C\exp\left\{\frac{1}{N}\sum_{k=1}^{m}\int_{kT-T}^{kT} y(t)s_0\,(t - kT + T - x_k)\,dt\right\} w_{pr}\,(x_1, \ldots, x_m). \tag{4}$$

If, in particular, $x_1 = \ldots = x_m$, we have

$$w_{pr}\,(x_1) = C\exp\left\{\frac{1}{N}\sum_{k=1}^{m}\int y\,(t' + kT - T)s_0\,(t' - x_1)\,dt'\right\} w_{pr}\,(x_1).$$

The last formula differs from (3) in that instead of $y(t) = s_0\,(t - x) + n(t)$ we must now take $\Sigma y(t + kT - T) = s_0\,(t - x) + n'(t)$, where $s_0\,(t - x) = ms_0\,(t - x)$ is the signal multiplied by m, and $n'(t) = \Sigma n(t - kT + T)$ is noise equal to the sum of the independent components, having a mean-square value $\sqrt{m}$ times greater than $n(t)$. The parameter

$$\alpha' = \frac{A_0'^2\tau_0}{mN} = m\frac{A_0'^2T}{N},$$

and so has been multiplied by m.

If the positions of the pulses in different periods are *a priori* completely independent, repeated observations will not increase the *a posteriori* knowledge given in formula (3). In intermediate cases, when the quantities $x_1, \dots, x_m$, though not identical, are correlated, the degree of accuracy of the *a posteriori* knowledge is described by some intermediate value of the parameter a $(A_0^2 r_0/N < a < mA_0^2 r_0/N)$ and tends to a certain finite limit when $m \to \infty$ in the case of a stationary process. This limit depends on the rate of the change of positions of x from one observation to another. When m is large, direct use of formula (4) is generally difficult, and it is better to utilize the properties of Markoff processes in finding the *a posteriori* probability density.

1. Equation for the probability of the pulse co-ordinates forming a Markoff chain

Let the co-ordinates of the pulses $x_1, \dots, x_m$ *a priori* form a Markoff chain. This means that the conditional distribution for x_{k+1} depends only on x_k and is independent of $x_{k-1}, x_{k-2}, \dots$ The multi-dimensional distribution $w_{pr}(x_1, \dots, x_m)$ entering into expression (4) can in this case be written in the form

$$w_{pr}(x_1, \dots, x_m) = w_T(x_m, x_{m-1}) \dots w_T(x_2, x_1)\, w_{pr}(x_1), \tag{5}$$

where $w_T(x_{k+1}, x_k) = w_T(x_{k+1} - x_k)$ is the transition probability, which we assume to depend only on the difference $x_{k+1} - x_k$ *.

Aiming at a more exact description of the real process, we can also take into account the dependence of the conditional probability for x_{k+1} not only on x_k but also on x_{k-1}, disregarding its dependence on $x_{k-2}, \dots$. Then the sequence $x_1, \dots, x_m$ will constitute a more complicated Markoff process, and the methods laid out below will be applicable with the necessary alterations. In general, the real process may be represented as a component of a multi-dimensional Markoff process with any previously assigned degree of accuracy. The complication of the process increases, or course, with increase of accuracy.

Confining our survey to the simple Markoff process, let us substitute distribution (5) in (4), and find

---

* This assumption is not essential. If greater generality is required, it need not be made, and we may even postulate the dependence of $w_T(x', x)$ on k.

$$w_{ps}(x_1, \ldots x_m) = C \prod_{k=0}^{m-1} \exp\left\{\frac{1}{N}\int_{kT}^{kT+T} y(t)\, s_0(t - kT - x_{k+1})\, dt\right\} \times$$
$$\times w_T(x_{k+1} - x_k) \exp\left\{\frac{1}{N}\int_0^T y(t)\, s_0(t - x_1)\, dt\right\} w_{pr}(x_1). \qquad (6)$$

From this it is easy to obtain a chain of equations determining the change of *a posteriori* probability,

$$w_k(x_k) = \int \ldots \int w_{ps}(x_1, \ldots, x_k) dx_1 \ldots dx_{k-1}, \qquad (7)$$

which describes the information concerning x_k after observation o' the signal $y(t)$ during $k-1$ preceeding periods. Assuming $m = k + 1$ in (6), ntegrating with respect to $x_1, \ldots, x_k$ and taking (7) into account, we find

$$w_{k+1}(x_{k+1}) = C \exp\left\{\frac{1}{N}\int_{kT}^{kT+T} y(t)\, s_0(t - kT - x_{k+1})\, dt\right\} \int w_T(x_{k+1} - x_k) \times$$
$$\times w_k(x_k)\, dx_k. \qquad (8)$$

If a computer device automatically realizes the successive transformations (8), then it also carries out optimal non-linear processing of the input signal $y(t)$. A block diagram of such a device is shown in Fig. 1. In it the functions w_k are realized as one function of time

$$w(t) = \begin{cases} w_1(t) & \text{when } 0 < t < T; \\ w_2(t - T) & \text{when } T < t < 2T; \\ \cdots\cdots & \\ w_k(t - kT + T) & \text{when } kT - T < t < kT; \\ \cdots\cdots & \end{cases} \qquad (9)$$

Replacing x_{k+1}, x_k by $t - kT$, $t' - kT$ in (8), and taking into account (9), we write transformation (8) in the form

$$w(t) = C \exp\left\{\frac{1}{N}\int y(t'')\, s_0(t'' - t)\, dt''\right\} \int w_T(t - t')\, w(t' - T)\, dt'. \qquad (10)$$

Block W in Fig. 1 realizes the Woodward-Davies optimal transformation, giving at the output the function

$$e(t) = \exp\left\{\frac{1}{N}\int s_0(t' - t)\, y(t')\, dt'\right\}. \tag{11}$$

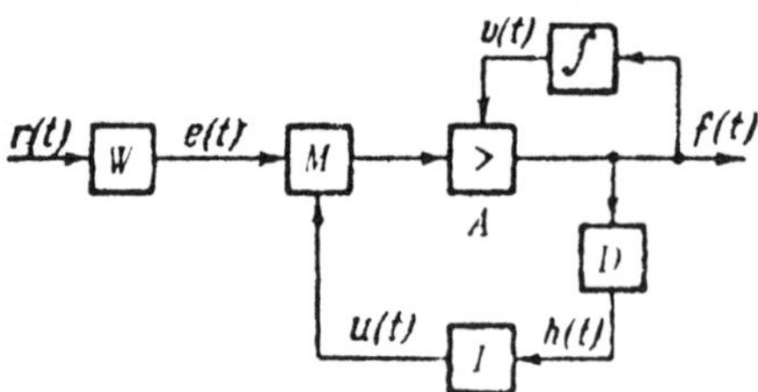

Fig. 1. Block-diagram of circuit for optimal non-linear processing of the input signal
W – Woodward-Davies block. M – multiplier block. A – Amplifier. D – delay. I – integral transformation ($r(t) = y(t)$)

The other blocks are so chosen as to achieve at the output of the whole device a function $f(t)$ proportional to $w(t)$ Block D brings about a delay T in time

$$h(t) = f(t - T). \tag{12}$$

Block I carries out the integral transformation

$$u(t) = \int w_T(t - t')\, h(t')\, dt', \tag{13}$$

and M the multiplication:

$$f(t) = e(t)\, u(t). \tag{14}$$

It is easy to see that in consequence of (11)-(14) the resulting function $f(t)$ is in fact proportional to the function $w(t)$ determined by transformation (10). The coefficient of proportionality does not play an important part. In order to approximately secure the condition of normalization, the amplifier A may be introduced into the circuit shown in Fig. 1; its coefficient of amplification $k(v)$ is controlled, let us say, by the function

$$v(t) = \frac{T}{T_0}\int_{-\infty}^{t} e^{-\frac{t-t'}{T_0}} f(t')\, dt' \quad (T_0 \gg T),$$

for instance, by the law $k(v) = k_0 + (1 - v)\, k_1$.

The circuit described is also suitable in the case when the useful pulse is a high frequency one. Here, as we know from [3], if the initial phase is completely random, the optimal Woodward-Davies block has to realize the transformation

$$e(t) = I_0\left(\frac{1}{N}\sqrt{z_1^2 + z_2^2}\right), \tag{15}$$

where

$$z_1 = \int s_0(t' - t) \sin \omega_0 (t - t')\, y(t')\, dt';$$

$$z_2 = \int s_0(t' - t) \cos \omega_0 (t - t')\, y(t')\, dt';$$

$I_0(z) = J_0(iz)$ being a Bessel function.

The transition probability $w_T(x - x')$ in (10) and (13) describes regular displacement of the pulse in time T and the dispersion of its co-ordinates. If the displacement equals qT the spreading is determined by the coefficient of diffusion b and the shift of the pulse in the period is relatively small, we can replace integration by differentiation, and assume

$$\int w_T(t - t')\, h(t')\, dt' = h(t) - gT\frac{dh(t)}{dt} + \frac{1}{2}bT\frac{d^2h(t)}{dt^2}. \tag{16}$$

In fact, the function $w_T(x - x')$ satisfies the Fokker-Planck equation

$$\frac{\partial w_T(x - x')}{\partial T} = -\frac{\partial}{\partial x}(g w_T) + \frac{b}{2}\frac{\partial^2 w_T}{\partial x^2}.$$

Consequently, the same equation is also satisfied by the function

$$\tilde{h}(x) = \int w_T(x - x')\, h(x')\, dx'.$$

Writing this equation, and replacing $d\tilde{h}(x)/dT$ by $(\tilde{h} - h)/T$ we obtain equation (16).

2. Determination of the phase of a narrow-band signal in white noise

For our second example, we shall consider the problem of the determination of the fluctuating phase $x(t)$ of a narrow-band signal

$$s(t) = A_0 \sin[\omega_0 t + x(t)] \tag{17}$$

with respect to the known realization of the sum $y(t) = s(t) + n(t)$. Here A_0, ω_0 are constant quantities known *a priori*; $n(t)$ is white noise described by the probability function (2).

The exactness of the determination of the phase $x(t)$ depends on the rapidity of its *a priori* change, and on the intensity of the noise. If the phase were constant, with increase of the observation time the error in determination of the phase would tend to zero. But since the phase is not constant, the error is finite, even with an unlimited observation period.

Let us consider an *a priori* phase variation process regarded as a Markoff process. Given a certain small time interval we shall write the *a priori* joint distribution density of the random values $x_1 = x(\Delta)$, $x_2 = x(2\Delta), \ldots, x_m = x(m\Delta)$ in the form

$$w_{pr}(x_1, \ldots, x_m) = w_\Delta(x_m, x_{m-1}) \ldots w_\Delta(x_2, x_1)\, w_{pr}(x_1), \tag{18}$$

where $w_\Delta(x_{k+1}, x_k)$ is, as before, the transition probability.

Bearing in mind (2), we find the multi-dimensional *a posteriori* distribution density

$$\begin{aligned} w_{ps}(x_1, \ldots, x_m) &= \text{const}\, W\,[y(t) - s(t)]\, w_{pr}(x_1, \ldots, x_m) = \\ &= \text{const}\exp\left\{-\frac{1}{2N}\int_0^{m\Delta}[y(t) - s(t)]^2\,dt\right\} w_\Delta(x_m, x_{m-1}) \ldots \\ &\quad \ldots w_\Delta(x_2, x_1)\, w_{pr}(x_1). \end{aligned} \tag{19}$$

This formula is similar to (6); therefore we may conduct our further investigation on similar lines to the previous one. For the one-dimensional *a posteriori* probabilities (7) we have a chain of transformations

$$\begin{aligned} w_{k+1}(x_{k+1}) &= C\exp\left\{-\frac{1}{2N}\int_{k\Delta}^{k\Delta+\Delta}[y(t) - A_0\sin(\omega_0 t + x_{k+1})]^2\,dt\right\} \times \\ &\quad \times \int w_\Delta(x_{k+1}, x_k)\, w_k(x_k)\,dx_k. \end{aligned} \tag{20}$$

In the case of the narrow-band signal (17), the expression under the integrand in the exponent is equal to

$$y^2(t) - 2y(t)A_0\sin(\omega_0 t + x) + \frac{A_0^2}{2}[1 - \cos 2(\omega_0 t + x)],$$

we can replace $2y(t)\, A_0 \sin(\omega_0 t + x)$ since we can refer the terms independent of $x - y^2$, $A_0^2/2$ – to the factor C, and the vibration function $\frac{1}{2} A_0^2 \cos 2(\omega_0 t + x)$ gives a small value on integration. Thus we can write instead of (20)

$$w_{k+1}(x_{k+1}) = C' \exp\left\{\frac{A_0}{N} \int_{k\Delta}^{k\Delta+\Delta} y(t) \sin(\omega_0 t + x)\, dt\right\} \times$$
$$\times \int w_\Delta(x_{k+1}, x_k)\, w_k(x_k)\, dx_k$$

or

$$w_k(x_k) = C' \exp\left\{\frac{A_0}{N} \sin x_k \int_{k\Delta-\Delta}^{k\Delta} y(t) \cos \omega_0 t dt + \right.$$

$$\left. + \frac{A_0}{N} \cos x_k \int_{k\Delta-\Delta}^{k\Delta} y(t) \sin \omega_0 t\, dt\right\} \int w_\Delta(x_k, x_{k-1})\, w_{k-1}(x_{k-1})\, dx_{k-1}. \quad (21)$$

This sequence of transformations can be realized by the aid of a device analogous to that already considered. To obtain the distribution density $w_k(x)$ as a function of time, it is convenient to scan with respect to x, with scanning period Δ. In order to do this, the voltage in the system must be worked out in saw-tooth form

$$\tilde{x}(t) = \frac{2\pi}{\Delta}(t - k\Delta) \text{ when } k\Delta < t < k\Delta + \Delta \quad (k = 0, 1, \ldots, m-1). (22)$$

The sequence of functions $w_1(x_1), \ldots, w_k(x_k)$ will be realized as one function of time

$$w(t) = w_k(\tilde{x}(t)) \text{ when } (k-1)\Delta < t < k\Delta \quad (k = 1, \ldots, m). \quad (23)$$

Substituting $x_k = \tilde{x}(t)$; $x_{k-1} = x(t - \Delta)$ in (21), and taking into account (22), (23), we obtain

$$w(t) = C' \exp\left\{\frac{A_0}{N} \sin \tilde{x}(t) \int_{k\Delta-\Delta}^{k\Delta} y(t') \cos \omega_0 t'\, dt' + \right.$$
$$\left. + \frac{A_0}{N} \cos \tilde{x}(t) \int_{k\Delta-\Delta}^{k\Delta} y(t') \sin \omega_0 t' dt'\right\} \int w_\Delta(\tilde{x}(t), \tilde{x}(t'))\, w(t' - \Delta)\, d\tilde{x}(t'). \quad (24)$$

The system shown in the block diagram realizes this transformation, and gives at its output a function $f(t)$ proportional to $w(t)$. Block W must now realize the transformation

$$e(t) = \exp\left\{\frac{A_0}{N}\sin\tilde{x}(t)\int_{k\Delta-\Delta}^{k\Delta} y(t')\cos\omega_0 t'dt' + \right.$$
$$\left. + \frac{A_0}{N}\cos\tilde{x}(t)\int_{k\Delta-\Delta}^{k\Delta} y(t')\sin\omega_0 t'\,dt'\right\}.$$

This may be replaced, with equal success, by the transformation

$$e(t) = \exp\left\{\frac{A_0}{N}\sin\tilde{x}(t)\int_{t-\Delta}^{t} y(t')\cos\omega_0 t'dt' + \right.$$
$$\left. + \frac{A_0}{N}\cos\tilde{x}(t)\int_{t-\Delta}^{t} y(t')\sin\omega_0\, t'dt'\right\} \tag{25}$$

or even

$$e(t) = \exp\left\{\frac{A_0}{N}\sin\tilde{x}(t)\int_{-\infty}^{t} e^{-\frac{t-t'}{\Delta}} y(t')\cos\omega_0 t'dt' + \right.$$
$$\left. + \frac{A_0}{N}\cos\tilde{x}(t)\int_{-\infty}^{t} e^{-\frac{t-t'}{\Delta}} y(t')\sin\omega_0 t'dt'\right\},$$

which simplifies the realization of the scheme. Block *I* carries out the transformation

$$u(t) = \int w_\Delta(\tilde{x}(t), \tilde{x}(t'))\, h(t')\, d\tilde{x}(t'). \tag{26}$$

The function $w_\Delta(x, x')$ statistically describes the *a priori* shift of the x co-ordinate in the interval Δ; this is the sum of the regular shift at a rate $g(x)$ and the diffusional spread, given by the diffusion coefficient $b(x)$. This *a priori* process is defined by a Fokker-Planck equation

$$\dot{w} = \frac{1}{2}\frac{\partial^2}{\partial x^2}[b(x)w] - \frac{\partial}{\partial x}[g(x)w].$$

With a sufficiently small Δ we get from this

$$w_k(x) = \int w_\Delta(x, x')\, w_{k-1}(x')\, dx' =$$
$$= w_{k-1}(x) + \frac{\Delta}{2}\frac{\partial^2}{\partial x^2}[b(x)w_{k-1}] - \Delta\frac{\partial}{\partial x}[g(x)w_{k-1}]. \tag{27}$$

Thus integration with the function $w_\Delta(x, x')$ can be replaced by a differential operation. Correspondingly, transformation (26) is replaced by:

$$u(t) = h(t) + \frac{\Delta}{2}\left(\frac{dt}{dx}\right)^2 \frac{d^2}{dt^2}[b(\tilde{x}(t))w] - \Delta\frac{dt}{dx}\frac{d}{dt}[g(\tilde{x}(t))w], \quad (28)$$

where dx/dt is the velocity of scanning. In our example, where x is the phase of the oscillations, we may assume that b is constant and $g = 0$. Then transformation (28) takes the form

$$u(t) = h(t) + \frac{1}{2}b\Delta\left(\frac{\Delta}{2\pi}\right)^2\frac{d^2h(t)}{dt^2}. \quad (29)$$

The other blocks in the diagram are the same as before.

The function $f(t)$ at the output of this system is proportional to the *a posteriori* probability, and carries the greatest possible amount of information on the value of x. The solution corresponding to it may be accepted.

3. Differential equations of optimal filter discrimination

Let us turn from relation (21) to the differential equation which is satisfied by the *a priori* distribution density. We shall bear in mind (27), and we shall expand the exponent in (21) in a Taylor series, and write the constant C' in the form $1 - \lambda\Delta + \Delta^2 \ldots$ Then, subtracting $w_k(x_{k-1})$ from both sides of (21), dividing by Δ and passing to the limit $\Delta \to 0$, we obtain

$$\dot{w}(x) = \frac{1}{2}\frac{\partial^2(bw)}{\partial x^2} - \frac{\partial(gw)}{\partial x} + [F(x, t) - \lambda]w. \quad (30)$$

In order to find the constant λ not dependent on x, we must integrate (3) with respect to x taking into account the fact that by virtue of the condition of normalization

$$\int \dot{w}dx = \frac{d}{dt}\int wdx = 0.$$

This gives

$$\lambda = \overline{F(x, t)} \equiv \int F(x, t)\,w(x)\,dx. \quad (31)$$

For the example already examined, the function $F(x, t)$ has the form

$$F(x, t) = \frac{A_0}{N} y(t) \sin(\omega_0 t + x) \tag{32}$$

and furthermore, equation (30) is somewhat simplified:

$$\dot{w}(x) = \frac{b}{2} \frac{\partial^2 w}{\partial x^2} + [F(x, t) - \overline{F}] w. \tag{33}$$

Instead of a differential equation of the type (30), (33), we can consider a set of differential equations with numerical parameters, which replace the probability density. We can take the semi-invariants of the *a posteriori* distribution for these parameters. Choosing a rather different way, we shall examine the differential equations for the parameters x_0, a_2, a_3, ...) which are determined by the equation

$$w(x, t) = \exp\left\{-\frac{C_0}{\mu^2} - \frac{1}{\mu^2} \sum_{j=2}^{\infty} \frac{1}{j!} a_j(t) [x - x_0(t)]^j\right\}. \tag{34}$$

Here C_0 is connected with the condition of normalization, and plays no further part. The parameter $x_0(t)$ is the co-ordinate corresponding to the maximum *a priori* probability. The quantity a_2/μ^2 determines the statistical dispersion or *a posteriori* error of the Δx co-ordinate:

$$\Delta x \sim \frac{\mu}{\sqrt{a_2}}. \tag{35}$$

The parameter μ is introduced simply for convenience of presentation. In another variant it can be assumed equal to unity. We shall select μ to equal a certain mean error $\overline{\Delta x}$ corresponding to stationary conditions. Then, in accordance with (35)

$$a_2 \sim 1. \tag{36}$$

Substituting (34) in (30), we can obtain differential equations defining the variation of the parameters x_0, a_2, a_3, ... If we were to succeed in creating a system to model these differential equations, then, taking $x_0(t)$ as the output signal, we would have an optimal non-linear system processing the input signal $y(t)$. However, since in practice it is impossible to model an infinite set of differential equations, we are faced with the problem of replacing this set by a finite one, by rejecting a number of irrelevant terms. An asymptotic method can be worked out for the solution of this problem, in which μ will play the part of a small parameter. The condition for applicability of this method can be formulated as the condition of relatively small *a posteriori* uncertainty.

We shall demonstrate this asymptotic method, using equation (33) as an example. Substituting $w = e^{-v/\mu^2}$ in (33), we find

$$\dot{v} = \frac{b}{2}\left[\frac{\partial^2 v}{\partial x^2} - \frac{1}{\mu^2}\left(\frac{\partial v}{\partial x}\right)^2\right] - \mu^2 F(x, t) + \mu^2 \overline{F}. \tag{37}$$

If here, in accordance with (34), we assume

$$v = C_0 + \sum_{j=2}^{\infty} \frac{1}{j!} a_j (x - x_0)^j,$$

write $F(x, t)$ in the form

$$F(x, t) = \sum_{l=0}^{\infty} \frac{1}{l!} F_l (x - x_0)^l \qquad \left(F_l = \left[\frac{\partial^l F}{\partial x^l}\right]_{x=x_0}\right) \tag{38}$$

and equate the terms with identical powers of the difference $x - x_0$, we shall obtain a chain of equations

$$-a_2 \dot{x}_0 = \frac{b}{2} a_3 - \mu^2 F_1; \tag{39}$$

$$\dot{a}_2 - a_3 \dot{x}_0 = \frac{b}{2\mu^2}[\mu^2 a_4 - 2a_2^2] - \mu^2 F_2; \tag{40}$$

$$\dot{a}_3 - a_4 \dot{x}_0 = \frac{b}{2\mu^2}[\mu^2 a_5 - 6a_2 a_3] - \mu^2 F_3; \tag{41}$$

. . . .

$$\dot{a}_n - a_{n+1}\dot{x}_0 = \frac{b}{2\mu^2}\left[\mu^2 a_{n+2} - n a_2 a_n - \binom{n}{2} a_3 a_{n-1} - \dots - n a_n a_2\right] - \mu^2 F_n, \tag{42}$$

. . . .

which is the mathematical equivalent of (33).

4. Breaking the chain of equations

The chain of equations (39) - (40) is broken if the terms with a_{n+1} and a_{n+2} are rejected from (42). To verify the possibility of this, we shall evaluate different terms of these equations with regard to the order of their magnitudes.

Let x_1 be the unit of variation rate of the function $f(x)$; then

$$F_l \sim \frac{F_2}{x_1^{l-2}} \quad (l = 3, 4, \ldots) \tag{43}$$

(in (32) we have $x_1 \sim 1$). Substituting this value in (41) and (42), and bearing in mind the fact that the equations thus obtained have identical structure, being distinguished only by the factor before F_2, we conclude that

$$a_l \sim \frac{a_2}{x_1^{l-2}} \tag{44}$$

or $a_l \sim x_1^{-l+2}$ by virtue of (36). As the basis for breaking the equation we shall take the condition

$$\mu \ll x_1 \quad \text{or} \quad \Delta x \ll x_1, \tag{45}$$

which we assume to be satisfied. As investigation shows, for the case in (32) the stationary *a posteriori* dispersion is expressed in terms of the initial parameters by the formula

$$\Delta x \sim \left(\frac{Nb}{A_0^2}\right)^{1/4}. \tag{46}$$

Thus it is convenient to suppose

$$\mu = \left(\frac{Nb}{A_0^2}\right)^{1/4}. \tag{47}$$

In this case condition (45) takes the form

$$\sqrt{Nb} \ll A_0. \tag{48}$$

Taking into account (44) and (36), we come to the conclusion that the term $\mu^2 a_{n+2}$ in (42) is smaller than the terms of the type $a_2 a_n$, $a_3 a_{n-1}, \ldots$ by a factor of about x_1^2/μ^2. In evaluating the term $a_{n+1}\dot{x}_0$ we shall take into account not its instantaneous value, but its integral operation over the time interval $T_1 = \mu^2/b$. In fact, equation (42), which we may write

$$\dot{a}_n + \frac{nb}{\mu^2} a_2 a_n = a_{n+1}\dot{x}_0 + \ldots,$$

has a solution

$$a_n^{\cdot}(t) = \int_{-\infty}^{t} \exp\left\{-\frac{nb}{\mu^2}\int_{t'}^{t} a_2 dt''\right\} [a_{n+1}\dot{x}_0 + \ldots] \, dt', \tag{49}$$

from which it is evident that the right side terms have only an integral effect on a_n, while the time constant of integration equals $\mu^2/nba_2 \sim \mu^2/b$. In evaluating with regard to order of magnitude, we can replace an integral of $a_{n+1}\dot{x}_0$ as in (49) by the expression

$$\int_{t-T_1}^{t} a_{n+1}\dot{x}_0 dt' \sim \varepsilon_{n+1} [x_0(t) - x_0(t - T_1)].$$

Thus the term $a_{n+1}\dot{x}_0$ has, in order of magnitude, the same integral operation as a_{n+1}

$$a_{n+1} \frac{x_0(t) - x_0(t - T_1)}{T_1}.$$

If the *a posteriori* error is not too great, the increment $a_0(t) - x_0(t - T)$ has the same order of magnitude as the *a priori* increment described by a purely diffusional law. Its dispersion equals bT_1, hence

$$a_{n+1} \frac{x_0(t) - x_0(t - T_1)}{T_1} \sim a_{n+1} \sqrt{\frac{b}{T_1}}.$$

Since $T_1 = \mu^2/b$ we have

$$a_{n+1} \frac{x_0(t) - x_0(t - T_1)}{T_1} \sim a_{n+1} \frac{b}{\mu} \left(\mu a_{n+1} \sim \frac{\mu}{x_1} a_n\right). \tag{50}$$

Consequently we may take it that the term $a_{n+1}\dot{x}_0$ is x_1/μ times greater than

$$\frac{b}{2\mu^2} \mu^2 a_{n+2}$$

and smaller by the same factor than terms of the type

$$\frac{b}{2\mu^2} a_2 a_n, \ldots,$$

i.e. its order of smallness is μ.

Condition (45) permits us to reject terms with a_{n+1} and a_{n+2} in (42). In this case a_n will have a relative error of the order of μ/x_1. In the previous equation

$$\dot{a}_{n-1} - a_n\dot{x}_0 = \frac{b}{2\mu^2}[\mu^2 a_{n+1} - (n-1) a_2 a_{n-1} - \ldots] - \mu^2 F_{n-1},$$

in order to determine a_{n-1}, we shall reject a_{n+1} and retain $a_n x_0$. Then a_{n-1} acquires an error of the order of $(\mu/x_1)^2$. In the equation for a_{n-2} we shall retain all the terms, but a_{n-2} will have a relative error of the order of $(\mu/x_1)^3$ resulting from the error of the functions a_{n-1}, a_n. Continuing the reasoning, we find that a_3 has a relative error of the order of $(\mu/x_1)^{n-2}$, and a_2 of the order of $(\nu x_1)^{n-1}$. Thus, if we are considering a finite set of n equations, the greater is n, the greater will be the accuracy of determination of the function $x_0(t)$.

This approach can be extended to the case of the more general equation (30). In this case (45) must be supplemented by a further condition

$$\mu \ll x_2, \tag{51}$$

where x_2 is the width unit of the *a priori* statistical dispersion in stationary conditions:

$$\frac{\partial}{\partial x}(bw_{st}) \sim \frac{bw_{st}}{x_2}, \text{ т. е. } x_2 \sim \frac{b}{g}. \tag{52}$$

Conditions (45) and (51) are satisfied when the function $F(x, t)$ is relatively large (the noise is small), or when b is relatively small (*a priori* instability small). Previously introducing the small parameter, we may conveniently assume

$$F(x, t) = \frac{1}{\mu^2} F^0(x, t); \; b(x) = \mu^2 b^0(x), \tag{53}$$

where $F^0(x, t)$, $b^0(x)$ are functions not containing the small parameter.* If we turn to F^0, b^0 in (30), and substitute (34), we shall obtain a set of equations in which the value of each term is determined by the power of the parameter μ in that term (assuming $x_2 \sim x_1$). The only exception is the combination $a_{n+1}(x_0 - g(x_0))$ to which an evaluation of type (50) is applicable, and which therefore has the order μ. Having assigned a definite degree of accuracy, it only remains to reject from the equations specified terms of a higher order of smallness.

* If another method is used, introduction of the small parameter would necessitate alteration of the time scale.

5. Equations of optimal filter discrimination in a Gaussian approximation

In the case of normal *a posteriori* distribution, all the coefficients in (34) except the first equal zero: $a_3 = a_4 = \ldots = 0$. If the distribution density is non-Gaussian, its degree of proximity to the Gaussian density can be evaluated from the relations of γ_n derivatives

$$\left[\frac{d^n \ln w}{dx^n}\right]_{x=x_0} = \frac{a_n}{\mu^2} \; (n = 3, 4, \ldots)$$

to the second derivative

$$-\frac{d^2 \ln w}{dx^2} = \frac{a_2}{\mu^2},$$

taken to the power $n/2$. Taking into account (36), (44), we obtain for these relations

$$\gamma_n = \frac{a_n}{a_2^{n/2}} \mu^{n-2} \sim \left(\frac{\mu}{x_1}\right)^{n-2} \quad (n = 3, 4, \ldots). \tag{54}$$

These correlations show that the distribution curve approaches the Gaussian with decrease of the *a posteriori* dispersion $\Delta x \sim \mu$ (condition (45)). By Gaussian approximation we mean the very simple case where, in equations (39) - (42), the functions $a_3, a_4, \ldots$ are assumed to equal zero. In this case we have a set of two equations:

$$\begin{aligned} a_2 \dot{x}_0 &= \mu^2 F_1; \\ \dot{a}_2 + \frac{b}{\mu^2} a_2^2 &= -\mu^2 F_2. \end{aligned} \tag{55}$$

The function a_2 is here determined with a relative error of the order of μ^2. In the first equation terms of order μ^2 are not taken into account; in the second equation, terms of order μ are also rejected; (we must remember that b/μ^2, $\mu^2 F_2$ have, by virtue of (53), the factual orders b^0, F_2^0, x_1^2, and $a_2 x_0$ has the order μ).

In this approximation, the equations corresponding to the more general equation (30) are

$$\begin{aligned} a_2 \dot{x}_0 &= g(x_0)\, a_2 + \mu^2 \frac{\partial F}{\partial x}(x_0), \\ \dot{a}_2 &= -2 \frac{\partial g}{\partial x}(x_0)\, a_2 - \frac{1}{\mu^2} b(x_0)\, a_2^2 - \mu^2 \frac{\partial^2 F}{\partial x^2}(x_0). \end{aligned} \tag{56}$$

Finally, let us consider the result of a multi-dimensional generalization of the theory we have expounded. Let an *a posteriori* process consisting of components x_a ($a = 1, \ldots, r$), be described by an equation [5]

$$\dot{w}(\vec{x}) = \frac{1}{2}\frac{\partial^2}{\partial x_\alpha \partial x_\beta}[b_{\alpha\beta} w] - \frac{\partial}{\partial x_\alpha}[g_\alpha w] + [F(\vec{x}, t) - \bar{F}]\, w. \tag{57}$$

Here summation with respect to the twice-occurring indices is implied. Substituting

$$\vec{w}(x) = \exp\left\{-\frac{C^0}{\mu^2} - \frac{1}{2\mu^2}\sum_{\alpha,\beta}{}' a_{\alpha\beta}(x_\alpha - x_\alpha^0)(x_\beta - x_\beta^0)\right\} \tag{58}$$

and rejecting terms of the same order of smallness as in (56), we obtain a set of equations in Gaussian approximation

$$\begin{aligned} a_{\alpha\beta}\dot{x}_\beta^0 &= a_{\alpha\beta} g_\beta(\vec{x}^0) + \mu^2 \frac{\partial F}{\partial x_\alpha}(x^0); \\ \dot{a}_{\alpha\beta} &= -\, a_{\alpha\gamma}\frac{\partial g_\gamma}{\partial x_\beta} - \frac{\partial g_\gamma}{\partial x_\alpha} a_{\gamma\beta} - \frac{1}{\mu^2} a_{\alpha\gamma} b_{\gamma\delta} a_{\delta\beta} - \mu^2 \frac{\partial^2 F}{\partial x_\alpha \partial x_\beta}. \end{aligned} \tag{59}$$

In the special case when $b_{\alpha\beta}$ is not dependent on x, when g_σ is a linear function of $\vec{x}$, and $F(\vec{x}, t)$ is, composed of a constant, a linear and a quadratic part, the *a posteriori* distribution is strictly Gaussian, and equations (59) are exact. Here in the stationary case, they are equivalent to the equations of optimal linear filtration derived from the Wiener-Kolmogorov theory [1-2]. In the more general case, they represent a generalization of this theory. The condition of their applicability is that of great *a posteriori* certainty; this is satisfied in most concrete problems, since reliable operation of the device is required in them.

The functions $x_0(t)$, $x_a^0(t)$ which are directly reproduced in a system modelling equations (55), (56) and (59) represent that same useful signal which was sought, filtered out of noise by optimal non-linear means.

Translated by J. Stuart

REFERENCES

1. A.N. Kolmogorov; *Izv. Akad. Nauk SSSR, Ser. Mat.* 5, (1941).
2. N. Wiener; *Extrapolation, Interpolation and Smoothing of Stationary Time Series.* John Wiley (1949).
3. F. Woodward and I. Davies; The principle of inverse probability in the theory of signal transmission. Included in a collection ed. N.A. Zheleznov, "*Teoriya peredachi*

elektricheskikh signalov pri nalichii pomekh", IL., (1953).
4. R.L. Stratonovich; *Teoriya verovatnosti i eyo primene-niya*, 4, 2, 239 (1959).
5. R.L. Stratonovich; *Teoriya veroyatnosti i eyo primeneniya*, 5, 2, 172 (1960).
6. R.L. Stratonovich; *Izv. vuzov MVO (Radiofizika)*, 2, 6, 892 (1959).

16

Reprinted from *J. Basic Eng.* **82**:35–45 (Mar. 1960)

A New Approach to Linear Filtering and Prediction Problems[1]

R. E. KALMAN
Research Institute for Advanced Study,[2]
Baltimore, Md.

The classical filtering and prediction problem is re-examined using the Bode-Shannon representation of random processes and the "state transition" method of analysis of dynamic systems. New results are:

(1) The formulation and methods of solution of the problem apply without modification to stationary and nonstationary statistics and to growing-memory and infinite-memory filters.

(2) A nonlinear difference (or differential) equation is derived for the covariance matrix of the optimal estimation error. From the solution of this equation the coefficients of the difference (or differential) equation of the optimal linear filter are obtained without further calculations.

(3) The filtering problem is shown to be the dual of the noise-free regulator problem.

The new method developed here is applied to two well-known problems, confirming and extending earlier results.

The discussion is largely self-contained and proceeds from first principles; basic concepts of the theory of random processes are reviewed in the Appendix.

Introduction

An important class of theoretical and practical problems in communication and control is of a statistical nature. Such problems are: (i) Prediction of random signals; (ii) separation of random signals from random noise; (iii) detection of signals of known form (pulses, sinusoids) in the presence of random noise.

In his pioneering work, Wiener [1][3] showed that problems (i) and (ii) lead to the so-called Wiener-Hopf integral equation; he also gave a method (spectral factorization) for the solution of this integral equation in the practically important special case of stationary statistics and rational spectra.

Many extensions and generalizations followed Wiener's basic work. Zadeh and Ragazzini solved the finite-memory case [2]. Concurrently and independently of Bode and Shannon [3], they also gave a simplified method [2] of solution. Booton discussed the nonstationary Wiener-Hopf equation [4]. These results are now in standard texts [5-6]. A somewhat different approach along these main lines has been given recently by Darlington [7]. For extensions to sampled signals, see, e.g., Franklin [8], Lees [9]. Another approach based on the eigenfunctions of the Wiener-Hopf equation (which applies also to nonstationary problems whereas the preceding methods in general don't), has been pioneered by Davis [10] and applied by many others, e.g., Shinbrot [11], Blum [12], Pugachev [13], Solodovnikov [14].

In all these works, the objective is to obtain the specification of a linear dynamic system (Wiener filter) which accomplishes the prediction, separation, or detection of a random signal.[4]

Present methods for solving the Wiener problem are subject to a number of limitations which seriously curtail their practical usefulness:

(1) The optimal filter is specified by its impulse response. It is not a simple task to synthesize the filter from such data.

(2) Numerical determination of the optimal impulse response is often quite involved and poorly suited to machine computation. The situation gets rapidly worse with increasing complexity of the problem.

(3) Important generalizations (e.g., growing-memory filters, nonstationary prediction) require new derivations, frequently of considerable difficulty to the nonspecialist.

(4) The mathematics of the derivations are not transparent. Fundamental assumptions and their consequences tend to be obscured.

This paper introduces a new look at this whole assemblage of problems, sidestepping the difficulties just mentioned. The following are the highlights of the paper:

(5) *Optimal Estimates and Orthogonal Projections.* The Wiener problem is approached from the point of view of conditional distributions and expectations. In this way, basic facts of the Wiener theory are quickly obtained; the scope of the results and the fundamental assumptions appear clearly. It is seen that all statistical calculations and results are based on first and second order averages; no other statistical data are needed. Thus difficulty (4) is eliminated. This method is well known in probability theory (see pp. 75–78 and 148–155 of Doob [15] and pp. 455–464 of Loève [16]) but has not yet been used extensively in engineering.

(6) *Models for Random Processes.* Following, in particular, Bode and Shannon [3], arbitrary random signals are represented (up to second order average statistical properties) as the output of a linear dynamic system excited by independent or uncorrelated random signals ("white noise"). This is a standard trick in the engineering applications of the Wiener theory [2–7]. The approach taken here differs from the conventional one only in the way in which linear dynamic systems are described. We shall emphasize the concepts of *state* and *state transition;* in other words, linear systems will be specified by systems of first-order difference (or differential) equations. This point of view is

---

[1] This research was supported in part by the U. S. Air Force Office of Scientific Research under Contract AF 49 (638)-382.

[2] 7212 Bellona Ave.

[3] Numbers in brackets designate References at end of paper.

[4] Of course, in general these tasks may be done better by nonlinear filters. At present, however, little or nothing is known about how to obtain (both theoretically and practically) these nonlinear filters.

Contributed by the Instruments and Regulators Division and presented at the Instruments and Regulators Conference, March 29–April 2, 1959, of The American Society of Mechanical Engineers.

Note: Statements and opinions advanced in papers are to be understood as individual expressions of their authors and not those of the Society. Manuscript received at ASME Headquarters, February 24, 1959. Paper No. 59—IRD-11.

natural and also necessary in order to take advantage of the simplifications mentioned under (5).

(7) *Solution of the Wiener Problem.* With the state-transition method, a single derivation covers a large variety of problems: growing and infinite memory filters, stationary and nonstationary statistics, etc.; difficulty (3) disappears. Having guessed the "state" of the estimation (i.e., filtering or prediction) problem correctly, one is led to a nonlinear difference (or differential) equation for the covariance matrix of the optimal estimation error. This is vaguely analogous to the Wiener-Hopf equation. Solution of the equation for the covariance matrix starts at the time t_0 when the first observation is taken; at each later time t the solution of the equation represents the covariance of the optimal prediction error given observations in the interval (t_0, t). From the covariance matrix at time t we obtain at once, without further calculations, the coefficients (in general, time-varying) characterizing the optimal linear filter.

(8) *The Dual Problem.* The new formulation of the Wiener problem brings it into contact with the growing new theory of control systems based on the "state" point of view [17–24]. It turns out, *surprisingly*, that the Wiener problem is the *dual* of the noise-free optimal regulator problem, which has been solved previously by the author, using the state-transition method to great advantage [18, 23, 24]. The mathematical background of the two problems is identical—this has been suspected all along, but until now the analogies have never been made explicit.

(9) *Applications.* The power of the new method is most apparent in theoretical investigations and in numerical answers to complex practical problems. In the latter case, it is best to resort to machine computation. Examples of this type will be discussed later. To provide some feel for applications, two standard examples from nonstationary prediction are included; in these cases the solution of the nonlinear difference equation mentioned under (7) above can be obtained even in closed form.

For easy reference, the main results are displayed in the form of theorems. Only Theorems 3 and 4 are original. The next section and the Appendix serve mainly to review well-known material in a form suitable for the present purposes.

Notation Conventions

Throughout the paper, we shall deal mainly with *discrete* (or *sampled*) dynamic systems; in other words, signals will be observed at equally spaced points in time (*sampling instants*). By suitable choice of the time scale, the constant intervals between successive sampling instants (*sampling periods*) may be chosen as unity. Thus variables referring to time, such as t, t_0, τ, T will always be integers. The restriction to discrete dynamic systems is not at all essential (at least from the engineering point of view); by using the discreteness, however, we can keep the mathematics rigorous and yet elementary. Vectors will be denoted by small bold-face letters: $\mathbf{a}, \mathbf{b}, \ldots, \mathbf{u}, \mathbf{x}, \mathbf{y}, \ldots$ A *vector* or more precisely an *n-vector* is a set of n numbers $x_1, \ldots x_n$; the x_i are the *co-ordinates* or *components* of the vector $\mathbf{x}$.

Matrices will be denoted by capital bold-face letters: $\mathbf{A}, \mathbf{B}, \mathbf{Q}, \mathbf{\Phi}, \mathbf{\Psi}, \ldots$; they are $m \times n$ arrays of elements $a_{ij}, b_{ij}, q_{ij}, \ldots$ The *transpose* (interchanging rows and columns) of a matrix will be denoted by the prime. In manipulating formulas, it will be convenient to regard a vector as a matrix with a single column.

Using the conventional definition of matrix multiplication, we write the *scalar product* of two *n*-vectors $\mathbf{x}, \mathbf{y}$ as

$$\mathbf{x}'\mathbf{y} = \sum_{i=1}^{n} x_i y_i = \mathbf{y}'\mathbf{x}$$

The scalar product is clearly a scalar, i.e., not a vector, quantity. Similarly, the *quadratic form* associated with the $n \times n$ matrix $\mathbf{Q}$ is

$$\mathbf{x}'\mathbf{Q}\mathbf{x} = \sum_{i,j=1}^{n} x_i q_{ij} x_j$$

We define the expression $\mathbf{x}\mathbf{y}'$ where $\mathbf{x}$ is an m-vector and $\mathbf{y}$ is an n-vector to be the $m \times n$ matrix with elements $x_i y_j$.

We write $E(\mathbf{x}) = E\mathbf{x}$ for the expected value of the random vector $\mathbf{x}$ (see Appendix). It is usually convenient to omit the brackets after E. This does not result in confusion in simple cases since constants and the operator E commute. Thus $E\mathbf{x}\mathbf{y}'$ = matrix with elements $E(x_i y_j)$; $E\mathbf{x}E\mathbf{y}'$ = matrix with elements $E(x_i)E(y_j)$.

For ease of reference, a list of the principal symbols used is given below.

Optimal Estimates

- t time in general; present time.
- t_0 time at which observations start.
- $x_1(t)$, $x_2(t)$ basic random variables.
- $y(t)$ observed random variable.
- $x_1^*(t_1|t)$ optimal estimate of $x_1(t_1)$ given $y(t_0), \ldots, y(t)$.
- L loss function (nonrandom function of its argument).
- ϵ estimation error (random variable).

Orthogonal Projections

- $\mathcal{Y}(t)$ linear manifold generated by the random variables $y(t_0), \ldots, y(t)$.
- $\bar{x}(t_1|t)$ orthogonal projection of $x(t_1)$ on $\mathcal{Y}(t)$.
- $\tilde{x}(t_1|t)$ component of $x(t_1)$ orthogonal to $\mathcal{Y}(t)$.

Models for Random Processes

- $\mathbf{\Phi}(t+1;\ t)$ transition matrix
- $\mathbf{Q}(t)$ covariance of random excitation

Solution of the Wiener Problem

- $\mathbf{x}(t)$ basic random variable.
- $\mathbf{y}(t)$ observed random variable.
- $\mathcal{Y}(t)$ linear manifold generated by $\mathbf{y}(t_0), \ldots, \mathbf{y}(t)$.
- $\mathcal{Z}(t)$ linear manifold generated by $\tilde{\mathbf{y}}(t|t-1)$.
- $\mathbf{x}^*(t_1|t)$ optimal estimate of $\mathbf{x}(t_1)$ given $\mathcal{Y}(t)$.
- $\tilde{\mathbf{x}}(t_1|t)$ error in optimal estimate of $\mathbf{x}(t_1)$ given $\mathcal{Y}(t)$.

Optimal Estimates

To have a concrete description of the type of problems to be studied, consider the following situation. We are given signal $x_1(t)$ and noise $x_2(t)$. Only the sum $y(t) = x_1(t) + x_2(t)$ can be observed. Suppose we have observed and know exactly the values of $y(t_0), \ldots, y(t)$. What can we infer from this knowledge in regard to the (unobservable) value of the signal at $t = t_1$, where t_1 may be less than, equal to, or greater than t? If $t_1 < t$, this is a *data-smoothing* (*interpolation*) problem. If $t_1 = t$, this is called *filtering*. If $t_1 > t$, we have a *prediction* problem. Since our treatment will be general enough to include these and similar problems, we shall use hereafter the collective term *estimation*.

As was pointed out by Wiener [1], the natural setting of the estimation problem belongs to the realm of probability theory and statistics. Thus signal, noise, and their sum will be random variables, and consequently they may be regarded as random processes. From the probabilistic description of the random processes we can determine the probability with which a particular sample of the signal and noise will occur. For any given set of measured values $\eta(t_0), \ldots, \eta(t)$ of the random variable $y(t)$ one can then also determine, in principle, the probability of simultaneous occurrence of various values $\xi_1(t)$ of the random variable $x_1(t_1)$. This is the conditional probability distribution function

$$Pr[x_1(t_1) \leqq \xi_1 | y(t_0) = \eta(t_0), \ldots, y(t) = \eta(t)] = F(\xi_1) \quad (1)$$

Evidently, $F(\xi_1)$ represents all the information which the measurement of the random variables $y(t_0), \ldots, y(t)$ has conveyed about the random variable $x_1(t_1)$. Any statistical estimate of the random variable $x_1(t_1)$ will be some function of this distribution and therefore a (nonrandom) function of the random variables $y(t_0), \ldots, y(t)$. This statistical estimate is denoted by $X_1(t_1|t)$, or by just $X_1(t_1)$ or X_1 when the set of observed random variables or the time at which the estimate is required are clear from context.

Suppose now that X_1 is given as a fixed function of the random variables $y(t_0), \ldots, y(t)$. Then X_1 is itself a random variable and its actual value is known whenever the actual values of $y(t_0), \ldots, y(t)$ are known. In general, the actual value of $X_1(t_1)$ will be different from the (unknown) actual value of $x_1(t_1)$. To arrive at a rational way of determining X_1, it is natural to assign a *penalty* or *loss* for incorrect estimates. Clearly, the loss should be a (i) positive, (ii) nondecreasing function of the *estimation error* $\epsilon = x_1(t_1) - X_1(t_1)$. Thus we define a *loss function* by

$$L(0) = 0$$

$$L(\epsilon_2) \geqq L(\epsilon_1) \geqq 0 \quad \text{when} \quad \epsilon_2 \geqq \epsilon_1 \geqq 0 \quad (2)$$

$$L(\epsilon) = L(-\epsilon)$$

Some common examples of loss functions are: $L(\epsilon) = a\epsilon^2, a\epsilon^4, a|\epsilon|, a[1 - \exp(-\epsilon^2)]$, etc., where a is a positive constant.

One (but by no means the only) natural way of choosing the random variable X_1 is to require that this choice should minimize the average loss or risk

$$E\{L[x_1(t_1) - X_1(t_1)]\} = E[E\{L[x(t_1) - X_1(t_1)]|y(t_0), \ldots, y(t)\}] \quad (3)$$

Since the first expectation on the right-hand side of (3) does not depend on the choice of X_1 but only on $y(t_0), \ldots, y(t)$, it is clear that minimizing (3) is equivalent to minimizing

$$E\{L[x_1(t_1) - X_1(t_1)]|y(t_0), \ldots, y(t)\} \quad (4)$$

Under just slight additional assumptions, optimal estimates can be characterized in a simple way.

Theorem 1. *Assume that L is of type* (2) *and that the conditional distribution function* $F(\xi)$ *defined by* (1) *is:*

(*A*) *symmetric about the mean* $\bar{\xi}$:

$$F(\xi - \bar{\xi}) = 1 - F(\bar{\xi} - \xi)$$

(*B*) *convex for* $\xi \leqq \bar{\xi}$:

$$F(\lambda\xi_1 + (1 - \lambda)\xi_2) \leqq \lambda F(\xi_1) + (1 - \lambda)F(\xi_2)$$

for all $\xi_1, \xi_2 \leqq \bar{\xi}$ *and* $0 \leqq \lambda \leqq 1$

Then the random variable $x_1^*(t_1|t)$ *which minimizes the average loss* (3) *is the conditional expectation*

$$x_1^*(t_1|t) = E[x_1(t_1)|y(t_0), \ldots, y(t)] \quad (5)$$

Proof: As pointed out recently by Sherman [25], this theorem follows immediately from a well-known lemma in probability theory.

Corollary. *If the random processes* $\{x_1(t)\}$, $\{x_2(t)\}$, *and* $\{y(t)\}$ *are gaussian, Theorem 1 holds.*

Proof: By Theorem 5, (*A*) (see Appendix), conditional distributions on a gaussian random process are gaussian. Hence the requirements of Theorem 1 are always satisfied.

In the control systems literature, this theorem appears sometimes in a form which is more restrictive in one way and more general in another way:

Theorem 1-a. *If* $L(\epsilon) = \epsilon^2$, *then Theorem 1 is true without assumptions* (*A*) *and* (*B*).

Proof: Expand the conditional expectation (4):

$$E[x_1^2(t_1)|y(t_0), \ldots, y(t)] - 2X_1(t_1)E[x_1(t_1)|y(t_0), \ldots, y(t)] + X_1^2(t_1)$$

and differentiate with respect to $X_1(t_1)$. This is not a completely rigorous argument; for a simple rigorous proof see Doob [15], pp. 77–78.

Remarks: (*a*) As far as the author is aware, it is not known what is the most general class of random processes $\{x_1(t)\}$, $\{x_2(t)\}$ for which the conditional distribution function satisfies the requirements of Theorem 1.

(*b*) Aside from the note of Sherman, Theorem 1 apparently has never been stated explicitly in the control systems literature. In fact, one finds many statements to the effect that loss functions of the general type (2) cannot be conveniently handled mathematically.

(*c*) In the sequel, we shall be dealing mainly with vector-valued random variables. In that case, the estimation problem is stated as: Given a vector-valued random process $\{\mathbf{x}(t)\}$ and observed random variables $\mathbf{y}(t_0), \ldots, \mathbf{y}(t)$, where $\mathbf{y}(t) = \mathbf{M}\mathbf{x}(t)$ ($\mathbf{M}$ being a singular matrix; in other words, not all co-ordinates of $\mathbf{x}(t)$ can be observed), find an estimate $\mathbf{X}(t_1)$ which minimizes the expected loss $E[L(\|\mathbf{x}(t_1) - \mathbf{X}(t_1)\|)]$, $\| \ \|$ being the norm of a vector.

Theorem 1 remains true in the vector case also, provided we require that the conditional distribution function of the n co-ordinates of the vector $\mathbf{x}(t_1)$,

$$Pr[x_1(t_1) \leqq \xi_1, \ldots, x_n(t_1) \leqq \xi_n | \mathbf{y}(t_0), \ldots, \mathbf{y}(t)] = F(\xi_1, \ldots, \xi_n)$$

be symmetric with respect to the n variables $\xi_1 - \bar{\xi}_1, \ldots, \xi_n - \bar{\xi}_n$ and convex in the region where all of these variables are negative.

Orthogonal Projections

The explicit calculation of the optimal estimate as a function of the observed variables is, in general, impossible. There is an important exception: The processes $\{x_1(t)\}$, $\{x_2(t)\}$ are gaussian.

On the other hand, if we attempt to get an optimal estimate under the restriction $L(\epsilon) = \epsilon^2$ and the additional requirement that the estimate be a linear function of the observed random variables, we get an estimate which is identical with the optimal estimate in the gaussian case, without the assumption of linearity or quadratic loss function. This shows that results obtainable by linear estimation can be bettered by nonlinear estimation only when (i) the random processes are nongaussian and even then (in view of Theorem 5, (*C*)) only (ii) by considering at least third-order probability distribution functions.

In the special cases just mentioned, the explicit solution of the estimation problem is most easily understood with the help of a geometric picture. This is the subject of the present section.

Consider the (real-valued) random variables $y(t_0), \ldots, y(t)$. The set of all linear combinations of these random variables with real coefficients

$$\sum_{i=t_0}^{t} a_i y(i) \quad (6)$$

forms a *vector space* (*linear manifold*) which we denote by $\mathcal{Y}(t)$. We regard, abstractly, any expression of the form (6) as "point" or "vector" in $\mathcal{Y}(t)$; this use of the word "vector" should not be confused, of course, with "vector-valued" random variables, etc. Since we do not want to fix the value of t (i.e., the total number of possible observations), $\mathcal{Y}(t)$ should be regarded as a finite-dimensional subspace of the space of all possible observations.

Given any two vectors u, v in $\mathcal{Y}(t)$ (i.e., random variables expressible in the form (6)), we say that u and v are *orthogonal* if $Euv = 0$. Using the Schmidt orthogonalization procedure, as described for instance by Doob [15], p. 151, or by Loève [16], p. 459, it is easy to select an *orthonormal basis* in $\mathcal{Y}(t)$. By this is meant a set of vectors $e_{t_0}, \ldots, e_t$ in $\mathcal{Y}(t)$ such that any vector in $\mathcal{Y}(t)$ can be expressed as a unique linear combination of $e_{t_0}, \ldots, e_t$ and

$$\left.\begin{aligned} Ee_ie_j &= \delta_{ij} = 1 \quad \text{if} \quad i = j \\ &= 0 \quad \text{if} \quad i \neq j \end{aligned}\right\} (i, j = t_0, \ldots, t) \qquad (7)$$

Thus any vector $\bar{x}$ in $\mathcal{Y}(t)$ is given by

$$\bar{x} = \sum_{i=t_0}^{t} a_i e_i$$

and so the coefficients a_i can be immediately determined with the aid of (7):

$$E\bar{x}e_j = E\left(\sum_{i=t_0}^{t} a_i e_i\right)e_j = \sum_{i=t_0}^{t} a_i Ee_ie_j = \sum_{i=t_0}^{t} a_i\delta_{ij} = a_j \qquad (8)$$

It follows further that any random variable x (not necessarily in $\mathcal{Y}(t)$) can be uniquely decomposed into two parts: a part $\bar{x}$ in $\mathcal{Y}(t)$ and a part $\tilde{x}$ orthogonal to $\mathcal{Y}(t)$ (i.e., orthogonal to every vector in $\mathcal{Y}(t)$). In fact, we can write

$$x = \bar{x} + \tilde{x} = \sum_{i=t_0}^{t} (Exe_i)e_i + \tilde{x} \qquad (9)$$

Thus $\bar{x}$ is uniquely determined by equation (9) and is obviously a vector in $\mathcal{Y}(t)$. Therefore $\tilde{x}$ is also uniquely determined; it remains to check that it is orthogonal to $\mathcal{Y}(t)$:

$$E\tilde{x}e_i = E(x - \bar{x})e_i = Exe_i - E\bar{x}e_i$$

Now the co-ordinates of $\bar{x}$ with respect to the basis $e_{t_0}, \ldots, e_t$ are given either in the form $E\bar{x}e_i$ (as in (8)) or in the form Exe_i (as in (9)). Since the co-ordinates are unique, $Exe_i = E\bar{x}e_i$ $(i = t_0, \ldots, t)$; hence $E\tilde{x}e_i = 0$ and $\tilde{x}$ is orthogonal to every base vector e_i and therefore to $\mathcal{Y}(t)$. We call $\bar{x}$ the *orthogonal projection* of x on $\mathcal{Y}(t)$.

There is another way in which the orthogonal projection can be characterized: $\bar{x}$ is that vector in $\mathcal{Y}(t)$ (i.e., that *linear* function of the random variables $y(t_0), \ldots, y(t)$) which minimizes the quadratic loss function. In fact, if w is any other vector in $\mathcal{Y}(t)$, we have

$$E(x - w)^2 = E(\tilde{x} + \bar{x} - w)^2 = E[(x - \bar{x}) + (\bar{x} - w)]^2$$

Since $\tilde{x}$ is orthogonal to every vector in $\mathcal{Y}(t)$ and in particular to $\bar{x} - w$ we have

$$E(x - w)^2 = E(x - \bar{x})^2 + E(\bar{x} - w)^2 \geqq E(x - \bar{x})^2 \qquad (10)$$

This shows that, if w also minimizes the quadratic loss, we must have $E(\bar{x} - w)^2 = 0$ which means that the random variables $\bar{x}$ and w are equal (except possibly for a set of events whose probability is zero).

These results may be summarized as follows:

Theorem 2. *Let $\{x(t)\}$, $\{y(t)\}$ random processes with zero mean (i.e., $Ex(t) = Ey(t) = 0$ for all t). We observe $y(t_0), \ldots, y(t)$.*

If either

(A) the random processes $\{x(t)\}$, $\{y(t)\}$ are gaussian; or

(B) the optimal estimate is restricted to be a linear function of the observed random variables and $L(\epsilon) = \epsilon^2$;

then

$$\begin{aligned} x^*(t_1|t) &= \textit{optimal estimate of } x(t_1) \textit{ given } y(t_0), \ldots, y(t) \\ &= \textit{orthogonal projection } \bar{x}(t_1|t) \textit{ of } x(t_1) \text{ on } \mathcal{Y}(t). \end{aligned} \qquad (11)$$

These results are well-known though not easily accessible in the control systems literature. See Doob [15], pp. 75–78, or Pugachev [26]. It is sometimes convenient to denote the orthogonal projection by

$$\bar{x}(t_1|t) \equiv x^*(t_1|t) = \hat{E}[x(t_1)|\mathcal{Y}(t)]$$

The notation $\hat{E}$ is motivated by part (*b*) of the theorem: If the stochastic processes in question are gaussian, then orthogonal projection is actually identical with conditional expectation.

Proof. (*A*) This is a direct consequence of the remarks in connection with (10).

(*B*) Since $x(t)$, $y(t)$ are random variables with zero mean, it is clear from formula (9) that the orthogonal part $\tilde{x}(t_1|t)$ of $x(t_1)$ with respect to the linear manifold $\mathcal{Y}(t)$ is also a random variable with zero mean. Orthogonal random variables with zero mean are uncorrelated; if they are also gaussian then (by Theorem 5 (*B*)) they are independent. Thus

$$\begin{aligned} 0 = E\tilde{x}(t_1|t) &= E[\tilde{x}(t_1|t)|y(t_0), \ldots, y(t)] \\ &= E[x(t_1) - \bar{x}(t_1|t)|y(t_0), \ldots, y(t)] \\ &= E[x(t_1)|y(t_0), \ldots, y(t)] - \bar{x}(t_1|t) = 0 \end{aligned}$$

Remarks. (*d*) A rigorous formulation of the contents of this section as $t \to \infty$ requires some elementary notions from the theory of Hilbert space. See Doob [15] and Loève [16].

(*e*) The physical interpretation of Theorem 2 is largely a matter of taste. If we are not worried about the assumption of gaussianness, part (*A*) shows that the orthogonal projection is the optimal estimate for all reasonable loss functions. If we do worry about gaussianness, even if we are resigned to consider only linear estimates, we know that orthogonal projections are *not* the optimal estimate for many reasonable loss functions. Since in practice it is difficult to ascertain to what degree of approximation a random process of physical origin is gaussian, it is hard to decide whether Theorem 2 has very broad or very limited significance.

(*f*) Theorem 2 is immediately generalized for the case of vector-valued random variables. In fact, we define the linear manifold $\mathcal{Y}(t)$ generated by $\mathbf{y}(t_0), \ldots, \mathbf{y}(t)$ to be the set of all linear combinations

$$\sum_{i=t_0}^{t} \sum_{j=1}^{m} a_{ij}y_j(i)$$

of all m co-ordinates of each of the random vectors $\mathbf{y}(t_0), \ldots, \mathbf{y}(t)$. The rest of the story proceeds as before.

(*g*) Theorem 2 states in effect that the optimal estimate under conditions (*A*) or (*B*) is a linear combination of all previous observations. In other words, the optimal estimate can be regarded as the output of a linear filter, with the input being the actually occurring values of the observable random variables; Theorem 2 gives a way of computing the impulse response of the optimal filter. As pointed out before, knowledge of this impulse response is not a complete solution of the problem; for this reason, no explicit formulas for the calculation of the impulse response will be given.

Models for Random Processes

In dealing with physical phenomena, it is not sufficient to give an empirical description but one must have also some idea of the underlying causes. Without being able to separate in some sense causes and effects, i.e., without the assumption of causality, one can hardly hope for useful results.

It is a fairly generally accepted fact that primary macroscopic sources of random phenomena are independent gaussian processes.[5] A well-known example is the noise voltage produced in a resistor due to thermal agitation. In most cases, *observed* random phenomena are not describable by independent random variables. The statistical dependence (correlation) between random signals observed at different times is usually explained by the presence of a dynamic system between the primary random source and the observer. *Thus a random function of time may be thought of as the output of a dynamic system excited by an independent gaussian random process.*

An important property of gaussian random signals is that they remain gaussian after passing through a linear system (Theorem 5(A)). Assuming independent gaussian primary random sources, if the observed random signal is also gaussian, we may assume that the dynamic system between the observer and the primary source is *linear*. This conclusion may be forced on us also because of lack of detailed knowledge of the statistical properties of the observed random signal: Given any random process with known first and second-order averages, we can find a gaussian random process with the same properties (Theorem 5(C)). Thus gaussian distributions and linear dynamics are natural, mutually plausible assumptions particularly when the statistical data are scant.

How is a dynamic system (linear or nonlinear) described? The fundamental concept is the notion of the *state*. By this is meant, intuitively, some quantitative information (a set of numbers, a function, etc.) which is the least amount of data one has to know about the past behavior of the system in order to predict its future behavior. The dynamics is then described in terms of *state transitions*, i.e., one must specify how one state is transformed into another as time passes.

A *linear* dynamic system may be described in general by the vector differential equation

$$\left.\begin{aligned} d\mathbf{x}/dt &= \mathbf{F}(t)\mathbf{x} + \mathbf{D}(t)\mathbf{u}(t) \\ \text{and} \qquad & \\ \mathbf{y}(t) &= \mathbf{M}(t)\mathbf{x}(t) \end{aligned}\right\} \qquad (12)$$

where $\mathbf{x}$ is an n-vector, the *state* of the system (the components x_i of $\mathbf{x}$ are called *state variables*); $\mathbf{u}(t)$ is an m-vector ($m \leq n$) representing the *inputs* to the system; $\mathbf{F}(t)$ and $\mathbf{D}(t)$ are $n \times n$, respectively, $n \times m$ matrices. If all coefficients of $\mathbf{F}(t)$, $\mathbf{D}(t)$, $\mathbf{M}(t)$ are constants, we say that the dynamic system (12) is *time-invariant* or *stationary*. Finally, $\mathbf{y}(t)$ is a p-vector denoting the outputs of the system; $\mathbf{M}(t)$ is an $n \times p$ matrix; $p \leq n$.

The physical interpretation of (12) has been discussed in detail elsewhere [18, 20, 23]. A look at the block diagram in Fig. 1 may be helpful. This is not an ordinary but a matrix block diagram (as revealed by the fat lines indicating signal flow). The integrator in Fig. 1 actually stands for n integrators such that the output of each is a state variable; $\mathbf{F}(t)$ indicates how the outputs of the integrators are fed back to the inputs of the integrators. Thus $f_{ij}(t)$ is the coefficient with which the output of the jth integrator is fed back to the input of the ith integrator. It is not hard to relate this formalism to more conventional methods of linear system analysis.

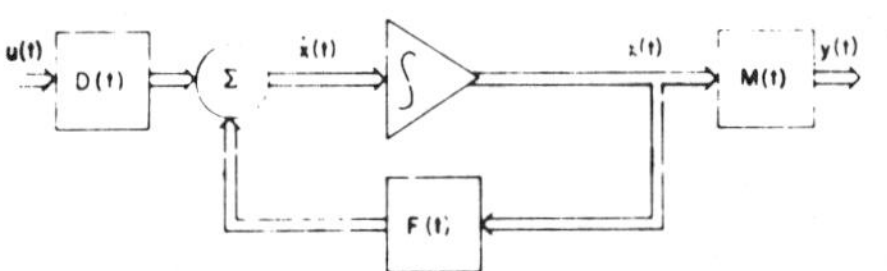

Fig. 1 Matrix block diagram of the general linear continuous-dynamic system

[5] The probability distributions will be gaussian because macroscopic random effects may be thought of as the superposition of very many microscopic random effects; under very general conditions, such aggregate effects tend to be gaussian, regardless of the statistical properties of the microscopic effects. The assumption of independence in this context is motivated by the fact that microscopic phenomena tend to take place much more rapidly than macroscopic phenomena; thus primary random sources would appear to be independent on a macroscopic time scale.

If we assume that the system (12) is stationary and that $\mathbf{u}(t)$ is constant during each sampling period, that is

$$\mathbf{u}(t + \tau) = \mathbf{u}(t); \quad 0 \leq \tau < 1, \quad t = 0, 1, \ldots \qquad (13)$$

then (12) can be readily transformed into the more convenient discrete form

$$\mathbf{x}(t + 1) = \mathbf{\Phi}(1)\mathbf{x}(t) + \mathbf{\Delta}(1)\mathbf{u}(t); \; t = 0, 1, \ldots$$

where [18, 20]

$$\mathbf{\Phi}(1) = \exp \mathbf{F} = \sum_{i=0}^{\infty} \mathbf{F}^i/i! \quad (\mathbf{F}^0 = \text{unit matrix})$$

and

$$\mathbf{\Delta}(1) = \left(\int_0^1 \exp \mathbf{F}\tau d\tau\right) \mathbf{D}$$

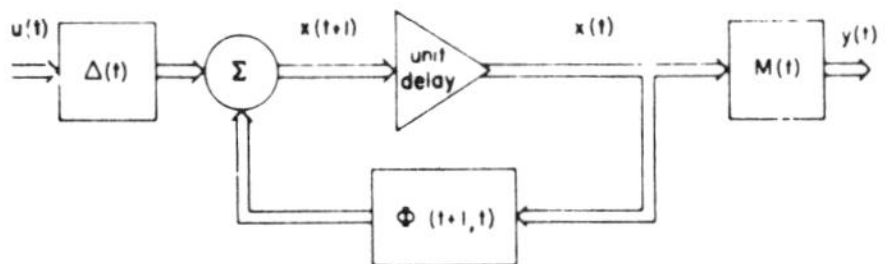

Fig. 2 Matrix block diagram of the general linear discrete-dynamic system

See Fig. 2. One could also express *exp* $\mathbf{F}\tau$ in closed form using Laplace transform methods [18, 20, 22, 24]. If $\mathbf{u}(t)$ satisfies (13) but the system (12) is nonstationary, we can write analogously

$$\left.\begin{aligned} \mathbf{x}(t + 1) &= \mathbf{\Phi}(t + 1;\, t)\mathbf{x}(t) + \mathbf{\Delta}(t)\mathbf{u}(t) \\ \mathbf{y}(t) &= \mathbf{M}(t)\mathbf{x}(t) \end{aligned}\right\} \; t = 0, 1, \ldots \qquad (14)$$

but of course now $\mathbf{\Phi}(t + 1;\, t)$, $\mathbf{\Delta}(t)$ cannot be expressed in general in closed form. Equations of type (14) are encountered frequently also in the study of complicated sampled-data systems [22]. See Fig. 2.

$\mathbf{\Phi}(t + 1;\, t)$ is the *transition matrix* of the system (12) or (14). The notation $\mathbf{\Phi}(t_2;\, t_1)$ (t_2, t_1 = integers) indicates transition from time t_1 to time t_2. Evidently $\mathbf{\Phi}(t;\, t) = \mathbf{I}$ = unit matrix. If the system (12) is stationary then $\mathbf{\Phi}(t + 1;\, t) = \mathbf{\Phi}(t + 1 - t) = \mathbf{\Phi}(1)$ = const. Note also the product rule: $\mathbf{\Phi}(t;\, s)\mathbf{\Phi}(s;\, r) = \mathbf{\Phi}(t;\, r)$ and the inverse rule $\mathbf{\Phi}^{-1}(t;\, s) = \mathbf{\Phi}(s;\, t)$, where t, s, r are integers. In a stationary system, $\mathbf{\Phi}(t;\, \tau) = \exp \mathbf{F}(t - \tau)$.

As a result of the preceding discussion, we shall represent random phenomena by the model

$$\mathbf{x}(t + 1) = \mathbf{\Phi}(t + 1;\, t)\mathbf{x}(t) + \mathbf{u}(t) \qquad (15)$$

where $\{\mathbf{u}(t)\}$ is a vector-valued, independent, gaussian random process, with zero mean, which is completely described by (in view of Theorem 5(C))

$$E\mathbf{u}(t) = \mathbf{0} \text{ for all } t;$$

$$E\mathbf{u}(t)\mathbf{u}'(s) = \mathbf{0} \quad \text{if} \quad t \neq s$$

$$E\mathbf{u}(t)\mathbf{u}'(t) = \mathbf{Q}(t).$$

Of course (Theorem 5 (A)), $\mathbf{x}(t)$ is then also a gaussian random process with zero mean, but it is no longer independent. In fact, if we consider (15) in the steady state (assuming it is a stable system), in other words, if we neglect the initial state $\mathbf{x}(t_0)$, then

$$x(t) = \sum_{r=-\infty}^{t-1} \Phi(t; r+1)u(r).$$

Therefore if $t \geq s$ we have

$$Ex(t)x'(s) = \sum_{r=-\infty}^{s-1} \Phi(t; r+1)Q(r)\Phi'(s; r+1).$$

Thus if we assume a linear dynamic model and know the statistical properties of the gaussian random excitation, it is easy to find the corresponding statistical properties of the gaussian random process $\{x(t)\}$.

In real life, however, the situation is usually reversed. One is given the covariance matrix $Ex(t)x'(s)$ (or rather, one attempts to estimate the matrix from limited statistical data) and the problem is to get (15) and the statistical properties of $u(t)$. This is a subtle and presently largely unsolved problem in experimentation and data reduction. As in the vast majority of the engineering literature on the Wiener problem, we shall find it convenient to start with the model (15) and regard the problem of obtaining the model itself as a separate question. To be sure, the two problems *should* be optimized jointly if possible; the author is not aware, however, of any study of the *joint* optimization problem.

In summary, the following assumptions are made about random processes:

Physical random phenomena may be thought of as due to primary random sources exciting dynamic systems. The primary sources are assumed to be independent gaussian random processes with zero mean; the dynamic systems will be linear. The random processes are therefore described by models such as (15). *The question of how the numbers specifying the model are obtained from experimental data will not be considered.*

Solution of the Wiener Problem

Let us now define the principal problem of the paper.

Problem I. *Consider the dynamic model*

$$x(t+1) = \Phi(t+1; t)x(t) + u(t) \quad (16)$$

$$y(t) = M(t)x(t) \quad (17)$$

where $u(t)$ is an independent gaussian random process of n-vectors with zero mean, $x(t)$ is an n-vector, $y(t)$ is a p-vector ($p \leq n$), $\Phi(t+1; t)$, $M(t)$ are $n \times n$, resp. $p \times n$, matrices whose elements are nonrandom functions of time.

Given the observed values of $y(t_0), \ldots, y(t)$ find an estimate $x^*(t_1|t)$ of $x(t_1)$ which minimizes the expected loss. (See Fig. 2, where $\Delta(t) = I$.)

This problem includes as a special case the problems of filtering, prediction, and data smoothing mentioned earlier. It includes also the problem of reconstructing all the state variables of a linear dynamic system from noisy observations of some of the state variables ($p < n!$).

From Theorem 2-*a* we know that the solution of Problem I is simply the orthogonal projection of $x(t_1)$ on the linear manifold $\mathcal{Y}(t)$ generated by the observed random variables. As remarked in the Introduction, this is to be accomplished by means of a linear (not necessarily stationary!) dynamic system of the general form (14). With this in mind, we proceed as follows.

Assume that $y(t_0), \ldots, y(t-1)$ have been measured, i.e., that $\mathcal{Y}(t-1)$ is known. Next, at time t, the random variable $y(t)$ is measured. As before let $\tilde{y}(t|t-1)$ be the component of $y(t)$ orthogonal to $\mathcal{Y}(t-1)$. If $\tilde{y}(t|t-1) \equiv 0$, which means that the values of all components of this random vector are zero for almost every possible event, then $\mathcal{Y}(t)$ is obviously the same as $\mathcal{Y}(t-1)$ and therefore the measurement of $y(t)$ does not convey any additional information. This is not likely to happen in a physically meaningful situation. In any case, $\tilde{y}(t|t-1)$ generates a linear manifold (possibly 0) which we denote by $\mathcal{Z}(t)$. By definition, $\mathcal{Y}(t-1)$ and $\mathcal{Z}(t)$ taken together are the same manifold as $\mathcal{Y}(t)$, and every vector in $\mathcal{Z}(t)$ is orthogonal to every vector in $\mathcal{Y}(t-1)$.

Assuming by induction that $x^*(t_1-1|t-1)$ is known, we can write:

$$\begin{aligned} x^*(t_1|t) &= \hat{E}[x(t_1)|\mathcal{Y}(t)] = \hat{E}[x(t_1)|\mathcal{Y}(t-1)] + \hat{E}[x(t_1)|\mathcal{Z}(t)] \\ &= \Phi(t+1; t)x^*(t_1-1|t-1) + \hat{E}[u(t_1-1)|\mathcal{Y}(t-1)] \\ &\quad + \hat{E}[x(t_1)|\mathcal{Z}(t)] \end{aligned} \quad (18)$$

where the last line is obtained using (16).

Let $t_1 = t + s$, where s is any integer. If $s \geq 0$, then $u(t_1-1)$ is independent of $\mathcal{Y}(t-1)$. This is because $u(t_1-1) = u(t+s-1)$ is then independent of $u(t-2), u(t-3), \ldots$ and therefore by (16–17), independent of $y(t_0), \ldots, y(t-1)$, hence independent of $\mathcal{Y}(t-1)$. Since, for all t, $u(t_0)$ has zero mean by assumption, it follows that $u(t_1-1)$ ($s \geq 0$) is orthogonal to $\mathcal{Y}(t-1)$. Thus if $s \geq 0$, the second term on the right-hand side of (18) vanishes; if $s < 0$, considerable complications result in evaluating this term. We shall consider only the case $t_1 \geq t$. Furthermore, it will suffice to consider in detail only the case $t_1 = t+1$ since the other cases can be easily reduced to this one.

The last term in (18) must be a linear operation on the random variable $\tilde{y}(t|t-1)$:

$$\hat{E}[x(t+1)|\mathcal{Z}(t)] = \Delta^*(t)\tilde{y}(t|t-1) \quad (19)$$

where $\Delta^*(t)$ is an $n \times p$ matrix, and the star refers to "optimal filtering."

The component of $y(t)$ lying in $\mathcal{Y}(t-1)$ is $\bar{y}(t|t-1) = M(t)x^*(t|t-1)$. Hence

$$\tilde{y}(t|t-1) = y(t) - \bar{y}(t|t-1) = y(t) - M(t)x^*(t|t-1) \quad (20)$$

Combining (18–20) (see Fig. 3) we obtain

$$x^*(t+1|t) = \Phi^*(t+1; t)x^*(t|t-1) + \Delta^*(t)y(t) \quad (21)$$

where

$$\Phi^*(t+1; t) = \Phi(t+1; t) - \Delta^*(t)M(t) \quad (22)$$

Thus optimal estimation is performed by a linear dynamic system of the same form as (14). The state of the estimator is the previous estimate, the input is the last measured value of the observable random variable $y(t)$, the transition matrix is given by (22). Notice that physical realization of the optimal filter requires only (*i*) the model of the random process (*ii*) the operator $\Delta^*(t)$.

The estimation error is also governed by a linear dynamic system. In fact,

$$\begin{aligned} \tilde{x}(t+1|t) &= x(t+1) - x^*(t+1|t) \\ &= \Phi(t+1; t)x(t) + u(t) - \Phi^*(t+1; t)x^*(t|t-1) \\ &\quad - \Delta^*(t)M(t)x(t) \end{aligned}$$

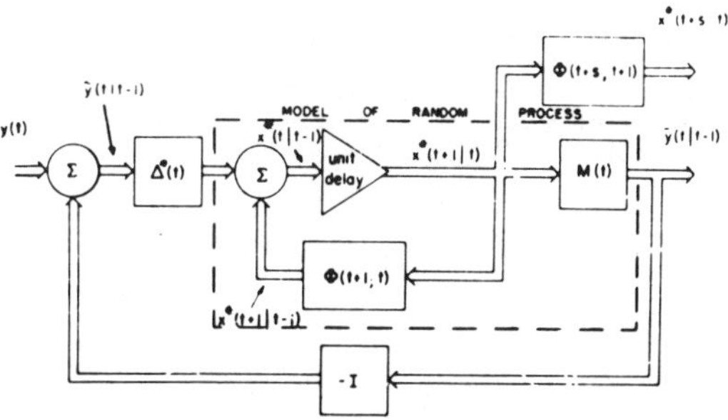

Fig. 3 Matrix block diagram of optimal filter

$$= \mathbf{\Phi}^*(t+1;\ t)\tilde{\mathbf{x}}(t|t-1) + \mathbf{u}(t) \tag{23}$$

Thus $\mathbf{\Phi}^*$ is also the transition matrix of the linear dynamic system governing the error.

From (23) we obtain at once a recursion relation for the covariance matrix $\mathbf{P}^*(t)$ of the optimal error $\tilde{\mathbf{x}}(t|t-1)$. Noting that $\mathbf{u}(t)$ is independent of $\mathbf{x}(t)$ and therefore of $\tilde{\mathbf{x}}(t|t-1)$, we get

$$\begin{aligned}\mathbf{P}^*(t+1) &= E\tilde{\mathbf{x}}(t+1|t)\tilde{\mathbf{x}}'(t+1|t)\\ &= \mathbf{\Phi}^*(t+1;\ t)E\tilde{\mathbf{x}}(t|t-1)\tilde{\mathbf{x}}'(t|t-1)\mathbf{\Phi}^{*\prime}(t+1;\ t) + \mathbf{Q}(t)\\ &= \mathbf{\Phi}^*(t+1;\ t)E\tilde{\mathbf{x}}(t|t-1)\tilde{\mathbf{x}}'(t|t-1)\mathbf{\Phi}'(t+1;\ t) + \mathbf{Q}(t)\\ &= \mathbf{\Phi}^*(t+1;\ t)\mathbf{P}^*(t)\mathbf{\Phi}'(t+1;\ t) + \mathbf{Q}(t)\end{aligned} \tag{24}$$

where $\mathbf{Q}(t) = E\mathbf{u}(t)\mathbf{u}'(t)$.

There remains the problem of obtaining an explicit formula for $\mathbf{\Delta}^*$ (and thus also for $\mathbf{\Phi}^*$). Since,

$$\tilde{\mathbf{x}}(t+1|\mathcal{Z}(t)) = \mathbf{x}(t+1) - \hat{E}[\mathbf{x}(t+1)|\mathcal{Z}(t)]$$

is orthogonal to $\tilde{\mathbf{y}}(t|t-1)$, it follows by (19) that

$$\begin{aligned}0 &= E[\mathbf{x}(t+1) - \mathbf{\Delta}^*(t)\tilde{\mathbf{y}}(t|t-1)]\tilde{\mathbf{y}}'(t|t-1)\\ &= E\mathbf{x}(t+1)\tilde{\mathbf{y}}'(t|t-1) - \mathbf{\Delta}^*(t)E\tilde{\mathbf{y}}(t|t-1)\tilde{\mathbf{y}}'(t|t-1).\end{aligned}$$

Noting that $\bar{\mathbf{x}}(t+1|t-1)$ is orthogonal to $\mathcal{Z}(t)$, the definition of $\mathbf{P}(t)$ given earlier, and (17), it follows further

$$\begin{aligned}0 &= E\tilde{\mathbf{x}}(t+1|t-1)\tilde{\mathbf{y}}'(t|t-1) - \mathbf{\Delta}^*(t)\mathbf{M}(t)\mathbf{P}^*(t)\mathbf{M}'(t)\\ &= E[\mathbf{\Phi}(t+1;\ t)\tilde{\mathbf{x}}(t|t-1) + \mathbf{u}(t|t-1)]\tilde{\mathbf{x}}'(t|t-1)\mathbf{M}'(t)\\ &\qquad - \mathbf{\Delta}^*(t)\mathbf{M}(t)\mathbf{P}^*(t)\mathbf{M}'(t).\end{aligned}$$

Finally, since $\mathbf{u}(t)$ is independent of $\mathbf{x}(t)$,

$$0 = \mathbf{\Phi}(t+1;\ t)\mathbf{P}^*(t)\mathbf{M}'(t) - \mathbf{\Delta}^*(t)\mathbf{M}(t)\mathbf{P}^*(t)\mathbf{M}'(t).$$

Now the matrix $\mathbf{M}(t)\mathbf{P}^*(t)\mathbf{M}'(t)$ will be positive definite and hence invertible whenever $\mathbf{P}^*(t)$ is positive definite, provided that none of the rows of $\mathbf{M}(t)$ are linearly dependent at any time, in other words, that none of the observed scalar random variables $y_1(t), \ldots, y_m(t)$ is a linear combination of the others. Under these circumstances we get finally:

$$\mathbf{\Delta}^*(t) = \mathbf{\Phi}(t+1;\ t)\mathbf{P}^*(t)\mathbf{M}'(t)[\mathbf{M}(t)\mathbf{P}^*(t)\mathbf{M}'(t)]^{-1} \tag{25}$$

Since observations start at t_0, $\tilde{\mathbf{x}}(t_0|t_0-1) = \mathbf{x}(t_0)$; to begin the iterative evaluation of $\mathbf{P}^*(t)$ by means of equation (24), we must obviously specify $\mathbf{P}^*(t_0) = E\mathbf{x}(t_0)\mathbf{x}'(t_0)$. Assuming this matrix is positive definite, equation (25) then yields $\mathbf{\Delta}^*(t_0)$; equation (22) $\mathbf{\Phi}^*(t_0+1;\ t_0)$, and equation (24) $\mathbf{P}^*(t_0+1)$, completing the cycle. If now $\mathbf{Q}(t)$ is positive definite, then all the $\mathbf{P}^*(t)$ will be positive definite and the requirements in deriving (25) will be satisfied at each step.

Now we remove the restriction that $t_1 = t+1$. Since $\mathbf{u}(t)$ is orthogonal to $\mathcal{Y}(t)$, we have

$$\mathbf{x}^*(t+1|t) = \hat{E}[\mathbf{\Phi}(t+1;\ t)\mathbf{x}(t) + \mathbf{u}(t)|\mathcal{Y}(t)] = \mathbf{\Phi}(t+1;\ t)\mathbf{x}^*(t|t)$$

Hence if $\mathbf{\Phi}(t+1;\ t)$ has an inverse $\mathbf{\Phi}(t;\ t+1)$ (which is always the case when $\mathbf{\Phi}$ is the transition matrix of a dynamic system describable by a differential equation) we have

$$\mathbf{x}^*(t|t) = \mathbf{\Phi}(t;\ t+1)\mathbf{x}^*(t+1|t)$$

If $t_1 \geq t+1$, we first observe by repeated application of (16) that

$$\begin{aligned}\mathbf{x}(t+s) &= \mathbf{\Phi}(t+s;\ t+1)\mathbf{x}(t+1)\\ &\quad + \sum_{r1}^{s-1}\mathbf{\Phi}(t+s;\ t+r)\mathbf{u}(t+r) \qquad (s \geq 1)\end{aligned}$$

Since $\mathbf{u}(t+s-1), \ldots, \mathbf{u}(t+1)$ are all orthogonal to $\mathcal{Y}(t)$,

$$\begin{aligned}\mathbf{x}^*(t+s|t) &= \hat{E}[\mathbf{x}(t+s)|\mathcal{Y}(t)]\\ &= \hat{E}[\mathbf{\Phi}(t+s;\ t+1)\mathbf{x}(t+1)\,|\mathcal{Y}(t)]\\ &= \mathbf{\Phi}(t+s;\ t+1)\mathbf{x}^*(t+1|t) \qquad (s \geq 1)\end{aligned}$$

If $s < 0$, the results are similar, but $\mathbf{x}^*(t-s|t)$ will have $(1-s)(n-p)$ co-ordinates.

The results of this section may be summarized as follows:

Theorem 3. (*Solution of the Wiener Problem*)

Consider Problem I. The optimal estimate $\mathbf{x}^*(t+1|t)$ of $\mathbf{x}(t+1)$ given $\mathbf{y}(t_0), \ldots, \mathbf{y}(t)$ is generated by the linear dynamic system

$$\mathbf{x}^*(t+1|t) = \mathbf{\Phi}^*(t+1;\ t)\mathbf{x}^*(t|t-1) + \mathbf{\Delta}^*(t)\mathbf{y}(t) \tag{21}$$

The estimation error is given by

$$\tilde{\mathbf{x}}(t+1|t) = \mathbf{\Phi}^*(t+1;\ t)\tilde{\mathbf{x}}(t|t-1) + \mathbf{u}(t) \tag{23}$$

The covariance matrix of the estimation error is

$$\operatorname{cov}\tilde{\mathbf{x}}(t|t-1) = E\tilde{\mathbf{x}}(t|t-1)\tilde{\mathbf{x}}'(t|t-1) = \mathbf{P}^*(t) \tag{26}$$

The expected quadratic loss is

$$\sum_{i=1}^{n} E\tilde{x}_i^2(t|t-1) = \operatorname{trace}\mathbf{P}^*(t) \tag{27}$$

The matrices $\mathbf{\Delta}^*(t)$, $\mathbf{\Phi}^*(t+1;\ t)$, $\mathbf{P}^*(t)$ are generated by the recursion relations

$$\left.\begin{aligned}&\mathbf{\Delta}^*(t) = \mathbf{\Phi}(t+1;\ t)\mathbf{P}^*(t)\mathbf{M}'(t)[\mathbf{M}(t)\mathbf{P}^*(t)\mathbf{M}'(t)]^{-1} &(28)\\ &\mathbf{\Phi}^*(t+1;\ t) = \mathbf{\Phi}(t+1;\ t) - \mathbf{\Delta}^*(t)\mathbf{M}(t) &(29)\\ &\mathbf{P}^*(t+1) = \mathbf{\Phi}^*(t+1;\ t)\mathbf{P}^*(t)\mathbf{\Phi}'(t+1;\ t) + \mathbf{Q}(t) &(30)\end{aligned}\right\}\ t \geq t_0$$

In order to carry out the iterations, one must specify the covariance $\mathbf{P}^*(t_0)$ of $\mathbf{x}(t_0)$ and the covariance $\mathbf{Q}(t)$ of $\mathbf{u}(t)$. Finally, for any $s \geq 0$, if $\mathbf{\Phi}$ is invertible

$$\begin{aligned}\mathbf{x}^*(t+s|t) &= \mathbf{\Phi}(t+s;\ t+1)\mathbf{x}^*(t+1|t)\\ &= \mathbf{\Phi}(t+s;\ t+1)\mathbf{\Phi}^*(t+1;\ t)\mathbf{\Phi}(t;\ t+s-1)\\ &\qquad \times \mathbf{x}^*(t+s-1|t-1)\\ &\quad + \mathbf{\Phi}(t+s;\ t+1)\mathbf{\Delta}^*(t)\mathbf{y}(t)\end{aligned} \tag{31}$$

so that the estimate $\mathbf{x}^*(t+s|t)$ $(s \geq 0)$ is also given by a linear dynamic system of the type (21).

Remarks. (*h*) Eliminating $\mathbf{\Delta}^*$ and $\mathbf{\Phi}^*$ from (28–30), a nonlinear difference equation is obtained for $\mathbf{P}^*(t)$:

$$\begin{aligned}\mathbf{P}^*(t+1) &= \mathbf{\Phi}(t+1;\ t)\{\mathbf{P}^*(t) - \mathbf{P}^*(t)\mathbf{M}'(t)[\mathbf{M}(t)\mathbf{P}^*(t)\mathbf{M}'(t)]^{-1}\\ &\qquad \times \mathbf{P}^*(t)\mathbf{M}(t)\}\mathbf{\Phi}'(t+1;\ t) + \mathbf{Q}(t) \qquad (t \geq t_0)\end{aligned} \tag{32}$$

This equation is linear only if $\mathbf{M}(t)$ is invertible but then the problem is trivial since all components of the random vector $\mathbf{x}(t)$ are observable $\mathbf{P}^*(t+1) = \mathbf{Q}(t)$. Observe that equation (32) plays a role in the present theory analogous to that of the Wiener-Hopf equation in the conventional theory.

Once $\mathbf{P}^*(t)$ has been computed via (32) starting at $t = t_0$, the explicit specification of the optimal linear filter is immediately available from formulas (29–30). Of course, the solution of Equation (32), or of its differential-equation equivalent, is a much simpler task than solution of the Wiener-Hopf equation.

(*i*) The results stated in Theorem 3 do not resolve completely Problem I. Little has been said, for instance, about the physical significance of the assumptions needed to obtain equation (25), the convergence and stability of the nonlinear difference equation (32), the stability of the optimal filter (21), etc. This can actually be done in a completely satisfactory way, but must be left to a future paper. In this connection, the principal guide and tool turns out to be the duality theorem mentioned briefly in the next section. See [29].

(*j*) By letting the sampling period (equal to one so far) approach zero, the method can be used to obtain the specification of a differential equation for the optimal filter. To do this, i.e., to pass from equation (14) to equation (12), requires computing the logarithm $\mathbf{F}^*$ of the matrix $\mathbf{\Phi}^*$. But this can be done only if $\mathbf{\Phi}^*$ is nonsingular—which is easily seen *not* to be the case. This is because it is sufficient for the optimal filter to have $n - p$ state variables, rather than n, as the formalism of equation (22) would seem to imply. By appropriate modifications, therefore, equation (22) can be reduced to an equivalent set of only $n - p$ equations whose transition matrix is nonsingular. Details of this type will be covered in later publications.

(*k*) The dynamic system (21) is, in general, nonstationary. This is due to two things: (1) The time dependence of $\mathbf{\Phi}(t+1; t)$ and $\mathbf{M}(t)$; (2) the fact that the estimation starts at $t = t_0$ and improves as more data are accumulated. If $\mathbf{\Phi}$, $\mathbf{M}$ are constants, it can be shown that (21) becomes a stationary dynamic system in the limit $t \to \infty$. This is the case treated by the classical Wiener theory.

(*l*) It is noteworthy that the derivations given are not affected by the nonstationarity of the model for $\mathbf{x}(t)$ or the finiteness of available data. In fact, as far as the author is aware, the only explicit recursion relations given before for the growing-memory filter are due to Blum [12]. However, his results are much more complicated than ours.

(*m*) By inspection of Fig. 3 we see that the optimal filter is a feedback system, and that the signal after the first summer is white noise since $\tilde{\mathbf{y}}(t|t-1)$ is obviously an orthogonal random process. This corresponds to some well-known results in Wiener filtering, see, e.g., Smith [28], Chapter 6, Fig. 6–4. However, this is apparently the first *rigorous* proof that every Wiener filter is realizable by means of a feedback system. Moreover, it will be shown in another paper that such a filter is always *stable*, under very mild assumptions on the model (16–17). See [29].

The Dual Problem

Let us now consider another problem which is conceptually very different from optimal estimation, namely, the noise-free regulator problem. In the simplest cases, this is:

Problem II. *Consider the dynamic system*

$$\mathbf{x}(t+1) = \hat{\mathbf{\Phi}}(t+1;\ t)\mathbf{x}(t) + \hat{\mathbf{M}}(t)\mathbf{u}(t) \tag{33}$$

where $\mathbf{x}(t)$ *is an n-vector,* $\mathbf{u}(t)$ *is an m-vector* ($m \leq n$), $\hat{\mathbf{\Phi}}$, $\hat{\mathbf{M}}$ *are* $n \times n$ *resp.* $n \times m$ *matrices whose elements are nonrandom functions of time. Given any state* $\mathbf{x}(t)$ *at time t, we are to find a sequence* $\mathbf{u}(t), \ldots, \mathbf{u}(T)$ *of control vectors which minimizes the performance index*

$$V[\mathbf{x}(t)] = \sum_{\tau=t}^{T+1} \mathbf{x}'(\tau)\hat{\mathbf{Q}}(\tau)\mathbf{x}(\tau)$$

where $\hat{\mathbf{Q}}(t)$ *is a positive definite matrix whose elements are nonrandom functions of time.* See Fig. 2, where $\mathbf{\Delta} = \hat{\mathbf{M}}$ and $\mathbf{M} = \mathbf{I}$.

Probabilistic considerations play no part in Problem II; it is implicitly assumed that every state variable can be measured exactly at each instant $t, t+1, \ldots, T$. It is customary to call $T \geq t$ the *terminal time* (it may be infinity).

The first general solution of the noise-free regulator problem is due to the author [18]. The main result is that the optimal control vectors $\mathbf{u}^*(t)$ are nonstationary linear functions of $\mathbf{x}(t)$. After a change in notation, the formulas of the Appendix, Reference [18] (see also Reference [23]) are as follows:

$$\mathbf{u}^*(t) = -\hat{\mathbf{\Delta}}^*(t)\mathbf{x}(t) \tag{34}$$

Under optimal control as given by (34), the "closed-loop" equations for the system are (see Fig. 4)

$$\mathbf{x}(t+1) = \hat{\mathbf{\Phi}}^*(t+1;\ t)\mathbf{x}(t)$$

and the minimum performance index at time t is given by

$$V^*[\mathbf{x}(t)] = \mathbf{x}'(t)\mathbf{P}^*(t-1)\mathbf{x}(t)$$

The matrices $\hat{\mathbf{\Delta}}^*(t)$, $\hat{\mathbf{\Phi}}^*(t+1;\ t)$, $\hat{\mathbf{P}}^*(t)$ are determined by the recursion relations:

$$\left.\begin{aligned} \hat{\mathbf{\Delta}}^*(t) &= [\hat{\mathbf{M}}'(t)\hat{\mathbf{P}}^*(t)\hat{\mathbf{M}}(t)]^{-1}\hat{\mathbf{M}}'(t)\hat{\mathbf{P}}^*(t)\hat{\mathbf{\Phi}}(t+1;\ t) && (35)\\ \hat{\mathbf{\Phi}}^*(t+1;\ t) &= \hat{\mathbf{\Phi}}(t+1;\ t) - \hat{\mathbf{M}}(t)\hat{\mathbf{\Delta}}^*(t) && (36)\\ \hat{\mathbf{P}}^*(t-1) &= \hat{\mathbf{\Phi}}'(t+1;\ t)\hat{\mathbf{P}}^*(t)\hat{\mathbf{\Phi}}^*(t+1;\ t) + \hat{\mathbf{Q}}(t) && (37) \end{aligned}\right\}\ t \leq T$$

Initially we must set $\hat{\mathbf{P}}^*(T) = \hat{\mathbf{Q}}(T+1)$.

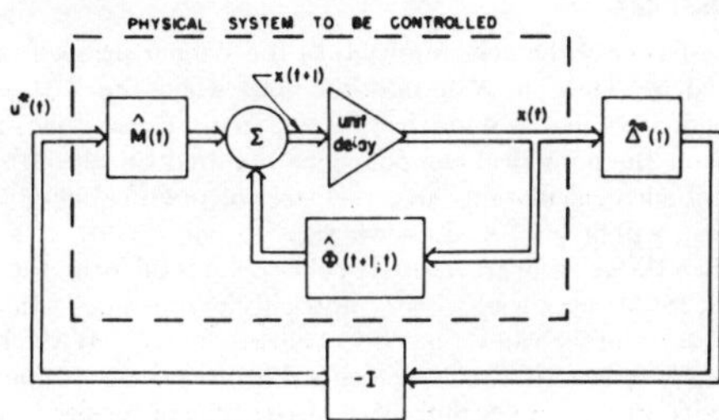

Fig. 4 Matrix block diagram of optimal controller

Comparing equations (35–37) with (28–30) and Fig. 3 with Fig. 4 we notice some interesting things which are expressed precisely by

Theorem 4. (*Duality Theorem*) *Problem I and Problem II are duals of each other in the following sense:*

Let $\tau \geq 0$. *Replace every matrix* $\mathbf{X}(t) = \mathbf{X}(t_0 + \tau)$ *in* (28–30) *by* $\hat{\mathbf{X}}'(t) = \hat{\mathbf{X}}'(T - \tau)$. *Then one has* (35–37). *Conversely, replace every matrix* $\hat{\mathbf{X}}(T - \tau)$ *in* (35-37) *by* $\mathbf{X}'(t_0 + \tau)$. *Then one has* (28–30).

Proof. Carry out the substitutions. For ease of reference, the dualities between the two problems are given in detail in Table 1.

Table 1

	Problem I	Problem II
1	$\mathbf{x}(t)$ (unobservable) state variables of random process.	$\mathbf{x}(t)$ (observable) state variables of plant to be regulated.
2	$\mathbf{y}(t)$ observed random variables.	$\mathbf{u}(t)$ control variables.
3	t_0 first observation.	T last control action.
4	$\mathbf{\Phi}(t_0+\tau+1;\ t_0+\tau)$ transition matrix.	$\hat{\mathbf{\Phi}}(T-\tau+1;\ T-\tau)$ transition matrix.
5	$\mathbf{P}^*(t_0+\tau)$ covariance of optimized estimation error.	$\hat{\mathbf{P}}^*(T-\tau)$ matrix of quadratic form for performance index under optimal regulation.
6	$\mathbf{\Delta}^*(t_0+\tau)$ weighting of observation for optimal estimation.	$\hat{\mathbf{\Delta}}^*(T-\tau)$ weighting of state for optimal control.
7	$\mathbf{\Phi}^*(t_0+\tau+1;\ t_0+\tau)$ transition matrix for optimal estimation error.	$\hat{\mathbf{\Phi}}^*(T-\tau+1;\ T-\tau)$ transition matrix under optimal regulation.
8	$\mathbf{M}(t_0+\tau)$ effect of state on observation.	$\hat{\mathbf{M}}(T-\tau)$ effect of control vectors on state.
9	$\mathbf{Q}(t_0+\tau)$ covariance of random excitation.	$\hat{\mathbf{Q}}(T-\tau)$ matrix of quadratic form defining error criterion.

Remarks. (*n*) The *mathematical* significance of the duality between Problem I and Problem II is that both problems reduce to the solution of the Wiener-Hopf-like equation (32).

(*o*) The *physical* significance of the duality is intriguing. Why are observations and control dual quantities?

Recent research [29] has shown that the essence of the Duality Theorem lies in the duality of constraints at the output (represented by the matrix $\hat{\mathbf{M}}(t)$ in Problem I) and constraints at the input (represented by the matrix $\hat{\mathbf{M}}(t)$ in Problem II).

(*p*) Applications of Wiener's methods to the solution of noise-free regulator problem have been known for a long time; see the recent textbook of Newton, Gould, and Kaiser [27]. However, the connections between the two problems, and in particular the duality, have apparently never been stated precisely before.

(*q*) The duality theorem offers a powerful tool for developing more deeply the theory (as opposed to the computation) of Wiener filters, as mentioned in Remark (*i*). This will be published elsewhere [29].

Applications

The power of the new approach to the Wiener problem, as expressed by Theorem 3, is most obvious when the data of the problem are given in numerical form. In that case, one simply performs the numerical computations required by (28–30). Results of such calculations, in some cases of practical engineering interest, will be published elsewhere.

When the answers are desired in closed analytic form, the iterations (28–30) may lead to very unwieldy expressions. In a few cases, $\mathbf{\Delta}^*$ and $\mathbf{\Phi}^*$ can be put into "closed form." Without discussing here how (if at all) such closed forms can be obtained, we now give two examples indicative of the type of results to be expected.

Example 1. Consider the problem mentioned under "Optimal Estimates." Let $x_1(t)$ be the signal and $x_2(t)$ the noise. We assume the model:

$$x_1(t + 1) = \phi_{11}(t + 1;\ t)x_1(t) + u_1(t)$$

$$x_2(t + 1) = u_2(t)$$

$$y_1(t) = x_1(t) + x_2(t)$$

The specific data for which we desire a solution of the estimation problem are as follows:

1 $t_1 = t + 1;\ t_0 = 0$
2 $Ex_1^2(0) = 0$, i.e., $x_1(0) = 0$
3 $Eu_1^2(t) = a^2$, $Eu^2(t) = b^2$, $Eu_1(t)u_2(t) = 0$ (for all t)
4 $\phi_{11}(t + 1;\ t) = \phi_{11} = \text{const.}$

A simple calculation shows that the following matrices satisfy the difference equations (28–30), for all $t \geqq t_0$:

$$\mathbf{\Delta}^*(t) = \begin{bmatrix} \phi_{11}C(t) \\ 0 \end{bmatrix}$$

$$\mathbf{\Phi}^*(t + 1;\ t) = \begin{bmatrix} \phi_{11}[1 - C(t)] & 0 \\ 0 & 0 \end{bmatrix}$$

$$\mathbf{P}^*(t + 1) = \begin{bmatrix} a^2 + \phi_{11}^2 b^2 C(t) & 0 \\ 0 & b^2 \end{bmatrix}$$

$$\text{where}\quad C(t + 1) = 1 - \frac{b^2}{a^2 + b^2 + \phi_{11}^2 b^2 C(t)}\qquad t \geqq 0 \tag{38}$$

Since it was assumed that $x_1(0) = 0$, neither $x_1(1)$ nor $x_2(1)$ can be predicted from the measurement of $y_1(0)$. Hence the measurement at time $t = 0$ is useless, which shows that we should set $C(0) = 0$. This fact, with the iterations (38), completely determines the function $C(t)$. The nonlinear difference equation (38) plays the role of the Wiener-Hopf equation.

If $b^2/a^2 \ll 1$, then $C(t) \cong 1$ which is essentially pure prediction. If $b^2/a^2 \gg 1$, then $C(t) \cong 0$, and we depend mainly on $x_1^*(t|t - 1)$ for the estimation of $x_1^*(t + 1|t)$ and assign only very small weight to the measurement $y_1(t)$; this is what one would expect when the measured data are very noisy.

In any case, $x_2^*(t|t - 1) = 0$ at all times; one cannot predict independent noise! This means that ϕ^*_{12} can be set equal to zero. The optimal predictor is a first-order dynamic system. See Remark (*j*).

To find the stationary Wiener filter, let $t = \infty$ on both sides of (38), solve the resulting quadratic equation in $C(\infty)$, etc.

Example 2. A number of particles leave the origin at time $t_0 = 0$ with random velocities; after $t = 0$, each particle moves with a constant (unknown) velocity. Suppose that the position of one of these particles is measured, the data being contaminated by stationary, additive, correlated noise. What is the optimal estimate of the position and velocity of the particle at the time of the last measurement?

Let $x_1(t)$ be the position and $x_2(t)$ the velocity of the particle; $x_3(t)$ is the noise. The problem is then represented by the model,

$$x_1(t + 1) = x_1(t) + x_2(t)$$

$$x_2(t + 1) = x_2(t)$$

$$x_3(t + 1) = \phi_{33}(t + 1;\ t)x_3(t) + u_3(t)$$

$$y_1(t) = x_1(t) + x_3(t)$$

and the additional conditions

1 $t_1 = t;\ t_0 = 0$
2 $Ex_1^2(0) = Ex_2(0) = 0$, $Ex_2^2(0) = a^2 > 0$;
3 $Eu_3(t) = 0$, $Eu_3^2(t) = b^2$.
4 $\phi_{33}(t + 1;\ t) = \phi_{33} = \text{const.}$

According to Theorem 3, $\mathbf{x}^*(t|t)$ is calculated using the dynamic system (31).

First we solve the problem of predicting the position and velocity of the particle one step ahead. Simple considerations show that

$$\mathbf{P}^*(1) = \begin{bmatrix} a^2 & a^2 & 0 \\ a^2 & a^2 & 0 \\ 0 & 0 & b^2 \end{bmatrix} \quad\text{and}\quad \mathbf{\Delta}^*(0) = \begin{bmatrix} 0 \\ 0 \\ 1 \end{bmatrix}$$

It is then easy to check by substitution into equations (28–30) that

$$\mathbf{P}^*(t) = \frac{b^2}{C_1(t - 1)} \times \begin{bmatrix} t^2 & t & -\phi_{33}t(t - 1) \\ t & 1 & -\phi_{33}(t - 1) \\ -\phi_{33}t(t - 1) & -\phi_{33}(t - 1) & \phi_{33}^2(t - 1)^2 + C_1(t - 1) \end{bmatrix}$$

is the correct expression for the covariance matrix of the prediction error $\tilde{\mathbf{x}}(t|t - 1)$ for all $t \geqq 1$, provided that we define

$$C_1(0) = b^2/a^2$$
$$C_1(t) = C_1(t - 1) + [t - \phi_{33}(t - 1)]^2,\ t \geqq 1$$

It is interesting to note that the results just obtained are valid also when ϕ_{33} depends on t. This is true also in Example 1. In conventional treatments of such problems there *seems* to be an essential difference between the cases of stationary and nonstationary noise. This misleading impression created by the conventional theory is due to the very special *methods* used in solving the Wiener-Hopf equation.

Introducing the abbreviation

$$C_2(0) = 0$$
$$C_2(t) = t - \phi_{33}(t - 1),\ t \geqq 1$$

and observing that

$$\begin{aligned} \operatorname{cov} \tilde{\mathbf{x}}(t + 1|t) &= \mathbf{P}^*(t + 1) \\ &= \mathbf{\Phi}(t + 1;\ t)[\operatorname{cov} \tilde{\mathbf{x}}(t|t)]\mathbf{\Phi}'(t + 1;\ t) + \mathbf{Q}(t) \end{aligned}$$

the matrices occurring in equation (31) and the covariance matrix of $\tilde{\mathbf{x}}(t|t)$ are found after simple calculations. We have, for all $t \geq 0$

$$\mathbf{\Phi}(t, t+1)\mathbf{\Delta}^*(t) = \frac{1}{C_1(t)}\begin{bmatrix} tC_2(t) \\ C_2(t) \\ C_1(t) - tC_2(t) \end{bmatrix}$$

$$\mathbf{\Phi}(t, t+1)\mathbf{\Phi}^*(t+1; t)\mathbf{\Phi}(t+1; t) = \frac{1}{C_1(t)}\begin{bmatrix} C_1(t) - tC_2(t) & C_1(t) - tC_2(t) & -\phi_{33}tC_2(t) \\ -C_2(t) & C_1(t) - C_2(t) & -\phi_{33}C_2(t) \\ -C_1(t) + tC_2(t) & -C_1(t) + tC_2(t) & +\phi_{33}tC_2(t) \end{bmatrix}$$

and

$$\operatorname{cov} \tilde{\mathbf{x}}(t|t) = E\tilde{\mathbf{x}}(t|t)\tilde{\mathbf{x}}'(t|t) = \frac{b^2}{C_1(t)}\begin{bmatrix} t^2 & t & -t^2 \\ t & 1 & -t \\ -t^2 & -t & t^2 \end{bmatrix}$$

To gain some insight into the behavior of this system, let us examine the limiting case $t \to \infty$ of a large number of observations. Then $C_1(t)$ obeys approximately the differential equation

$$dC_1(t)/dt \simeq C_2^2(t) \qquad (t \gg 1)$$

from which we find

$$C_1(t) \simeq (1 - \phi_{33})^2t^3/3 + \phi_{33}(1 - \phi_{33})t^2 + \phi_{33}^2t + b^2/a^2 \qquad (t \gg 1) \quad (39)$$

Using (39), we get further

$$\mathbf{\Phi}^{-1}\mathbf{\Phi}^*\mathbf{\Phi} \cong \begin{bmatrix} 1 & 1 & 0 \\ 0 & 1 & 0 \\ -1 & -1 & 0 \end{bmatrix} \text{ and } \mathbf{\Phi}^{-1}\mathbf{\Delta}^* \cong \begin{bmatrix} 0 \\ 0 \\ 1 \end{bmatrix} \qquad (t \gg 1)$$

Thus as the number of observations becomes large, we depend almost exclusively on $x_1^*(t|t)$ and $x_2^*(t|t)$ to estimate $x_1^*(t+1|t+1)$ and $x_2^*(t+1|t+1)$. Current observations are used almost exclusively to estimate the noise

$$x_3^*(t|t) \cong y_1(t) - x_1^*(t|t) \qquad (t \gg 1)$$

One would of course expect something like this since the problem is analogous to fitting a straight line to an increasing number of points.

As a second check on the reasonableness of the results given, observe that the case $t \gg 1$ is essentially the same as prediction based on continuous observations. Setting $\phi_{33} = 0$, we have

$$E\tilde{x}_1^2(t|t) \cong \frac{a^2b^2t^2}{b^2 + a^2t^3/3} \qquad (t \gg 1;\ \phi_{33} = 0)$$

which is identical with the result obtained by Shinbrot [11], Example 1, and Solodovnikov [14], Example 2, in their treatment of the Wiener problem in the finite-length, continuous-data case, using an approach entirely different from ours.

Conclusions

This paper formulates and solves the Wiener problem from the "state" point of view. On the one hand, this leads to a very general treatment including cases which cause difficulties when attacked by other methods. On the other hand, the Wiener problem is shown to be closely connected with other problems in the theory of control. Much remains to be done to exploit these connections.

References

1 N. Wiener, "The Extrapolation, Interpolation and Smoothing of Stationary Time Series," John Wiley & Sons, Inc., New York, N. Y., 1949.

2 L. A. Zadeh and J. R. Ragazzini, "An Extension of Wiener's Theory of Prediction," *Journal of Applied Physics*, vol. 21, 1950, pp. 645-655.

3 H. W. Bode and C. E. Shannon, "A Simplified Derivation of Linear Least-Squares Smoothing and Prediction Theory," *Proceedings IRE*, vol. 38, 1950, pp. 417-425.

4 R. C. Booton, "An Optimization Theory for Time-Varying Linear Systems With Nonstationary Statistical Inputs," *Proceedings IRE*, vol. 40, 1952, pp. 977-981.

5 J. H. Laning and R. H. Battin, "Random Processes in Automatic Control," McGraw-Hill Book Company, Inc., New York, N.Y., 1956.

6 W. B. Davenport, Jr., and W. L. Root, "An Introduction to the Theory of Random Signals and Noise," McGraw-Hill Book Company, Inc., New York, N. Y., 1958.

7 S. Darlington, "Linear Least-Squares Smoothing and Prediction, With Applications," *Bell System Tech. Journal*, vol. 37, 1958, pp. 1221-1294.

8 G. Franklin, "The Optimum Synthesis of Sampled-Data Systems," Doctoral dissertation, Dept. of Elect. Engr., Columbia University, 1955.

9 A. B. Lees, "Interpolation and Extrapolation of Sampled Data," *Trans. IRE Prof. Group on Information Theory*, IT-2, 1956, pp. 173-175.

10 R. C. Davis, "On the Theory of Prediction of Nonstationary Stochastic Processes," *Journal of Applied Physics*, vol. 23, 1952, pp. 1047-1053.

11 M. Shinbrot, "Optimization of Time-Varying Linear Systems With Nonstationary Inputs," Trans. ASME, vol. 80, 1958, pp. 457-462.

12 M. Blum, "Recursion Formulas for Growing Memory Digital Filters," *Trans. IRE Prof. Group on Information Theory*, IT-4, 1958, pp. 24-30.

13 V. S. Pugachev, "The Use of Canonical Expansions of Random Functions in Determining an Optimum Linear System," *Automatics and Remote Control* (USSR), vol. 17, 1956, pp. 489-499; translation pp. 545-556.

14 V. V. Solodovnikov and A. M. Batkov, "On the Theory of Self-Optimizing Systems (in German and Russian)," Proc. Heidelberg Conference on Automatic Control, 1956, pp. 308-323.

15 J. L. Doob, "Stochastic Processes," John Wiley & Sons, Inc., New York, N. Y., 1955.

16 M. Loève, "Probability Theory," Van Nostrand Company, Inc., New York, N. Y., 1955.

17 R. E. Bellman, I. Glicksberg, and O. A. Gross, "Some Aspects of the Mathematical Theory of Control Processes," RAND Report R-313, 1958, 244 pp.

18 R. E. Kalman and R. W. Koepcke, "Optimal Synthesis of Linear Sampling Control Systems Using Generalized Performance Indexes," Trans. ASME, vol. 80, 1958, pp. 1820-1826.

19 J. E. Bertram, "Effect of Quantization in Sampled-Feedback Systems," *Trans. AIEE*, vol. 77, II, 1958, pp. 177-182.

20 R. E. Kalman and J. E. Bertram, "General Synthesis Procedure for Computer Control of Single and Multi-Loop Linear Systems," *Trans. AIEE*, vol. 77, II, 1958, pp. 602-609.

21 C. W. Merriam, III, "A Class of Optimum Control Systems," *Journal of the Franklin Institute*, vol. 267, 1959, pp. 267-281.

22 R. E. Kalman and J. E. Bertram, "A Unified Approach to the Theory of Sampling Systems," *Journal of the Franklin Institute*, vol. 267, 1959, pp. 405-436.

23 R. E. Kalman and R. W. Koepcke, "The Role of Digital Computers in the Dynamic Optimization of Chemical Reactors," Proc. Western Joint Computer Conference, 1959, pp. 107-116.

24 R. E. Kalman, "Dynamic Optimization of Linear Control Systems, I. Theory," to appear.

25 S. Sherman, "Non-Mean-Square Error Criteria," *Trans. IRE Prof. Group on Information Theory*, IT-4, 1958, pp. 125-126.

26 V. S. Pugachev, "On a Possible General Solution of the Problem of Determining Optimum Dynamic Systems," *Automatics and Remote Control* (USSR), vol. 17, 1956, pp. 585-589.

27 G. C. Newton, Jr., L. A. Gould, and J. F. Kaiser, "Analytical Design of Linear Feedback Controls," John Wiley & Sons, Inc., New York, N. Y., 1957.

28 O. J. M. Smith, "Feedback Control Systems," McGraw-Hill Book Company, Inc., New York, N. Y., 1958.

29 R. E. Kalman, "On the General Theory of Control Systems," Proceedings First International Conference on Automatic Control, Moscow, USSR, 1960.

APPENDIX
RANDOM PROCESSES: BASIC CONCEPTS

For convenience of the reader, we review here some elementary definitions and facts about probability and random processes. Everything is presented with the utmost possible simplicity; for greater depth and breadth, consult Laning and Battin [5] or Doob [15].

A *random variable* is a function whose values depend on the outcome of a chance event. The *values* of a random variable may be any convenient mathematical entities; real or complex numbers, vectors, etc. For simplicity, we shall consider here only real-valued random variables, but this is no real restriction. Random variables will be denoted by $x, y, \ldots$ and their values by $\xi, \eta, \ldots$. Sums, products, and functions of random variables are also random variables.

A random variable x can be explicitly defined by stating the probability that x is less than or equal to some real constant ξ. This is expressed symbolically by writing

$$Pr(x \leq \xi) = F_x(\xi);\ F_x(-\infty) = 0,\ F_x(+\infty) = 1$$

$F_x(\xi)$ is called the *probability distribution function* of the random variable x. When $F_x(\xi)$ is differentiable with respect to ξ, then $f_x(\xi) = dF_x(\xi)/d\xi$ is called the *probability density function* of x.

The *expected value* (*mathematical expectation, statistical average, ensemble average, mean*, etc., are commonly used synonyms) of any nonrandom function $g(x)$ of a random variable x is defined by

$$Eg(x) = E[g(x)] = \int_{-\infty}^{\infty} g(\xi)dF_x(\xi) = \int_{-\infty}^{\infty} g(\xi)f_x(\xi)d\xi \quad (40)$$

As indicated, it is often convenient to omit the brackets after the symbol E. A sequence of random variables (finite or infinite)

$$\{x(t)\} = \ldots, x(-1), x(0), x(1), \ldots \quad (41)$$

is called a *discrete* (or *discrete-parameter*) *random* (or *stochastic*) *process*. One particular set of observed values of the random process (41)

$$\ldots, \xi(-1), \xi(0), \xi(1), \ldots$$

is called a *realization* (or a *sample function*) of the process. Intuitively, a random process is simply a set of random variables which are indexed in such a way as to bring the notion of time into the picture.

A random process is *uncorrelated* if

$$Ex(t)x(s) = Ex(t)Ex(s) \quad (t \neq s)$$

If, furthermore,

$$Ex(t)x(s) = 0 \quad (t \neq s)$$

then the random process is *orthogonal*. Any uncorrelated random process can be changed into orthogonal random process by replacing $x(t)$ by $x'(t) = x(t) - Ex(t)$ since then

$$\begin{aligned} Ex'(t)x'(s) &= E[x(t) - Ex(t)]\cdot[x(s) - Ex(s)] \\ &= Ex(t)x(s) - Ex(t)Ex(s) = 0 \end{aligned}$$

It is useful to remember that, if a random process is orthogonal, then

$$E[x(t_1) + x(t_2) + \ldots]^2 = Ex^2(t_1) + Ex^2(t_2) + \ldots \quad (t_1 \neq t_2 \neq \ldots)$$

If $\mathbf{x}$ is a vector-valued random variable with components $x_1, \ldots, x_n$ (which are of course random variables), the matrix

$$[E(x_i - Ex_i)(x_j - Ex_j)] = E(\mathbf{x} - E\mathbf{x})(\mathbf{x}' - E\mathbf{x}') = \text{cov}\ \mathbf{x} \quad (42)$$

is called the *covariance matrix* of x.

A random process may be specified explicitly by stating the probability of simultaneous occurrence of any finite number of events of the type

$$x(t_1) \leq \xi_1, \ldots, x(t_n) \leq \xi_n;\ (t_1 \neq \ldots \neq t_n), \text{ i.e.,}$$

$$Pr[(x(t_1) \leq \xi_1, \ldots, x(t_n) \leq \xi_n)] = F_{x(t_1), \ldots, x(t_n)}(\xi_1, \ldots, \xi_n) \quad (43)$$

where $F_{x(t_1), \ldots, x(t_n)}$ is called the *joint probability distribution function* of the random variables $x(t_1), \ldots, x(t_n)$. The *joint probability density function* is then

$$f_{x(t_1), \ldots, x(t_n)}(\xi_1, \ldots, \xi_n) = \partial^n F_{x(t_1), \ldots, x(t_n)}/\partial\xi_1, \ldots, \partial\xi_n$$

provided the required derivatives exist. The expected value $Eg[x(t_1), \ldots, x(t_n)]$ of any nonrandom function of n random variables is defined by an n-fold integral analogous to (40).

A random process is *independent* if for any finite $t_1 \neq \ldots \neq t_n$, (43) is equal to the product of the first-order distributions

$$Pr[x(t_1) \leq \xi_1] \ldots Pr[x(t_n) \leq \xi_n]$$

If a set of random variables is independent, then they are obviously also uncorrelated. The converse is not true in general. For a set of more than 2 random variables to be independent, it is not sufficient that any pair of random variables be independent.

Frequently it is of interest to consider the probability distribution of a random variable $x(t_{n+1})$ of a random process given the actual values $\xi(t_1), \ldots, \xi(t_n)$ with which the random variables $x(t_1), \ldots, x(t_n)$ have occurred. This is denoted by

$$Pr[x(t_{n+1}) \leq \xi_{n+1} | x(t_1) = \xi_1, \ldots, x(t_n) = \xi_n] = \frac{\int_{-\infty}^{\xi_{n+1}} f_{x(t_1), \ldots, x(t_{n+1})}(\xi_1, \ldots, \xi_{n+1})d\xi_{n+1}}{f_{x(t_1), \ldots, x(t_n)}(\xi_1, \ldots, \xi_n)} \quad (44)$$

which is called the *conditional probability distribution function* of $x(t_{n+1})$ given $x(t_1), \ldots, x(t_n)$. The *conditional expectation*

$$E\{g[x(t_{n+1})]|x(t_1), \ldots, x(t_n)\}$$

is defined analogously to (40). The conditional expectation is a random variable; it follows that

$$E[E\{g[x(t_{n+1})]|x(t_1), \ldots, x(t_n)\}] = E\{g[x(t_{n+1})]\}$$

In all cases of interest in this paper, integrals of the type (40) or (44) need never be evaluated explicitly; only the *concept* of the expected value is needed.

A random variable x is *gaussian* (or *normally distributed*) if

$$f_x(\xi) = \frac{1}{[2\pi E(x - Ex)^2]^{1/2}} \exp\left[-\frac{1}{2}\frac{(\xi - Ex)^2}{E(x - Ex)^2}\right]$$

which is the well-known bell-shaped curve. Similarly, a random vector $\mathbf{x}$ is *gaussian* if

$$f_x(\xi) = \frac{1}{(2\pi)^{n/2}(\det \mathbf{C})^{1/2}} \exp\left[-\frac{1}{2}(\xi - E\mathbf{x})'\mathbf{C}^{-1}(\xi - E\mathbf{x})\right]$$

where $\mathbf{C}^{-1}$ is the inverse of the covariance matrix (42) of $\mathbf{x}$. A *gaussian random process* is defined similarly.

The importance of gaussian random variables and processes is largely due to the following facts:

Theorem 5. (*A*) *Linear functions (and therefore conditional expectations) on a gaussian random process are gaussian random variables.*

(*B*) *Orthogonal gaussian random variables are independent.*

(*C*) *Given any random process with means $Ex(t)$ and covariances $Ex(t)x(s)$, there exists a unique gaussian random process with the same means and covariances.*

17

Reprinted from pp. 285–300 of *Proc. 2nd AGARD Guided Missiles Seminar (Guidance and Control)*, AGARDograph 21, NATO Advanced Groups for Aerospace R & D, 1956

RECENT DEVELOPMENTS IN FIXED AND ADAPTIVE FILTERING

A. G. Carlton and J. W. Follin, Jr.*

SUMMARY

Classical optimal filtering methods have been extended to a large class of problems in which the input has incompletely specified characteristics. By minimax principles the optimal filter and the best input are determined. Two problems of time-varying filters are considered, first the optimal settling of filters to steady state and second the design of adaptive filters which adjust to varying or unknown environment.

SOMMAIRE

Les méthodes classiques de filtrage optimum ont été étendues à une large classe de problèmes, dans laquelle la grandeur d'entrée a des caractéristiques incomplètement spécifiées. En se basant sur des principes "minimax", le filtre optimum et la grandeur d'entrée la plus intéressante sont déterminés. Deux problèmes de filtres en fonction du temps, sont considérés, en premier l'ajustement optimum de filtre au régime permanent et deuxièmement l'étude de filtres adaptés qui ajustent au milieu inconnu ou variable.

1. INTRODUCTION

Linear filtering theory is largely based on the fundamental work of Wiener, "Extrapolation, Interpolation, and Smoothing of Stationary Time Series," 1950, which in many respects paralleled the independent work of Kolmogorov.

It may be well to review the problem considered by Wiener. He assumed that there was available the entire past history of a time series consisting of signal-plus-noise possibly correlated with the signal, that all processes involved were stationary and indeed ergodic to the second order, with known autocorrelation and cross-correlation functions. He wished to determine the realizable linear filter to apply to the signal in order to minimize the mean-square difference between the output and the message translated by an assigned positive or negative time interval. Wiener solved this problem by using variational analysis on the weighting function to obtain an integral equation, then by using subtle Fourier analysis to solve the integral equation. Multiple time series were handled by an extension of this technique.

It will be noted that the problem solved by Wiener contains two restrictions beyond the assumptions: first, optimization is restricted to linear functions; second, the loss function whose expectation is minimized is the squared error.

In attempting to extend filtering theory, it is appropriate to modify or eliminate various of these assumptions or restrictions. Various investigators have widened the field of permissible filters and dealt with alternative loss functions. Nonstationary processes have been considered; some trivial obvious results have been obtained, and some adaptive filters

*Applied Physics Laboratory, The Johns Hopkins University.

appear suitable, but little has been done that is both significant and rigorous. In this paper we shall remove the assumption of complete knowledge of the correlation functions and also indicate some minor extensions of the basic theory and techniques, consider the optimum filter with only portions of the signal history available, and attempt to classify the types of adaptive filters.

2. THE FREQUENCY APPROACH

The reader of Wiener's work will note that although his basic problem is formulated in terms of time series, correlation functions, and weighting functions, his solutions are expressed in terms of spectral densities and transfer functions. It appears reasonable, consequently, to set up the problem in the frequency domain.

From the elementary properties of spectral densities, we have

$$\sigma^2 = \int \left\{ \left| e^{i\omega\alpha} - F \right|^2 M + \left| F \right|^2 N \right\} d\omega . \tag{1}$$

where σ^2 is the mean-square error, $F(\omega)$ the filter transfer function, α the time translation, and M and N the signal-noise spectral densities, so normalized that the signal power is $\int M(\omega)\, d\omega$. All integrals are taken over the entire real frequency axis unless otherwise indicated, and the dependence of the variables on ω will usually not be indicated. It is assumed here and henceforth that the signal and noise are independent; this entails no loss in generality in the classical developments, where M can be regarded as the sum of the signal and signal-on-noise spectral densities, N as the sum of the noise and noise-on-signal spectral densities.

It is useful to consider the spectral densities as resolutions of the signal power into a continuum of frequency components. From this standpoint it is clear that ergodic properties are not relevant to the problem of <u>linear</u> filtering, although the optimal filter may be nonlinear if the second-order characteristics are not ergodic.

Before applying variations to minimize σ^2 by choice of F, let us indicate some extensions of this relation to problems beyond the original one. In the first place it will be noted that Eq. (1) is valid even though the power of signal, noise, or both is unbounded; it is not necessary that the correlation function exist. It will be seen that this may be of importance.

A trivial generalization of Eq. (1) is to replace $e^{i\omega\alpha}$ by the Fourier transform $Y(\omega)$ of any desired linear operator on the signal, e.g. by $i\omega$ for the derivative. Thus,

$$\sigma^2 = \int \left\{ \left| Y - F \right|^2 M + \left| F \right|^2 N \right\} d\omega . \tag{2}$$

Another easy extension is to the case in which it is desired to weight errors unequally for various frequency components. A symmetric nonnegative function $W(\omega)$, so normalized that $\int W(\omega)\, d\omega = 1$, could be inserted to obtain

$$\sigma_*^2 = \int \left\{ \left| Y - F \right|^2 M + \left| F \right|^2 N \right\} W \, d\omega , \tag{3}$$

which is of course formally equivalent to Eq. (2) with M and N replaced by MW and NW.

The next extension is to minimize σ^2 subject to a restriction on the mean power of the output or some other linear function of the output. As an example, if σ^2 must be minimized subject to the restriction that the output acceleration power must be less than β, i.e., that

$$\int (M + N) |F|^2 \omega^4 \, d\omega \leq \beta, \tag{4}$$

we should minimize

$$\sigma^2 + \lambda\beta = \int \left\{ |Y - F|^2 M + |F|^2 N + \lambda(M + N)|F|^2 \omega^4 \right\} d\omega, \tag{5}$$

and select λ to satisfy Eq. (4). In certain cases, especially with non-Gaussian processes, restraints such as Eq. (4) may preferably be applied only to the noise. In this case the integral to be minimized can be reduced to Eq. (2) by suitable definition of N. This type of side condition can be introduced formally as here, or can be used in the definition of the class over which F is optimized. Several simultaneous side conditions can be introduced in the same way.

The final extension to be considered is to replace the class of realizable transfer functions by other classes, say $\mathcal{F}$, as appropriate.

3. MINIMIZATION OF σ^2 BY $F \subset \mathcal{F}$

We now consider the selection of F within the class $\mathcal{F}$ to minimize σ^2. An increment J or F will produce an incremental σ^2 of

$$\sigma^2_{J + F} - \sigma^2_F = \int |J|^2 (M + N) \, d\omega + \int \bar{J} \left[(M + N) F - MY \right] d\omega, \tag{6}$$

a relation obtained by noting that every transfer function has an even real part and an odd imaginary part.

The transfer function F will be optimal in $\mathcal{F}$ if the right hand integral of Eq. (6) is nonnegative for every J such that $J + F \subset \mathcal{F}$ and $\int |J|^2 (M + N) \, d\omega$ is finite.

The absolutely optimal F is, from Eq. (6), evidently

$$F_0 = \frac{MY}{M + N} \tag{7}$$

The optimal realizable F is given by

$$F_R = \frac{1}{(M + N)^+} \left[\frac{MY}{(M + N)^-} \right]_+ \tag{8}$$

where the new symbols denote factorization and decomposition of a meromorphic function as

$$H \simeq H^+ H^- = H_+ + H_- , \tag{9}$$

H^+ being analytic and without poles or zeroes in the lower half plane, H^- the conjugate of H^+; H_+ and H_- have no poles in the lower half plane and upper half plane, respectively. Polynomial terms of H appear in H_+. If M + N is not factorable, $F_R = F_0$.

To check the validity of Eq. (8), substitute it in Eq. (6), obtaining

$$\sigma^2_{J + F_R} - \sigma^2_{F_R} = \int |J|^2 (M + N) \, d\omega + \int \bar{J} (M + N)^- \left[\frac{MY}{(M + N)^-} \right]_- d\omega ; \tag{10}$$

the latter integral is zero, by contour integration over the upper half plane, for J sufficiently convergent, i.e., for $\int |J|^2 (M + N) \, d\omega$ finite.

As an example let us consider the spectra

$$M = \theta/\omega^4 , \quad N = \phi , \quad \text{with } Y = 1 , \tag{11}$$

the absolute optimum (realizable with infinite delay) is then

$$F_0(p) = \frac{1}{1 + \phi p^4/\theta} \tag{12}$$

and the optimal realizable F is

$$F_R(p) = \frac{1 + \sqrt{2}\, p\,(\phi/\theta)^{1/4}}{1 + \sqrt{2}\, p\,(\phi/\theta)^{1/4} + p^2\,(\phi/\theta)^{1/2}} . \tag{13}$$

This transfer function is the zero velocity lag loop with .7 critical damping which will be discussed below. The corresponding errors are

$$\sigma_0^2 = .3535\, \phi^{3/4}\, \theta^{1/4}$$

$$\sigma_R^2 = 1.414\, \phi^{3/4}\, \theta^{1/4}. \tag{14}$$

A very useful relation can be obtained by manipulation of the second integral of Eq. (6) as follows:

$$\int \bar{J} \left[(M + N) F - MY \right] d\omega$$

$$= \int \bar{J}/\bar{F} \left[(M + N) |F|^2 - MY\bar{F} \right] d\omega$$

$$= \int \bar{J}/\bar{F} \left[(M + N) |F|^2 - \text{Re}\, MY\bar{F} - \text{Im}\, MY\bar{F} \right] d\omega. \tag{15}$$

This integral is zero for sufficiently convergent realizable J if

$$|F_R|^2 (M + N) = \text{Re}\, MY\bar{F} + \text{Re:Im}\, MY\bar{F}, \tag{16}$$

where Re:ImH represents the real complement to the given imaginary function such as to render H realizable.

This can be more formally expressed by the Bode relation

$$|F_R|^2 (M + N) = \frac{1}{i\pi} \int_* \frac{w (MY\bar{F})_w\, dw}{\omega^2 - w^2} , \tag{17}$$

where $\int_*$ indicates disregard of poles at $w = \pm\, \omega$,* provided MY does not diverge for large ω.

Solutions of Eq. (16) for simple forms of MY are readily obtained; for example,

$$|F_R|^2 (M + N) = \begin{cases} MY, & \text{if } MY \text{ is constant} \\ MY\bar{F}(ia), & \text{if } MY = \dfrac{c \cdot a^2}{a^2 + \omega^2} \\ MY + k, & \text{if } MY = c_0 + c_2 \omega^2 + c_4 \omega^4 \end{cases} \tag{18}$$

$\left[K \text{ determined by } \int \log |F_R|^2\, d\omega = 0 \right].$

By symmetry under the interchange of $M \leftrightarrow N$ and $F \leftrightarrow (Y - F)$ we obtain

$$|Y - F_R|^2 (M + N) = NY, \tag{19}$$

if NY is constant.

For ordinary filtering, with $Y = 1$, these define the optimal spectral density of the error signal in a servo type filter, and can be used to construct an adaptive filter which will become optimal for any signal spectrum.

Expressions for $|F_R|^2$, based on Eq. (16) are quite useful in optimizing filters with side conditions such as limited mean-square output, since the expressions for $|F_R|^2$ are frequently more convenient to use than those for F_R in determining the Lagrange multipliers.

4. MINIMAX FILTERING

Let us consider now some typically statistical problems, in which one has incomplete knowledge of the spectrum of noise, of signal, or both. We discuss below adaptive quasi-linear filters which appear suitable for cases in which one spectrum is completely unknown, and which can cope with cases involving a spectral density of known form but unknown magnitude.

The problems considered by Wiener were essentially probabilistic, i.e., the system is completely described in terms of appropriate probability measures; the problems we are now considering are statistical, in that we are dealing with a system defined by probabilities, some of which are unknown. Our problem in this case is one of statistical estimation of a function of the signal. Our estimating function should be optimal in some sense. One of the most logical criteria for an estimate, developed by Abraham Wald, is that it minimizes the maximum expected loss; that is, each filter is assessed on the basis of the expected loss with the possible system which is least favorable for the given filter, and the optimal filter is that filter for which the maximum expected loss is minimum. This formulation of statistical decision theory is very similar to two-person game theory, independently developed by von Neuman. We adopt this criterion and consider as optimal the minimax filter, with the loss function proportional to the squared error. In a completely prescribed system, the minimax linear filter is the Wiener optimal linear filter.

Minimax theory offers a strong justification for the use of linear filters. If the distribution functions of the processes are not known but the class of possible distributions includes Gaussian distributions, the minimax filter is linear, since with a linear filter the mean-square error is independent of the form of the distribution function, and with a nonlinear filter the mean-square error exceeds that with a linear filter when the processes are Gaussian.

Typical problems encountered in practice involve situations in which the noise process is known to be limited in power, in mean square velocity, etc.:

$$\int N \, d\omega = C_0, \int N\omega^2 \, d\omega = C_1, \text{ etc.} \tag{20}$$

Such restrictions can be put in the general form

$$\int N\theta \, d\omega = 1, \tag{21}$$

where θ is a prescribed symmetric nonnegative function. It can be shown straightforwardly that the variation in σ^2 due to variation n in N is a nonnegative function of n, zero if and only if n = 0, plus

$$\int n \left[|F_{M,N}|^2 - \lambda\theta \right] d\omega, \tag{22}$$

where $F_{M,N}$ is the optimal transfer function with M and N, and λ is a Lagrange multiplier to be selected to satisfy Eq. (21). From Eq. (22) and the fact that n + N must be nonnegative, it follows that the maximum N is N_0 given by

$$\begin{aligned} |F_{M,N_0}|^2 &= \lambda\theta & \omega: N_0 > 0 \\ &\le \lambda\theta & \omega: N_0 = 0. \end{aligned} \tag{23}$$

This result is easily generalized to the case of several inequality restrictions

$$K_j \int N\theta_j \, d\omega = 1 \qquad (j = 1, 2 \ldots, k) \tag{24}$$

with the K_j not specified but ≥ 1.

The maximum N is N_0 satisfying

$$|F_{M,N_0}|^2 = \sum \lambda_j K_j \theta_j \qquad \omega : N_0 > 0$$

$$\leq \sum \lambda_j K_j \theta_j \qquad \omega : N_0 = 0. \tag{25}$$

with the λ_j, K_j satisfying Eq. (24) and also

$$\lambda_j (K_j - 1) = 0, \qquad (j = 1, 2 \ldots, k). \tag{26}$$

Similar results are obtained for cases in which the signal spectral density is subject to one or more equalities or inequalities, and where both spectra are limited only by such restrictions.

The results just given have derived the maximin spectrum or spectra, but our object is to determine the minimax filter. For this purpose we now prove that

$$\underset{F}{\text{Min}} \ \underset{N}{\text{Max}} \ \sigma^2_{N,F} = \underset{N}{\text{Max}} \ \underset{F}{\text{Min}} \ \sigma^2_{N,F}$$

and that the minimax F is F_{M,N_0}.

To prove this, we observe that

$$\underset{F}{\text{Min}} \ \underset{N}{\text{Max}} \ \sigma^2_{N,F} \leq \text{Max} \ \sigma^2_{N,F_{M,N_0}}$$

$$= \underset{N}{\text{Max}} \int \left\{ N |F_{M,N_0}|^2 + M |Y - F_{M,N_0}|^2 \ d\omega \right.$$

$$= \sum \lambda_j + \int M |Y - F_{M,N_0}|^2 \ d\omega = \underset{N}{\text{Max}} \ \underset{F}{\text{Min}} \ \sigma^2_{N,F}.$$

We have thus shown that

$$\underset{F}{\text{Min}} \ \underset{N}{\text{Max}} \ \sigma^2_{N,F} \leq \underset{N}{\text{Max}} \ \underset{F}{\text{Min}} \ \sigma^2_{N,F}. \tag{27}$$

But by the fundamental theorem of game theory,

$$\underset{F}{\text{Min}} \ \underset{N}{\text{Max}} \ \sigma^2_{N,F} \geq \underset{N}{\text{Max}} \ \underset{F}{\text{Min}} \ \sigma^2_{N,F}. \tag{28}$$

from which it follows at once that the "game" is determined, with the minimax filter being F_{M,N_0} and the maximin N being the N_0 previously defined.

The basic result of the minimax approach to optimum filtering is that the errors depend in the second order on the spectra and the form of the filter. As a consequence, if a suitable approximation to the optimal form is used and the parameters are adjusted properly the resulting system will be satisfactory.

5. TIME-VARYING FILTERS

Let us now consider the problem of filtering when only a finite and perhaps fragmentary history of the signals is available. In this case the filter parameters are variable, and we must assume a particular form of transfer function. In general, the steady-state optimal filter with variable band-pass and damping is best. We may attack this problem in the time domain by considering the rate of change of the filtering errors and adjusting parameters to maximize the rate of decrease of the error.

For a linear system the tracking accuracy may be described in terms of the variances of the tracking error and error rate. As an example let us consider the simple zero velocity lag feedback system in Fig. 1. The input consisting of signal and noise $x(t)+x_n(t)$ is at the left, the output, x_c, at the right. The equations of the system are

$$\frac{dx_c}{dt} = \dot{x}_c + b(x - x_c) + bx_n$$

$$\frac{dx_{\dot{c}}}{dt} = a(x - x_c) + ax_n \; .$$

(29)

Note, in explanation of the notation, that $x_{\dot{c}} \neq \dot{x}_c$.

Actually the principal interest centers upon the errors $\epsilon_c = x - x_c$, $\epsilon_{\dot{c}} = \dot{x} - x_{\dot{c}}$. In the second diagram of Fig. 1 is shown the error loop equivalent to the original signal loop. The signal x now appears as an acceleration input to the first integrator. This is a significant advantage when, as is often the case, the acceleration spectrum of the signal is known.

So far a, b are unrestricted. If $\ddot{x}$ and x_n both have flat spectral densities θ, ϕ respectively then this filter is optimal with the values of a and b previously derived. The present purpose is to extend the optimization to the transient period. The gains a, b in this case are time functions and the resulting system, while not necessarily optimum among all possible systems, is the best obtainable with a given structure.

The key to the solution lies in setting up the differential equations relating the variances and covariances of the integrator outputs ϵ_c , $\epsilon_{\dot{c}}$. Write Eqs. (29) in terms of the errors and express the solution in the neighborhood of t as a power series in Δt. Terms beyond the first degree in Δt are not required. The result is most easily obtained by direct use of the second figure

$$\epsilon_c = \epsilon_{c_0} + \epsilon_{\dot{c}_0}\Delta t - b\epsilon_{c_0}\Delta t + b\int_t^{t+\Delta t} x_n\, dt + O(\Delta t^2)$$

(30)

$$\epsilon_{\dot{c}} = \epsilon_{\dot{c}_0} - a\epsilon_{c_0}\Delta t + \int_t^{t+\Delta t} \ddot{x}\, dt + a\int_t^{t+\Delta t} x_n\, dt + O(\Delta t^2).$$

(31)

Square Eq. (30) and average over the ensemble of inputs $\ddot{x}$, x_n. Then denoting resulting variances and covariances by $\widehat{\epsilon_c}$, $\widehat{\epsilon_{\dot{c}}}$, $\widehat{\epsilon_{c\dot{c}}}$, respectively

$$\widehat{\epsilon_c} - \widehat{\epsilon_{c_0}} = 2\widehat{\epsilon_{c\dot{c}_0}}\Delta t - 2b\widehat{\epsilon_{c_0}}\Delta t + b^2\phi\Delta t + O(\Delta t^2)$$

(32)

where ϕ is the spectral density (assumed constant of x_n).

The term ϕ is derived as follows: $\phi(a)$ is the autocorrelation function of x_n.

$$\left[\int_t^{t+\Delta t} x_n\, dt \int_t^{t+\Delta t} x_n\, dt\right]$$

$$= \int_t^{t+\Delta t}\int_t^{t+\Delta t} \left[x_n(u)\, x_n(v)\right] du\, dv$$

$$= \int_0^{\Delta t}\int_0^{\Delta t} \phi(u-v)\, du\, dv$$

$$= \phi\int_0^{\Delta t}\int_0^{\Delta t} \delta(u-v)\, du\, dv$$

$$= \phi\,\Delta t.$$

Dividing Eq. (32) by Δt and letting $\Delta t \to 0$

$$\frac{d\widehat{\epsilon_c}}{dt} = 2\widehat{\epsilon_{c\dot{c}}} - 2b\widehat{\epsilon_c} + b^2\phi .$$

(33)

Similarly, assuming the spectral density θ of $\ddot{x}$ is flat

$$\frac{d\hat{\epsilon}_{\dot{c}}}{dt} = -2a\,\hat{\epsilon}_{c\dot{c}} + a^2\phi - \theta \tag{34}$$

$$\frac{d\hat{\epsilon}_{c\dot{c}}}{dt} = \hat{\epsilon}_{\dot{c}} - a\hat{\epsilon}_{c} - b\hat{\epsilon}_{c\dot{c}} + a\,b\,\phi . \tag{35}$$

These variance and covariance equations may be used to adjust the gains to get optimum tracking. This will surely result if functions a, b can be found making the right hand members of Eqs. (33), (34), and (35) simultaneously minimum for this will make $\hat{\epsilon}_c$ and $\hat{\epsilon}_{\dot{c}}$ decrease at maximum rate.

This simultaneous minimum does occur and at

$$a = \frac{\hat{\epsilon}_{c\dot{c}}}{\phi}, \qquad b = \frac{\hat{\epsilon}_{c}}{\phi}, \tag{36}$$

as can be seen by setting the partial derivatives of all three right members with respect to a and b equal to zero. The resulting system has optimum tracking and rate of settling and the variances facilitate evaluating performance of the system.

This optimization of the transient behavior has a byproduct, the known steady-state result. For in this case the left hand members of Eqs. (33), (34), and (35) vanish when using Eq. (36)

$$a = \left(\frac{\theta}{\phi}\right)^{1/2}, \qquad b = \sqrt{2}\left(\frac{\theta}{\phi}\right)^{1/4} . \qquad \text{(steady-state)} \tag{37}$$

For simplicity the input acceleration spectrum has been assumed flat. There are several ways to deal with nonflat spectra. For example it can be shown that for a general $\theta(\omega)$ with autocorrelation function $\phi(t)$, θ in Eq. (34) would be replaced by $2\int_0^t W(t,\tau)\,\phi(t-\tau)\,d\tau$. Here $W(t,\tau)$ is the impulsive response of $\epsilon_{\dot{c}}$ to $\ddot{x}$. Or a flat sprectrum could be filtered by $\theta(\omega)^+$ $\left[\theta(\omega)^- = \theta(\omega)^+\,\theta(\omega)^-\right]$ to give a signal of spectrum $\theta(\omega)$ into the integrator.

This last is particularly simple in the Markoff case $\theta = \theta_0/(C^2 + \omega^2)$ where $\theta^+ = 1/(C + j\omega)$. The original error tracking loop could be modified as shown in the third figure thus introducing one additional integrator. Proceeding as before, six variance-covariance equations result. In general, with a system involving n integrators (n'th order differential operation), there will be one-half n(n + 1) variance equations although generally some are trivial.

The transient filtering problem has been discussed on the basis that the noise and signal spectra are known. This leads to a solution of the optimum settling time of a filter and to the best combination of parameters even if the transfer function is not optimal. If the noise and signal spectra are not known then the above technique of computing variances fails and other methods must be used. If sufficient time is available it is possible to measure the spectral densities and use the variance methods, but less cumbersome methods are desirable.

6. ADAPTIVE FILTERS

The discussion which follows will concern adaptive systems, i.e., those which change parameters or adapt as a function of the environment. In general, the rate of change of parameters is slow compared to the data rate of the input so that they may be treated as time-varying rather than nonlinear systems. If the response of the adapting loop is fast, paper analysis is impossible and simulation techniques are needed. In designing an adaptive system it is necessary to consider the response and stability of the loop as well as the source of intelligence to be employed in adjusting the system.

Adaptive systems may be classified according to several distinct criteria as follows:

a. Object of adaptive loop

   (1) Setting of gain or transfer function.

   (2) Adjusting output level or other parameter.

   (3) Adjusting stability margins of main loop.

b. Source of information

   (1) Measurements of normal input or output.

   (2) Injection of tracer signal outside band-pass of normal input.

   (3) Time sharing tracer signal.

   (4) Amplitude or phase of self-excited oscillations.

c. Type of system

   (1) Open loop, i.e., system adjusted according to measurements on the input.

   (2) Closed loop, i.e., system measures signal or tracer output.

In addition, all adaptive systems may be classified according to standard servo practice as electrical or mechanical, digital or analog, etc., but such distinctions are not desirable for the present purpose. The class of adaptive servos ranges from standard AGC and AFC loops to servo driven auto-transformers for voltage regulation to more sophisticated optimal filtering loops.

Let us now look at examples of the three different adaptive systems listed under the first criterion (a). These are the zero velocity lag tracking loop mentioned earlier, a similar filter with limited output acceleration, and a system for maintaining a servo loop as tight as possible without instability when the loop gain is slowly varying or not known accurately.

Fig. 2 shows a simple servo where only the gain is varied and the loop gain is to be maximized (possibly to minimize the effects of a variable back torque from the load). The unknown gain is assumed to be in the servo. The band-pass filter passes the frequency at which instability is expected and this signal is detected and used to adjust the gain to damp the oscillations.

Let λ_0 be the gain at which the loop is neutrally stable. Then the rate of buildup or decay of oscillations is proportioned to $(\lambda/\lambda_0 - 1)$. The amplitude, z, of the oscillations satisfies the equation

$$\dot{z} = k_1 z(\lambda - \lambda_0)/\lambda_0 \doteq k_1 z_0 (\lambda - \lambda_0)/\lambda_0 \tag{38}$$

If the signal is picked off at point A, the control equation is

$$\dot{\lambda} = -g(s)(z - z_0) \tag{39}$$

while if it is picked off at point B, we have

$$\dot{\lambda} = g(s)\,\lambda\,(z - z_0)$$

$$\doteq -\lambda_0\, g(s)\,(z - z_0). \tag{40}$$

Combining these equations we find that in case A

$$\left[s^2 + \frac{k_1 z_1}{\lambda_0}\, g(s)\right]\lambda = -k_1 z_1 \lambda_0\, g(s)\left(\frac{1}{\lambda_0}\right), \tag{41}$$

while for case B

$$\left[s^2 + k_1 z_1\, g(s)\right]\lambda = k_1 z_1\, g(s)\,\lambda_0\,. \tag{42}$$

While both systems can be made stable, and both have $\lambda = \lambda_0$ as the steady-state solution, the transient response of the adaptive loop depends on the required gain in case A, but in case B the transient is fixed and the system has zero velocity lag with respect to variations of λ_0. If λ_0 increases linearly with time then $\lambda = \lambda_0$ but $z > z_1$ with $\dot{z} = 0$. Thus a closed-loop adaptive system is better here. These statements are subject to modification if there exists dead space or friction in the main or adaptive loop; actually simulation is then required to determine the behavior.

If the main loop shaping network f(s) is properly chosen then a very good servo response is obtainable but input signals at the loop resonance frequency must be avoided. This design is especially useful for a regulator, i.e., when $x_i = 0$ and the servo is designed to counter the back torque.

Fig. 3 is a diagram of an adaptive filter in which the r.m.s. output acceleration due to noise is limited. The adaptive loop is very simple and, with the gains as shown, has a response which is independent of input noise spectral density. This can be seen from the fact that the error in r.m.s. acceleration is proportional to $\Delta\lambda/\lambda$ and hence the control equation is

$$\dot{\lambda} \sim \Delta\lambda, \tag{43}$$

if the filter g(s) is unity.

If it is desired to have the adaptive loop time constant proportional to the main loop time constant then the λ must be replaced by λ^2 and the control equation is

$$\dot{\lambda} \sim \lambda\;\Delta\lambda. \tag{44}$$

If the noise is flat or of known spectral shape it is possible to measure its amplitude at the input to the filter and compute the r.m.s. acceleration from the known transfer function of the filter. It is then possible to use an open loop method of adjusting λ but, while this eliminates adaptive loop stability problems it does not have the accuracy in adjusting λ. If the noise does not have the expected spectrum the filter f(s) will not be optimal in shape; however the performance of the system will deviate in the second order if λ is correct, but errors in λ may give first order effects on system performance due to violation of the constraints.

While this loop is very simple we have discussed it because some simulator studies of the effects of nonlinearities may be of interest. The nonlinearity concerned is that due to a fixed limiter inserted in the loop as shown at the bottom of Fig. 3. The analysis was carried out using various fixed values of λ and noise rather than closing the adaptive loop as above.

Fig. 4 shows the results obtained. The solid upper curve shows σ^2 vs. λ for an unlimited system. If the limiter is inserted then the dashed curve is appropriate; the minimum is only a few percent above the optimum at the minimum. The lower half of the figure shows the effect of the limiter on the output acceleration before and after the limit. The result of the simulator study was that the minimum in σ^2 occurs almost exactly at $a_1 = L$, hence the simple adaptive system just described forms a very sophisticated tracking loop.

Fig. 5 is a block diagram of the zero velocity lag tracking loop in which we do not know the signal or noise spectra although we assume the noise to be nearly flat. In order to adjust band-pass we may use the result, Eq. (18) that the error signal spectrum is proportional to the noise spectrum when the loop is optimal. While the transfer of the loop is not correct if the noise is not flat or if the signal is not that assumed, it is still true that adjusting the loop band-pass so that the error signal spectrum is flat is nearly optimum.

The method of measurement is to use two filters f_1 (s), a low-pass filter covering the band-pass of the main loop, and $f_2(s)$ covering an equal band-pass just above the main loop and take the ratio of the outputs. The optimum shape of such filters has not been determined but simulator runs show suitable performance for simple filters. $f_2(s)$ should have a finite band-pass because the actual high frequency noise is unimportant; only the noise in the vicinity of the main loop band-pass is important.

In the loop as shown the obvious scale factors have been inserted to make the response frequency (relative to the main loop) independent of the value of the spectra. The filter $f_3(s)$ determines the band-pass of the adaptive loop and, in order to have minimum r.m.s. errors in λ, $f_3(s)$ must be adjusted so that the lags in following changes in the spectra are balanced by the fluctuations in the noise out of the detectors. From this criterion we can determine the adaptive loop band-pass as $\omega_A \sim \sqrt{\lambda/T}$, where $1/T \sim \dot{\phi}/\phi$ is the effective time constant relating to the change in input spectra. However, if step changes in the input signals are contemplated then the adaptive loop should be as tight as stability dictates and the gain settings in the figure are correct.

The ratio of the filter outputs minus one is proportional to $\delta\lambda/\lambda$ so that

$$\dot{\lambda} \sim \lambda \, \Delta \lambda \tag{45}$$

and the band-pass in the adaptive loop is proportional to that in the main loop.

In all of the adaptive systems considered it is easy to specify the gain changes to keep the loop dynamically similar for different inputs, but it is harder to specify the exact band-pass or the shape of the filters in the adaptive loop. It is always possible to make a linear stability analysis, if the inputs are fixed, and the noise out of the squaring circuits can be computed for Markovian noise but no general theory of optimal design exists.

There are many ways of instrumenting adaptive servos which give adequate performance and the effects of nonlinearities and complexity must be considered carefully if a satisfactory design is to be obtained. For example, the use of smoothed absolute value instead of r.m.s. leads to only a few percent more noise in the adaptive loop. As another example the division in the last example may be replaced by a subtraction if the dynamic range is not too great. At low input signals the loop is then sluggish but the tracking error is small due to the small input.

Underlying the design of adaptive servos is the assumption of relatively slow, or only occasional, changes in environment. If rapid changes occur a different main loop, possibly nonlinear, is required. Adaptive loops may be called quasi-linear but because they are nonlinear no general method of analysis has emerged to determine optimal performance as a standard of comparison with specific loops, or to check instrumentation approximations.

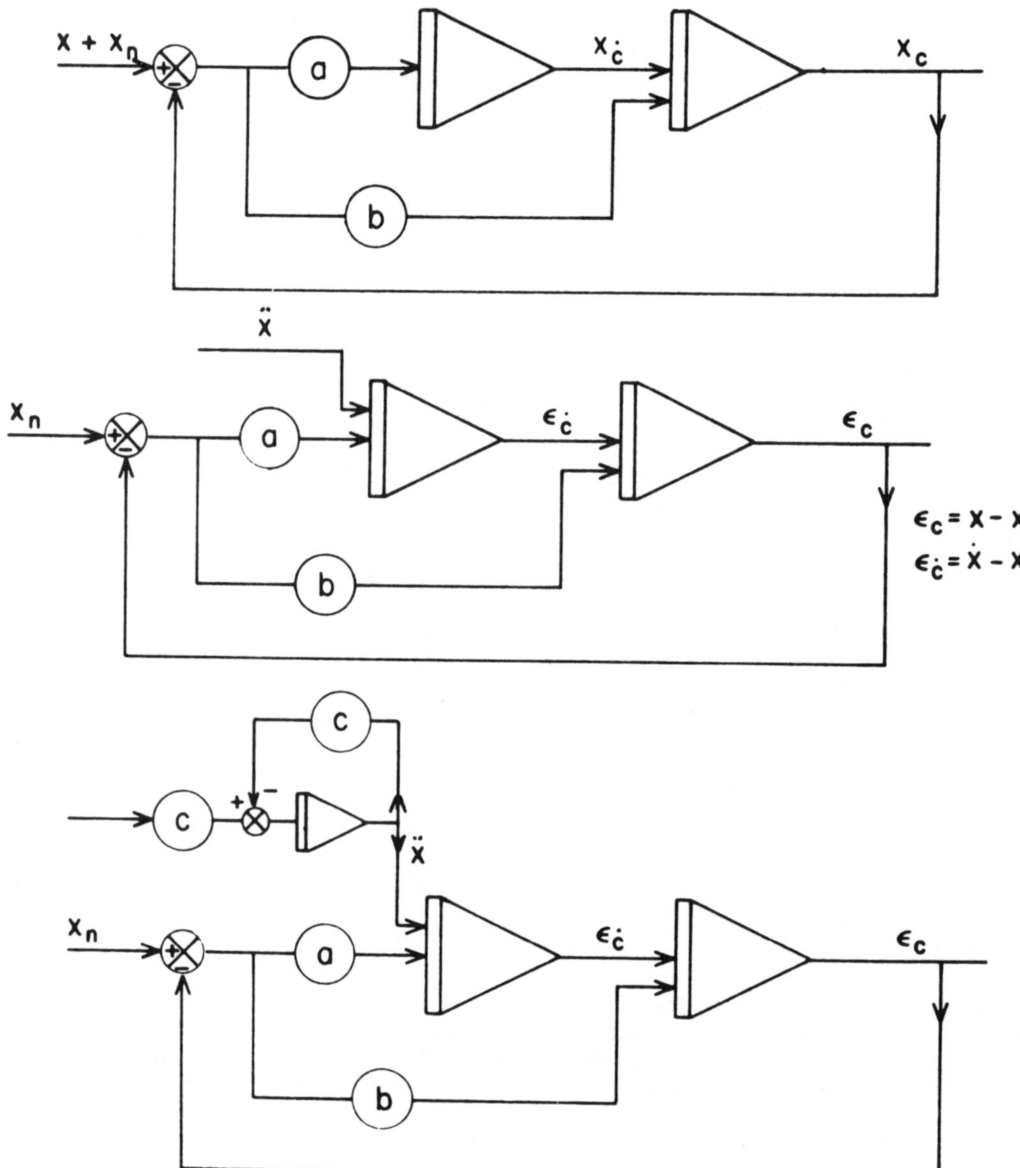

Fig. 1. Zero velocity lag feedback system.

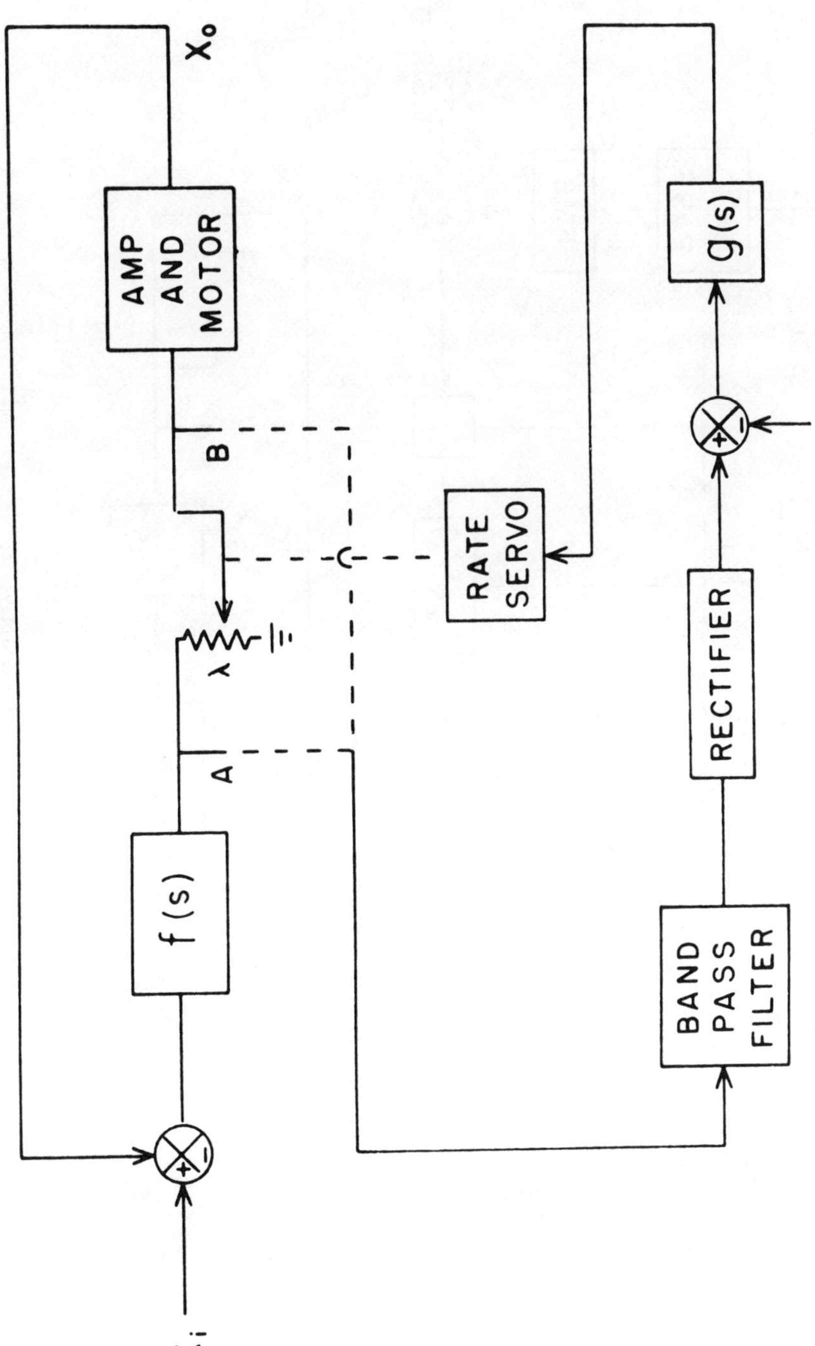

Fig. 2. Maximized loop gain servo.

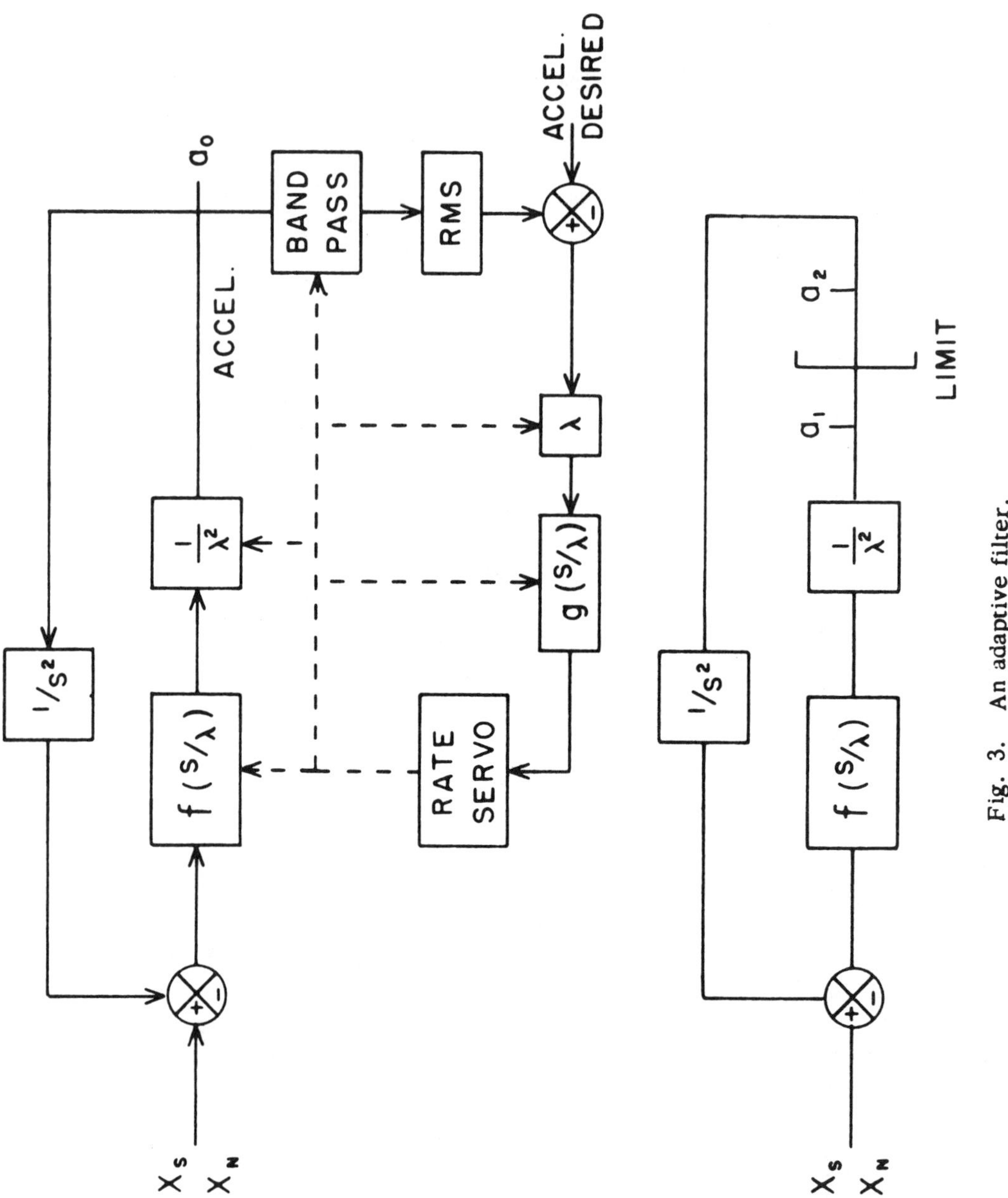

Fig. 3. An adaptive filter.

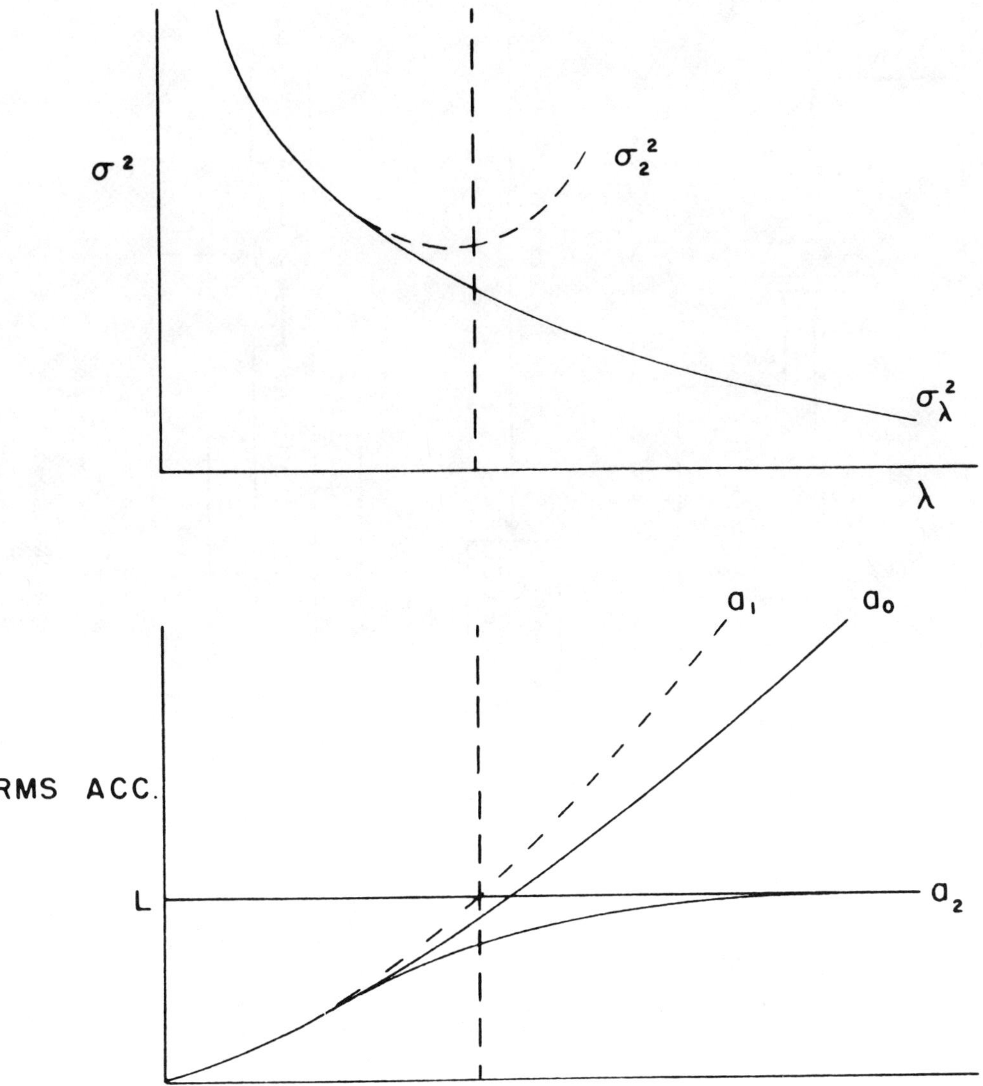

Fig. 4. Results of system of Fig. 3

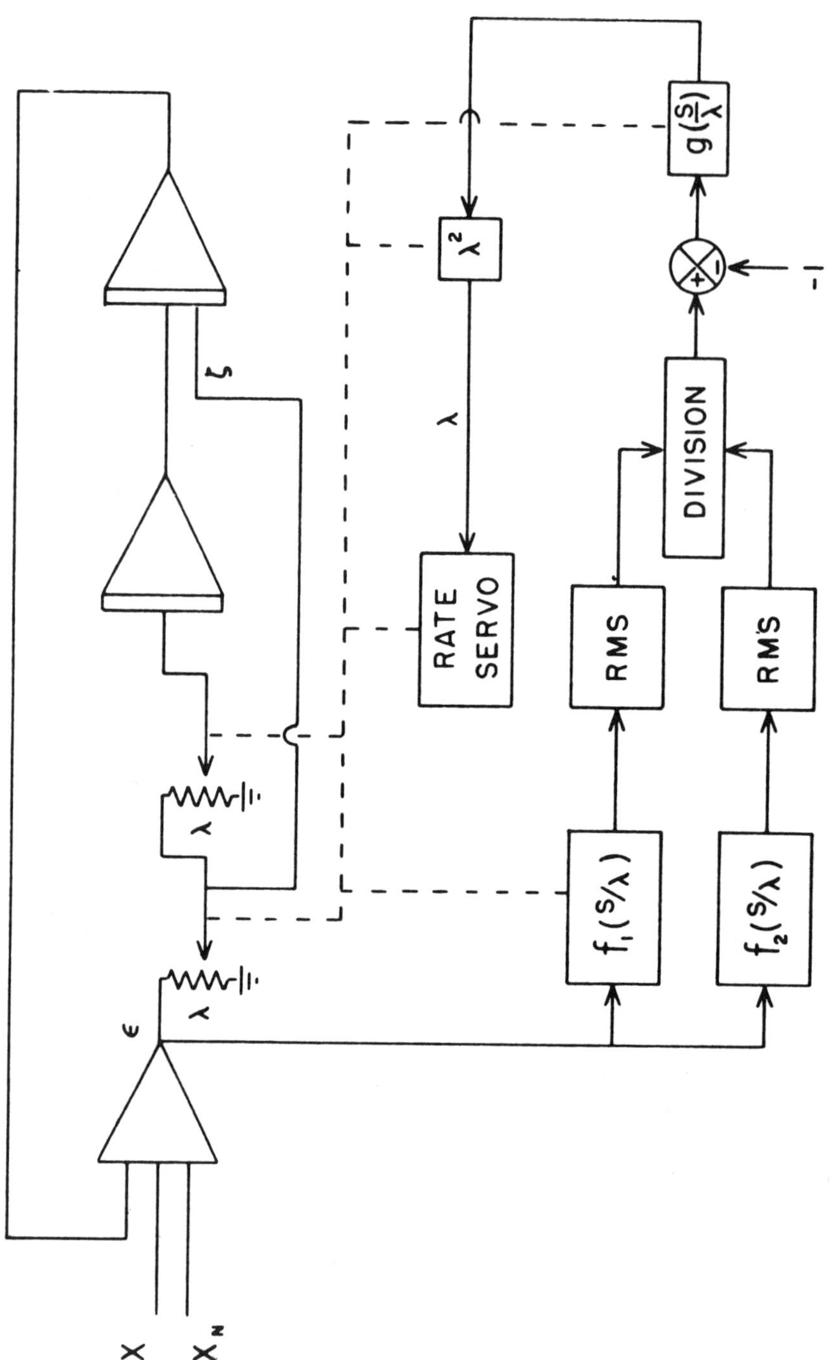

Fig. 5. Zero velocity lag tracking loop.

18

Reprinted from pp. 55–63 of *Theory and Applications of Kalman Filtering*, AGARDOgraph 139, C. T. Leondes, ed., NATO Advanced Groups for Aerospace R & D, 1970

FURTHER COMMENTS ON THE DERIVATION OF KALMAN FILTERS, SECTION II: GAUSSIAN ESTIMATES AND KALMAN FILTERING

Y. Genin

1. INTRODUCTION

The aim of this contribution is twofold. First, it will give the reader, who is familiar with mean squares techniques, sufficient background to gain easy access to Kalman filter theory. To this end, the intimate connection between the Kalman and the Gaussian estimation theory will be pointed out: more precisely, it will be shown that the application of results obtained by Gauss, at least in essence, to discrete stochastic linear processes leads to the Kalman filter.

The first part recalls the main points of the Gauss theory[6], as slightly extended by Plackett[12]. The core of this section is a restatement of two classical theorems, which are of primary importance in modern linear filtering techniques: the first constitutes the cornerstone of recursive filtering, while the second offers its dual form, i.e. accumulative filtering.

The second part is devoted to the application of these theorems to discrete stochastic processes and shows how the Kalman filter and its dual form fit naturally in the framework provided by Gauss.

2. THE GAUSSIAN LEAST SQUARES THEORY

2.1 Statement of the Problem

Consider the following equation of measurements:

$$z = Hx + v , \qquad (1)$$

where z is a N-dimensional vector of measurements, x a n-dimensional constant state vector to be determined ($n < N$), H a $N \times n$ matrix of maximal rank and v a N-dimensional vector of errors with zero mean and positive definite covariance matrix R.

The problem is to find an estimate $\hat{x}$ of the unknown state x as a linear combination of the measurements z,

$$\hat{x} = Az \qquad (2)$$

so that the estimate $\hat{x}$ be unbiased and have the smallest variance for each of its elements.

2.2 A Deterministic Problem

Let us consider a particular problem, in which Equation (1) holds[6]. The z vector is known and is assumed to take the value $\bar{z}$. Let us temporarily disregard probabilistic and statistical considerations and seek a "good" approximation $\bar{x}$ of the state x, in the form of a linear combination of the known measurements $\bar{z}$. If we arbitrarily decide to measure the quality of the approximation by adopting the usual mean squares criterion

$$(\bar{z} - H\bar{x})^T W(\bar{z} - H\bar{x}) = \text{minimum} , \qquad (3)$$

with W a positive definite weighting matrix, the best approximation is readily found to be

$$\bar{x} = (H^T WH)^{-1} H^T W\bar{z} , \qquad (4)$$

which solves this simple deterministic problem.

2.3 Theorem 1

Coming back to our main problem (Section 2.1), we shall now prove the following fundamental theorem. *The unbiased estimate* $\hat{x}$, *which is a linear combination of the measurements* z *and has the smallest variance for each of its components, is given by*

$$\hat{x} = (H^T R^{-1} H)^{-1} H^T R^{-1} z . \qquad (5)$$

It has the same form as Equation (4), the solution of the deterministic problem in which the weighting matrix W *is set equal to* R^{-1} .

The estimate is readily proved to be unbiased, for its mean value

$$E(\hat{x}) = (H^TR^{-1}H)^{-1}H^TR^{-1}E(z)$$

may be written in view of Equation (1),

$$E(\hat{x}) = (H^TR^{-1}H)^{-1}H^TR^{-1}[Hx + E(v)]$$

$$= x \,. \tag{6}$$

The covariance matrix P of the estimate $\hat{x}$ is defined by

$$P = E[(\hat{x} - x)(\hat{x} - x)^T]$$

which, in this case, has the form

$$P = (H^TR^{-1}H)^{-1}H^TR^{-1}E(vv^T)R^{-1}H(H^TR^{-1}H)^{-1}$$

$$= (H^TR^{-1}H)^{-1} \,. \tag{7}$$

Before proving that each diagonal element of the P matrix is minimized by Equation (5), let us first establish the following important lemma. *Lemma. Let* M *be any* $n \times n$ *non-negative definite matrix. Then the unbiased estimate* $\hat{x}$ *, which is a linear combination of the measurements* z *and minimizes the quadratic form*

$$E[(\hat{x} - x)^TM(\hat{x} - x)] \tag{8}$$

is again given by Equation (5).

Suppose this proposition were false and let $\hat{x} = Bz$, the solution with a matrix B different from $(H^TR^{-1}H)^{-1}H^TR^{-1}$. Without restriction, M may be assumed to be symmetric, for (8) is a homogeneous quadratic form. If it has multiple eigenvalues, it may be submitted to the Gauss-Schmidt orthogonalization procedure[3] so that, in any case, (8) may be written

$$\sum_{j=1}^{n} \lambda_j q_j^T E[(\hat{x} - x)(\hat{x} - x)^T] q_j \,, \tag{9}$$

with q_j the eigenvector corresponsing to the eigenvalue λ_j of the matrix M . Note that the eigenvalues λ_j are all real and non-negative, for M is a non-negative definite real symmetric matrix. On the other hand, since the estimate $\hat{x}$ is unbiased, we have

$$BH = I \,, \tag{10}$$

with I the identity matrix. The quadratic form (9) is therefore equivalent to

$$\sum_{j=1}^{n} \lambda_j q_j^T BRB^T q_j \,,$$

which may be transformed into

$$\sum_{j=1}^{n} \lambda_j q_j^T BRB^T q_j = \sum_{j=1}^{n} \lambda_j q_j^T (H^TR^{-1}H)^{-1} q_j + \sum_{j=1}^{n} \lambda_j q_j^T [B - (H^TR^{-1}H)^{-1}H^TR^{-1}] R [B - (H^TR^{-1}H)^{-1}H^TR^{-1}]^T q_j \tag{11}$$

in view of Equation (10) and the identity

$$B = (H^TR^{-1}H)^{-1}H^TR^{-1} + [B - (H^TR^{-1}H)^{-1}H^TR^{-1}] \,. \tag{12}$$

Since in the summation (11) each term is non-negative, we are left with

$$B = (H^TR^{-1}H)^{-1}H^TR^{-1} \,, \tag{13}$$

which proves the lemma.

A direct consequence of the lemma is that the estimate Equation (5) has the smallest variance for each of its elements, as can be seen by choosing all elements of M equal to zero, except for a unit element anywhere on the diagonal.

2.4 Theorem 2 (Recursive Estimation)

Suppose now a partition of the measurements vector z in two subvectors z_{k-1}, z_k. Partitioning v and H accordingly, Equation (1) may be written

$$\left.\begin{aligned} z_{k-1} &= H_{k-1}x + v_{k-1} \\ z_k &= H_k x + v_k \end{aligned}\right\} \tag{14}$$

and we assume the set v_{k-1} to be uncorrelated with the set v_k, so that R has the form

$$R = \begin{bmatrix} R_{k-1} & 0 \\ 0 & R_k \end{bmatrix} . \tag{15}$$

Let $\hat{x}_{k-1}$ and P_{k-1} be the minimum variance unbiased estimate and its associated covariance matrix, defined on the subset z_{k-1} only. Then *the minimum variance unbiased estimate* $\hat{x}_k$ *defined on the set of measurements* $[z_{k-1}, z_k]$ *may be obtained without reprocessing the subset* z_{k-1} *and is given by*

$$\hat{x}_k = \hat{x}_{k-1} + K_k(z_k - H_k\hat{x}_{k-1}) , \tag{16}$$

with K_k a gain matrix defined by

$$K_k = P_{k-1}H_k^T[H_kP_{k-1}H_k^T + R_k]^{-1} , \tag{17}$$

while the covariance matrix P_k is obtained by

$$P_k = (I - K_kH_k)P_{k-1} . \tag{18}$$

To demonstrate the above relations, note that, in view of Equation (7), the inverse covariance matrix is

$$\begin{aligned} P_k^{-1} &= H^TR^{-1}H \\ &= H_{k-1}^TR_{k-1}^{-1}H_{k-1} + H_k^TR_k^{-1}H_k \\ &= P_{k-1}^{-1} + H_k^TR_k^{-1}H_k . \end{aligned} \tag{19}$$

This expression has a classical form in matrix algebra, known as the Frobenius form, so that the inverse may be readily obtained:

$$\begin{aligned} P_k &= P_{k-1} - P_{k-1}H_k^T[H_kP_{k-1}H_k^T + R_k]^{-1}H_kP_{k-1} \\ &= (I - K_kH_k)P_{k-1} . \end{aligned} \tag{20}$$

On the other hand, in view of Equation (5), the new estimate $\hat{x}_k$ has the form

$$\begin{aligned} \hat{x}_k &= P_kH^TR^{-1}z \\ &= P_k[H_{k-1}^TR_{k-1}^{-1}z_{k-1} + H_k^TR_k^{-1}z_k] . \end{aligned} \tag{21}$$

Combination of Equations (20) and (21) yields the result

$$\begin{aligned} \hat{x}_k &= (I - K_kH_k)P_{k-1}[H_{k-1}^TR_{k-1}^{-1}z_{k-1} + H_k^TR_k^{-1}z_k] \\ &= (I - K_kH_k)\hat{x}_{k-1} + (I - K_kH_k)P_{k-1}H_k^TR_k^{-1}z_k \\ &= (I - K_kH_k)\hat{x}_{k-1} + P_{k-1}H_k^TR_k^{-1}z_k - \\ &\quad - P_{k-1}H_k^T[H_kP_{k-1}H_k^T + R_k]^{-1}[H_kP_{k-1}H_k^T + R_k]R_k^{-1}z_k + \\ &\quad + P_{k-1}H_k^T[H_kP_{k-1}H_k^T + R_k]^{-1}z_k \\ &= \hat{x}_{k-1} + K_k(z_k - H_k\hat{x}_{k-1}) . \end{aligned} \tag{22}$$

which completes the proof of the theorem.

The generalization to an arbitrary number of subsets of measurements z_k uncorrelated with each other is obvious.

2.5 Theorem 3 (Accumulative Estimation)

With the same definitions as above for z_{k-1}, z_k, R_{k-1}, R_k, $\hat{x}_{k-1}$ and P_{k-1}, let us now denote by $\hat{x}_k$ the minimum variance unbiased estimate defined on the subset z_k only, and let P_k be the corresponding covariance matrix; then, *the minimum variance unbiased estimate* $\hat{x}$ *defined on the two subsets* $[z_{k-1}, z_k]$ *is directly obtained by*

$$\hat{x} = P[P_{k-1}^{-1}\hat{x}_{k-1} + P_k^{-1}\hat{x}_k] \quad . \tag{23}$$

with a covariance matrix P computed from the relation

$$P^{-1} = P_{k-1}^{-1} + P_k^{-1} \quad . \tag{24}$$

The proof of this theorem is a direct consequence of Equations (5) and (7); in fact, we have

$$\begin{aligned} P^{-1} &= H^TR^{-1}H \\ &= H_{k-1}^TR_{k-1}^{-1}H_{k-1} + H_k^TR_k^{-1}H_k \\ &= P_{k-1}^{-1} + P_k^{-1} \end{aligned} \tag{25}$$

and

$$\begin{aligned} \hat{x} &= PH^TR^{-1}z \\ &= P[H_{k-1}^TR_{k-1}^{-1}z_{k-1} + H_k^TR_k^{-1}z_k] \\ &= P[P_{k-1}^{-1}\hat{x}_{k-1} + P_k^{-1}\hat{x}_k] \quad . \end{aligned} \tag{26}$$

The generalization to an arbitrary number of subsets of measurements uncorrelated with each other is again obvious.

2.6 Observability

From Equation (7) the existence condition for a minimum variance unbiased estimate is readily deduced: the unknown state vector is observable if, and only if, the matrix H has maximal rank.

2.7 Comments

Theorem 2 and its dual form (Theorem 3) have received numerous applications in the field of satellite orbit determination, especially Reference 4; the filtering procedures of Swerling[14], Claus[5] and Battin[1,2] are direct consequences of them. Theorem 2 is of primary importance in modern filtering theory, for it constitutes the cornerstone of the Kalman recursive filter.

It can be proved[11] that, for normal processes, linear least squares estimation yields the same solution as non-linear least squares estimation. Moreover, in that case, the least squares estimate turns out to be identical to the maximum likelihood estimate, which requires to maximize the conditional probability density of measuring z, assuming x.

The minimum variance unbiased estimate is often called Gaussian estimate, Markov estimate, for obvious reasons in view of Theorem 2 or weighted least squares estimate, underlining the occurrence of the covariance matrix R^{-1} in the quadratic form (9).

Let us finally emphasize that the minimum variance unbiased estimate requires no *a priori* information concerning the error distribution function.

3. THE KALMAN-BUCY FILTERING THEORY

3.1 The Kalman Recursive Filter

It will be shown in this section that the Kalman filter, almost simultaneously derived by Kalman and Bucy[9], Battin[1] and Swerling[14] is a direct consequence of Theorems 1 and 2, when applied to first-order discrete linear systems.

Consider an n-dimensional non-constant state vector, taking the value x_i at time t_i and obeying the following first-order discrete linear equation ($i = 1, \ldots, N$):

$$x_i = \Phi_{i,i-1}x_{i-1} + w_{i-1} \quad . \tag{27}$$

where w_{i-1} is a white noise random vector sequence with zero mean and positive definite covariance matrix Q_{i-1}

$$E(w_{i-1}) = 0$$

$$E(w_{i-1}w_{i-1}^T) = Q_{i-1}$$

$$E(w_i w_k^T) = \delta_{ki} Q_i \ ,$$

while $\Phi_{i,i-1}$ is the system state transition matrix from time t_{i-1} to time t_i . At each time t_i , the state vector x_i is observed through the measurement equation

$$z_i = H_i x_i + v_i \ , \tag{28}$$

with the same definitions for z_i , H_i and v_i as in Theorem 2. It is further assumed that the two white noise random sequences v_i and w_{i-1} are uncorrelated.

Then, the problem may be formulated as follows. Find from the measurements $(z_i, x_{i-1}, \ldots, z_1)$ the minimum variance unbiased estimate $\hat{x}_i$ (covariance matrix P_i) of the state x_i , depending linearly on the measurements, assuming the minimum variance unbiased estimate $\hat{x}_{i-1}$ (covariance matrix P_{i-1}) to be known from the measurements $(z_{i-1}, z_{i-2}, \ldots, z_1)$.

In order to apply the theorems of the preceeding sections to the problem at hand, let us introduce the vectors x^i , z^i , w^{i-1} , v^i defined by the following recurrence relations:

$$x^i = \begin{bmatrix} x_i \\ x^{i-1} \end{bmatrix}, \qquad z^i = \begin{bmatrix} z_i \\ z^{i-1} \end{bmatrix}$$

$$w^{i-1} = \begin{bmatrix} w_{i-1} \\ w^{i-2} \end{bmatrix}, \qquad v^i = \begin{bmatrix} v_i \\ v^{i-1} \end{bmatrix} .$$

If we similarly* define the matrices Q^{i-1} , R^i , H^i , Φ^i , F^i ,

$$Q^{i-1} = E[(w^{i-1})(w^{i-1})^T]$$

$$R^i = E[(v^i)(v^i)^T]$$

$$H^i = \begin{bmatrix} H_i & 0 \\ 0 & H^{i-1} \end{bmatrix}$$

$$\Phi^i = \begin{bmatrix} I \\ \Phi^{i-1}\Phi^{-1}_{i,i-1} \end{bmatrix}$$

$$F^i = \begin{bmatrix} 0 & 0 \\ \Phi^{i-1}\Phi^{-1}_{i,i-1} & F^{i-1} \end{bmatrix} ,$$

the whole set of Equations (27) and (28) for i running from 1 up to i , may then be globally written

$$z^i = H^i \Phi^i x_i - H^i F^i w^{i-1} + v^i \ , \tag{29}$$

or equivalently

$$z^i = H^i \Phi^i x_i + e_i \ , \tag{30}$$

with e_i the random vector variable

$$e_i = - H^i F^i w^{i-1} + v^i \ , \tag{31}$$

with zero mean and covariance matrix C_i :

$$C_i = H^i F^i Q^{i-1} (H^i F^i)^T + R^i \ . \tag{32}$$

---

* The matrix $\Phi_{i,i-1}$ is assumed to have an inverse $\Phi^{-1}_{i,i-1}$.

Note that Equation (30) has the form of Equation (1) so that, by Theorem 1, the estimate $\hat{x}_{i-1}$ may be written

$$\hat{x}_{i-1} = P_{i-1}(H^{i-1}\Phi^{i-1})^T C_{i-1}^{-1} z^{i-1} \tag{33}$$

$$P_{i-1} = [(H^{i-1}\Phi^{i-1})^T C_{i-1}^{-1}(H^{i-1}\Phi^{i-1})]^{-1} \ , \tag{34}$$

in view of Equations (5) and (7).

Before calculating $\hat{x}_i$, let us first compute the minimum variance unbiased estimate $\hat{x}_i'$ (covariance matrix P_i') of the state x_i, based upon the measurements $(z_{i-1}, z_{i-2}, \ldots, z_1)$ only.

Introducing (27) in the generalized measurement equation

$$z^{i-1} = H^{i-1}\Phi^{i-1}x_{i-1} + e_{i-1} \ , \tag{35}$$

one obtains

$$z^{i-1} = H^{i-1}\Phi^{i-1}\Phi_{i,i-1}^{-1}x_i + e_i' \ , \tag{36}$$

where e_i' is a new random vector variable:

$$e_i' = e_{i-1} - H^{i-1}\Phi^{i-1}\Phi_{i,i-1}^{-1}w_{i-1} \ , \tag{37}$$

with zero mean and covariance matrix C_i':

$$C_i' = C_{i-1} + H^{i-1}\Phi^{i-1}\Phi_{i,i-1}^{-1}Q_{i-1}(H^{i-1}\Phi^{i-1}\Phi_{i,i-1}^{-1})^T \ .$$

As a Frobenius form is recognized in the above equation, $(C_i')^{-1}$ is known, i.e.

$$(C_i')^{-1} = C_{i-1}^{-1} - C_{i-1}^{-1}H^{i-1}\Phi^{i-1}\Phi_{i,i-1}^{-1}[(\Phi_{i,i-1}^{-1})^T P_{i-1}^{-1}\Phi_{i,i-1}^{-1} + Q_{i-1}^{-1}]^{-1}(H^{i-1}\Phi^{i-1}\Phi_{i,i-1}^{-1})^T C_{i-1}^{-1} \tag{38}$$

so that the covariance matrix P_i', which may be written, in view of Theorem 1,

$$(P_i')^{-1} = (H^{i-1}\Phi^{i-1}\Phi_{i,i-1}^{-1})^T(C_i')^{-1}(H^{i-1}\Phi^{i-1}\Phi_{i,i-1}^{-1}) \ , \tag{39}$$

reduces to

$$\begin{aligned}(P_i')^{-1} &= (\Phi_{i,i-1}^{-1})^T P_{i-1}^{-1}\Phi_{i,i-1}^{-1} - (\Phi_{i,i-1}^{-1})^T P_{i-1}^{-1}\Phi_{i,i-1}^{-1}[(\Phi_{i,i-1}^{-1})^T P_{i-1}^{-1}\Phi_{i,i-1}^{-1} + Q_{i-1}^{-1}]^{-1}(\Phi_{i,i-1}^{-1})^T P_{i-1}^{-1}\Phi_{i,i-1}^{-1} \\ &= [\Phi_{i,i-1}P_{i-1}\Phi_{i,i-1}^T + Q_{i-1}]^{-1} \ ,\end{aligned} \tag{40}$$

as can be easily verified, using Equations (34) and (38). Thus, P_i' is equal to

$$P_i' = \Phi_{i,i-1}P_{i-1}\Phi_{i,i-1} + Q_{i-1} \ . \tag{41}$$

A similar manipulation of $\hat{x}_i'$, which by Theorem 1 is defined to be

$$\hat{x}_i' = P_i'(H^{i-1}\Phi^{i-1}\Phi_{i,i-1}^{-1})^T(C_i')^{-1}z^{i-1} \ , \tag{42}$$

leads via Equations (33), (34), (38) and (41) to the result

$$\hat{x}_i' = \Phi_{i,i-1}\hat{x}_{i-1} \ . \tag{43}$$

We are now in position to simply derive the Kalman filter equations, for all the conditions are satisfied to apply Theorem 2 with

$$\begin{aligned} z_k &= z_i \\ P_{k-1} &= P_i' \\ \hat{x}_{k-1} &= \hat{x}_i' \\ H_k &= H_i \\ R_k &= R_i \ , \end{aligned}$$

so that the estimate $\hat{x}_i$ is immediately given by Equations (16), (17) and (18):

$$\hat{x}_i = \hat{x}'_i + K_i(z_i - H_i\hat{x}'_i) \tag{44}$$

$$K_i = P'_iH_i^T(H_iP'_iH_i^T + R_i)^{-1} \tag{45}$$

$$P_i = (I - K_iH_i)P'_i \ . \tag{46}$$

Equations (41), (43), (44), (45) and (46) constitute the Kalman filter. The physical meaning of these equations is evident: Equations (41) and (43) extrapolate the estimate $\hat{x}_{i-1}$ from time t_{i-1} to time t_i through the state transition equation (27) while Equations (44), (45) and (46) update this extrapolated estimate on account of the new available measurements z_i .

The dynamical system model (27) may be readily extended to include deterministic forcing terms f_{i-1} :

$$x_i = \Phi_{i,i-1}x_{i-1} + f_{i-1} + w_{i-1} \ . \tag{47}$$

This clearly affects the extrapolated estimate $\hat{x}'_i$ only and Equation (43) must be replaced by

$$\hat{x}'_i = \Phi_{i,i-1}\hat{x}_{i-1} + f_{i-1} \ , \tag{48}$$

the other filtering equations remaining unchanged.

The Kalman filter is capable of other important extensions; although out of the scope of the present contribution let us quote the simultaneous estimation of imprecisely known parameters, entering the dynamical system linearly[13], the replacement of white noise sequences for w_{i-1} , v_i by wide-sense Markov sequences[13] or by sequences correlated with respect to each other[10], stochastic optimization, and so on ...

3.2 The Accumulative Filter

It has been shown in the preceding section, that the Kalman filter is a direct application of Theorem 2 to first-order discrete linear systems; it is therefore natural to ask whether Theorem 3 may be applied to the same problem[7,8].

Consider Equations (27) and (28) of Section 3.1 with the same definitions and assumptions and let us introduce a n-dimensional vector y_i such that at any time t_i the following relation holds:

$$\hat{x}_i = P_iy_i \ . \tag{49}$$

It is easily verified from Theorems 1 and 2 that such a vector may always be defined. Furthermore, suppose that the vector y_{i-1} and the inverse covariance matrix P_{i-1}^{-1} are available at time t_{i-1} and let S_i be a matrix defined by

$$S_i = (\Phi_{i,i-1}^{-1})^T P_{i-1}^{-1}\Phi_{i,i-1}^{-1} \ , \tag{50}$$

Equations (40) and (41) may then be written

$$P'_i = S_i^{-1} + Q_{i-1} \tag{51}$$

$$(P'_i)^{-1} = D_iS_i \ , \tag{52}$$

with D_i a matrix given by the relation

$$D_i = I - S_i[S_i + Q_{i-1}^{-1}]^{-1} \ . \tag{53}$$

Similarly, the extrapolated y_{i-1} vector, i.e. y'_i , can be written

$$y'_i = D_i(\Phi_{i,i-1}^{-1})^T y_{i-1} \ , \tag{54}$$

in view of Equation (43).

Use of Theorem 3 yields the new vector y_i at time t_i ,

$$y_i = y'_i + H_i^TR_i^{-1}z_i \ , \tag{55}$$

and the inverse estimate covariance matrix

$$P_i^{-1} = (P'_i)^{-1} + H_i^TR_i^{-1}H_i \ . \tag{56}$$

The estimate $\hat{x}_i$ is then available via

$$\hat{x}_i = P_i y_i \ . \tag{57}$$

Equations (50), (52), (53), (54), (55) and (56) constitute the accumulative filter equations for the problem at hand. It is important, to note that the calculation of the estimate $\hat{x}_i$ is made independently of the accumulative filter: it means that any error in the inversion of P_i^{-1} does not affect the succeeding estimates. For the general case, however, the accumulative filter is not a practical tool to solve the problem, in view of the $n \times n$ matrix inversion required in Equation (53). The situation is completely different in the absence of plant noise; in such cases ($w_{i-1} = 0$ for any i), the M_i matrix reduces to the identity matrix so that the accumulative filter equations simply become

$$S_i = (\Phi_{i,i-1}^{-1})^T P_{i-1}^{-1} \Phi_{i,i-1} \tag{50$'$}$$

$$(P_i')^{-1} = S_i \tag{52$'$}$$

$$M_i = I \tag{53$'$}$$

$$y_i' = (\Phi_{i,i-1}^{-1})^T y_{i-1} \tag{54$'$}$$

$$y_i = y_i' + H_i^T R_i^{-1} z_i \tag{55$'$}$$

$$(P_i)^{-1} = (P_i')^{-1} + H_i^T R_i^{-1} H_i \ . \tag{56$'$}$$

Any numerical matrix inversion has completely disappeared in the scheme, which is not the case in the Kalman filter: the computation procedure is made particularly stable and this may be of primary importance when weakly observable systems are dealt with.

3.3 Observability Condition

In view of Theorem 1 and Equation (29), an observability condition is easily obtained: the state x_i is observable if, and only if, the matrix $(H^i\Phi^i)^T C_i^{-1}(H^i\Phi^i)$ has an inverse or, more compactly, if and only if the matrix $H^i\Phi^i$ has a generalized inverse (C_i is positive definite).

4. CONCLUSION

The modern recursive filtering theory has been shown to proceed from an idea originally due to Gauss, at least in essence. More precisely, simple theorems arising from the minimum variance estimation of a constant state, linearly depending on the measurements, have been applied to discrete first-order stochastic linear systems, by the way of a restatement of the problem. As a result, the Kalman filter and its dual form have been obtained by simple algebraic manipulations.

REFERENCES

1. Battin, R.H. — *A Statistical Optimizing Navigation Procedure for Space Flight.* American Rocket Society Journal, Vol.32, 1962, pp.1681-1682.

2. Battin, R.H. — *Astronautical Guidance.* McGraw-Hill, New York, 1967.

3. Bellman, R. — *Introduction to Matrix Analysis.* McGraw-Hill, New York, 1960.

4. Blackmann, R.E. — *Methods of Orbit Refinement.* Bell System Technical Journal, Vol.43, 1965, pp.885-909.

5. Claus, A.J. — *Orbit Determination for Communication Satellites from Angular Data Only.* American Rocket Society 17th Annual Meeting, Los Angeles, 1962.

6. Gauss, K.F. — *Theory of the Motion of the Heavenly Bodies Moving About the Sun in Conic Sections.* Dover, New York, 1963.

7. Genin, Y. — *A Note on Linear Minimum Variance Fstimation Problems.* Institute of Electrical and Electronic Engineers, Transactions on Automatic Control, Vol.AC-13, 1968, p.103.

8. Genin, Y. — *Gaussian Estimates and Modern Linear Discrete Filtering Methods.* Revue A, Vol.13, 1968, pp.70-78.

9. Kalman, R.E.
   Bucy, R.S. — *New Results in Linear Filtering and Prediction Theory.* Journal of Basic Engineering, Vol.83D, 1961, pp.95-108.

10. Kalman, R.E. — *Fundamental Study of Adaptive Control Systems.* Technical Report ASD-TR 61, Vol.1, (NASA N 62-15355).

11. Papoulis, A. — *Probability, Random Variables and Stochastic Processes.* McGraw-Hill, New York, 1965.

12. Rosenbroek, H.H. — *On the Connection Between Discrete Linear Filters and Some Formulae of Gauss.* Reprints of AFRA Congress, Paper 1, Paris, May 4-7, 1965.

13. Sorenson, H.W. — *Kalman Filtering Techniques.* Advances in Control Systems, Vol.3, Chapter 5, Academic Press, 1966.

14. Swerling, P. — *A Proposed Stagewise Differential Correction Procedure for Satellite Tracking and Prediction.* Rand Corporation, Paper P-1292, 1958.

19

Reprinted from pp. 354–364 of *Proc. Optimum System Synthesis Conference, 1962*, Tech. Doc. Rep. No. ASD-TDR-63-119, Aeronautical Systems Division, Wright-Patterson AFB, 1963, 395 pp.

SMOOTHING FOR LINEAR AND NONLINEAR DYNAMIC SYSTEMS

Arthur E. Bryson* and Malcolm Frazier**

ABSTRACT

A method is presented for making maximum likelihood estimates of the state variables of linear and nonlinear dynamic systems over a finite time interval, utilizing measurements made during this time interval and, if available, a priori estimates of the initial conditions and the forcing functions. For a linear system, the filtering results are identical to those of Kalman and Bucy; the smoothing result is an extension of their work.

ACKNOWLEDGEMENT

The first author wishes to thank the Boeing Company and the Raytheon Company for support of this work.

STATEMENT OF THE SMOOTHING PROBLEM

Given: (1) A dynamic system governed by nonlinear differential equations,

$$\dot{x} = f(x, u, t)$$

where

$x(t)$ = $n \times 1$ matrix of state variables,
$u(t)$ = $m \times 1$ matrix of forcing functions,
f = $n \times 1$ matrix of known functions of x, u, t
t = independent variable (often time)
$(\dot{\ }) = \frac{d}{dt}(\ )$

(2) An a priori estimate of initial conditions,

$$E[x(t_o)] = \bar{x}_o \text{, and}$$

an a priori estimate of errors in $\bar{x}_o$,

$$E\left\{[x(t_o) - \bar{x}_o][x(t_o) - \bar{x}_o]^T\right\} = \bar{P}_o \text{, an } n \times n \text{ matrix}$$

(symmetric).

---

*Professor of Mechanical Engineering, Harvard University, Cambridge, Mass.

**Senior Engineer, Raytheon Company, Bedford, Mass.

(3) An a priori estimate of the forcing functions,

$$E[u(t)] = \overline{u}(t), \text{ and}$$

an a priori estimate of errors in $\overline{u}(t)$,

$$E\left\{[u(t) - \overline{u}(t)]\,[u(\tau) - \overline{u}(\tau)]^T\right\} = \overline{Q}(t)\,\delta(t-\tau)$$

(4) A set of measurements for $t_o \le t \le t_f$,

$$E[z(t)] = \overline{z}(t), \text{ a } p \times 1 \text{ matrix of functions,}$$

where the theoretical relations between "exact" measurements, $z(t)$, and state variables $x(t)$, is $h[z(t), x(t), t] = 0$, a $p \times 1$ matrix of known functions, and an a priori estimate of errors in $\overline{z}(t)$,

$$E\left\{[z(t) - \overline{z}(t)]\,[z(\tau) - \overline{z}(\tau)]^T\right\} = \overline{R}(t)\,\delta(t-\tau)$$

In the above equations E[ ] means "expected value of [ ]", i.e. an ensemble average over many identical systems performing simultaneously. The sumbol $\delta(t-\tau)$ is the Dirac delta function and, appearing in the correlation equations above, it indicates "white noise" in the forcing functions and the measurements. In effect what this means is that we assume the correlation times of the noise are short compared to times of interest in the problem. The sumbol $(\ )^T$ indicates the transpose of the matrix ( ).

Find: The maximum likelihood estimate of $u(t) = \hat{u}(t)$ and the initial conditions $x(t_o) = \hat{x}_o$. Note this gives the maximum likelihood estimate of $x(t) = \hat{x}(t)$ from

$$\dot{\hat{x}} = f(\hat{x}, \hat{u}, t)$$

$$x(t_o) = x_o$$

SOLUTION TO THE SMOOTHING PROBLEM

The maximum likelihood estimate will be obtained by minimizing

(1) $$J = \frac{1}{2}\tilde{x}_o^T \overline{P}_o^{-1} \tilde{x}_o + \frac{1}{2}\int_{t_o}^{t_f} (\tilde{z}^T \overline{R}^{-1} \tilde{z} + \tilde{u}^T \overline{Q}^{-1} \tilde{u})\, dt,$$

where

$$\tilde{(\ )} = \hat{(\ )} - \overline{(\ )}$$

with the constraints:

(2) $$h(\hat{z}, \hat{x}, t) = 0$$

(3) $$\dot{\hat{x}} - f(\hat{x}, \hat{u}, t) = 0.$$

This is a standard problem in the calculus of variations.* It may be solved by adjoining the constraints (2) and (3) to the expression for J in (1):

(4) $$J = \frac{1}{2}\tilde{x}_o^T \overline{P}_o^{-1} \tilde{x}_o + \int_{t_o}^{t_f} \left\{ \frac{1}{2}\tilde{z}^T \overline{R}^{-1} \tilde{z} + \frac{1}{2}\tilde{u}^T \overline{Q}^{-1} \tilde{u} + \lambda^T [\dot{\hat{x}} - f(\hat{x}, \hat{u}, t)] + \mu^T h(\hat{z}, \hat{x}, t) \right\} dt$$

Now consider variations of $\hat{u}$ and $\hat{x}_o$ from their optimum values:

(5) $$\delta J = \tilde{x}_o^T \overline{P}_o \delta\hat{x}_o + \int_{t_o}^{t_f} \left\{ \tilde{z}^T \overline{R}^{-1} \delta\hat{z} + \tilde{u}^T \overline{Q}^{-1} \delta\hat{u} + \lambda^T [\delta\dot{\hat{x}} - \frac{\partial f}{\partial x}\delta\hat{x} - \frac{\partial f}{\partial u}\delta\hat{u}] + \mu^T [\frac{\partial h}{\partial z}\delta\hat{z} + \frac{\partial h}{\partial x}\delta\hat{x}] \right\} dt$$

or,

(6) $$\delta J = \tilde{x}_o^T \overline{P}_o^{-1} \delta\hat{x}_o + [\lambda^T \delta x]_{t_o}^{t_f} + \int_{t_o}^{t_f} \left\{ (\tilde{z}^T \overline{R}^{-1} + \mu^T \frac{\partial h}{\partial z})\delta\hat{z} + (-\dot{\lambda}^T - \lambda^T \frac{\partial f}{\partial x} + \mu^T \frac{\partial h}{\partial x})\delta\hat{x} + (\tilde{u}^T \overline{Q}^{-1} - \lambda^T \frac{\partial f}{\partial u})\delta\hat{u} \right\} dt$$

Let us now choose $\lambda(t)$ and $\mu(t)$ so that

(7) $$\dot{\lambda}^T + \lambda^T \frac{\partial f}{\partial x} - \mu^T \frac{\partial h}{\partial x} = 0; \quad \lambda(t_f) = 0, \quad \tilde{x}_o^T \overline{P}_o^{-1} - \lambda^T(t_o) = 0,$$

(8) $$\tilde{z}^T \overline{R}^{-1} + \mu^T \frac{\partial h}{\partial z} = 0.$$

Then Eqn. (6) becomes,

(9) $$\delta J = \int_{t_o}^{t_f} (\tilde{u}^T \overline{Q}^{-1} - \lambda^T \frac{\partial f}{\partial u})\delta\hat{u}\, dt.$$

*See for example Bryson, A. E., and Denham, W. F., "A Steepest Ascent Method for Solving Optimum Programming Problems," Jour. Appl. Mech., June 1962, pp. 247-257.

If $\hat{u}(t)$ is the optimal estimate of the forcing functions then $\delta J = 0$ for arbitrary $\delta\hat{u}$. For this to be true, it is necessary that

(10) $$\tilde{u}^T \bar{Q}^{-1} - \lambda^T \frac{\partial f}{\partial u} = 0$$

Summarizing, the optimal estimate of u(t) and x(t) will be obtained by solving the two point boundary value problem,

Diff. Eqns.

(11) $$\dot{\hat{x}} = f(\hat{x}, \hat{u}, t) \quad \text{with} \quad \hat{u} = \bar{u} + \bar{Q}\left(\frac{\partial f}{\partial u}\right)^T \lambda ,$$

(12) $$\dot{\lambda} = -\left(\frac{\partial f}{\partial x}\right)^T \lambda - \left(\frac{\partial h}{\partial x}\right)^T \left(\frac{\partial h}{\partial z}\right)^{-T} \bar{R}^{-1}(\hat{z} - \bar{z}), \quad \text{with } h(\hat{z}, \hat{x}, t) = 0.$$

Boun. Cond.

(13) $$\hat{x}(t_o) = \bar{x}(t_o) + \bar{P}_o \lambda(t_o)$$

(14) $$\lambda(t_f) = 0$$

Note 1: If no a priori estimates of $x(t_o)$ and u(t) are given, this essentially means $\bar{P}_o \to \infty$, $\bar{Q} \to \infty$. The conditions involving $\bar{x}_o$, $\bar{u}$, $\bar{P}_o$ and $\bar{Q}$ above are simply replaced by,

(15) $$\lambda^T \frac{\partial f}{\partial u} = 0$$

(16) $$\lambda(t_o) = 0$$

Note 2: A simple extension of this analysis would be to make $\bar{R}$ and $\bar{Q}$ functions of x and u.

Note 3: Another simple extension would be to include estimation of some parameters of the system.

SMOOTHING FOR LINEAR DYNAMIC SYSTEMS

If the system is linear, it can be put into the form,

(17) $$\dot{x} = F(t)\,x + G(t)\,u$$

(18) $$z = H(t)\,x + w(t)$$

where w(t) is a known set of bias functions. The smoothing problem then simplifies to the following:

$$\frac{d}{dt}\begin{bmatrix}\hat{x}\\ \lambda\end{bmatrix} = \begin{bmatrix}F, & G\bar{Q}G^T\\ H^T\bar{R}^{-1}H, & -F^T\end{bmatrix}\begin{bmatrix}\hat{x}\\ \lambda\end{bmatrix} + \begin{bmatrix}G\bar{u}\\ -H^T\bar{R}^{-1}(\bar{z}-w)\end{bmatrix} \quad \begin{matrix}(19)\\(20)\end{matrix}$$

with boundary conditions,

(21) $\hat{x}(t_o) - \bar{P}_o\lambda(t_o) = \bar{x}_o$

(22) $\lambda(t_f) = 0$

This two-point boundary value problem can be solved in a straightforward manner due to the linearity of eqns. (19) and (20) as follows:

Let $\hat{x}_p(t)$ and $\lambda_p(t)$ be solutions of Eqns. (19) and (20) with initial conditions

(23) $\hat{x}_p(t_o) = \bar{x}(t_o)$,

(24) $\lambda_p(t_o) = 0$.

Let $\Phi_x(t)$ and $\Phi_\lambda(t)$ be $n \times n$ matrices which are solutions to the homogeneous version of Eqns. (19) and (20), i.e., with $\bar{u} = \bar{z} - w = 0$, with initial conditions,

(25) $\Phi_x(t_o) = \bar{P}_o$,

(26) $\Phi_\lambda(t_o) =$ unit matrix $= I$.

It follows that

(27) $\hat{x}(t) = \hat{x}_p(t) + \Phi_x(t)\lambda(t_o)$

(28) $\lambda(t) = \lambda_p(t) + \Phi_\lambda(t)\lambda(t_o)$

In order to satisfy the boundary condition (22), we must have

(29) $\lambda(t_o) = -[\Phi_\lambda(t_f)]^{-1}\lambda_p(t_f)$

Combining (29) with (27), (28) and the relation,

(30) $\hat{u}(t) = \bar{u}(t) + \bar{Q}G^T\lambda(t)$

we have finally,

$$\hat{x}(t) = \hat{x}_p(t) - \Phi_x(t)\,[\Phi_\lambda(t_f)]^{-1}\,\lambda_p(t_f) \tag{31}$$

$$\hat{u}(t) = \bar{u}(t) + \bar{Q}G^T\left\{\lambda_p(t) - \Phi_\lambda(t)\,[\Phi_\lambda(t_f)]^{-1}\,\lambda_p(t_f)\right\} \tag{32}$$

$$\hat{z}(t) = H\hat{x} + w \tag{33}$$

This is the solution to the smoothing problem, i.e., the problem of finding the maximum likelihood estimates of x(t), u(t), and z(t) over an entire interval $t_o \leq t \leq t_f$.

The filtering problem is concerned with finding the maximum likelihood estimate of $x(t_f)$ only, and the prediction problem is concerned with finding the maximum likelihood estimate of x(t) for $t > t_f$. Elegant solutions to the filtering and prediction problems have been given by Kalman and Bucy*. Solutions to these latter two problems by the present method are given in the next section.

It is also possible to find the covariance matrix of errors in the smoothing estimate $\hat{x}(t)$ by considering small variations in $\hat{x}$, $\bar{x}$, $\hat{u}$, $\bar{u}$, $\hat{z}$, $\bar{z}$ in Eqns. (19) through (22):

$$\frac{d}{dt}\begin{bmatrix} \delta\hat{x} \\ \delta\lambda \end{bmatrix} = \begin{bmatrix} F\,, & G\bar{Q}G^T \\ H^T\bar{R}^{-1}H, & -F^T \end{bmatrix}\begin{bmatrix} \delta\hat{x} \\ \delta\lambda \end{bmatrix} + \begin{bmatrix} G\,\delta\bar{u} \\ -H^T\bar{R}^{-1}\delta\bar{z} \end{bmatrix} \tag{34, 35}$$

$$\delta\lambda(t_f) = 0 \tag{36}$$

$$\delta\hat{x}(t_o) - \bar{P}_o\,\delta\lambda(t_o) = \delta\bar{x}_o \tag{37}$$

Let

$$\left\{\begin{aligned} P_{xx}(t) &= E(\delta\hat{x}\,\delta\hat{x}^T) \\ P_{x\lambda}(t) &= E(\delta\hat{x}\,\delta\lambda^T) \\ P_{\lambda\lambda}(t) &= E(\delta\lambda\;\delta\lambda^T) \end{aligned}\right. \tag{38}$$

*Kalman, R. E., and Bucy, R. S., "New Results in Linear Filtering and Prediction Theory," Jour. Basic Engr., March 1961.

It then follows immediately from Eqns. (34) to (37) and the relations

(39) $$E[\delta\bar{u}(t)\,\delta\bar{u}(\tau)^T] = \bar{Q}(t)\,\delta(t-\tau),$$

(40) $$E[\delta\bar{z}(t)\,\delta\bar{z}(\tau)^T] = \bar{R}(t)\,\delta(t-\tau),$$

(41) $$E[\delta\bar{u}(t)\,\delta\bar{z}(\tau)^T] = 0,$$

that the covariance matrices (38) are solutions to

(42) $$\dot{P} = \mathcal{F}P + P\mathcal{F}^T + \mathcal{G}$$

where

(43, 44) $$P = \begin{bmatrix} P_{xx}, & P_{x\lambda} \\ P_{x\lambda}^T, & P_{\lambda\lambda} \end{bmatrix}, \quad \mathcal{F} = \begin{bmatrix} F & , & G\bar{Q}G^T \\ H^T\bar{R}^{-1}H, & & -F^T \end{bmatrix},$$

$$\mathcal{G} = \begin{bmatrix} G\bar{Q}G^T, & 0 \\ 0 \;, & H^T\bar{R}^{-1}H \end{bmatrix}.$$

The boundary conditions for (42) are

(45) $$\begin{cases} P_{\lambda\lambda}(t_f) = 0 \\ P_{x\lambda}(t_f) = 0 \\ P_{xx}(t_o) - \bar{P}_o P_{x\lambda}^T(t_o) - P_{x\lambda}(t_o)\bar{P}_o + \bar{P}_o P_{\lambda\lambda}(t_o)\bar{P}_o = \bar{P}_o \end{cases}$$

Thus, finding the covariance matrices is also a two-point boundary value problem.

It can be shown that the correlation matrix of errors in $\hat{u}(t)$ is given by,

(46) $$GQG^T = G\bar{Q}G^T + G\bar{Q}G^T P_{x\lambda}^T + P_{x\lambda}G\bar{Q}G^T,$$

where

(47) $$E[\delta\hat{u}(t)\,\delta\hat{u}(\tau)^T] = Q(t)\,\delta(t-\tau).$$

Similarly,

(48) $$H^T\bar{R}^{-1}R\bar{R}^{-1}H = -H^T\bar{R}^{-1}HP_{x\lambda} - P_{x\lambda}^T H^T\bar{R}^{-1}H,$$

where

(49) $$E[\delta\hat{z}(t)\,\delta\hat{z}(\tau)^T] = R(t)\,\delta(t-\tau).$$

FILTERING FOR LINEAR SYSTEMS

If only $\hat{x}(t_f)$ is wanted, the two point boundary value problem of Eqs. (19) - (22) may be converted to an initial value problem. Referring to Eq. (31), we define a new quantity

(50) $$y(t) = \hat{x}_p(t) - \Phi_x(t)[\Phi_\lambda(t)]^{-1}\lambda_p(t)$$

Note that

(51) $$y(t_f) = \hat{x}(t_f)$$

Let us also define another quantity

(52) $$P_y(t) = \Phi_x(t)[\Phi_\lambda(t)]^{-1}$$

Differentiating (52) with respect to t and using the relations

(53) $$\frac{d}{dt}\begin{bmatrix}\Phi_x \\ \Phi_\lambda\end{bmatrix} = \begin{bmatrix} F, & G\bar{Q}G^T \\ H^T\bar{R}^{-1}H, & -F^T\end{bmatrix}\begin{bmatrix}\Phi_x \\ \Phi_\lambda\end{bmatrix}$$

(54) $$\begin{cases}\Phi_x(t_o) = \bar{P}_o \\ \Phi_\lambda(t_o) = I\end{cases}$$

It is straightforward to show that

(55) $$\dot{P}_y = FP_y + P_yF^T - P_yH^T\bar{R}^{-1}HP_y + G\bar{Q}G^T$$

(56) $$P_y(t_o) = \bar{P}_o$$

Differentiating (50) with respect to t and using (19), (20), (23), (24), (52), and (55), it is also straightforward to show that

(57) $$\dot{y} = Fy + G\bar{u} + P_yH^T\bar{R}^{-1}(\bar{z} - Hy - w)$$

(58) $$y(t_o) = \bar{x}_o$$

Eqs. (55) through (58) are precisely those obtained earlier by Kalman and Bucy (Loc. Cit.). Fig. 1 shows a block diagram of this filter. Note that $P_y(t)$ may be precalculated or calculated simultaneously with y(t).

The solution to the smoothing problem (linear) of Eqs. (19) through (22) can be obtained by "filtering" to obtain $y(t_f) = x(t_f)$, then integrating Eqs. (19) and (20) backward with boundary conditions

(59, 60) $$\hat{x}(t_f) = y(t_f) \; ; \quad \lambda(t_f) = 0 \, .$$

A block diagram of this "smoother" is shown in Fig. 2.

Similarly it can be shown that

(61) $$P_y(t_f) = P_{xx}(t_f) \, .$$

Hence, the solution to the covariance Eqs. (42) - (45) for the smoothing problem can be obtained by "filtering" to obtain $P_y(t_f) = P_{xx}(t_f)$, then integrating Eqs. (42) - (44) backward with boundary conditions

(62) $$P_{xx}(t_f) = P_{yy}(t_f) \, ,$$

(63) $$P_{x\lambda}(t_f) = 0 \, ,$$

(64) $$P_{\lambda\lambda}(t_f) = 0 \, .$$

The impulse response functions (kernels) for the linear filter are easily found using the equations adjoint to equations (19) and (20),

(65)
(66) $$\frac{d}{dt}\begin{bmatrix} \Psi_x \\ \Psi_\lambda \end{bmatrix} = \begin{bmatrix} -F^T \, , & -H^T \bar{R}^{-1} H \\ -G\bar{Q}G^T, & F \end{bmatrix} \begin{bmatrix} \Psi_x \\ \Psi_\lambda \end{bmatrix}$$

with boundary conditions,

(67) $$\Psi_x(t_f) = I = \text{unit matrix}$$

(68) $$\Psi_x(t_o)\bar{P}_o + \Psi_\lambda(t_o) = 0 \, .$$

It follows immediately that

(69) $$\hat{x}(t_f) = \int_{t_o}^{t_f} [\Psi_x^T G\bar{u} - \Psi_\lambda^T H^T \bar{R}^{-1}(\bar{z} - w)] \, dt + \Psi_x^T(t_o)\bar{x}_o \, .$$

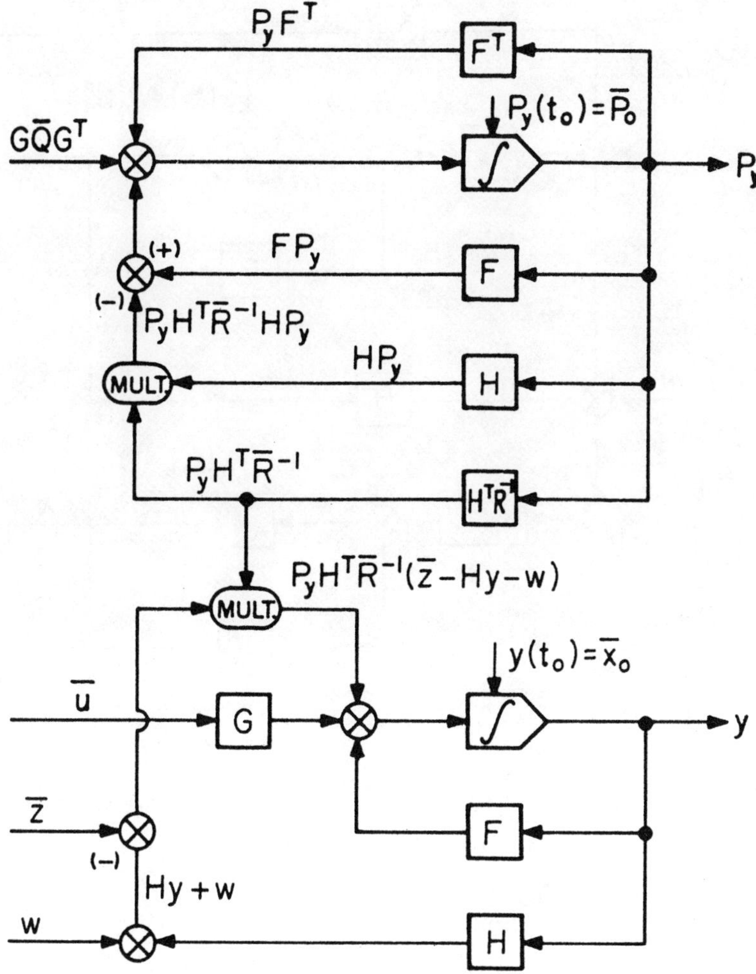

FIGURE 1. OPTIMAL FILTER FOR LINEAR MULTIVARIABLE SYSTEM

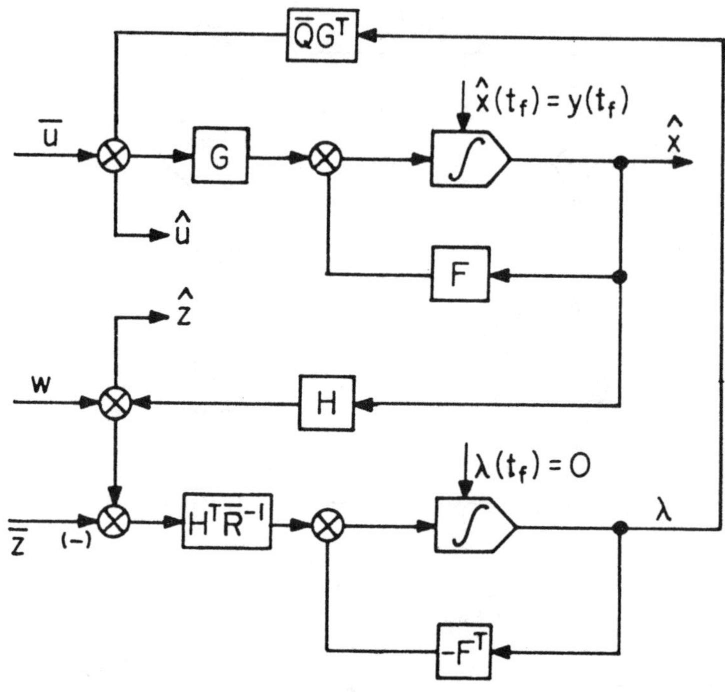

FIGURE 2. OPTIMAL SMOOTHER FOR LINEAR MULTIVARIABLE SYSTEM. (TO BE RUN BACKWARD AFTER $\hat{x}(t_f)$ OBTAINED FROM OPTIMAL FILTER)

20

Reprinted with permission from *SIAM J. Control* **10**(3):477–486 (1972)

A NOTE ON LEAST SQUARES ESTIMATION BY THE INNOVATIONS METHOD*

THOMAS KAILATH†

Abstract. A rigorous proof by the innovations method is given for certain results on linear least squares estimation, especially of the Stratonovich–Kalman–Bucy formulas. Some discussion is also given of the nonlinear problem.

1. Introduction. In [1]–[2] it was stated that informal arguments were to be used for the derivation of the Stratonovich–Kalman–Bucy least squares estimation formulas by the innovations method. There have been several queries as to a rigorous presentation; we shall provide one such derivation here. It will be seen that the rigorous derivation follows the spirit of the informal one.

The idea of the innovations method is to replace the given observation process by a simpler process that contains the same "information" as the given process. The point is that the estimation problem with the simpler process is usually much easier to solve than the original problem. This method was first used for discrete-time stationary prediction problems by Wold (1938) and Kolmogorov (1939) and then extended to continuous-time stationary processes by Bode and Shannon (1950) and Zadeh and Ragazzini (1950) (these references and several related ones are cited in [1] and [3]). The simpler process has been called the innovations process by Wiener and Masani. For linear least squares problems, we shall use a white-noise process or rather its integrated version, a process with orthogonal increments, as the innovations process; however, other choices are possible (cf. [3], [4]).

In our presentation, we start with the simple but basic problem of finding the least squares estimate of a random variable x, given observations of a related orthogonal-increments process $\upsilon(\cdot)$. If we now specialize x to being the state of a linear dynamical system and if we take $\upsilon(\cdot)$ as the innovations process of noisy observations of the state of this system, then we can readily obtain the Stratonovich–Kalman–Bucy equations [5, p. 675], [6]. Finally in § 4 we note that by using certain recent results the innovations method can also be extended to certain nonlinear problems.

2. Linear estimation of a random variable from an orthogonal-increments process. Let x be a random variable with

$$Ex = 0, \quad E|x|^2 < \infty, \tag{1}$$

and let $\{\upsilon(\tau), 0 \leqq \tau \leqq t\}$ be a related process of orthogonal increments with

$$E\upsilon(t) = 0, \quad E\upsilon(t)\upsilon'(s) = \int_0^{t\wedge s} I\, d\tau, \tag{2}$$

where $t \wedge s$ denotes the minimum of t and s and the prime denotes transpose.

* Received by the editors July 19, 1971, and in revised form October 8, 1971.

† Department of Electrical Engineering, Stanford University, Stanford, California 94305. This work was supported by the Applied Mathematics Division of the Air Force Office of Scientific Research under Contract AF44-620-69-C-0101 and by the JSEP at Stanford University, Contract N-00014-67-A-0112-0044.

We could have assumed that the covariance of $v(\cdot)$ was the integral of a positive-definite function, but the normalization to the identity function is convenient and entails no loss of generality. We wish to find the least squares estimate, $\hat{x}(t)$, in the form

$$\hat{x}(t) = \int_0^t g_t(s)\, dv(s), \tag{3}$$

where

$$\int_0^t \operatorname{tr}\,[g_t(s)g_t'(s)]\, ds < \infty. \tag{4}$$

There is no loss of generality here, since the so-called Wiener stochastic integral in (3) is the most general form of a finite-variance linear functional of the process $\{v(\tau), 0 \leqq \tau \leqq t\}$ (see, for example, Doob [7]).

To determine the weighting function $g_t(s)$, we shall use the well-known criterion that

$$\tilde{x}(t) \triangleq x - \hat{x}(t) \perp v(\tau), \qquad 0 \leqq \tau \leqq t, \tag{5}$$

where $\perp$ denotes orthogonality (zero-correlation). This yields

$$Exv'(\tau) = E\left[\int_0^t g_t(s)\, dv(s) \int_0^\tau 1' \cdot dv'(u)\right] = \int_0^\tau g_t(s)\, ds \tag{6}$$

by using the properties of Wiener stochastic integrals [7, Chap. IX]. Therefore,

$$g_t(\tau) = \frac{\partial}{\partial \tau} Exv'(\tau), \qquad \tau \leqq t, \tag{7}$$

and

$$\hat{x}(t) = \int_0^t \left[\frac{\partial}{\partial \tau} Exv'(\tau)\right] dv(\tau). \tag{8}$$

To go further, we need more assumptions on x. A common set of assumptions is treated in the next section.

3. The Stratonovich–Kalman–Bucy problem. We now assume that

$$x = x(t), \tag{9}$$

where $x(\cdot)$ is the solution of a linear stochastic differential equation

$$dx(\tau) = F(\tau)x(\tau)\, d\tau + G(\tau)\, dU(\tau), \quad x(0) = x_0, \tag{10}$$

and $U(\cdot)$ is a process of orthogonal increments such that

$$EU(\tau) = 0, \quad EU(t)U'(s) = \int_0^{t \wedge s} Q(\tau)\, d\tau, \qquad Q \geqq 0, \tag{11}$$

and

$$Ex_0 = 0, \quad Ex_0x_0' = \Pi_0, \quad EU(t)x_0' = 0, \qquad t \geqq 0. \tag{12}$$

We also assume that we have observations of the form

(13) $$dY(\tau) = z(\tau)\,d\tau + dW(\tau), \quad z(\tau) = H(\tau)x(\tau),$$

where $W(\cdot)$ is an orthogonal-increments process with

(14) $$EW(t) = 0, \quad EW(t)W'(s) = \int_0^{t\wedge s} I\,d\tau,$$

(15) $$EW(t)x_0' \equiv 0, \quad EW(t)U'(s) = \int_0^{t\wedge s} C(\tau)\,d\tau.$$

We shall also assume that

(16) $$F(\cdot), G(\cdot), H(\cdot) \quad \text{are continuous functions.}$$

As will be seen from the proofs below, the assumptions (16) are stronger than really necessary, but they are the conventional ones.

We may remark that in derivations using white-noise and delta functions, the presence of the correlation term $C(\cdot)$ usually leads to various unsatisfactory discussions about "partial" delta functions and to the use of relations like (see [1], [6])

$$\int_{-\infty}^{t} \delta(t-\tau)\,d\tau = \tfrac{1}{2}.$$

In the author's opinion, the need to clarify this problem is perhaps a more important motivation for going to a stochastic integral treatment than just the desire to obtain a formal mathematical proof.

Next, we introduce the fundamental matrix of $F(\cdot)$, which is defined as the unique solution of the equation

$$d\Phi(t,s)/dt = F(t)\Phi(t,s), \quad \Phi(0,0) = I.$$

In terms of $\Phi(\cdot,\cdot)$, we can write the solution $x(\cdot)$ of (10) as

(17) $$x(t) = \Phi(t,0)x_0 + \int_0^t \Phi(t,s)G(s)\,dU(s).$$

We also note the properties

(18) $$E[W(t) - W(s)]x'(\tau) = 0, \qquad \tau \leqq s, t,$$

and

(19) $$E\int_0^T \operatorname{tr}\,[H(s)x(s)x'(s)H'(s)]\,ds < \infty.$$

(a) *The innovations process.* For the solution, the first step is to obtain the innovations process. From (13), (18), (19) it can be shown that the process

(20) $$\nu(t) = Y(t) - \int_0^t H(s)\hat{x}(s)\,ds = \int_0^t H(s)\tilde{x}(s)\,ds + W(t)$$

is a process of orthogonal increments with

$$Ev(t) = 0, \quad Ev(t)v'(s) = \int_0^{t \wedge s} I \, d\tau \tag{21}$$

and that

$$\sigma\{v(\tau), \tau \leqq t\} = \sigma\{Y(\tau), \tau \leqq t\}, \qquad 0 \leqq t \leqq T, \tag{22}$$

where $\sigma\{\cdot\}$ denotes the sigma fields generated by the specified random variables. Therefore, $v(\cdot)$ is the innovations process of the observations $Y(\cdot)$. Proofs of the fundamental equations (20)–(22) are outlined in the Appendix.

Next, in order to calculate $\hat{x}(t)$, we shall need (cf. (8)) to evaluate

$$g_t(s) = \frac{\partial}{\partial s} E[x(t)v'(s)].$$

We shall find, as will be shown presently, that the assumptions (10)–(15) on $x(\cdot)$ easily yield

$$g_t(s) = \Phi(t, s)K(s), \qquad 0 \leqq s \leqq t, \tag{23}$$

where

$$K(t) = P(t)H'(t) + G(t)C'(t) \tag{24}$$

and

$$P(t) = E\tilde{x}(t)\tilde{x}'(t), \quad \tilde{x}(t) = x(t) - \hat{x}(t). \tag{25}$$

Therefore, by (8),

$$\hat{x}(t) = \int_0^t \Phi(t, s)K(s) \, dv(s), \tag{26}$$

or in differential form,

$$d\hat{x}(t) = F(t)\hat{x}(t) \, dt + K(t) \, dv(t), \quad \hat{x}(0) = 0, \tag{27}$$

which is the celebrated equation of Stratonovich, Kalman and Bucy.

Proof of (23)–(24). We have, by (20),

$$\begin{aligned} g_t(s) &= \frac{\partial}{\partial s} E[x(t)v'(s)] \qquad (s \leqq t) \\ &= \frac{\partial}{\partial s} E\left[\int_0^s x(t)\tilde{x}'(u)H'(u) \, du\right] + \frac{\partial}{\partial s} E[x(t)W'(s)]. \end{aligned} \tag{28}$$

By the use of Schwarz' inequality and the Fubini theorem, we can show that the first term on the right-hand side of (28) is

$$\begin{aligned} \frac{\partial}{\partial s} \int_0^s E[x(t)\tilde{x}'(u)H'(u)] \, du &= E[x(t)\tilde{x}'(s)]H'(s) \\ &= \Phi(t, s)E[x(s)\tilde{x}'(s)]H'(s) \\ &\quad + E\left[\int_s^t \Phi(t, \tau)G(\tau) \, dU(\tau) \cdot \tilde{x}'(s)\right]H'(s) \\ &= \Phi(t, s)P(s)H'(s) + 0, \end{aligned}$$

while the second term is equal to

$$\frac{\partial}{\partial s}E\left[\Phi(t,0)x_0W'(s)+\int_0^t \Phi(t,\tau)G(\tau)\,dU(\tau)\int_0^s dW(\sigma)\right]$$

$$=\frac{\partial}{\partial s}\int_0^s \Phi(t,\tau)G(\tau)C(\tau)\,d\tau=\Phi(t,s)G(s)C(s), \qquad s\leqq t.$$

The sum of these two terms yields the desired relations (23)–(24).

(b) *Derivation of Riccati equation for* $P(\cdot)$. The mean square error $P(t)$ obeys a differential equation, which is also easily obtained via the innovations. We first observe that for the process $x(\cdot)$ described by the stochastic differential equation (10)–(12), a direct calculation using (17) yields

$$\Pi(t)\triangleq Ex(t)x'(t)=\Phi(t,0)\Pi_0\Phi'(t,0)+\int_0^t \Phi(t,\tau)G(\tau)Q(\tau)G'(\tau)\Phi'(t,\tau)\,d\tau.$$

Equivalently, we have the differential equation

$$\dot{\Pi}(t)=F(t)\Pi(t)+\Pi(t)F'(t)+G(t)Q(t)G'(t),\quad \Pi(0)=\Pi_0. \tag{29}$$

Since $\nu(\cdot)$ has orthogonal increments, a similar calculation applied to $\hat{x}(\cdot)$, as given by the stochastic differential equation (27), yields

$$\Sigma(t)\triangleq E\hat{x}(t)\hat{x}'(t),\quad \dot{\Sigma}(t)=F(t)\Sigma(t)+\Sigma(t)F'(t)+K(t)IK'(t),\quad \Sigma(0)=0. \tag{30}$$

But since $\hat{x}(\cdot)$ and $\tilde{x}(\cdot)$ are orthogonal, we have

$$\Pi(t)=Ex(t)x'(t)=E\hat{x}(t)\hat{x}'(t)+E\tilde{x}(t)\tilde{x}'(t)=\Sigma(t)+P(t), \tag{31}$$

and therefore,

$$\begin{aligned}\dot{P}(t)&=\dot{\Pi}(t)-\dot{\Sigma}(t)\\ &=F(t)P(t)+P(t)F'(t)-K(t)K'(t)+G(t)Q(t)G'(t),\\ P(0)&=\Pi_0.\end{aligned} \tag{32}$$

This is the well-known matrix Riccati differential equation of the Stratonovich–Kalman–Bucy theory; the above derivation was first given in [3, footnote 7].

As far as the author is aware, the only previous direct rigorous proof of the Stratonovich–Kalman–Bucy result is the one due to Wonham [8], which proceeds via a careful limiting argument from the discrete-time case. Some rigorous proofs have been proposed via specializations of certain nonlinear filtering formulas, but until recently all such proofs have needed fairly strong conditions on the process $x(\cdot)$. Recently, however, Fujisaki, Kallianpur and Kunita [9] have given a derivation (partly based on innovations) under fairly general conditions; their results can be specialized to the linear case, but such a proof will be less direct than the one we have given here. Even more recently, Balakrishnan [23] has given a proof based on martingale arguments, but it does not seem to be as direct as ours.

4. Some extensions; nonlinear estimation. Similar arguments can be used to justify the results obtained by the innovations method for linear smoothing [2], and linear filtering and smoothing based only on covariance functions [10], [11];

various extensions, for example, to infinite-dimensional and distributed parameter systems, can also be made.

Somewhat surprisingly the innovations method can also be used [12] for nonlinear least squares estimation given observations of the form

$$dY(t) = z(t)\,dt + dW(t),$$

where $W(\cdot)$ is a Wiener process but $z(\cdot)$ is not necessarily Gaussian. If

$$E\int |z(t)|\,dt < \infty$$

and the increments of $W(\cdot)$ are independent of the sigma fields generated by past $z(\cdot)$ and $W(\cdot)$, then it can be shown [13], [14] that $\upsilon(\cdot)$ defined by

$$d\upsilon = dY - \hat{z}(t)\,dt, \quad \hat{z}(t) = E[z(t)|\sigma\{Y(\tau), \tau \leqq t\}]$$

is a Wiener process with the same covariance function as $W(\cdot)$. The difficult problem is to show that there is no loss of information in working with $\upsilon(\cdot)$, i.e., $\sigma\{\upsilon(\tau), \tau \leqq t\} = \sigma\{Y(\tau), \tau \leqq t\}$. In [12], the rather strong hypothesis

$$|z(t)| \leqq M$$

was necessary. Recently, however, Yerzhov [15] has shown that a sufficient condition for the equality of the sigma fields associated with $\upsilon(\cdot)$ and $Y(\cdot)$ is that the measures induced in the space of continuous functions by the processes $Y(\cdot)$ and $W(\cdot)$ be mutually absolutely continuous. This will be the case, for instance, in the widely studied problem in which it is assumed that $z(\cdot)$ and $W(\cdot)$ are completely independent and $z(\cdot)$ is square-integrable almost surely. (Several other assumptions will also yield mutual absolute continuity.) Given the innovations process, a derivation very similar to that of the linear problem can be used (as informally described in [12], see also [9]) to obtain the differential equations of Stratonovich, Kushner and others for the conditional density function. Here we shall only quote the analogue of the linear estimation result in § 2. [*Note added in proof*. Shepp and P. Varaiya have pointed out to me that Yerzhov's proof appears to be inadequate.]

The least squares estimate of a finite-variance random variable x given observations $\{\upsilon(\tau), 0 \leqq \tau \leqq t\}$ of a Wiener process $\upsilon(\cdot)$ can be written

$$\begin{aligned} \hat{x}(t) &= E[x(t)|\,\mathfrak{F}_t], \\ &= \int_0^t \left\{\frac{\partial}{\partial\tau} E[x\upsilon'(\tau)|\,\mathfrak{F}_\tau]\right\} d\upsilon(\tau), \qquad \mathfrak{F}_t = \sigma\{\upsilon(\tau), \tau \leqq t\}. \end{aligned} \tag{33}$$

The analogy to (8) is striking. If we now make further assumptions on x, for example, that x is the value at time t of a diffusion process or a Markov jump process, etc., then from the representation (33), we can obtain equations for the evolution of $\hat{x}(t)$ or more generally of the conditional characteristic function or the conditional probability density function.

Appendix: the innovations results of (20)–(22). There are many ways to prove that the process $\upsilon(\cdot)$ defined by (20) of the main text is a process of orthogonal increments. For example, one could specialize the more general result of [13], [14], quoted in §4, which is obtained by using certain martingale theorems. Or one could use the simpler arguments of (61)–(62) of [16]. It is somewhat more difficult to show that the process $Y(\cdot)$ can be uniquely recovered from the process $\upsilon(\cdot)$. Hitsuda [17] has used martingale theory to give a rigorous derivation of (20)–(22). However, here we shall outline a proof based on a factorization theorem of Gohberg and Krein [18]; this proof is the rigorous counterpart of the informal arguments used in [1] and [3]. The result of Gohberg and Krein is for operators on certain general Hilbert spaces, but here we shall need only the result for the space L_2.

Let K be an integral operator on $L_2[0, 1]$, with a real symmetric square-integrable kernel (for notational simplicity, we assume everything scalar in this Appendix)

$$K(t, s) = K(s, t), \quad \int_0^1 \int_0^1 K^2(t, s)\, dt\, ds < \infty. \tag{A.1}$$

Suppose also that the operator $I + K$ is positive definite, or equivalently that the eigenvalues of K are all greater than -1. Then Gohberg and Krein [18] show that the equation

$$h(t, s) + \int_0^t h(t, \tau)K(\tau, s)\, d\tau = K(t, s), \qquad 0 \leqq s \leqq t \leqq 1, \tag{A.2}$$

$$h(t, s) = 0, \qquad s > t, \tag{A.3}$$

has a unique square-integrable solution and that

$$(I + K)^{-1} = (I - h^*)(I - h), \tag{A.4}$$

where h^* denotes the adjoint of the Volterra operator h that has kernel $h(t, s)$.

Furthermore, if $k(t, s)$ is the unique square-integrable solution of

$$h(t, s) + \int_0^t k(t, \tau)h(\tau, s)\, d\tau = k(t, s), \qquad 0 \leqq s \leqq t \leqq 1, \tag{A.5}$$

$$k(t, s) = 0, \qquad s > t, \tag{A.6}$$

then we can also write

$$I + K = (I + k)(I + k^*). \tag{A.7}$$

We use these results as follows. Suppose that

$$dY = z(t)\, dt + dW,$$

where

$$EW(t)W(s) = \int_0^{t \wedge s} d\tau, \quad E \int_0^1 z^2(t)\, dt < \infty,$$

and the increments of $W(\cdot)$ are uncorrelated with past values of $z(\cdot)$. It is easy to verify that the assumptions (10)–(16) of the main text yield a process $Y(\cdot)$ that satisfies these conditions. The first thing to do is to show that a process $Y(\cdot)$ defined as above has covariance function

$$EY(t)Y(s) = t \wedge s + \int_0^t \int_0^s K(u, v)\, du\, dv,$$

where $K(u, v)$ meets the conditions of the Gohberg–Krein theorem. This fact can be established by a direct calculation—see, for example, the discussion of (48)–(49) in [19]. (A more elegant method is to assume temporarily that $z(\cdot)$ and $W(\cdot)$ are Gaussian processes and then to note that, by any one of a variety of criteria, the process $Y(\cdot)$ is mutually absolutely continuous with respect to the Wiener process $W(\cdot)$. Therefore, by a theorem of Shepp [20] (see also [21], [22]), the covariance of Y must have the stated properties.)

The next step is to observe that the linear least squares estimate of $z(\cdot)$ can be written as

$$\hat{z}(t) = \int_0^t h(t, s)\, dY(s) = \int_0^t h(t, s) z(s)\, ds + \int_0^t h(t, s)\, dW(s),$$

where $h(t, s)$ is defined by (A.2)–(A.3). This follows easily by using the projection definition of $\hat{z}(\cdot)$. Now the fact that

$$\upsilon(t) = Y(t) - \int_0^t d\tau \int_0^\tau h(\tau, u)\, dY(u)$$

has covariance function $t \wedge s$ follows by a direct but tedious calculation. The basic idea is symbolically that

$$\begin{aligned} \dot{\upsilon}_t &= \dot{Y}_t - \int_0^t h(t, u)\, dY(u) \\ &= [(I - h)Y]_t \end{aligned}$$

so that

$$\begin{aligned} E\dot{\upsilon}_t\dot{\upsilon}_s &= (I - h)E\dot{Y}_t\dot{Y}_s(I - h^*) \\ &= (I - h)(I + k)(I + k^*)(I - h^*) = I. \end{aligned}$$

Finally, since $(I - h)$ is invertible, we expect that $Y(\cdot)$ can be recovered from $\upsilon(\cdot)$ by

$$\dot{Y}_t = (I - h)^{-1}\dot{\upsilon}_t = (I + k)\dot{\upsilon}_t$$

or equivalently,

$$Y(t) = \upsilon(t) + \int_0^t d\tau \int_0^\tau k(\tau, u)\, d\upsilon(u).$$

This last equation can be directly verified without much difficulty. The right-hand side is equal to

$$Y(t) - \int_0^t d\tau \int_0^\tau h(\tau, u)\, dY(u) + \int_0^t d\tau \int_0^\tau k(\tau, u)\, dY(u)$$

$$- \int_0^t d\tau \int_0^\tau k(\tau, u) \int_0^u h(u, \sigma)\, dY(\sigma)\, du.$$

By the use of a Fubini theorem for Wiener integrals [7, pp. 430–431], the last term can be rewritten as

$$\int_0^t d\tau \int_0^\tau dY(\sigma) \int_0^\tau k(\tau, u)h(u, \sigma)\, du.$$

But now the sum of the last three terms is

$$-\int_0^t d\tau\left[\int_0^\tau \left[h(\tau, \sigma) - k(\tau, \sigma) + \int_0^\tau k(\tau, u)h(u, \sigma)\, du\right] dY(\sigma)\right],$$

which is zero because the term in square brackets is zero almost everywhere by the definition of h and k.

Acknowledgment. It is a pleasure to acknowledge discussions on § 2 with G. Kallianpur.

REFERENCES

[1] T. KAILATH, *An innovations approach to least-squares estimation, Part I: Linear filtering in additive white noise*, IEEE Trans. Automatic Control, AC-13 (1968), pp. 646–655.

[2] T. KAILATH AND P. FROST, *An innovations approach to least-squares estimation, Part II: Linear smoothing in additive white noise*, Ibid., AC-13 (1968), pp. 655–660.

[3] T. KAILATH, *The innovations approach to detection and estimation theory*, Proc. IEEE, 58 (1970), pp. 680–695.

[4] D. L. DUTTWEILER, *Reproducing kernel Hilbert space techniques for detection and estimation problems*, Doctoral thesis, Department of Electrical Engineering, Stanford University, Stanford, Calif., 1970.

[5] R. L. STRATONOVICH, *Detection and estimation of signals in non-Gaussian noise*, Proc. IEEE, 58 (1970), pp. 670–679.

[6] R. E. KALMAN AND R. S. BUCY, *New results in linear filtering and prediction theory*, Trans. ASME J. Basic Engrg., Ser. A, 83 (1961), pp. 95–107.

[7] J. L. DOOB, *Stochastic Processes*, John Wiley, New York, 1953.

[8] W. M. WONHAM, *Random differential equations in control theory*, Probabilistic Methods in Applied Mathematics, vol. 2, A. T. Bharucha-Reid, ed., Academic Press, New York, 1969, pp. 131–212.

[9] M. FUJISAKI, G. KALLIANPUR AND H. KUNITA, *Stochastic differential equations for the nonlinear filtering problem*, Osaka J. Math., to appear.

[10] T. KAILATH AND R. GEESEY, *An innovations approach to least-squares estimation, Part IV: Recursive estimation given the covariance function*, IEEE Trans. Automatic Control, AC-16 (1971), pp. 720–727.

[11] R. GEESEY AND T. KAILATH, *An innovations approach to least-squares estimation, Part V: Recursive estimation in colored noise*, submitted for publication.

[12] P. FROST AND T. KAILATH, *An innovations approach to least-squares estimation, Part III: Nonlinear least-squares estimation*, IEEE Trans. Automatic Control, AC-16 (1971), pp. 217–226.

[13] T. KAILATH, *Some extensions of the innovations theorem*, Bell System Tech. J., 50 (1971), pp. 1487–1494.

[14] R. LIPTSER AND A. SHIRYAEV, *On absolutely continuous measures, corresponding to processes of diffusion type, relative to Wiener measure*, Izv. Akad. Nauk., to appear.

[15] M. YERZHOV, *Existence and uniqueness of solutions of stochastic functional equations*, Teor. Veroyatnost. i Primenen., to appear.

[16] T. KAILATH, *A general likelihood-ratio formula for random signals*, IEEE Trans. Information Theory, IT-15 (1969), pp. 350–361.

[17] M. HITSUDA, *Representation of Gaussian processes equivalent to Wiener processes*, Osaka J. Math., 5 (1968), pp. 299–312.

[18] I. Ts. Gohberg and M. G. Krein, *The factorization of operators in Hilbert space*, Amer. Math. Soc. Transl. (2), 51 (1966), pp. 155–188.

[19] T. Kailath, *Likelihood ratios for Gaussian processes*, IEEE Trans. Information Theory, IT-16 (1970), pp. 276–288.

[20] L. Shepp, *Radon–Nikodym derivatives of Gaussian measures*, Ann. Math. Statist., 37 (1966), pp. 321–354.

[21] T. Kailath, *On measures equivalent to Wiener measure*, Ibid., 38 (1967), pp. 261–263.

[22] G. Kallianpur and H. Oodaira, *The equivalence and singularity of Gaussian measures*, Proc. Symp. on Time Series Analysis, M. Rosenblatt, ed., John Wiley, New York, 1963, pp. 000–000.

[23] A. V. Balakrishnan, *Stochastic differential systems, I*, Lecture notes, Dept. of System Science, UCLA, 1971.

AUTHOR CITATION INDEX

SUBJECT INDEX

About the Editor

After receiving his Sc.D. degree from the Massachusetts Institute of Technology in 1961, THOMAS KAILATH worked at the Jet Propulsion Laboratories and the California Institute of Technology before joining Stanford University in 1963. There he became Professor of Electrical Engineering in 1968 and Director of the Information Systems Laboratory in 1971. His research and teaching have covered a wide field, including multipath channels, feedback communications, signal detection, least-squares estimation, operator theory, and linear system theory. Among other honors, he has held Guggenheim (1970) and Churchill (1977) Fellowships, and was elected a Fellow of the Institute of Electrical and Electronic Engineers in 1970 and of the Institute of Mathematical Statistics in 1975. In 1975 he served as President of the IEEE Information Theory Group.